SOURCE BOOK

OF

MEDICAL HISTORY

COMPILED WITH NOTES BY
LOGAN CLENDENING, M.D.

DOVER PUBLICATIONS, INC.
NEW YORK
HENRY SCHUMAN
NEW YORK

This Dover edition, first published in 1960, is an
unabridged and unaltered republication of the work
originally published in 1942.

International Standard Book Number

ISBN-13: 978-0-486-20621-9
ISBN-10: 0-486-20621-1

Library of Congress Catalog Card Number: 60-2873

Manufactured in the United States by LSC Communications
20621120   2017
www.doverpublications.com

For

RALPH H. MAJOR, M.D.

Inspiring and generous chief, delightful companion, loyal friend, in the fellowship of science and history and books.

This new Dover edition is dedicated to the memory of

LOGAN CLENDENING
May, 1884 — January, 1945
Doctor and Teacher

For his contribution to medical education and medical history.

PREFACE

THE IDEA, AS WELL AS THE NAME "SOURCE BOOK," IS TAKEN FROM THE
collections of the same general kind which have long been used as
an adjunct to teaching in the field of general history. A number of
anthologies of original contributions in the field of medicine have
already appeared—"Syllabus of Lectures in Medical History," by
John D. Comrie (circa 1880), used at the University of Edinburgh;
"Epoch-Making Contributions to Medicine, Surgery and Allied Sciences," by C. N. B. Camac (1909); "Greek Medicine," by Arthur J.
Brock (1929); "Selected Readings in Pathology," by Esmond R.
Long (1929); "Selected Readings in Physiology," by John R. Fulton
(1930); "Classic Descriptions of Disease," by Ralph H. Major (1932);
"Classical Contributions to Obstetrics and Gynaecology," by Herbert
Thoms (1935); "A Mirror for Surgeons—Selected Readings in Sur-
gery," by Sir D'Arcy Power (1939); "Pioneers in Abdominal Sur-
gery," by Zachary Cope (1939); "Cardiac Classics," by Fredrick A.
Willius and Thomas E. Keys (1941).

These are naturally my models, but whereas they deal with special
subjects, or, as in the case of Dr. Camac's selections, with special
enthusiasms of the editor, the present work attempts to give a more
comprehensive survey of classical medical writings in chronological
order. The introductory notes to each section or author have been
made as brief and plain as possible. For more complete discussions
it is desirable for the student to consult the standard histories of
medicine.

The sole responsibility for the selections included in this book rests
with the editor. I regret as much as anyone that many solid contributions had to be omitted for lack of space and reasons of expediency.
The problem of spelling also presented a problem, for which I
undertook the onus. I arbitrarily decided that the spelling should,
where any doubt existed, give a flavor of the period when the original
work appeared. I am totally unfitted for textual criticism and have
made no attempts in that direction.

A special feature is the inclusion of selections from lay literature
that show a contemporary view of medical life at different ages.
Thus the accounts of Greek medicine by Herondas, Aristophanes,

Plato, and Thucydides; of Arabian medicine from the Arabian
Nights; Chaucer's "Doctor of Physic"; the seventeenth century doctors from Molière; the barber-surgeon from "Gil Blas"; the doctors
of Dickens and Thackeray and pre-anaesthetic surgery from "Rab
and His Friends," give a panorama of medical life that the student
of history cannot afford to ignore.

Few of the selections give the original in its entirety, but it is
hoped that the general spirit has been preserved. In several instances, the editor has felt that an abstract of an entire book would
clarify understanding.

I wish to express here my thanks to the translators who contributed original drafts to this collection: Dr. W. P. Hotchkiss, who
made the translation of the Preface of the Fabrica; Julian F. Smith,
of the Hooker Scientific Library, Central College, Fayette, Missouri,
who made the translations of Paracelsus, d'Amato, and Vierordt, and
to my wife who made the translations from Pasteur, Itard, and
Claude Bernard.

Many scholars have permitted me to use their translations: Dr.
Samuel Lambert, of Vesalius; Dr. Esmond R. Long, of Benivieni;
Dr. John F. Fulton, of Galvani and Malpighi; Dr. J. M. Hayman,
of Malpighi; Dr. Otto Glasser, of Roentgen; Dr. Wilmer Cave
Wright, of Ramazzini; Dr. Herbert Thoms, of Credé and Semmel-
weiss. The University of Chicago Press kindly allowed me to use
excerpts from the late Dr. Breasted's translation of the Edwin H.
Smith Papyrus. C. C. Thomas, publishers, Springfield, Illinois;
Columbia University Press; Wehman Brothers Publishing Co.; the
Charaka Club, Paul B. Hoeber, Inc., New York; the *Annals of Medical History*; the New York Academy of Medicine; the National
Tuberculosis Association; Dr. Solomon Claiborne Martin; the Philmar Press; and the Urologic and Cutaneous Press (Fracastorius'
"Origin of Syphilis"), have permitted me to use material published
by them. To all of these generous persons I am most grateful; as I
am to the various translations on which I ventured to base my own
adaptations.

To Dr. Leona Baumgartner and to Dr. Ralph H. Major I am
obligated for general and special advice. To Mr. Paul B. Hoeber
and his staff, whose patience I have repeatedly tried, my apologies
and thanks.

LOGAN CLENDENING

Kansas City, Missouri

CONTENTS

ix

CONTENTS

CONTENTS   xi

CONTENTS xiii

SOURCE BOOK OF
MEDICAL HISTORY

I

THE EGYPTIAN PAPYRI

THE FIRST *written* ACCOUNTS OF MEDICAL EXPERIENCE ARE FOUND IN THE Egyptian Papyri. According to the dates assigned by modern scholarship (more recent than were at first estimated) they are: *Kahun* Papyrus, 1900 B.C., three columns, gynecology, discovered by Sir Flinders Petri, 1889, at Fayum; *Edwin H. Smith* Papyrus, 1600 B.C., surgical, forty-eight descriptions of disease, incomplete, discovered at Luxor in 1862, now at New York; *Ebers* Papyrus, 1550 B.C., medical prescriptions and charms, one hundred and eight columns, apparently complete, largest and best preserved, discovered at Thebes, in 1862, by Georg Ebers, now at Leipzig; *Hearst* Papyrus, subject same as Ebers, calculated to be about the same date as the *Ebers*, because many of the prescriptions are identical, fifteen columns, discovered at Der-el-Ballas in 1899, now at the University of California; *Berlin* Papyrus, late but probably a copy of an earlier document, two hundred and four columns, mostly obstetrical, very magical; *London* Papyrus, late, largely magical, now in the British Museum.

Ebers edited a magnificent reproduction and hieratic translation, with glossary, of the Ebers Papyrus. Joachim published a German translation in 1890. Cyril P. Bryan published a partial, rather unsatisfactory, translation of the Ebers Papyrus in 1930. B. Ebbell published a full, excellent translation of the entire papyrus in 1937. Breasted translated into English the Smith Papyrus, and this was published along with a facsimile reproduction of the Papyrus by the University of Chicago Press. A translation of the Hearst Papyrus is being prepared by faculty members of the University of California. Bayard Holmes and P. Gad Kitterman (Medicine in Ancient Egypt, Cincinnati, 1914) have brief translations of parts of the Kahun and Berlin papyri.

REFERENCE

DAWSON, W. R. The Beginnings: Egypt and Assyria. Clio Medica Series. New York, Paul B. Hoeber, 1930.

KAHUN PAPYRUS

Knowledge of a woman whose back aches, and the inside of her thighs are painful. Say to her, it is the falling of the womb. Do thou for her thus: uah grains; shasha fruit 1-64, hekt, cow's milk I henu, cook, let it cool, make it into gruel, drink four mornings.

EDWIN H. SMITH PAPYRUS

A series of case histories all arranged in a formal stereotyped manner. First the surgical condition is described. Then advice as to whether the surgeon should or should not attempt to treat it. If treatment is advised it is then described. Typical of these histories are those subjoined:

CASE ELEVEN—A BROKEN NOSE*

TITLE

Instructions concerning a break in the column of his nose.

EXAMINATION

If thou examinest a man having a break in the column of his nose, his nose being disfigured, and a [depression] being in it, while the swelling that is on it protrudes, [and] he had discharged blood from both his nostrils, [conclusion in diagnosis].

DIAGNOSIS

Thou shouldst say concerning him: "One having a break in the column of his nose. An ailment which I will treat."

TREATMENT

Thou shouldst cleanse [it] for him [with] two plugs of linen. Thou shouldst place two [other] plugs of linen saturated with grease in the inside of his two nostrils. Thou shouldst put [him] at his mooring stakes until the swelling is reduced [lit. drawn out]. Thou shouldst apply for him stiff rolls of linen by which his nose is held fast. Thou shouldst treat him afterward [with] lint, every day until he recovers.

GLOSS A

As for: "The column of his nose," it means the outer edge of his nose as far as its side[s] on the top of his nose, being the inside of his nose in the middle of his two nostrils.

* Translation by Breasted. By permission of the University of Chicago Press.

GLOSS B

As for: "His two nostrils," [it means] the two sides of his nose extending to his [two] cheek[s], as far as the back of his nose; the top of his nose is loosened.

CASE TWENTY-FIVE—A DISLOCATION OF THE MANDIBLE

TITLE

Instructions concerning a dislocation of his mandible.

EXAMINATION

If thou examinest a man having a dislocation in his mandible, shouldst thou find his mouth open [and] his mouth can not close for him, thou shouldst put thy thumb[s] upon the ends of the two rami of the mandible in the inside of his mouth,. [and] thy two claws [meaning two groups of fingers] under his chin, [and] thou shouldst cause them to fall back so that they rest in their places.

DIAGNOSIS

Thou shouldst say concerning him "one having a dislocation in his mandible. An ailment which I will treat."

TREATMENT

Thou shouldst bind it with *ymrw*, [and] honey every day until he recovers.

EBERS PAPYRUS

The Ebers Papyrus is a materia medica, or rather a formulary. Under the heading of diseases are given a number of prescriptions with some hints (not always very definite) of how and when to apply them. The first section is entitled "Internal Medical Diseases"; it gives directions on how to open the bowels, to close the bowels, to expel worms, to expel fevers, etc. Other sections are "Diseases of the Eye," "Diseases of the Skin," "Diseases of the Teeth," "Diseases of Women," "Diseases of the Ears." At the beginning of the Papyrus is an incantation to be recited by the physician or the priest when giving the remedies. Thus the Papyrus represents both the magical and the empirical forms of *primitive* medicine.

RECITAL ON DRINKING A REMEDY*

Here is the great remedy. Come thou who expellest evil things in this my stomach and drives them out from these my limbs! Horus and Seth have been conducted to the big palace at Heliopolis, where they consulted over the connection between Seth's testicles with Horus, and Horus shall get well like one who is on earth. He who drinks this shall be cured like these gods who are above. . . . These words should be said when drinking a remedy. Really excellent, proven many times!

INTERNAL MEDICAL DISEASES

The beginning of a compilation of remedies.

To expel *diseases in the belly*: thwj mixed with beer, is drunk by the man.

Another [remedy] for the belly, when it is ill: cumin ½ ro, goosefat 4 ro, are boiled, strained and taken.

Another: figs 4 ro, sebesten 4 ro, sweet beer 20 ro likewise.

Remedy *to open the bowels*: milk 25 ro, sycamore-fruit 8 ro, honey 8 ro, are boiled, strained and taken for four days.

Another to cause evacuation: Honey—are made into suppository.

Remedy for dejection: colocynth 4 ro, are mixed together, eaten and swallowed with beer 10 ro or wine 5 ro.

Another: colocynth 4 ro, honey 1 ro, are mixed together and eaten by the man in one day.

To kill *roundworm* [Ascaris Lumbriocoides] root of pomegranate 5 ro, water 10 ro, remains during the night in the dew, is strained and taken in 1 day.

Another to expel *the rose* [? erysipelas] in the belly: sory, wax, turpentine, are ground, mixed together and [it] is anointed herewith. Then thou shalt prepare remedy for evacuation after obstruction of his bowels: colocynth, senna, fruit of sycamore, are ground, mixed together and shaped into 4 cakes and let him eat it.

DISEASES OF THE EYE

The beginning of a compilation on the eye.

This is to be done for the accumulation of *purulency* in the blood of the eye: honey, balm of Mecca, gum ammoniac [?]. To treat its

* Translation based on Ebers' transcript, Joachim's translation, and Ebbell's English translation, made under the supervision of the editor.

discharge: frankincense, myrrh, yellow ochre. To treat the growth [?] red ochre, malachite, honey. Afterwards thou shalt prepare for him: oil, sagapen hntt in frankincense, yellow ochre, frankincense, goose-fat, atibium, oil, herewith [it] is bandaged for 4 days. Thou shalt not disturb much.

Another for *night-blindness* in the eyes: liver of ox, roasted and crushed out, is given for this trouble. Really excellent!

Another for *blindness*: pig's eyes, its humour is fetched, real stibium, red ochre, less of honey, are ground fine, mixed together and poured into the ear of the man, so that he may be cured immediately. Do [it] and thou shalt see. Really excellent! Thou shalt recite as a spell: I have brought this which was applied to the seat of yonder trouble: it will remove the horrible suffering. Twice.

DISEASES OF THE SKIN

For *bites* by men: shell of ndw-vessel, leek, are crushed, mixed together and [it] is bandaged therewith.

The beginning of remedies to expel *discharging exanthema of the scalp* [achor]: seed of ricinus, grease, balanites-oil, are mixed together and [it] is anointed therewith.

Another:—is crushed with honey and [it] is bandaged therewith.

Another to make the hair of a bald-headed person grow: fat of lion, fat of hippopotamus, fat of crocodile, fat of cat, fat of serpent, fat of ibex, are mixed together and the head of a bald person is anointed therewith.

II

GREEK MEDICINE
THE CULT OF AESCULAPIUS

AESCULAPIUS, OR ASKLEPIOS, THE GREEK GOD OF HEALTH, WAS WORSHIPPED in temples, to which patients repaired to be cured. Many of the ruins of these temples have been uncovered and partially restored by modern archeological research. They were usually built in salubrious spots near springs. The most famous, at least of those revealed by modern excavation, were at Cos, the island where Hippocrates reputedly was born, at Pergamos, the birthplace of Galen, at Athens, at Epidaurus, and at Corinth.

The priests trained to preside in the temples, were instructed in treatment by diet, baths, mineral springs, psychotherapy, and probably in the use of drugs and manipulation. Animal sacrifices were made—cocks, swine, rams, and rarely goats (in some temples goat sacrifices were forbidden). The serpent was sacred to Aesculapius, whose symbol was the staff and coiled serpent. Classic Greek literature gives us many references and descriptions of Aesculapius and the temple rites.

HOMER

[From the *Iliad*, Book II]

Aesculapius is mentioned in the Iliad, but not in the Odyssey. It should be noted that he is referred to as a man, not as a god. Like the Egyptian, Im-ho-tep, from whom, according to some historians, the Greeks borrowed Aesculapius, he was apparently first a living man, and was afterwards raised to the status of a god.

The two sons of Asklepios led him, goodly physicians, Machaon and Podalirius. [There follows a catalogue of the ships—Asklepios is described as the leader of the forces from Trikka, Ithomè and Orchala and therefore a man.]

[From the *Iliad*, Book IV]

King Agamemnon said: "Dear Menelaus! Would that it were so, yet the physician must explore thy wound, and with his balsams

6

soothe the bitter pain. Call Machaon hither, the son of Asklepios, the blameless physician and bring him to the Achaian general, the warlike Menelaus."

Machaon's heart was touched and forth he went through the great throng, the army of the Greeks and came to where Atreus' warlike son was wounded. Without delay he drew the arrow from the fairly-fitted belt. The barbs were bent in drawing. Then he loosed the embroidered belt and carefully o'erlooked the wound where fell the bitter shaft, cleaned it from blood and sprinkled over it with skill the soothing balsam which of yore the friendly Chiron to his father gave.

HERONDAS

THE WOMEN SACRIFICING TO ASKLEPIOS

[From the *Mimes*]

Herondas (300-250 B.C.) was a poet living on the island of Cos. He wrote mimes, a poetic form which described simple life and simple people. "The Women Sacrificing to Asklepios" gives us an idea of the decoration of the temples and of the ritual used.

Characters:
  Phile
  Kynno
  Kydilla
  Kokkale

Phile: Sovereign Paion, who governest Trikka and who livest in lovely Kos and Epidauros! Hail to Koronis who gave thee birth, and to Apollo! also to her whom, with thy right hand, thou touchest, Hygieia! and to ye whose venerable altars are here: Panacea, Epione, and Iaso, Hail also! and to ye who laid waste the houses and walls of Laomedon; the healers of savage ills, Podalirius and Machaon, Hail! and to all the gods who dwell at thy hearth, and the goddesses, Father Paion! Accept propitiously what is good in my sacrifice of this cock, herald of domestic walls. We can offer little from our house: far less than we would; for we would have brought an ox or a fatted sow instead of a cock, as an offering for the healing of our ills which we wouldst thou remove, O King, laying upon us thy gentle hands!

To the right, Kokkale; place the tablet on the right hand of Hygieia.

Ah, dear Kynno; what lovely statues. Who wrought this stone and who set it up here? *Kynno:* The sons of Praxiteles: don't you see those letters on the base? And it was Euthies, the son of Praxon, who had it set there. May Paion be kind to them and to Euthies for their lovely work. See, Phile, that child looking up to the apple; wouldn't you say she would faint if she didn't get it? *Phile:* And that old man, Kynno. And, by the Fates, how that little boy is strangling the fox-goose! If it were not plainly stone before us, you might believe him to speak. *Kynno:* But the day has come and the crowd is pressing. You— remain here; the door has opened and I see the sanctuary.

[The sacrifice is offered and the temple further admired.]

Phile: Do you see, Kynno? What works! Wouldn't you say that a new Athena had graven all these beautiful things?—Hail to thee, Mistress!—and that naked boy: if I scratched him wouldn't it leave a mark, Kynno? for he has flesh on him that quivers with life on the panel. And these silver tongs: if Myellos saw them, or Pataikiskos, the son of Lamprion, wouldn't their eyes pop out, believing them really made of silver? And that ox with the man leading him and another with them; and the man with the hook-nose, and the one with the snub-nose; aren't they the living day! If it wer'n't so unwomanly, I would have cried out for fear of that ox injuring me; one of his eyes, Kynno, glares at me so! *Kynno:* Yes, Phile, the hands of the Ephesian Apelles were truthful in everything they did, and one could not say: "That man saw some things while other things were hidden." Whatever it occurred to him to touch, even the gods, he succeeded with. A man who has seen him or his works, without being properly amazed ought to be hung up and beaten like clothes in a fuller's shop. *Priest:* Your sacrifice, women, has been well accomplished and with favorable presages. No one has evoked Paion more efficaciously than you have.

*ARISTOPHANES**

[From the *Plutus*]

Aristophanes' comedy, *Plutus, the God of Riches,* furnishes us another contemporary and doubtless first-hand description of the temple mysteries. The plot concerns the eternal question—Why are the virtuous poor and the wicked rich? The answer is that Plutus, the God of Riches, must be blind. So he is taken to the temple of Aesculapius to have his sight restored.

The dialogue which concerns the ritual is thus:

Chremylus: We two must give back sight—
Blepsidemus: Give sight? To whom?
Chremylus: To Plutus—by some one device or other.
Blepsidemus: So then, he's really blind?
Chremylus: He is, by Heaven.
Blepsidemus: No wonder that he never came to me!
Chremylus: But now—so please the gods—he'll make amends.
Blepsidemus: Come then—a leech! A leech—shouldst not have fetched one?
Chremylus: What leech has Athens now? They're gone together.
The art and its rewards—no fee no physic!
But listen, I was thinking
To lay him down at Aesculapius' shrine.
That were the way—
Blepsidemus: Far best, by all the powers: Away—delay not—something do, and quickly.

Carion: The master, boys, has prosper'd gloriously,
Or rather Plutus' self: instead of blind,
His eyes are clear—clean'd out, and fairly—whiten'd,
A kindly leech in Aesculapius finding.
Chorus: O lucky day!
Hurra! Huzza!
Carion: Like it or not, rejoicing-time is come.
Attend. The whole affair will I relate from first to last—
Soon as we reach'd the god,
Guiding a man, most miserable then,
Most happy now, if happy man there be;

* Translation by John Hookham Frere, 1847.

First to the salt sea sand we led him down,
And there we—duck'd him.
Wife: Happy he, by Jupiter!
A poor fellow, duck'd in the cold brine.
Carion: Thence to the sanctuary hied we; and
When on the altar cakes and corn-oblations
Were dedicated—to Vulcan's flame a wafer—
We laid our Plutus down, as meet it was,
While each of us fell to, to patch a bed up.
Wife: And were there other suitors to the god?
Carion: Why, one was Neoclides, blind is he,
Yet at thieving he will surpass our best eyes;
And many a one besides, with all diseases
Laden;—but when the sexton gave
The word to sleep, the lamps extinguishing,
And strictly charged *"if any hear a noise,
Mute let him be"*—we squatted around in order.
Well:
Sleep I could not, but a certain pot
Of porridge hugely took my fancy; 'twas lying there
Some small space distant from an old wife's head,
Towards which I felt a wondrous motion draw me;—
So, venturing a peep, I spy the priest
Our offerings—scones and figs—snatching away
From off the holy table; after this,
Round every altar, one by one, he grop'd
If any where a single cake were left;
Then these he *bless'd*—into a sort of satchel.
So, thinking 'twas a deed of vast devotion,
Bent on the pot of porridge, up get I.
Wife: Wretch! Fear'st thou not the god?
Carion: By the gods, I did,
Lest he should get before me to the pot,
Garlands and all;—his priest had tutor'd me.
When this was past, forthwith I muffled up,
Cowering with dread; but he, most doctor-like,
Perform'd his rounds; inspecting case by case,
Then placed a lad beside him his stone mortar
Pestle and chest.

For Neoclides first he took in hand
To pound a cataplasm—throwing in
Three heads of Tenian garlic; these he bruised,
Commixing in the mortar benjamin
And mastic; drenching all with Sphettian vinegar,
He plaster'd o'er his eyelids, inside out,
To give him greater torment;—squalling, bawling,
The wretch sprung up to flee; then laugh'd the god,
And cried, "Now sit ye down beplaster'd there,
And take thine oath I keep thee from the sessions."

Wife: O what a patriot and a prudent god!

Carion: He next sat down by Plutus;
And handled first his head; then with a cloth
Of linen, clean and napless, wiped the eyelids
Quite round and round; then Panacea
Wrapped in a purple petticoat his head,
And all his face; then Aesculapius whistled—
With that out darted from the shrine two serpents
Of most prodigious size.

Wife: Merciful heavens!

Carion: And these, smooth gliding underneath the petticoat,
Lick'd with their tongues—so seem'd to me—his eyelids,
And, ere you'd toss me off ten half-pint bumpers,
Plutus—O mistress!—up rose Plutus seeing.
Loud clapp'd I then both hands for ecstasy,
And fell to waking master; but the god
Vanish'd into the temple, self and serpents.
Then those that couch'd beside him—canst thou guess
How they *did* fondle Plutus, and all night
Slept not, but watch'd till morning glimmer'd through?
While I was lauding lustily the god,
That in a twinkling he gave sight to Plutus,
And Neoclides blinded worse than ever.

PLATO

DEATH OF SOCRATES

[From the *Phaedo*]

Crito made a sign to the servant who was standing by: and he went out and having been absent for some time, returned with the jailer carrying the cup of poison. Socrates said: "You my good friend who are experienced in these matters, shall give me directions how I am to proceed." The man answered: "You have only to walk about until your legs are heavy and there to lie down and the poison will act!" Then raising the cup to his lips readily and cheerfully he drank off the poison and he walked about until, as he said, his legs began to fail, and then he lay down, and the man who gave him the poison now and then looked at his feet and legs: and after a while he pressed his leg hard and asked him if he could feel and he said No. Then he felt them himself and said: when the poison reaches my heart that will be the end. He was beginning to grow cold about the groin, when he uncovered his face and said: "Crito, I owe a cock to Asclepius; will you remember to pay the debt?" "The debt shall be paid," said Crito. "Is there anything else?" There was no answer; but in a minute or two a movement was heard, and the attendants uncovered him; his eyes were set, and Crito closed his eyes and mouth.

III

HIPPOCRATES

HIPPOCRATES—THE "FATHER OF MEDICINE"—IS MENTIONED BY PLATO AS "A professional trainer of medical students." There are three ancient biographies of Hippocrates which state that he was born on Cos in 460 B.C., that he belonged to the guild of physicians called Asclepiadae, that he traveled all over Greece, that he stayed the plague at Athens and at other places, that his life was long but of uncertain length, tradition making him 85 to 109 at his death. There is a good deal of fable in these accounts, and modern scholarship is inclined to be skeptical as to whether Hippocrates was really an actual person, or whether from vague beginnings, the name came to be applied to all the writings of the Corpus, the product of numerous disciples of a school of thought. Still, Plato's evidence is usually reliable, and would indicate he was a living man. There is nothing improbable about the accounts except the reference to staying the plague at Athens: Thucydides says that medical treatment was unavailing.

Whether the Asclepiadae were priests in the temples of Aesculapius, or a guild bound by rules, whose members swore the Hippocratic Oath, is debatable.

The Works of Hippocrates are varied: they are a library of medical practice probably written by a school rather than one man. "Airs, Waters and Places," "Epidemics," "Regimen," "Prognostics," "The Sacred Disease" (refutation of the idea that disease was sent by a god), "Aphorisms," "Ancient Medicine," are some of the books in the Corpus.

The Hippocratic Oath is traditionally the obligation which Hippocrates imposed upon his pupils. Actually its origin is unknown. From internal evidence it can be dated as originating between the fifth century B.C. and the first century A.D.: Aesculapius was not invoked as a god before the fifth century B.C. and Apollo would not have been invoked as a god after the first century A.D. Again studying from internal evidence, it is not an oath but an indenture between master and pupil. That it was widely used in this way up to the Middle Ages is assumed, since there is a Christian and an Arabic form of oath; in these, in the invocation, the names of the pagan deities are replaced by Christian and Moslem gods. Galen wrote that ancient physicians comprised "a clan of Aesclepiadae, who taught their sons anatomy, who, in turn, transmitted their learning to the next generation." Many believe the Oath was the pledge required of all who were admitted to this clan of the Aesclepiadae.

There are many puzzling clauses in the Oath: (1) The prohibition against cutting for the stone: this is omitted from the Christian oath: it is suggested that cutting for the stone involved castration, which was abhorrent to the pagan Greeks, acceptable to Christians with their notions of celibacy. (2) The prohibition against giving poisons—the words are really *pharmacon oudeni*—deadly drugs: in modern practice this would prevent the giving of any drug with a lethal range.

REFERENCES

ADAMS, F. The Works of Hippocrates Translated from the Greek with a Preliminary Discourse and Annotations. London, Sydenham Society, 1849. 2 Vols.

JONES, W. H. S. The Doctor's Oath. Cambridge University Press, 1924.

NITTIS, S. The Hippocratic Oath in Reference to Lithotomy. *Bull. Med. Hist.*, 7:719 (July), 1939.

JONES, W. H. S. Hippocrates, with an English translation. Loeb Classical Library. London, William Heinemann; New York, G. P. Putnam's Sons, 1923.

LITTRÉ, E. Hippocrates, with French translation, 1839-61. 10 Vols. [Still probably to be considered the final authority.]

WITHINGTON, E. T. Medical History. London, 1894.

THE OATH

I swear by Apollo Physician, by Asclepius, by Health, by Panacea and by all the gods and goddesses, making them my witnesses, that I will carry out, according to my ability and judgement, this oath and this indenture. To hold my teacher in this art equal to my own parents; to make him partner in my livelihood; when he is in need of money to share mine with him; to consider his family as my own brothers, and to teach them this art, if they want to learn it, without fee or indenture; to impart precept, oral instruction, and all other instruction to my own sons, the sons of my teacher, and to indentured pupils who have taken the physician's oath, but to nobody else. I will use treatment to help the sick according to my ability and judgment, but never with a view to injury and wrong-doing. Neither will I administer a poison to anybody when asked to do so, nor will I suggest such a course. Similarly I will not give a woman a pessary to cause abortion. But I will keep pure and holy both my life and my art. I will not use the knife, not even, verily, on sufferers from stone, but I will give place to such as are craftsmen therein. Into whatsoever houses I enter, I will enter to help the sick, and I will abstain from all intentional wrong-doing and harm, especially from abusing the bodies of man or woman, bond or free. And whatsoever I shall see or hear in the course of my profession, as well as outside my profession in my intercourse with men, if it be what should not be published abroad, I will never divulge, holding such things to be holy secrets. Now if I carry out this oath, and break it not, may I gain for ever reputation among all men for my life and for my art; but if I transgress it and forswear myself, may the opposite befall me.

Aphorisms*

[From the *Aphorisms*]

SECTION I

1. Life is short, and the Art long; the occasion fleeting; experience fallacious, and judgement difficult. The physician must not only be prepared to do what is right himself, but also to make the patient, the attendants, and externals co-operate.

4. A slender and restricted diet is always dangerous in chronic diseases, and also in acute diseases, where it is not requisite. And again, a diet brought to the extreme point of attenuation is dangerous; and repletion, when in the extreme, is also dangerous.

12. The exacerbations and remissions will be indicated by the diseases, the seasons of the year, the reciprocation of the periods, whether they occur every day, every alternate day, or after a longer period, and by the supervening symptoms; as, for example, in pleuritic cases, expectoration, if it occur at the commencement, shortens the attack, but if it appear later, it prolongs the same; and in the same manner the urine, and alvine discharges, and sweats, according as they appear along with favorable or unfavorable symptoms, indicate diseases of a short or long duration.

13. Old persons endure fasting most easily; next, adults; young persons not nearly so well; and most especially infants, and of them such as are of a particularly lively spirit.

19. Neither give nor enjoin anything to persons during periodical paroxysms, but abstract from the accustomed allowance before the crisis.

* Translated from the Greek by Francis Adams. Sydenham Society, 1849.

24. Use purgative medicines sparingly in acute diseases, and at the commencement, and not without proper circumspection.

SECTION II

1. In whatever disease sleep is laborious, it is a deadly symptom; but if sleep does good, it is not deadly.

2. When sleep puts an end to delirium, it is a good symptom.

3. Both sleep and insomnolency, when immoderate, are bad.

19. In acute diseases it is not quite safe to prognosticate either death or recovery.

23. Acute diseases come to a crisis in fourteen days.

24. The fourth day is indicative of the seventh; the eighth is the commencement of the second week; and hence, the eleventh being the fourth of the second week, is also indicative; and again, the seventeenth is indicative, being the fourth from the fourteenth, and the seventh from the eleventh.

25. The summer quartans are, for the most part, of short duration; but the autumnal are protracted, especially those occurring near the approach of winter.

26. It is better that a fever succeed to a convulsion, than a convulsion to a fever.

32. For the most part, all persons in ill health, who have a good appetite at the commencement, but do not improve, have bad appetite again towards the end; whereas, those who have a very bad appetite at the commencement, and afterwards acquire a good appetite, get better off.

41. Persons who have had frequent and severe attacks of swooning, without any manifest cause, die suddenly.

42. It is impossible to remove a strong attack of apoplexy, and not easy to remove a weak attack.

43. Of persons who have been suspended by the neck, and are in a state of insensibility, but not quite dead, those do not recover who have foam at the mouth.

44. Persons who are naturally very fat are apt to die earlier than those who are slender.

47. Pains and fevers occur rather at the formation of pus than when it is already formed.

48. In every movement of the body, whenever one begins to endure pain, it will be relieved by rest.

SECTION III

24. In the different ages the following complaints occur: to little and new born children, aphthae, vomiting, coughs, sleeplessness, frights, inflammation of the navel, watery discharges from the ears.
25. At the approach of dentition, pruritus of the gums, fevers, convulsions, diarrhoea, especially when cutting the canine teeth, and in those who are particularly fat, and have constipated bowels.
26. To persons somewhat older, affections of the tonsils, incurvation of the spine at the vertebra next the occiput, asthma, calculus, round worms, ascarides, acrochordon, satyriasmus, struma, and other tubercles (phymata), but especially the aforesaid.
27. To persons of more advanced age, and now on the verge of manhood, the most of these diseases, and, moreover, more chronic fevers, and epistaxis.
28. Young people for the most part have a crisis in their complaints, some in forty days, some in seven months, some in seven years, some at the approach of puberty; and such complaints of children as remain, and do not pass away about puberty, or in females about the commencement of menstruation, usually become chronic.
29. To persons past boyhood, haemoptysis, phthisis, acute fevers, epilepsy, and other diseases, but especially the aforementioned.
30. To persons beyond that age, asthma, pleurisy, pneumonia, lethargy, phrenitis, ardent fevers, chronic diarrhoea, cholera, dysentery, lientery, haemorrhoids.
31. To old people dyspnoea, catarrhs accompanied with coughs, dysuria, pains of the joints, nephritis, vertigo, apoplexy, cachexia, pruritus of the whole body, insomnolency, defluxions of the bowels, of the eyes, and of the nose, dimness of sight, cataract (glaucoma), and dullness of hearing.

SECTION IV

76. When small fleshy substances like hairs are discharged along with thick urine, these substances come from the kidneys.
78. In those cases where there is a spontaneous discharge of bloody urine, it indicates rupture of a small vein in the kidneys.
79. In those cases where there is a sandy sediment in the urine, there is calculus in the bladder (or kidneys).

80. If a patient pass blood and clots in his urine, and have strangury, and if a pain seize the hypogastric region and perineum, the parts about the bladder are affected.

SECTION V

8. In pleuritic affections, when the disease is not purged off in fourteen days, it usually terminates in empyema.

9. Phthisis most commonly occurs between the ages of eighteen and thirty-five years.

10. Persons who escape an attack of quinsy, and when the disease is turned upon the lungs, die in seven days; or if they pass these they become affected with empyema.

11. In persons affected with phthisis, if the sputa which they cough up have a heavy smell when poured upon coals, and if the hairs of the head fall off, the case will prove fatal.

13. In persons who cough up frothy blood, the discharge of it comes from the lungs.

SECTION VI

12. When a person has been cured of chronic hemorrhoids, unless one be left, there is danger of dropsy or phthisis supervening.

13. Sneezing coming on, in the case of a person afflicted with hiccup, removes the hiccup.

14. In a case of dropsy, when the water runs by the veins into the belly, it removes the disease.

19. When a bone, cartilage, nerve, the slender part of the jaw, or prepuce, are cut out, the part is neither restored, nor does it unite.

20. If blood be poured out preternaturally into a cavity, it must necessarily become corrupted.

28. Eunuchs do not take the gout, nor become bald.

29. A woman does not take the gout, unless her menses be stopped.

30. A young man does not take the gout until he indulges in coition.

35. Hiccup supervening in dropsical cases is bad.

41. When pus formed anywhere in the body does not point, this is owing to the thickness of the part.

42. In cases of jaundice, it is a bad symptom when the liver becomes indurated.

43. When persons having large spleens are seized with dysentery, and if the dysentery pass into a chronic state, either dropsy or lientery supervenes, and they die.

46. Such persons as become hump-backed from asthma or cough before puberty, die.

54. In acute diseases, complicated with fever, a moaning respiration is bad.

57. Persons are most subject to apoplexy between the ages of forty and sixty.

SECTION VII

1. In acute diseases, coldness of the extremities is bad.

4. A chill supervening on a sweat is not good.

5. Dysentery, or dropsy, or ecstasy coming on madness is good.

15. From a spitting of blood there is a spitting of pus.

16. From spitting of pus arise phthisis and a flux; and when the sputa are stopped, they die.

34. When bubbles settle on the surface of the urine, they indicate disease of the kidneys, and that the complaint will be protracted.

35. When the scum on the surface is fatty and copious, it indicates acute diseases of the kidneys.

37. Haematemesis, without fever, does not prove fatal, but with fever it is bad; it is to be treated with refrigerant and styptic things.

47. If a dropsical patient be seized with hiccup the case is hopeless.

51. Sneezing arises from the head, owing to the brain being heated, or the cavity (ventricle?) in the head being filled with humours; the air confined in it then is discharged, and makes a noise, because it comes through a narrow passage.

54. In those cases where phlegm is collected between the diaphragm and the stomach, and occasions pain, as not finding a passage into either of the cavities, the disease will be carried off if the phlegm be diverted to the bladder by the veins.

56. Anxiety, yawning, rigor,—wine drunk with an equal proportion of water, removes these complaints.

57. When tubercles (phymata) form in the urethra, if they suppurate and burst, the pain is carried off.

87. Those diseases which medicines do not cure, iron [the knife?] cures; those which iron cannot cure, fire cures; and those which fire cannot cure, are to be reckoned wholly incurable.

METHOD OF PROGNOSIS*

[From the *Prognostics*]

SECTION I

It appears to me a most excellent thing for the physician to cultivate Prognosis; for by foreseeing and foretelling, in the presence of the sick, the present, the past, and the future, and explaining the omissions which patients have been guilty of, he will be the more readily believed to be acquainted with the circumstances of the sick; so that men will have confidence to intrust themselves to such a physician. And he will manage the cure best who has foreseen what is to happen from the present state of matters. For it is impossible to make all the sick well; this, indeed, would have been better than to be able to foretell what is going to happen; but since men die, some even before calling the physician, from the violence of the disease, and some die immediately after calling him, having lived, perhaps, only one day or a little longer, and before the physician could bring his art to counteract the disease; it therefore becomes necessary to know the nature of such affections, how far they are above the powers of the constitution; and, moreover, if there be anything divine in the diseases, and to learn a foreknowledge of this also. Thus a man will be the more esteemed to be a good physician, for he will be the better able to treat those aright who can be saved, from having long anticipated everything; and by seeing and announcing beforehand those who will live and those who will die, he will thus escape censure.

SECTION II

He should observe thus in acute diseases; first, the countenance of the patient, if it be like itself, for this is the best of all; whereas the most opposite to it is the worst, such as the following: a sharp nose, hollow eyes, collapsed temples; the ears cold, contracted, and their lobes turned out; the skin about the forehead being rough, distended, and parched; the colour of the whole face being green, black, livid, or lead-coloured.† If the countenance be such at the commencement of the disease, and if this cannot be accounted for

* Translation by Francis Adams, 1849.
† The Hippocratic facies.

from the other symptoms, inquiry must be made whether the patient has long wanted sleep; whether his bowels have been very loose; and whether he has suffered from want of food; and if any of these causes be confessed to, the danger is to be reckoned so far less; and it becomes obvious, in the course of a day and a night, whether or not the appearance of the countenance proceed from these causes. But if none of these be said to exist, and if the symptoms do not subside in the aforesaid time, it is to be known for certain that death is at hand. And, also, if the disease be in a more advanced stage either on the third or fourth day, and the countenance be such, the same inquiries as formerly directed are to be made, and the other symptoms are to be noted, those in the whole countenance, those on the body, and those in the eyes; for if they shun the light, or weep involuntarily, or squint, or if the one be less than the other, or if the white of them be red, livid, or has black veins in it; if there be a gum upon the eyes, if they are restless, protruding, or are become very hollow; and if the countenance be squalid and dark, or the colour of the whole face be changed—all these are to be reckoned bad and fatal symptoms. The physician should also observe the appearance of the eyes from below the eyelids in sleep; for when a portion of the white appears owing to the eyelids not being closed together, and when this is connected with diarrhoea or purgation from medicine, or when the patient does not sleep thus from habit, it is to be reckoned an unfavorable and very deadly symptom; but if the eyelid be contracted, livid, or pale, or also the lip, or nose, along with some of the other symptoms, one may know for certain that death is close at hand. It is a mortal symptom also, when the lips are relaxed, pendent, cold, and blanched.

CASE HISTORIES*

[From the *Epidemics*, Book I, Section III]

CASE I

Philiscus, who lived by the Wall, took to bed on the first day of acute fever; he sweated; towards night was uneasy. On the second day all the symptoms were exacerbated; late in the evening had a proper stool from a small clyster, the night quiet. On the third day, early in the morning and until noon, he appeared to be free

* Translation by Francis Adams, 1849.

from fever; towards evening, acute fever, with sweating, thirst, tongue parched; passed black urine; night uncomfortable, no sleep; he was delirious on all subjects. On the fourth, all the symptoms exacerbated, urine black; night more comfortable, urine of a better colour. On the fifth, about mid-day, had a slight trickling of pure blood from the nose; urine varied in character, having floating in it round bodies, resembling semen, and scattered, but which did not fall to the bottom; a suppository having been applied, some scanty, flatulent matters were passed; night uncomfortable, little sleep, talking incoherently; extremities altogether cold, and could not be warmed; urine black; slept a little towards day; loss of speech, cold sweats; extremities livid; about the middle of the sixth day he died. *The respiration throughout, like that of a person recollecting himself, was rare, and large,* the spleen was swelled up in a round tumour, the sweats cold throughout, the paroxysms on the even days.

CASE II

Silenus lived on the Broad-way, near the house of Evalcidas. From fatigue, drinking, and unseasonable exercises, he was seized with fever. He began with having pain in the loins; he had heaviness of the head, and there was stiffness of the neck. On the first day the alvine discharges were bilious, unmixed, frothy, high coloured, and copious; urine black, having a black sediment; he was thirsty, tongue dry; no sleep at night. On the second, acute fever, stools more copious, thinner, frothy; urine black, an uncomfortable night, slight delirium. On the third, all symptoms exacerbated; an oblong distension, of a softish nature, from both sides of the hypochondrium to the navel; stools thin, and darkish; urine muddy, and darkish; no sleep at night; much talking, laughter, singing, he could not restrain himself. On the fourth, in the same state. On the fifth, stools bilious, unmixed, smooth, greasy; urine thin, and transparent; slight absence of delirium. On the sixth, slight perspiration about the head; extremities cold, and livid; much tossing about; no passage of the bowels, urine suppressed, acute fever. On the seventh, loss of speech; extremities could no longer be kept warm; no discharge of urine. On the eighth, a cold sweat all over; red rashes with sweat, of a round figure, small, like vari, persistent, not subsiding; by means of a slight stimulus, a copious discharge from the bowels, of a thin and undigested character, with pain; urine acrid, and passed with pain; extremities slightly heated; sleep slight, and comatose; speechless; urine thin, and transparent. On the ninth, in the same state. On the tenth, no drink taken; comatose, sleep slight; alvine discharges the same; urine abundant, and thickish; when allowed to stand, the sediment farinaceous and white; extremities again cold. On the eleventh he died. At the commencement, and throughout, the respiration was slow and large; there was a constant throbbing in the hypochondrium; his age was about twenty.

CASE V

The wife of Epicrates, who was lodged at the house of Archigetes, being near the term of delivery, was seized with a violent rigor, and, as was said, she did not become heated; next day the same. On the third, she was delivered of a daughter, and everything went on properly. On the day following her delivery, she was seized with acute fever, pain in the cardiac region of the stomach, and in the genital parts. Having had a suppository, was in so far relieved; pain in the head, neck, loins; no sleep; alvine discharges scanty, bilious, thin, and unmixed; urine thin, and blackish. Towards the night of the sixth day from the time she was seized with the fever, became delirious. On the seventh, all the symptoms exacerbated; insomnolency, delirium, thirst; stools bilious, and high coloured. On the eighth, had a rigor; slept more. On the ninth, the same. On the tenth, her limbs painfully affected; pain again of the cardiac region of the stomach; heaviness of the head; no delirium; slept more; bowels constipated. On the eleventh, passed urine of a better colour, and having an abundant sediment, felt lighter. On the fourteenth, had a rigour; acute fever. On the fifteenth, had a copious vomiting of bilious and yellow matters; sweated; fever gone; at night acute fever; urine thick, sediment white. On the seventeenth, an exacerbation; night uncomfortable; no sleep; delirium. On the eighteenth, thirsty; tongue parched; no sleep; much delirium; legs painfully affected. About the twentieth, in the morning, had a slight rigor; was comatose; slept tranquilly; had slight vomiting of bilious and black matters; towards night deafness. About the twenty-first, weight generally in the left side, with pain; slight cough; urine thick, muddy, and reddish; when allowed to stand, had no sediment; in other respects felt lighter; fever not gone; fauces painful from the commencement, and red; uvula retracted; defluxion remained acrid, pungent, and saltish throughout. About the twenty-seventh, free of fever; sediment in the urine; pain in the side. About the thirty-first was attacked with fever, bilious diarrhoea; slight bilious vomiting on the fortieth. Had a complete crisis, and was freed from the fever on the eightieth day.

CASE IX

Criton, in Thasus, while still on foot, and going about, was seized with a violent pain in the great toe; he took to bed the same day, had rigors and nausea, recovered his heat slightly, at night was delirious. On the second, swelling of the whole foot, and about the ankle erythema, with distension, and small bullae (phlyctaenae); acute fever; he became furiously deranged; alvine discharges bilious, unmixed, and rather frequent. He died on the second day from the commencement.

CASE X

The Clazomenian who was lodged by the Well of Phrynichides was seized with fever. He had pain in the head, neck, and loins from the beginning, and immediately afterwards deafness; no sleep, acute fever, hypochondria elevated with a swelling, but not much distension; tongue dry. On the fourth, towards night, he became delirious. On the fifth, in an uneasy state. On the sixth, all the symptoms exacerbated. About the eleventh a slight remission; from the commencement to the fourteenth day the alvine discharge thin, copious, and of the colour of water, but were well supported; the bowels then became constipated. Urine throughout thin, and well coloured, and had many substances scattered through it, but no sediment. About the sixteenth, urine somewhat thicker, which had a slight sediment; somewhat better, and more collected. On the seventeenth, urine again thin; swellings about both his ears, with pain; no sleep, some incoherence; legs painfully affected. On the twentieth, free of fever, had a crisis, no sweat, perfectly collected. About the twenty-seventh, violent pain of the right hip; it speedily went off. The swellings about the ears subsided, and did not suppurate, but were painful. About the thirty-first, a diarrhoea, attended with a copious discharge of watery matters, and symptoms of dysentery; passed thick urine; swellings about the ears gone.

About the fortieth day, had pain in the right eye, sight dull. It went away.

PROGNOSIS FROM SIGNS*

[From the *Prognostics*]

SECTION VII

That state of the hypochondrium [abdomen] is best when it is free from pain, soft, and of equal size on the right and the left. But if inflamed, or painful, or distended; or when the right and left sides are of disproportionate sizes;—all these appearances are to be dreaded. And if there be also pulsation in the hypochondrium, it indicates perturbation or delirium; and the physician should examine the eyes of such persons; for if their pupils be in rapid motion, such persons may be expected to go mad. A swelling in the hypochondrium, that is hard and painful, is very bad, provided it occupy the whole hypochondrium; but if it be on either side, it is less dangerous when on the left. Such swellings at the commencement of the disease prognosticate speedy death; but if the fever has passed twenty days, and the swelling had not subsided, it turns to a suppuration. A discharge of blood from the nose occurs to such in the first period, and proves very useful; but inquiry should be made if they have headache or indistinct vision; for if there be such, the disease will be determined thither. The discharge of blood is rather to be expected in those who are younger than thirty-five years.

SECTION XII

The urine is best when the sediment is white, smooth, and consistent during the whole time, until the disease come to a crisis, for it indicates freedom from danger, and an illness of short duration; but if deficient, and if it be sometimes passed clear, and sometimes with a white and smooth sediment, the disease will be more protracted, and not so void of danger. But if the urine be reddish, and the sediment consistent and smooth, the affection, in this case, will be more protracted than the former, but still not fatal. But farinaceous sediments in the urine are bad, and still worse are the leafy; the white and thin are very bad, but the furfuraceous are still worse

* Translation by Francis Adams, 1849.

than these. Clouds carried about in the urine are good when white, but bad if black. When the urine is yellow and thin, it indicates that the disease is unconcocted; and if it (the disease) should be protracted, there may be danger lest the patient should not hold out until the urine be concocted. But the most deadly of all kinds of urine are the fetid, watery, black, and thick; in adult men and women the black is of all kinds of urine the worst, but in children, the watery. In those who pass thin and crude urine for a length of time, if they have otherwise symptoms of convalescence, an abscess may be expected to form in the parts below the diaphragm. And fatty substances floating on the surface are to be dreaded, for they are indications of melting. And one should consider respecting the kinds of urine, which have clouds, whether they tend upwards or downwards, and the colors which they have and such as fall downwards, with the colors as described, are to be reckoned good and commended; but such as are carried upwards, with the colors as described, are to be held as bad, and are to be distrusted. But you must not allow yourself to be deceived if such urine be passed while the bladder is diseased; for then it is a symptom of the state, not of the general system, but of a particular viscus.

SECTION XVII

Empyema may be recognized in all cases by the following symptoms: In the first place, the fever does not go off, but is slight during the day, and increases at night, and copious sweats supervene, there is a desire to cough, and the patients expectorate nothing worth mentioning, the eyes become hollow, the cheeks have red spots on them, the nails of the hands are bent, the fingers are hot, especially their extremities, there are swellings in the feet, they have no desire of food, and small blisters (phlyctaenae) occur over the body.

IV

THUCYDIDES

THUCYDIDES (471-400 B.C.) IS THE FIRST ACCURATE HISTORIAN. HIS DESCRIPTION of the plague at Athens is, he tells us, that of an eye witness. As you study it, ask yourself the question, "How would I describe an epidemic that was new and strange and struck my own city? What facts would I record and in what order?" Then notice how completely Thucydides covers the ground.

In spite of the careful description of the signs and symptoms, however, it is not possible to identify positively Thucydides' plague. W. H. S. Jones thought it was malaria. More likely it was bubonic plague. The ancient and medieval world was the stage for many destructive epidemics. Both bubonic plague and influenza cause intense cyanosis so either could have been "the black death." After Thucydides, the great plagues were (1) that following the eruption of Vesuvius, A.D. 79, which destroyed Pompeii; (2) the plague of Orosius, A.D. 125, preceded by an invasion of grasshoppers which spread over Italy and Northern Africa, Carthage, and Utica; (3) the plague of Antoninus, A.D. 164-180, all over the Roman Empire; (4) the pestilence of Cyprian, A.D. 251-266, probably smallpox; (5) the black death appeared in Europe about 1348, and repeated its visitations too often to mention.

REFERENCE

CRAWFORD, R. Plague and Pestilence in Literature and Art. London, Oxford University Press, 1914.

THE PLAGUE AT ATHENS*

[From *History of the Peloponnesian War*, Book II, Chapters 47-54]

As soon as summer returned (this was in 430 B.C.), the Peloponnesian army, comprising as before two thirds of the force of each confederate state, under the command of the Lacedaemonian king Archidamus, the son of Zeuxidamus, invaded Attica, where they established themselves and ravaged the country. They had not been there many days when the plague broke out at Athens for the first

* Translation by Jowett.

time. A similar disorder is said to have previously smitten many places, particularly Lemnos, but there is no record of such a pestilence occurring elsewhere, or of so great a destruction of human life. For a while physicians, in ignorance of the nature of the disease, sought to apply remedies; but it was in vain, and they themselves were among the first victims, because they oftenest came into contact with it. No human art was of any avail, and as to supplications in temples, inquiries of oracles, and the like, they were utterly useless, and at last men were overpowered by the calamity and gave them all up.

The disease is said to have begun south of Egypt in Aethiopia; thence it descended into Egypt and Libya, and after spreading over the greater part of the Persian empire, suddenly fell upon Athens. It first attacked the inhabitants of the Piraeus, and it was supposed that the Peloponnesians had poisoned the cisterns, no conduits having as yet been made there. It afterwards reached the upper city, and then the mortality became far greater. As to its probable origin or the causes which might or could have produced such a disturbance of nature, every man, whether a physician or not, will give his own opinion. But I shall describe its actual course, the symptoms by which any one who knows them beforehand may recognize the disorder should it ever reappear. For I was myself attacked, and witnessed the sufferings of others.

The season was admitted to have been remarkably free from ordinary sickness; and if anybody was already ill of any other disease, it was absorbed in this. Many who were in perfect health, all in a moment, and without any apparent reason, were seized with violent heats in the head and with redness and inflammation of the eyes. Internally the throat and the tongue were quickly suffused with blood, and the breath became unnatural and fetid. There followed sneezing and hoarseness; in a short time the disorder, accompanied by a violent cough, reached the chest; then fastening lower down, it would move the stomach and bring on all the vomits of bile to which physicians have ever given names; and they were very distressing. An ineffectual retching, producing violent convulsions, attacked most of the sufferers; some as soon as the previous symptoms had abated, others not until long afterwards. The body externally was not so very hot to the touch, nor yet pale; it was of a livid color inclining to red, and breaking out in pustules and ulcers. But the internal fever was intense; the sufferers could not bear to have on them even the finest linen garment; they insisted on being naked, and there was nothing which they longed for more eagerly than to throw themselves into cold water. And many of those who had no one to look after them actually plunged into the cisterns, for they were tormented by unceasing thirst, which was not in the least assuaged whether they drank little or much. They could not sleep; a restlessness which was intolerable never left them. While the disease was at its height the body, instead of wasting away, held out amid these sufferings in a marvellous manner, and either they died on the seventh or ninth day, not of weakness, for their strength was not exhausted, but of internal fever, which was the end of most; or, if they survived, then the disease descended into the bowels and there produced violent ulceration; severe diarrhoea at the same time set in, and at a later stage caused exhaustion, which finally, with few exceptions, carried them off. For the disorder which had originally settled in the head passed gradually through the whole body, and, if a person got over the worst, would often seize the extremities and leave its mark, attacking the fingers and the toes; and some escaped with the loss of these, some with the loss of their eyes. Some again had no sooner recovered than they were seized with forgetfulness of all things and knew neither themselves nor their friends.

The malady took a form not to be described, and the fury with which it fastened upon each sufferer was too much for human nature to endure. There was one circumstance in particular which distinguished it from ordinary diseases. The birds and animals which feed on human flesh, although so many bodies were lying unburied, either never came near them, or died if they touched them. This was proved by a remarkable disappearance of the birds of prey, who were not to be seen either about the bodies or anywhere else; while in the case of the dogs the fact was even more obvious, because they live with men.

Such was the general nature of the disease: I omit many strange peculiarities which characterized individual cases. None of the ordinary sicknesses attacked any one while it lasted, or, if they did, they ended in the plague. Some of the sufferers died from want of care, others equally who were receiving the greatest attention. No single remedy could be deemed a specific; for that which did good to one did harm to another. No constitution was of itself strong enough to resist or weak enough to escape the attacks; the disease carried off all alike, and defied every mode of treatment. Most appalling was the despondency which seized upon any one who felt himself sickening; for he instantly abandoned his mind to despair, and, instead of holding out, absolutely threw away his chance of life. Appalling too was the rapidity with which men caught the infection; dying like sheep if they attended on one another; and this was the principal cause of mortality. When they were afraid to visit one another, the sufferers died in their solitude, so that many houses were empty because there had been no one left to take care of the sick; or if they ventured they perished, especially those who aspired to heroism. For they went to see their friends without thought of themselves and were ashamed to leave them, even at a time when the very relations of the dying were at last growing weary and ceased to make lamentations, overwhelmed by the vastness of the calamity. But whatever instances there may have been of such devotion, more often the sick and the dying were tended by the pitying care of those who had recovered, because they knew the course of the disease and were themselves free from apprehension. For no one was ever attacked a second time, or not with a fatal result. All men congratulated them, and they themselves, in the excess of their joy at the moment, had an innocent fancy that they could not die of any other sickness.

The crowding of the people out of the country into the city aggravated the misery, and the newly arrived suffered most. For, having no houses of their own, but inhabiting in the height of summer stifling huts, the mortality among them was dreadful, and they perished in wild disorder. The dead lay as they had died, one upon another, while others hardly alive wallowed in the streets and crawled about every fountain craving for water. The temples in which they lodged were full of corpses of those who died in them; for the violence of the calamity was such that men, not knowing where to turn, grew reckless of all law, human and divine. The customs which had hitherto been observed at funerals were universally violated, and they buried their dead each one as best he could. Many, having no proper appliances, because the deaths in their household had been so frequent, made no scruple of using the burial-place of others. When one man had raised a funeral pile, others would come, and throwing on their dead first, set fire to it; or when some other corpse was already burning, before they could be stopped would throw their own dead upon it, and depart.

There were other and worse forms of lawlessness which the plague introduced at Athens. Men who had hitherto concealed their indulgence in pleasure now grew bolder. For, seeing the sudden change—how the rich died in a moment, and those who had nothing immediately inherited their property—they reflected that life and riches were alike transitory, and they resolved to enjoy themselves while they could, and to think only of pleasure. Who would be willing to sacrifice himself to the law of honor when he knew not whether he would ever live to be held in honor? The pleasure of the moment and any sort of thing which conduced to it took the place both of honor and of expediency. No fear of God or law of man deterred a criminal. Those who saw all perishing alike, thought that the worship or neglect of the Gods made no difference. For offences against human law no punishment was to be feared; no one would live long enough to be called to account. Already a far heavier sentence had been passed and was hanging over a man's head; before that fell, why should he not take a little pleasure?

Such was the grievous calamity which now afflicted the Athenians; within the walls their people were dying, and without, their country was being ravaged. In their troubles they naturally called to mind a verse which the elder men among them declared to have been current long ago:—

"A Dorian War will come and a plague with it."

There was a dispute about the precise expression; some saying that *limos*, a famine, and not *loimos*, a plague, was the original word. nevertheless, as might have been expected, for men's memories reflected their sufferings, the argument in favor of *loimos* prevailed at the time. But if ever in future years another Dorian War arises which happens to be accompanied by a famine, they will probably repeat the verse in the other form. The answer of the oracle to the Lacedaemonians when the God was asked "whether they should go to war or not," and he replied "that if they fought with all their might, they would conquer, and that he himself would take their part," was not forgotten by those who had heard of it, and they quite imagined that they were witnessing the fulfilment of his words. The disease certainly did set in immediately after the invasion of the Peloponnesians, and did not spread into Peloponnesus in any degree worth speaking of, while Athens felt its ravages most severely, and next to Athens the places which were most populous. Such was the history of the plague.

V

ARISTOTLE

ARISTOTLE (384-322 B.C.), THE MOST COMPREHENSIVE MIND OF CLASSICAL antiquity, may be said to have applied to general science the objective method of observation used by Hippocrates in clinical medicine. The other Greek scientists or natural philosophers, such, for instance, as Heraclitus and Parmenides, indulged in vague speculation about the nature of the universe, without bothering to verify their fundamental data. Aristotle obviously made direct observations. He covered an immense field—physics, astronomy, meteorology, psychology, natural history. In biology, his work on embryology (*De generatione animalium*) is full of acute observation; *Historia animalium* and *De partibus animalium* furnish a rough, fairly good idea of animal and human anatomy, as may be seen from the excerpt that follows. His ideas of physiology were more speculative and less sound.

REFERENCES

ARISTOTLE. Selections. Scribner's Philosophy series, with introduction. New York, Scribner's, 1927.
SINGER, C. Greek Biology and Greek Medicine. London, Oxford University Press, 1922.
THOMPSON, W. D'A. On Aristotle as a Biologist. London, Oxford University Press, 1913.
The Works of Aristotle. Translated into English under the editorship of J. A. Smith and W. D. Ross. London, Oxford University Press, *circa* 1930. 11 Vols.

DESCRIPTION OF THE ORGANS OF THE ANIMAL BODY*

[From *Parts of Animals*]

1. The external parts of the body are arranged in this manner; and, as I have said, are for the most part named and known from habit. But the internal parts are not so well known, and those of the human body are the least known. So that in order to explain them

---

* From De Partibus animalium and Historia animalium. Translation by Ogle, 1882. These selections were taken from different parts of these works to give an idea of Aristotle's anatomy.

we must compare them with the same parts of those animals which are most nearly allied.

2. First of all, the brain is placed in the fore-part of the head, and it occupies the same position in all animals that have this part, which belongs to all sanguineous and cephalopodous animals. In proportion to his size, man has the largest brain of all animals, and the moistest. Two membranes enclose the brain: that outside the skull is the strongest; the inner membrane is slighter than the outer one. In all animals the brain is in two portions. The cerebellum is placed upon the brain at its lowest extremity. It is different from the brain both to the touch and in appearance.

3. The back of the head is empty and hollow in all animals in proportion to their size, for some have a large head, but the part lying under the face is less in those animals which have round faces; others have a small head and large jaws, as the whole tribe of Lophuri. In all animals the brain is without blood, nor does it contain any veins, and it is naturally cold to the touch. The greater number of animals have a small cavity in the centre of the brain. And round this a membrane filled with veins: this membrane is like skin, and encloses the brain. Above the brain is the smoothest and weakest bone in the head—it is called sinciput.

4. Three passages lead from the eye to the brain; the largest and the middle-sized to the cerebellum, the least to the brain itself. The least is that which is nearest the nostril; the greater are parallel, and do not meet; but the middle-sized passages meet: this is most evident in fishes, and these passages are nearer to the brain than the larger, but the least separate from each other, and do not meet.

5. Within the neck is the oesophagus, which also derives its additional name, the isthmus, from its length and narrowness, and the trachea. The trachea lies in front of the oesophagus in all animals which possess this part, that is, all animals which breathe from the lungs. The trachea is cartilaginous in its nature, and contains but little blood: it is surrounded with many smooth rings of cartilage, and it lies upon the upper part towards the mouth, opposite the passage from the nostril to the mouth, wherefore, also, if any liquid is drawn into it in drinking, it passes out of the mouth through the nostrils.

6. Between the passages is the epiglottis, which can be folded over the passage which extends from the trachea to the mouth; by the

epiglottis the passage of the tongue is closed, at the other extremity the trachea reaches to the middle of the lungs, and afterwards divides to each side of the lungs. For the lung is double in all animals which possess this part, though the division is not so marked in viviparous animals, and least of all in man. The human lungs are anomalous, neither being divided into many lobes, as in other animals, nor being smooth.

7. In oviparous animals, such as birds and the oviparous quadrupeds, the parts are very widely separated, so that they appear to have two lungs; they are, however, only two divisions of the trachea extending to each side of the lungs; the trachea is also united with the great vein and with the part called the aorta. When the trachea is filled with air it distributes the breath into the cavities of the lungs, which have cartilaginous interstices ending in a point; the passages of these interstices go through the whole lungs, always dividing from greater into less.

8. The heart is connected with the trachea by fatty and cartilaginous muscular bands. There is a cavity near the junction, and in some animals, when the trachea is filled with breath, this cavity is not always distinguishable, but in larger animals it is evident that the breath enters in. This then is the form of the trachea, which only inhales and exhales breath, and nothing else either dry or moist, or it suffers pain till that which has passed down is coughed up.

9. The oesophagus is joined to the mouth from above, near the trachea, being united both to the spine and the trachea by membranaceous ligaments. It passes through the diaphragm into the cavity of the stomach, is fleshy in its nature, and is extensible both in length and breadth. The human stomach is like that of a dog, not a great deal larger than the entrail, but like a wide bowel; after this there is an entrail simply rolled together, then an entrail of moderate width. The lower part of the abdomen is like that of a hog, for it is wide, and from this to the seat it is short and thick.

10. The omentum is united to the abdomen in the middle, and is in its nature a fatty membrane, as in other animals with a single stomach and teeth in both jaws. The mesenterium is over the bowels; it is membranaceous, broad, and fat; it is united to the great vein and the aorta: through it extend many numerous veins at its junction with the intestines, reaching from above downwards. This

is the nature of the oesophagus, trachea, and the parts of the abdominal cavity.

[From *History of Animals*, Chapter XIV]

1. The heart has three cavities: it lies above the lungs, near the division of the trachea. It has a fat and thick membrane, by which it is united to the great vein and the aorta, and it lies upon the aorta near the apex; and the apex is placed in all animals, whether they have or have not a chest, the apex of the heart is forwards, though it often escapes notice by the change of position in the parts when dissected. The gibbous portion of the heart is upwards; its apex is generally fleshy and thick, and there is a sinew in the cavities.

2. In all other animals which have a chest the heart is placed in the centre; in man it is rather on the left side, inclining a little from the division of the mammae towards the left breast in the upper part of the chest; it is not large; its whole form is not long, but rather round, except that the extremity ends in a point. It has three cavities, as I have said. The greatest is that on the right, the least on the left, the middle one is of intermediate size. They are all perforated towards the lungs. It has both the two smaller, and all of them perforated towards the lungs, and this is evident in one of the cavities downwards from its point of attachment.

3. Near the principal cavity it is attached to the great vein to which also the mesenterium is united, and in the middle it is attached to the aorta. Passages lead from the lungs to the heart, and they are divided in the same way as the trachea, following the passages from the trachea throughout the whole lungs, and the passages leading from the heart are on the upper part. There is no passage which is common to them both, but by their union they receive the breath and transmit it through the heart; for one of the passages leads to the right cavity, and the other to the left. We will hereafter speak of the great vein and the aorta in the portion of our work which treats of these parts.

4. In all animals which have lungs and are viviparous, either internally or externally, the lung has more blood than all the other parts; for the whole lung is spongy, and through each perforation branches of the great vein proceed. Those persons are deceived who say that the lungs are empty, drawing their conclusion from dis-

sected animals, from which all the blood has escaped. Of all the viscera the heart alone contains blood, and in the lungs the blood is not in the lungs themselves, but in the veins by which they are perforated. But in the heart itself the blood is in each of the cavities, but the thinnest blood is in the middle cavity.

5. Beneath the lungs is that division of the trunk which is called the diaphragm. It is united to the ribs, the hypochondriac region, and the spine. In the centre is a smooth membranous part, and there are veins extending through it. The human veins are thick in proportion to the size of the body. Under the diaphragm, on the right side is the liver, on the left the spleen, alike in all animals which are furnished with these parts in their natural form and without monstrosity, for already there has been observed an altered order in some quadrupeds. They are joined to the abdomen near the omentum.

6. The appearance of the human spleen is narrow and long, like that of the hog. Generally speaking, and in most animals, the liver is not furnished with a gall, though this is found in some animals. The human liver is round, like that of the ox. This is the case also in animals offered for sacrifice, as in the district of Chalcis, in Euboea, where the sheep have no gall, and in Naxos it is so large in nearly all the animals, that strangers who come to sacrifice are surprised, and think that it is ominous, and not at all natural. The liver is united with the great vein, but has no part in common with the aorta. For a vein branches off from the great vein through the liver, at the place where the gates of the liver, as they are called, are situated. The spleen also is only connected with the great vein, for a vein extends from this to the spleen.

7. Next to these are the kidneys, which lie close to the spine. In their nature they are like the kidneys of oxen. In all animals that have kidneys the right kidney lies higher than the left, and is covered with less fat, and is more dry than the left. This is the same in all animals. Passages lead from them to the great vein and to the aorta, but not to the cavity; for all animals, except the seal, have a cavity in their kidneys, though it is greater in some than in others. The human kidneys, though similar to those of oxen, are more solid than in other animals, and the passages that lead to them end in the body of the kidney; and this is a proof that they do not pass through them, that they contain no blood in the living animal, nor

is it coagulated in them when dead; but they have a small cavity, as I said before. From the cavity of the kidneys two strong passages lead to the bladder, and two others, strong and continuous, lead to the aorta.

8. A hollow, sinewy vein is attached to the middle of each kidney, which extends from the spine through small branches, and disappears towards the hip, though it afterwards appears again upon the hip. The branches of these veins reach to the bladder; for the bladder is placed lowest of all, being united to the passages which proceed from the kidneys by the neck which reaches to the urethra; and nearly all round its circumference it is united by smooth and muscular membranes, very similar in form to those upon the diaphragm of the chest.

9. The human bladder is moderately large in size, and the pudendum is united to the neck of the bladder, having a strong passage above and a small one below. One of these passages leads to the testicles; the other, which is sinewy and cartilaginous, to the bladder. From this are appended the testicles of the male, concerning which we will treat in the part devoted to their consideration. These parts are the same in the female, who differs in none of the internal parts except the womb, the appearance of which may be learned from the drawings in the books on anatomy. Its position is upon the entrails. The bladder is above the uterus. In a future book we will speak of the nature of the uterus generally; for it is not alike, nor has it the same nature in them all.

These are the internal and external parts of the human body, and this is their nature and their manner.

VI

THE FOUR HUMOURS

THE HUMORAL DOCTRINE INFLUENCED MEDICAL THOUGHT FOR TWO THOUSAND years. In the form of various systems—van Helmont, Cullen, Hahnemann, Rokitansky—it is found down to very modern times. Indeed in our doctrines of constitutions (constitutional inadequacy, sthenic, hypersthenic types of body) and in our casual references to psychological constitutional types (a phlegmatic, sanguine, choleric, melancholic man) it survives today.

Empedocles, the philosopher of Sicily, may be said to have first suggested the theory of the four humours. Hippocrates and Galen applied the idea to medicine. Empedocles' dates are uncertain. He died later than 444 B.C. His theory of nature was expounded in a poem only fragments of which have been preserved.

The universe is made up of four elements—earth, air, water, fire. In the human body these correspond to cold, dry, moist, and hot humours. When they are in balance the result is health. When one gets the upper hand, good regimen demands that its opposite be applied. Blood is hot and moist; phlegm is cold and moist; yellow bile is hot and dry; black bile is cold and dry. Men are likely to show a balance towards one humour. They are sanguine, phlegmatic, choleric, or melancholy.

EMPEDOCLES

[Fragments from *Poem on Nature*]

From all of these—sun, earth, sky and sea—are united in all their parts, according to their affinities for one another. Those things that differ most in origin, and mixture and the forms imprinted on each, are most hostile, being altogether unaccustomed to unite.

Out of water and earth, and air and fire mingled together arose the forms and colours of all mortal things.

HIPPOCRATES

[From *On the Constitution of Man. The Four Humours*, Part IV*]

Concerning the composite parts of man's body, it has blood, phlegm, yellow bile and melancholy bile (black bile): these make

* ΠΕΡΙ ΦΥΣΙΟΣ ΑΝΘΡΩΠΟγ. From Littré.

up his parts and through them he feels illness or enjoys health. When all of these elements are truly balanced and mingled, he feels the most perfect health. Illness occurs when one of these qualities is in excess or is lessened in amount or is entirely thrown out of the body. Because when one of these elements is isolated so that it has no balance, by one of the others, the particular part of the body where it is supposed to make balance naturally becomes diseased.

GALEN

[From *On Habits*]

A clear proof of this may be found in the different humours produced by each article of food: some produce a melancholic (atrabiliary) blood, others a phlegmatic blood, or blood containing a fair amount of pale so-called yellow bile: some again produce pure blood.

VII

GALEN

GALEN WAS BORN IN 130 A.D. AT PERGAMUM, A CITY IN NORTHWEST ASIA Minor which rivalled Alexandria in culture. His father, Nicon, was an architect and mathematician. Galen was well educated in mathematics, logic, and philosophy at the university of his native town. When he was seventeen, his father had "vivid dreams" that his son was to study medicine, and he placed him at the Aesculapian at Pergamum. Afterwards Galen prosecuted his medical studies at Corinth, Smyrna, and Alexandria. After four years of practice in his native town, as physician to the gladiators, he went, at the age of thirty-one, to Rome where he soon made a great reputation. There he attended the Emperor Marcus Aurelius, gave lectures in anatomy and had an enormous practice. He fell into controversies with the two predominant medical sects, the Methodists and Erisistratists, and because of his self-assurance and methods of winning notoriety his fellow practitioners made things so uncomfortable for him that he was forced to leave the city in 166. He visited the copper mines of Cyprus where he obtained ores of medical value; Palestine, where he procured some of the juice-balsam of the scriptural "balm of Gilead," and was scarcely back in Pergamum before he was hastily summoned to Rome, 168 A.D., by the order of the Emperors Marcus Aurelius and Verus. The plague was making rapid progress and Verus died of it. Marcus Aurelius urged Galen to accompany him on the German campaign, but Galen pleaded a warning from Aesculapius in a dream, and was excused. He was given medical charge of the young heir apparent, Commodus. At this time he began his medical writings. He continued his connection with the court until the reign of Septimus Severus. He died about 201 A.D., whether at Rome or Pergamum is unknown.

Galen's writings are in Greek, are very voluminous and cover every field of practice. They were regarded during the Middle Ages as the ultimate authority, and were taught as a rigid system, a practice which would have been distasteful to Galen himself.

REFERENCES

WALSH, J. Galen's Second Sojourn in Italy and His Treatment of the Family of Marcus Aurelius. *Medical Life*, Vol. 37, No. 9, New Series No. 120, September, 1930.

Loeb Classical Library. Galen on the Natural Faculties, with an English translation by Arthur John Brock. London, William Heinemann, 1916.

BROCK, A. J. Greek Medicine. J. M. Dent & Sons, New York, 1929. Pp. 130-244.

FINLAYSON, J. Bibliographical Demonstrations of Classical Medical Writers—Galen. *British Medical Journal*, 1:573; 730-771, 1892.

ON THE PULSE*

The heart and all the arteries pulsate with the same rhythm, so that from one you can judge of all; not that it is possible to feel the pulsations of all to the same extent, but only in those areas where the artery is close to the surface. For you could not perceive the pulsations of arteries which are deep in the body or under muscles, or which lie within bones, or which have other bodies in front of them in an animal in natural health. When the body is emaciated, the pulsations felt in the artery that lies along the vertebral column frequently indicate the throbbing of the abdomen, and arteries in the limbs previously indistinguishable have been felt.

But in all cases the pulsations of the arteries in the soles of the feet and the wrist are easily felt. Not so distinct, yet by no means indistinguishable, are the pulsations of the arteries behind the ears and in the arms, and others that do not lie deep.

But you could not find any arteries more convenient or more suitable for taking the pulse than those in the wrists, for they are easily visible, as there is little flesh over them, and it is not necessary to strip any part of the body of clothing for them, as is necessary with many others, and they run in a straight course; and this is of no small help in the accuracy of diagnosis.

The artery will seem to the touch to be distended in every dimension.

In an animal in a normal state of health you will find the artery quite moderately distended; but in abnormal conditions sometimes the tension is too low, sometimes too great in every dimension. Now you must remember what a normal pulse is like, and if you find an abnormal pulse of excessive breadth, you should term it "broad," and if of excessive length "long," and if of excessive depth "deep,"

---

* Galen wrote in several treatises on the pulse, viz: (1) Libellus de Pulsibus ad Tirones, (2) Libri Quatuor de Pulsum Differentiis, (3) Libri Quatuor de Pulsibus Dignoscendis, (4) Libri Quatuor de Causis Pulsuum, (5) Libri Quatuor de Praesagitione ex Pulsibus, (6) Synopsis Sexdecim Librorum de Pulsibus, (7) Pulsuum Compendium. Excerpts from various parts of these treatises give some idea of Galen's pulse lore.

and in like manner the opposite of these "narrow," "short," and "shallow." And a pulse that is in all these dimensions abnormally diminished is termed "small," and one that is abnormally augmented "large." Such, then, are the varieties of pulse as far as dimension goes.

As regards special characteristics, there is swiftness and slowness. The strength of the pulse or the reverse is determined by the force with which it repels the touch; if it repels violently it is strong, if weakly the reverse.

And there are variations in the softness or hardness of the arterial coat; it is soft when the artery appears, so to speak, flesh-like to the touch; hard when it seems dry and hard, like leather.

For the speed or slowness of the pulse depends, we said, on the rate of movement, and the strength or feebleness on the character of the pulsation, and the largeness or smallness on the length of the diastole. But the diastole is not devoid of movement, and there is no need of movement in a soft or hard body for it to be such. These four variations in pulses you will find according to the beat.

Besides, there is a fifth variety depending on the pauses between the beats. For such is the term usually given by medical men to the space of time between the beats, within which the artery expands and contracts. Moreover, I think that beginners should practise themselves as though the systole could not be felt. The two terms I shall use are the pulsation and the pause. By the pulsation I mean the feeling of the artery striking against the finger as it is expanded; by the pause I mean the period of quiescence between the pulsations, according to the length of which normal pulses are rapid, slow, or medium. These you will determine by the length of the pause. For a pulse is rapid when the interval of quiescence is short, slow when the interval is long. You may call it indifferently quiescence or pause between the pulsations or systole.

Regularity and irregularity occur in the above-mentioned variations. By regularity is meant an even and unbroken series. For example, when the dimension of a series of pulsations continues the same, the pulse would be termed regular in size; and if the rate were unaltered, regular in rate. The same holds good in speaking of violence, feebleness, and frequency of pulse. Irregularity means the destruction of even rhythm in whatever varieties of pulse it occurs.

For one may be irregular in size, another in rate, another in violence, feebleness, and frequency, and so on.

Sometimes, too, when a number of beats are definite and regular, an uneven pulsation occurs in the midst of the even ones; and this may happen in various ways. For there may be three regular beats, then the fourth irregular, and so on continuously; or there may be four regular and the fifth irregular. The same thing may occur with any other number, for frequently the sixth is irregular after five regular beats, or the seventh after six. So, then, in these cases, a normal rhythm is not preserved, and so the pulse is not normal; and yet, as a certain fixed order of beat is maintained, it is regular.

For though the number is always constant, yet an irregular beat occurring in the midst of regular beats destroys the normal rhythm; but the recurring cycle insures a certain regularity. But if no period recurs, such a pulse is termed irregular.

An abnormality may occur even in a single pulsation, owing to the different relations which the parts of the artery bear to one another in rest and in movement, and owing to the special movement of each separate pulsation.

When the parts are at rest the abnormality consists in the artery seeming to have been drawn out of position upwards and downwards, and forwards and backwards, and to the right and to the left; but when in movement from the movement of the parts being too quick or too slow, or too soon or too late, too violent or too feeble, too long in duration or too short, being in perpetual movement or not moving at all.

The pulses are arranged, as far as it is possible for one to be taken with another, one with many, many with many, and some of them have a name; for instance, the worm-like (vermiform), the ant-like (formicans), and the hectic pulses. The worm-like pulse is a condition in which it seems as though a worm were creeping along the artery which is in waves of pulsation, the whole of the artery not being distended at the same time. If this takes place, accompanied by a short relaxation, it is called worm-like; but if with a long interval merely wave-like. The worm-like pulse, too, is readily seen to be feeble and beating quickly. But the pulse that has sunk to the extreme limits of feebleness, frequency, and smallness is called ant-like, and this, though it appears to be swift, is not really so.

So, too, the pulse is termed hectic, just as we apply the term to a fever, when it does not vary greatly, but remains much the same continuously, being entangled and never getting free, as the whole condition is one of disease in fevers and pulses of this sort. I think I have said enough for beginners on the subject of varieties of pulses.

Let me then now sum up shortly what I have been speaking of, and then proceed to the subject next in order.

An excessive pulse is that which occurs when the artery is greatly distended in length, depth, and breadth; a pulse is long when the artery is distended only in length, broad when distended in breadth, deep when in depth. A violent pulse is one that strikes strongly against the finger; a soft gentle pulse occurs when the coat of the artery is soft. The pulse is rapid when the artery is distended in a short space of time; frequent when there is little interval; regular when each successive beat is the same; constant when each recurring cycle of beats is the same; a pulse that is uneven in one beat is termed irregular in one beat.

A CASE HISTORY
[From *De Locis Affectis*,* Book II]

Upon the occasion of my first visit to Rome I completely won the admiration of the philosopher Glaucon by the diagnosis which I made in the case of one of his friends. Meeting me one day in the street he shook hands with me and said: "I have just come from the house of a sick man, and I wish that you would visit him with me. He is a Sicilian physician, the same person with whom I was walking when you met me the other day." "What is the matter with him?" I asked. Then coming nearer to me he said, in the frankest manner possible: "Gorgias and Apelas told me yesterday that you had made some diagnoses and prognoses which looked to them more like acts of divination than products of the medical art pure and simple. I would therefore like very much to see some proof, not of your knowledge but of this extraordinary art which you are said to possess." At this very moment we reached the entrance of the patient's house, and so, to my regret, I was prevented from having any further conversation with him on the subject and from explaining to him how the element of good luck often renders it possible for a physician to give, as it were offhand, diagnoses and prognoses of this exceptional

* Lectures on Bibliographic Demonstrations. Translation by James Finlayson. *Brit. M. J.*, 1892.

character. Just as we were approaching the first door, after entering the house, we met a servant who had in his hand a basin which he had brought from the sick room and which he was on his way to empty upon the dung heap. As we passed him I appeared not to pay any attention to the contents of the basin, but at a mere glance I perceived that they consisted of a thin sanio-sanguinolent fluid, in which floated excrementitious masses that resembled shreds of flesh —an unmistakable evidence of disease of the liver. Glaucon and I, not a word having been spoken by either of us, passed on into the patient's room. When I put out my hand to feel of the latter's pulse, he called my attention to the fact that he had just had a stool, and that, owing to the circumstance of his having gotten out of bed, his pulse might be accelerated. It was in fact somewhat more rapid than it should be, but I attributed this to the existence of an inflammation. Then, observing upon the window sill a vessel containing a mixture of hyssop and honey and water, I made up my mind that the patient, who was himself a physician, believed that the malady from which he was suffering was a pleurisy; the pain which he experienced on the right side in the region of the false ribs (and which is also associated with inflammation of the liver) confirming him in this belief, and thus inducing him to order for the relief of the slight accompanying cough the mixture to which I have called attention. It was then that the idea came into my mind that, as fortune had thrown the opportunity in my way, I would avail myself of it to enhance my reputation in Glaucon's estimation. Accordingly, placing my hand on the patient's right side over the false rib, I remarked: "This is the spot where the disease is located." He, supposing that I must have gained this knowledge by simply feeling his pulse, replied with a look which plainly expressed admiration mingled with astonishment, that I was entirely right. "And," I added simply to increase his astonishment, "you will doubtless admit that at long intervals you feel impelled to indulge in a shallow, dry cough, unaccompanied by any expectoration." As luck would have it, he coughed in just this manner almost before I had got the words out of my mouth. At this Glaucon, who had hitherto not spoken a word, broke out into a volley of praises. "Do not imagine," I replied, "that what you have observed represents the utmost of which medical art is capable in the matter of fathoming the mysteries of disease in a living person. There still remain one or two other symptoms to which I will direct your attention." Turning then to the patient I remarked: "When you draw a longer breath you feel a more marked pain, do you not, in the region which I indicated; and with this pain there is associated a sense of weight in the hypochondrium?" At these words the patient expressed his astonishment and admiration in the strongest possible terms. I wanted to go a step farther and announce to my audience still another symptom which is sometimes observed in the more serious maladies of the liver (scirrhus, for example), but I was afraid that I might compromise the laudation which had been bestowed upon me. It then occurred to me that I might safely make the announcement if I put it somewhat in the form of a prognosis. So I remarked to the patient: "You will probably soon experience, if you have not already done so, a sensation of something pulling upon the right clavicle." He admitted that he had already noticed this symptom. "Then I will give just one more evidence of this power of divination which you believe that I possess. You, yourself, before I arrived on the scene, had made up your mind that your ailment was an attack of pleurisy, etc."

Glaucon's confidence in me and in the medical art, after this episode, was unbounded.

On the Functions of the Human Body*
[From *On the Natural Faculties*, Book I, Section XIII]

ON THE KIDNEYS

The doctrines of Hippocrates may be judged as to their truth and exactitude not only from a study of his commentators and opponents, but by going directly to Nature and observing the functions of animals, the subjects of natural research. Anyone who does not believe, after such a study, that there exists in the parts of animals the faculty for doing the work for which they are most capable must continuously deny the facts. Take as an example the study made by Asclepiades, the physician, of the kidneys. That they are the organs for separating out the urine was the belief of such physicians as Hippocrates, Diocles, Erisistratus and Praxagoras, but in truth any butcher knows this from the fact that he sees every day the position of the kidneys and the duct (called the ureter) which runs from each kidney into the bladder and by studying this anatomy he reasons what their use is and the nature of their functions. But aside from

* From ΓΑΛΗΝΟΤ ΠΕΡΙ ΦΤΣΙΚΩΝ ΔΤΝΑΜΕΩΝ. Translated under the supervision of the editor.

the butchers, all patients who suffer from painful urination or retention of urine call themselves nephritics or kidney patients when they have a colic in the loins or pass sandy matter in the urine.

ON THE UTERUS

In the case of the eliminative function, the os opens, whilst the whole fundus approaches as near as possible to the os, expelling the foetus as it does so: and along with the fundus the nearby parts—which form an encirclement round the whole organ—co-operate in the work; they squeeze upon the foetus and propel it bodily outwards. And, in many women who have had a long labor, violent pains cause forcible prolapse of the whole womb; here almost the same thing happens as frequently occurs in wrestling-bouts and struggles, when in our eagerness to overturn and throw others we are ourselves upset along with them; for similarly when the uterus is forcing the foetus forward it sometimes becomes entirely prolapsed, and particularly when the ligaments connecting it with the spine happen to be lax.

A wonderful device of Nature's also is this—that, when the foetus is alive, the os uteri is closed with perfect accuracy, but if it dies, the os at once opens up to the extent which is necessary for the foetus to make its exit. The midwife, however, does not make the parturient woman get up at once and sit down on the (obstetric) chair, but she begins by palpating the os as it gradually dilates, and the first thing she says is that it has dilated "enough to admit the little finger," then that "it is bigger now," and as we make enquiries from time to time, she answers that the size of the dilation is increasing. And when it is sufficient to allow of the passage of the foetus, she then makes the patient get up from her bed and sit on the chair, and bids her make every effort to expel the child. Now, this additional work which the patient does of herself is no longer the work of the uterus but of the abdominal muscles, which also help us in defaecation and micturition.

ON BONES
[From *On Anatomical Procedure*,* Book i, Chapter ii]

The human bones are subjects of study with which you should first become perfectly familiar. You cannot merely read about the

* Translation by the editor.

bones in one of these books which are called by some "Osteology," by others "The Skeleton," and by others simply "On Bones," such as this my own book; which is much more reliable and exact than any previously written on the subject. Pursue by hard study, then, not only the descriptions of the bones in the book, but also acquaint yourself with the appearance of each of the bones, by the use of your own eyes handling each bone by itself so that you become a first-hand observer.

At Alexandria this is very easy, since the physicians in that country accompany the instruction they give to their students with opportunities for personal inspection (at autopsy). Hence you must try to get to Alexandria for this reason alone, if for no other. But if you cannot manage this, still it is not impossible to obtain a view of human bones. Personally I have very often had a chance to do this where tombs or monuments have become broken up. On one occasion a river, having risen to the level of a grave which had been carelessly constructed a few months previously, easily disintegrated this; then by the force of its current it swept right over the dead man's body, of which the flesh had already putrefied while the bones were still closely attached to one another. This it carried away downstream for the distance of a league, till, coming to a lake-like stretch with sloping banks, it here deposited the corpse. And here the latter lay ready for inspection, just as though prepared by a doctor for his pupil's lesson.

Once also I examined the skeleton of a robber, lying on a mountain-side a short distance from the road. This man had been killed by some traveller whom he had attacked, but who had been too quick for him. None of the inhabitants of the district would bury him, but in their detestation of him they were delighted when his body was eaten by birds of prey: the latter, in fact, devoured the flesh in two days and left the skeleton ready, as it were, for anyone who cared to enjoy an anatomical demonstration.

As regards yourself, then, even if you do not have the luck to see anything like this, still you can dissect an ape, and learn each of the bones from it, by carefully removing the flesh. For this purpose you must choose apes which most resemble man. Such are those in whom the jaws are not prominent nor the canine teeth large. In such apes you will also find the other parts as in man, whence they walk and run on two legs. Those of them, again, that are like the dog-headed baboons (*cynocephalae*) have longer muzzles and large canine teeth; they have difficulty in standing upright on two legs, let alone walking about or running.

CASE HISTORY

[From *On Prognostics*,* Chapter xi]

The case of the Emperor Marcus Aurelius himself was quite wonderful. He believed as did the physicians of his entourage who had gone abroad with him that he was beginning to have a feverish paroxysm. But they all proved wrong both on the second and third day. He had on the preceding day taken a draught of bitter aloes at the first hour, and then some theriac as was his daily custom. Next he took some food about the sixth hour, washed at sunset and had a small meal. During the whole night there ensued colicky pains with intestinal evacuations. This made him feverish, and when his attendant physicians observed this, they gave orders that he should be kept quiet; then they prescribed slop diet at the ninth hour. After this I was myself also summoned to come and sleep in the Palace. Then, when the lamps were newly lit, a messenger came to call me at the Emperor's orders. Three doctors had been observing him since about daybreak, and two of them felt his pulse, and they all considered this the beginning of a febrile attack. I stood by, however, without saying anything; so the Emperor, looking at me first, asked why, when the others felt his pulse, I alone did not do so. I said to him, "Two of these gentlemen have already done this, and probably when they were abroad with you they already learned by experience the characteristics of your pulse; hence I expect they will be better able to judge its present condition." On my saying this he bade me also feel his pulse. It seemed to me that, taking his age and constitution into account, the pulse was far from indicating the beginning of a febrile attack. I declared that this was not the onset of fever, but that his stomach was overloaded by the food he had taken, which had turned to phlegm prior to ejection.

My diagnosis seemed praiseworthy to the Emperor, and he repeated three times in succession: "That's it. It is just what you say. I feel I am weighed down by chilling food." He, therefore, asked from all who were expert in the study of the pulse, one of whom

* Translation by the editor.

was even Archigenes, some peculiar sign of an attack. Some answered that the systole of the artery could be found at the beginning, others that the systole was by no means evident, while I said that an accurate touch perceived differences even when these were very slight.

He now asked what was to be done. The other physicians suggested sending him to the bath, or giving him nourishment. When he asked me I answered as I felt saying, "If another was so affected I would give him some wine to which he was accustomed with a dash of pepper in it. In the case of a king for whom physicians make special effort to do even the little extras which add to immediate comfort, I would in addition apply to the orifice of the stomach a wool bandage with warm spikenard ointment." He asserted that it was his custom when anything was wrong with his stomach to apply a warm spikenard ointment on purple wool. Ordering Pitholaus to prepare it, and dismissing us he had his feet rubbed by warm hands, and asked for Sabine wine with pepper.

After the drink he said to Pitholaus, that there was one physician who was not hide-bound by rules, and from this time he never stopped lauding me. He is the First of Physicians, said he, and also of Philosophers. For Marcus had already had experience with many, not only desirous of money, but contentious, vain-glorious, envious and malignant.

VIII

ARETAEUS THE CAPPADOCIAN

ARETAEUS, THE CAPPADOCIAN (SECOND TO THIRD CENTURY A.D.) WAS OF THE late Greek period. He gave good accounts of pneumonia, empyema, diabetes, tetanus, and paralysis of the hemiplegic form, pointing out that the paralysis occurred on the opposite side to the lesion in the brain.

ON ULCERATIONS ABOUT THE TONSILS

[From *On the Causes and Symptoms of Acute Diseases,** Book I,
Chapter IX]

Ulcers occur on the tonsils, some indeed of an ordinary nature, mild and innocuous; but others of an unusual kind, pestilential and fatal. Such as are clean, small, superficial, without inflammation and without pain, are mild; but such as are hard, hollow, foul and covered with a white, livid or black concretion, are pestilential. . . . And if the disease spread outwardly to the mouth and reach the columella (uvula) and divide it asunder, and if it extend to the tongue, the gums, and the alveoli, the teeth also become loosened and black; and the inflammation seizes the neck; and these die within a few days from the inflammation, fever, foetid smell and want of food. But, if it spread to the thorax by the windpipe, it occasions death by suffocation within the space of a day.

. . . Wherefore the children, until puberty especially suffer, for children have large and cold respiration; for there is moist heat in them; . . . and they bawl loud both in anger and in sport. . . .

The land of Egypt especially engenders it, the air thereof being dry from respiration, and the food diversified consisting of roots, herbs of many kinds, acrid seeds, and thick drink; namely the water of the Nile, and the sort of ale prepared from barley. Syria also, and more especially Coelosyria, engenders these diseases, and hence they have been named Egyptian and Syrian ulcers.

*ΑΡΕΤΑΙΟΤ ΚΑΠΠΑΔΟΚΟΤ ΤΑΣΩΖΟΜΕΝΑ. Translated from the Greek by Francis Adams, Sydenham Society, 1856.

CURE OF AFFECTIONS ABOUT THE COLUMELLA (OR UVULA)

[From *On the Therapeutics of Acute Diseases,** Book I, Chapter VIII]

Of the affections which form about the columella, some require to be treated by excision; but the surgical treatment of such cases does not come within the design of this work. Some are to be treated as acute affections; for some of them readily prove fatal by suffocation and dyspnoea. . . . If, then, the patient be young, we must open the vein at the elbow, and evacuate copiously by a larger incision than usual; for such an abstraction frees one from suffocation, as it were, from strangulation. It is necessary, also, to inject with a mild clyster, but afterwards with an acrid one again and again, until one has drawn from the parts above by revulsion, and let ligatures be applied to the extremities above the ankles and knees, and above the wrists and forearms to the arms. But if the suffocation be urgent, we must apply a cupping instrument to the occiput and to the thorax, with some scarifications, and also do everything described by me under synanche; for the mode of death is the same in both. We must also use the same medicines to the mouth, both astringents and emollients, with fomentations of the external parts, cataplasms, and liniments to the mouth. For the forms named columella and uva, as an astringent medicine take the juice of pomegranate, acacia dissolved in honey or water, hypocists, Samian, Lemnian, or Sinopic earth and the inspisated juice of sour grapes. But if the diseased part be ulcerated, gum and starch moistened in the decoction of roses or of dates, and the juice of ptisan or of spelt. But in columella let there be more of the stronger medicines, from myrrh, costus and cyperus; for the columella endures these acrid substances. But should the part suffocate, in some instances, even the bones of the palate become diseased, and the patients have died, wasted by a protracted consumption.

*ΑΡΕΤΑΙΟΥ ΚΑΠΠΑΔΟΚΟΥ ΤΑΣΩΖΟΜΕΝΑ. Translated from the Greek by Francis Adams. Sydenham Society, 1856.

On Paralysis

[From *On the Causes and Symptoms of Chronic Diseases,** Book I,

Chapter VII]

. . . Wherefore, the parts are sometimes paralyzed singly, as one eyebrow, or a finger, or still larger a hand, or a leg; and sometimes more together; and sometimes the right or the left only, or each by itself, or all together either entirely or in a less degree; and the parts only which are distant, homonymous and in pairs—the eyes, hands and legs; and also the parts which cohere, as the nose on one side, the tongue to the middle line of separation, and the one tonsil, the isthmus faucium, and the parts concerned in deglutition to one half. I fancy also, that sometimes the stomach, the bladder, and the rectum, as far as its extremity, suffers in like manner. But the internal parts when in a paralytic state, are concealed from the sight. . . .

If, therefore, the commencement of the affection be below the head, such as the membrane of the spinal marrow, the parts which are homonymous and connected with it are paralyzed: the right on the right side and the left on the left side. But if the head be primarily affected on the right side, the left side of the body will be paralyzed: and the right if on the left side. The cause of this is the interchange in the origins of the nerves; for they do not pass along the same side, the right on the right side, until their terminations; but each of them passes over to the other side from that of its origin, decussating each other in the form of the letter X.

On Diabetes

[From *On the Causes and Symptoms of Chronic Diseases,*† Book II,

Chapter II]

Diabetes is a wonderful affection, not very frequent among men, being a melting down of the flesh and limbs into urine. Its cause is of a cold and humid nature as in dropsy. The course is the common one, namely the kidneys and bladder; for the patients never stop making water, but the flow is incessant, as if from the opening of aqueducts. The nature of the disease then is chronic, and it takes

* Translated from the Greek by Francis Adams. Sydenham Society, 1856.
†ΑΡΕΤΑΙΟΤ ΚΑΠΠΑΔΟΚΟΤ ΤΑΣΩΟΜΕΝΑ.   Translated from the Greek by Francis Adams. Sydenham Society, 1856.

a long period to form, but the patient is short-lived if the constitution of the disease be completely established; for the melting is
rapid, the death speedy. Moreover life is disgusting and painful;
thirst unquenchable; excessive drinking, which however is disproportionate to the large quantity of urine for more urine is passed;
and one cannot stop them either from drinking or making water.
Or if for a time they abstain from drinking, their mouth becomes
parched and their body dry; the viscera seem as if scorched up; they
are affected with nausea, restlessness, and a burning thirst; and at
no distant term they expire. Thirst as if scorched up with fire. But
by what method could they be restrained from making water? Or
how can shame become more potent than pain? And even if they
were to restrain themselves for a short time they become swelled in
the loins, scrotum and hips; and when they give vent they discharge
the collected urine, and the swellings subside for the overflow passes
to the bladder.

If the disease be fully established, it is strongly marked; but if it
be merely coming on, the patients have the mouth parched, saliva
white, frothy, as if from thirst (for the thirst is not yet confirmed),
weight in the hypochondriac region. A sensation of heat or of cold
from the stomach to the bladder is, as it were, the advent of the
approaching disease; they now make a little more water than usual
and there is thirst but not yet great.

But, if it increase still more, the heat is small indeed, but pungent,
and seated in the intestines; the abdomen shrivelled, veins pro-
tuberant, general emaciation when the quantity of urine and thirst
have already increased; and when, at the same time, the sensation
appears in the extremity of the member, the patients immediately
make water. Hence the disease appears to me to have got the name
diabetes, as if from the Greek word which signifies a siphon, because
the fluid does not remain in the body, but uses the man's body as a
bladder whereby to leave it. They stand out for a certain time,
though not very long for they pass urine with pain, and the emaciation is dreadful; nor does any great portion of the drink get into
the system, and many parts of the flesh pass out along with the
urine.

The cause of it may be that some of the acute diseases may have
terminated in this; and during the crisis of the disease may have left
some malignity lurking in the part. It is not improbable also, that

something pernicious, derived from the other diseases which attack the bladder and kidneys, may sometimes prove the cause of this affection. But if anyone is bitten by the dipsas [a species of viper] the affection induced by the wound is of this nature; for the reptile, the dipsas, if it bite one, kindles up an unquenchable thirst. For they drink copiously, not as a remedy for thirst, but so as to produce repletion of the bowels, by the insatiable desire of drink. But if one be pained by the distention of the bowels and feel uncomfortable, and abstain from drink for a little, he again drinks copiously from thirst, and thus the evils alternate; for the thirst and the drink conspire together. Others do not pass urine, nor is there any relief from what is drank. Wherefore, what from insatiable thirst, an overflow of liquids, and distention of the belly, the patients have suddenly burst.

TREATMENT

The affection of diabetes is a species of dropsy, both in cause and in condition, differing only in the place by which the humour runs. For, indeed, in ascites the receptacle is the peritoneum, and it has no outlet, but remains there and accumulates. But in diabetes, the flow of the humour from the affected part and the melting are the same, but the defluxion is determined to the kidneys and the bladder; and in dropsical cases this is the outlet when the disease takes a favorable turn; and it is good when it proves a solution of the cause, and not merely a lightening of the burden. In the latter disease the thirst is greater; for the fluid running off dries the body.

But the remedies for the stoppage of the melting are the same as those for dropsy. For the thirst there is need of a powerful remedy for in kind it is the greatest of all sufferings; and when a fluid is drunk it stimulates the discharge of urine; and sometimes as it flows off it melts and carries away with it particles of the body. Medicines then which cure thirst are required, for the thirst is great with an insatiable desire of drink, so that no amount of fluid would be sufficient to cure the thirst. We must, then, by all means strengthen the stomach, which is the fountain of the thirst. When, therefore, you have purged with the hiera, use as Epithemes the nard, mastich, dates, and raw quinces; the juice of these with nard and rose oil is very good for lotions; their pulp with mastich and dates, form a cataplasm. . . .

But the water used as drink is to be boiled with autumn fruit. The food is to be milk, and with it cereals, starch, groats of spelt, gruels. Astringent wines to give tone to the stomach, and these but little diluted, in order to dissipate and clear away the other humours; for thirst is engendered by saltish things.

IX

CELSUS

AURELIUS CORNELIUS CELSUS IS A SOMEWHAT SHADOWY FIGURE WHO LIVED in the reign of Tiberius Caesar (42 B.C. to 37 A.D.). He wrote in Latin. He was a noble, a great landowner, and compiled his books in order to record the best that was known about agriculture, veterinary medicine, and human illnesses. *De re medicinae,* his medical treatise, was partly based on the Hippocratic canon and other classical works, and partly a record of his own experience. Neglected by the narrow pedants of the Middle Ages, it became very popular with the revival of learning, and was one of the first medical books to be printed (1478). It consists of eight books, the first four being descriptions of diseases that can be treated by diet and regimen, the last four diseases that can be treated by diet and surgery. The fourth book contains a description of inflammation, giving the four classical signs—rubor, dolor, color, tumor.

REFERENCE
CELSUS. De Re Medicinae. Translated by W. G. Spencer. Loeb Classical Library. Cambridge, Harvard University Press, 1935. 3 vols.

PREFACE
[From *On Medical Subjects**]

As agriculture promises food to the healthy, so medicine promises health to the sick. There is no place in the world where this art is not found; for even the most barbarous nations are acquainted with herbs, and other easy remedies for wounds and diseases. However, it has been more improved by the Greeks than any other people; though not from the infancy of that nation, but only a few ages before our own times.

Those, then, who declare for a theory in medicine, look upon the following things as necessary: the knowledge of the occult and constituent causes of distemper; next, of the evident ones; then of the natural actions; and, lastly, of the internal parts. They call these causes occult in which we inquire of what principles our bodies are

* De Re Medicinae. Translated from the Latin by James Grieve, M.D. London, 1838.

composed, what constitutes health, and what sickness. For they hold it impossible that anyone should know how to cure disease, if he be ignorant of the causes whence they proceed; and that it is not to be doubted but one method of cure is required, if redundancy or deficiency in any of the four principles* be the cause of diseases, as some philosophers have maintained; another, if the fault be wholly in the humours, as Herophilus thought; another if in the inspired air, as Hippocrates believed; another, if the blood be transfused into those vessels which are designed only for air, and occasion an inflammation, which the Greeks call phlegmon, and that inflammation causes such a commotion as we observe in a fever, which was the opinion of Erasistratus; another if the corpuscles, passing through the invisible pores should stop, and obstruct the passage as Asclepiades maintained; that he will proceed in the proper method of curing a disease, who is not deceived in its original cause.

On the other hand, those who from experience style themselves empirics, admit indeed the evident causes as necessary; but affirm the inquiry after the occult causes and natural actions to be fruitless, because nature is incomprehensible.

Of Disease of the Large Intestine and Its Cure

[From *On Medical Subjects*, Book iv, Chapter xii]

That Distemper, which is seated in the large intestine, principally affects that part where I mentioned the caecum to be situated. There is a violent inflation: vehement pains, especially on the right side; the intestine seems to be inverted, which almost forces out the wind. In most people it comes after colds and crudity, then ceases; and while they live it often returns, and torments but does not shorten life.

When this pain has begun, it is proper to apply dry and warm fomentations, but first of all mild, and then stronger; and at the same time by friction to make a derivation of the matter to the extremities, that is, of the legs and arms; if it is not removed, to make use of dry cupping, where the pain is. There is also a medicine calculated for this distemper which is called colicon. Cassius claimed the glory of this invention. It has the best effect given by way of

---

* Four principles air, earth, fire, and water.

potion; but even externally applied, by dispersing the wind, it eases the pain.

Neither food nor drink should be given, till the pain be over. The regimen for such patients I have already mentioned. The composition which is called colicon, consists of the following ingredients: of costos, anise, castor each p. Xiii, parsley p. Xiv, long pepper and round each p. Xv, tears of poppy, round cyperus myrrh, nard, of each p. Xvi. These are incorporated in honey. Now this may be swallowed alone, and taken with warm water.*

OF DISEASES OF THE FAUCES AND THEIR CURES

[From *On Medical Subjects*, Book IV, Chapter IV]

Of a spitting of blood. A spitting of blood may strike a greater terror. But that sometimes is less, and sometimes more dangerous. For it issues sometimes from the gums, and without expectorating anything; but breaks out from the mouth in the same manner as from the nose. . . . But sometimes, when the throat and windpipe are ulcerated, a frequent cough forces out blood too. Neither is it uncommon for it to come from the lungs, or the breast or the side of the liver. Women whose menses are suppressed, often have these spittings. . . . And it often happens that the blood is followed by pus.

Now, sometimes, stopping the blood is alone sufficient to effect a cure. But if ulcers have followed, if pus, if there be a cough, diseases are formed, which differ in nature and danger according to the parts they possess.

OF A PERIPNEUMONY, AND ITS CURE

[From *On Medical Subjects*, Book IV, Chapter VII]

From the frame of the body we must proceed to the bowels; and first of all to the lungs. Whence a violent and acute distemper arises, which the Greeks call peripneumonia. The nature of it is this. The whole lungs are affected. And their disorder is followed by a cough bringing up bile, or pus a weight of the precordia and the whole breast, difficulty of breathing, violent fevers, continual watching,

---

* Note on Celsus' weights and measures: The symbol }(is the denarius. It probably is about 62 grains Troy wt. The exact symbol is as above but we have used the Roman X. The small numerals following probably represent fractions of the denarius.

prostration of appetite, and a consumption. This kind of a distemper is attended with more danger than pain.

It is fit if the strength will admit of it to let blood; if not to make use of dry cupping to the praecordia; and if the patient can endure it by gestation to dissipate; if he cannot bear that, to move him gently within the house; to give him, in drink, hyssop boiled with a dry fig; or a decoction of hyssop or rue in hydromel; to use friction longest upon the shoulders, a little shorter on the arms and feet and legs, gentle over the lungs and to do this twice every day.

As to the diet, he ought neither to have salt things, nor acrid, nor bitters, nor astringents; but what is of the milder kind. . . .

And it is not improper during the violence of a distemper to keep the windows close upon the patient; when it has a little abated, three or four times a day to open them a little and let in fresh air. Then when he begins to recover, for several days to abstain from wine.

X

DIOSCORIDES

DIOSCORIDES (CA. 40-CA. 90 A.D.) WAS A GREEK WHO SERVED IN THE ROMAN armies as a surgeon in the time of Nero. He collected plant specimens all over the known world, as he traveled with the armies, and wrote a work variously known as *Materia Medica* or *Herbal*. There is a good English translation called *Greek Herbal* (Gunther, Oxford Press, 1934). Dioscorides followed a plan which differed from that of the Ebers Papyrus—the Papyrus named a disease and listed all the remedies for it: Dioscorides described a plant and told all the diseases it would cure. His work was authoritatively used until the seventeenth century (see Chaucer's *Doctor of Physic*).

ON WILLOW

[From the *Materia Medica*,* Book 1]

The willow is a tree known by every one. Its fruit, leaves, bark and sap possess astringent properties. The leaves rubbed down with a little pepper and wine are good for patients with diseased intestines. It prevents conception if taken by itself with water. The fruit helps blood-spitting; the bark does the same. If burned and applied mixed with vinegar it removes knots and warts. The juice from the leaves and bark warmed in a pomegranate shell with rose oil is good for earache. The decoction is a fine fomentation for gout. It clears away scruf too. Juice is got from it by cutting into the bark when the tree is in flower. It is found solidified inside. Lastly, it has the property of clearing away opacities from the pupils of the eyes.

ON MUSTARD

[From the *Materia Medica*, Book 11]

It has heating and digestive properties, it causes shrinkage, and if chewed induces phlegm. Its juice mixed with hydromel makes a

* Translation by John D. Comrie, 1880.

good gargle for swollen tonsils and chronic thickening and roughness of the air-passages. If held to the nostrils bruised it induces
sneezing. It helps epileptics and arouses women stifled by the womb.
Lethargic people have their heads shaved and this smeared over.
Mixed with figs and applied till the part is red it is good for lum-
bago and diseased spleens, and in short for every chronic pain whenever we wish to bring anything from below to the surface by the
principle of counter-irritation. If applied as a plaster it cures bald-
ness. It clears the complexion, and in honey or tallow or salve removes puffiness under the eyes. It is used in vinegar as an ointment
for leprosy and severe eruptions. It is swallowed dry in recurring
fevers, sprinkled on the drink like barley-meal. It can be usefully
mixed with blistering or desquamating plasters. It helps dullness
of hearing and noises in the head when ground with figs and put
into the ears. Its juice mixed with honey is good to smear on dim
eyes and rough eyelids. The juice is pressed out from fresh seed and
dried in the sun.

On Artemisia Maritima (Santonica)

[From the *Materia Medica*, Book iii]

Artemisia maritima—some too call it seriphos—which grows on the
Taurus range throughout Cappadocia and in Taphosiris in Egypt,
and which the worshippers of Isis use in place of palm-leaves, is a
light-stemmed grass like small absinth, full of little seeds, slightly bitter, bad for the stomach, of heavy smell, astringent with heating
power. Boiled by itself or with rice and taken with honey, it kills
ascarides and round worms, expelling them easily. Cooked with soup
or porridge it can do the same. The sheep grazing in Cappadocia are
especially fat.

On Castor Oil

[From the *Materia Medica*, Book v]

Cici or croton some call wild sesame, others Cyprian sesele, the
Egytians systhamna, others trixis, soothsayers fever-blood, the
Romans ricinum, and others lupa. It is called croton (a tick) on account of the resemblance of the seed to this animal. It is a tree the
size of a small fig; the leaves like those of a plane but bigger,

smoother, and darker. Its trunk and stems are hollow like a reed. The fruit is in rough bunches, and when skinned is like the tick animals, from it is pressed castor-oil, as it is called. It is not edible, but is useful for lamps and plasters. Thirty berries cleaned, pounded, and swallowed bring phlegm, bile, and water through the bowels. They also cause vomiting. Purging by this means is very disagreeable and troublesome, upsetting the stomach violently. Berries pounded and made into a plaster cure acne and freckles. The leaves crushed with barley-flour check swellings and inflammation of the eyes and swollen breasts, and relieve attacks of erysipelas either alone or made into a plaster with vinegar.

XI

ARABIC MEDICINE

THE ARABIANS ALONE TENDED THE LAMP OF THE INTELLECT IN EUROPE DURING the Middle Ages. Filtered across the Pyrenees from Spain, Arabic culture was a stimulating force. The introduction of arithmetic alone was exciting (Arabic numerals made possible addition, subtraction, multiplication, division, which cannot be done with Roman numerals; the Arabs also introduced zero), but Islam's influence in medicine was deep and lasting.

REFERENCES

Legacy of Islam. London, Oxford University Press, 1931.
BROWNE, E. G. Arabian Medicine. London, Cambridge University Press, 1921.

THE ARABIAN NIGHTS

In *The Arabian Nights* is found the slave girl Tawaddud's account of Arabian medicine. In the story of the spendthrift heir (Burton's translation, Vol. v) an accomplished slave girl recites the foundation of Mohammedan theology, law, philosophy, medicine, astronomy, and other arts and sciences. We get some idea of what Arabian education must have been if this is what a slave girl knew. It is generally believed that the Thousand and One Nights is a product of the tenth century, that it was derived from Indian or Persian sources, but was thoroughly Arabianized. Lane says it is related to its Persian sources only as the Aeneid is related to the Odyssey. Some scholars date it as late as the sixteenth century. It appeared first in the western world in the French translation of Antoine Galland (1704-1717).

NOW WHEN IT WAS THE FOUR HUNDRED AND FORTY-NINTH NIGHT*

She said, it hath reached me, O auspicious King, that when the damsel defeated the Koranist and took off his clothes and sent him away confused, then came forward the skilled physician and said to her,

*Translation by Richard Burton, 1885.

"We are free of theology and come now to physiology. Tell me, therefore, how is man made; how many veins, bones and vertebrae are there in his body; which is the first and chief vein and why Adam was named Adam?" She replied, "Adam was called Adam, because of his udmah, that is, the wheaten colour of his complexion and also (it is said) because he was created of the adim of the earth, that is to say, of the surface-soil. His breast was made of the earth of the Ka'abah, his head of earth from the East and his legs of earth from the West. There were created for him seven doors in his head, viz., the eyes, the ears, the nostrils and the mouth, and two passages, before and behind. The eyes were made the seat of the sight-sense, the ears the seat of the hearing-sense, the nostrils the seat of the smell-sense, the mouth the seat of the taste-sense and the tongue to utter what is in the heart of man. Now Adam was made of a compound of the four elements, which be water, earth, fire and air. The yellow bile is the humour of fire, being hot-dry; the black bile that of the earth, being cold-dry; the phlegm that of water, being cold-moist. There were made in men three hundred and sixty veins, two hundred and forty-nine bones, and three souls or spirits, the animal, the rational and the natural, to each of which is allotted its proper function. Moreover, Allah made him a heart and a spleen and lungs and six intestines and a liver and two kidneys and buttocks and brain and bones and skin and five senses; hearing, seeing, smell, taste, touch. The heart He set on the left side of the breast and made the stomach the guide and governor thereof. He appointed the lungs for a fan to the heart and established the liver on the right side, opposite thereto. Moreover, He made, besides this, the diaphragm and the viscera and set up the bones of the breast and latticed them with the ribs." "How many ventricles are there in a man's head?" —Three, which contain five faculties, styled the intrinsic senses, to wit, common sense, imagination, the thinking faculty, perception, and memory. . . .

"Describe to me the configuration of the bones." She replied, "Man's frame consists of two hundred and forty bones, which are divided into three parts, the head, the trunk and the extremities. The head is divided into calvarium and face. The skull is constructed of eight bones, and to it are attached the four osselets of the ear. The face is furnished with an upper jaw of eleven bones and a lower jaw of one; and to these are added the teeth two-and-thirty in number, and the os hyoides. The trunk is divided into spinal column, breast and basin. The spinal column is made up of four-and-twenty bones, called Fikar or vertebrae; the breast, of the breastbone and the ribs, which are four-and-twenty in number, twelve on each side; and the basin of the hips, the sacrum and the os coccygis. The extremities are divided into upper and lower, arms and legs. The arms are again divided firstly into shoulder, comprising shoulder blades and collar bone; secondly into the upper arm which is one bone; thirdly into fore-arm, composed of two bones, the radius and the ulna, and fourthly into the hand, consisting of the wrist, the metacarpus of five and the fingers, which number five, of three bones each, called the phalanges, except the thumb, which hath but two. The lower extremities are divided, firstly into thigh, which is one bone, secondly into leg, composed of three bones, the tibia, the fibula and the patella, and thirdly into the foot, divided, like the hand, into tarsus, metatarsus and toes; and is composed of seven bones, ranged in two rows, two in one and five in the other; and the metatarsus is composed of five bones and the toes number five, each of three phalanges except the big toe which hath only two," (?) "Which is the root of the veins?" The aorta, from which they ramify, and they are many, none knoweth the tale of them save He who created them; but I repeat, it is said that they number three hundred and sixty. The liver is the seat of pity, the spleen of laughter and the kidneys of craft; the lungs are ventilators, the stomach the storehouse and the heart the prop and pillar of the body. When the heart is sound, the whole body is sound, and when the heart is corrupt, the whole body is corrupt. (?) "What are the outward signs and symptoms evidencing disease in the members of the body, both external and internal?" A physician, who is a man of understanding, looketh into the state of the body and is guided by the feel of the hands, according as they are firm, or flabby, hot, cool, moist or dry. Internal disorders are also indicated by external symptoms, such as yellowness of the white of the eyes, which denoteth jaundice, and bending of the back, which denoteth disease of the lungs. And Shahrazad perceived the dawn of day and ceased saying her permitted say.

NOW WHEN IT WAS THE FOUR HUNDRED AND FIFTY-FIRST NIGHT

She said, It hath reached me, O auspicious King, that when the damsel had described to the doctor the outer signs and symptoms quoth he, "Thou hast replied aright! now what are the internal symptoms of disease?" The science of the diagnosis of disease by internal symptoms is founded upon six canons, (1) the patient's actions; (2) what is evacuated from his body; (3) the nature of the pain and (4) the site thereof; (5) swelling and (6) the effluvia given off his person. "How cometh hurt to the head?" By the ingestion of food upon food, before the first be digested, and by fullness upon fullness; this it is that wasteth peoples. He who would live long, let him be early with the morning-meal and not late with the evening-meal; let him be sparing of commerce with women and chary of such depletory measures as cupping and blood-letting; and let him make of his belly three parts, one for food, one for drink and the third for air; for that a man's intestines are eighteen spans in length and it befitteth that he appoint six for meat, six for drink, and six for breath. If he walk, let him go gently; it will be wholesomer for him and better for his body and more in accordance with the saying of the Almighty, "Walk not proudly on the earth." "What are the symptoms of yellow bile and what is to be feared therefrom?" The symptoms are sallow complexion and bitter taste in the mouth with dryness; failure of appetite, venereal and other, and rapid pulse; and the patient hath to fear high fever and delirium and eruptions and jaundice and tumour and ulcers of the bowels and excessive thirst. "What are the symptoms of black bile and what hath the patient to fear from it, an it get the mastery of the body?" The symptoms are false appetite and great mental disquiet and cark and care; and it behoveth that it be evacuated, else it will generate melancholia and leprosy and cancer and disease of the spleen and ulceration of the bowels.

"Thou hast replied aright! What sayest thou of the Hammam?"

"Let not the full man enter it. Quoth the Prophet, "The bath is the blessing of the house, for that it cleanseth the body and calleth to mind the Fire.""

"What Hammams are best for bathing in?" Those whose waters are sweet and whose space is ample and which are kept well aired; their atmosphere representing the four seasons—autumn, and summer and winter and spring.

"What kind of food is the most profitable?" That which women make and which hath not cost overmuch trouble and which is readily digested. The most excellent of food is brewis or bread sopped in broth; according to the saying of the Prophet, "Brewis excelleth other food, even as Avishah excelleth other women."

"What kind of meat is the most profitable?" Mutton; but jerked meat is to be avoided, for there is no profit in it.

"What of fruits?" Eat them in their prime and quit them when their season is past.

"What sayest thou of drinking water?" Drink it not in large quantities nor swallow it by gulps, or it will give thee head-ache and cause divers kinds of harm; neither drink it immediately after leaving the Hammam nor after carnal copulation or eating (except it be after the lapse of fifteen minutes for a young man and forty for an old man), nor after waking from sleep.

"What of drinking fermented liquors?" Doth not the prohibition suffice thee in the Book of Almighty Allah, where He saith, "Verily, wine and lots and images, and the divining arrows are an abomination, of Satan's work: therefore avoid them, that ye may prosper."

"What sayest thou of cupping?" It is for him who is over full of blood and who hath no defect therein; and whoso would be cupped, let it be during the wane of the moon, on a day without cloud, wind or rain and on the seventeenth of the month. If it fall on a Tuesday, it will be the more efficacious, and nothing is more salutary for the brain and eyes and for clearing the intellect than cupping.

"Four things kill and ruin the body: entering the Hammam on a full stomach, eating salt food; copulation on a plethora of blood and lying with an ailing woman; for she will weaken thy strength and infect thy frame with sickness; and an old woman is deadly poison."

AVICENNA

Avicenna (980-1037 A.D.) was physician-in-chief to the hospital at Bagdad. Widely learned in the Greek scientific classics he exerted a great influence on contemporary thought. He was court physician to a succession of caliphs, and this eminent position enhanced his authority. Besides his medical writings he made significant contributions to geology in his theory of the formation of mountains. His influence on medical thought in the Middle Ages was second only to Galen's. The *Canon Medicinae* is an attempt to codify all medical knowledge. LeClerc wrote, "Avicenna is an intellectual phenomenon. Never perhaps has an example been seen of so precocious, quick and wide an intellect extending and asserting itself with so strange and indefatigable an activity." Osler wrote, "The touch of the man never reached me until I read some of his mystical and philosophical writings translated by Mehren. It is Plato over again."

REFERENCE

GRUNER, O. C. A Treatise on the Canon of Medicine of Avicenna, Incorporating a translation of the first book. London, Luzac & Co., 1930.

OF FRICTION

[From *Canon of Medicine*,* Lib. i, Fen. 3, Doct. 2, Cap. 4]

One kind of friction is hard, which enlarges or thickens: another is gentle, which loosens. One is prolonged, which causes thinness; another is moderate, which fattens. When these are combined, corresponding results will be produced. There is also a friction which is rough, as with rough clothes, which quickly draws the blood to the outward parts, and one which is smooth, as with the palm or with soft cloths, which collects the blood and keeps it in the part. Now, the objects of friction are:—the thickening of thin bodies, the softening of hard ones. Besides there is a friction of preparation, which comes before exercise, and is begun gently; this, when the desire to rise is felt, braces and hardens. Then there is a friction of restoration, which comes after exercise and is called rest-inducing friction. The object of this is the resolution of superfluities retained in the muscles, not evacuated by exercise, that they may be evaporated, and that fatigue may not occur. This friction must be done smoothly and gently.

OF JABROL

[From *Canon of Medicine*, Lib. ii, Tract. ii, Cap. 365]

What is Jabrol? It is the root of the woodland mandrake, or of any kind of mandrake, large and of human shape; on this account it is called mandrake. For mandrake is the name of any natural object, for instance, a growing plant, in human shape. It is immaterial whether the meaning of this name was discovered or not, for many names have now meanings which were not at first discovered; moreover, a form of mandrake has been found like spongy wood which is

* Canon Medicinae of Avicenna. From a Latin translation of 1608, adapted by the editor.

crumbling down like great costus. Nature.—It is cold in the third degree, dry in itself, and it has a slight amount of heat in the opinion of some. The root is strong and drying, while the rind of the root is weak. The dried leaves are also used, though they are sometimes applied moist, and in the mandrake itself there is moisture. Uses and properties.—It is narcotic, it exudes gum, and has juice, but the juice is stronger than the gum. Whoever wishes to have anything cut from his limbs may drink three measures of it in wine: this will produce insensibility.

OF THE SIGNS OF PLEURISY

[From *Canon of Medicine*, Lib. iii. Tract. iv, Cap. 2]

The signs of simple pleurisy are quite clear. They are fever invariably present because of the vicinity of the heart. Secondly, a stabbing pain under the ribs, since the membrane is shaggy, and often this is not apparent except when breathing. And sometimes there is tension with stabbing; sometimes this is very great. Tension signifies great extent and stabbing great severity in its penetration. The third sign is difficulty of breathing owing to compression caused by effusion, also the shallowness and frequency of breathing. The fourth is a saw-like pulse. The difference from the usual condition of the pulse is increased by the want of strength and the extent of the cause. The fifth sign is cough; sometimes the cough at the beginning of this illness is dry, and later with sputum; and whenever the cough is with sputum at the beginning it is matter for congratulation. The cough, however, does not occur unless the lung is affected by its proximity.

RHAZES

Rhazes (860-932) was born in Persia, studied at Bagdad, and in Palestine, Egypt, and Spain. He was a celebrated teacher and physician in the cities of the Eastern Caliphate. His descriptions of smallpox and measles are the first authentic accounts in literature.

REFERENCE

Medical Classics. Vol. 4, No. 1, September, 1939.

AUTHOR'S PREFACE

[From the *Treatise on the Smallpox and Measles**]

In the name of God, the Compassionate, the Merciful—Abu Ben Mohammed Ibu Zacariya says—It happened on a certain night at a meeting in the house of a nobleman of great goodness and excellence, and very anxious for the explanation and facilitating of useful sciences for the good of mankind, that mention being made of the smallpox, I then spoke what came into my mind on the subject. Whereupon our host (may God favor all men by prolonging the remainder of his life) wished me to compose a suitable, solid, and complete discourse on this disease, because there has not appeared up to the present time either among the ancients or the moderns an accurate and satisfactory account of it. And therefore I composed this discourse, hoping to receive my reward from the Almighty and Glorious God, and awaiting his good pleasure.

ON THE SYMPTOMS WHICH INDICATE THE APPROACHING ERUPTION OF THE SMALLPOX AND MEASLES

[From the *Treatise on the Smallpox and Measles*, Chapter III]

The eruption of the smallpox is preceded by a continued fever, pain in the back, itching in the nose, and terrors in sleep. These are the more peculiar symptoms of its approach, especially a pain in the back with fever; then also a pricking which the patient feels all over his body; a fullness of the face, which at times goes and comes; an inflamed colour, and vehement redness in both the cheeks; a redness of both the eyes; a heaviness of the whole body; great uneasiness; the symptoms of which are stretching and yawning; a pain in the throat and chest; with a slight difficulty in breathing, and cough; a dryness of the mouth, thick spittle, and hoarseness of the voice; pain and heaviness of the head; inquietude, distress of mind, nausea, and anxiety; (with the difference that the inquietude, nausea and anxiety are more frequent in the measles than in the smallpox) while on the other hand, the pain in the back is more peculiar to the smallpox than to the measles; heat of the whole body, an inflamed colour, and shining redness, and especially an intense redness of the gums.

* Translated from the Arabic by William Alexander Greenhill. London, Sydenham Society, 1848.

When, therefore, you see these symptoms, or some of the worst of them (such as pain of the back, and the terrors of sleep, with the continued fever) then you may be assured that the eruption of one or other of these diseases in the patient is nigh at hand; except that there is not in the measles so much pain of the back as in smallpox; nor in the smallpox so much anxiety and nausea as in the measles, unless the smallpox be of a bad sort; and this shows that the measles came from a very bilious blood.

[From the *Treatise on Smallpox and Measles*, Chapter 1]

We will now begin therefore by mentioning the efficient cause of this distemper, and why hardly anyone escapes it; and then we will treat of the other things, that relate to it section by section; and we will (with God's assistance) speak on every one of these points with what we consider to be sufficient copiousness.

I say then that every man, from the time of his birth till he arrives at old age is continually tending to dryness; and for this reason the blood of children and infants is much moister than the blood of young men, and still more so than that of old men. And besides that it is much hotter; as Galen testifies in his Commentary on the Aphorisms in which he says, "The heat of children is greater in quantity than the heat of young men, and the heat of young men is more intense in quality." And this is evident from the force with which the natural processes such as digestion and growth of the body are carried on in children. For this reason the blood of infants and children may be compared to *must* in which the coction leading to perfect ripeness has not yet begun, nor the movement towards fermentation taken place; the blood of young men may be compared to *must*, which has already fermented and made a hissing noise, and has thrown out abundant vapors and its superfluous parts, like wine which is now still and quiet and arrived at its full strength; and as to the blood of old men, it may be compared to wine which has now lost its strength and is beginning to grow vapid and sour.

Now the smallpox arises when the blood putrefies and ferments, so that the superfluous vapors are thrown out of it, and it is changed from the blood of infants which is like *must*, into the blood of young men, which is like wine perfectly ripened; and the smallpox itself may be compared to the fermentation, and the hissing noise which takes place in *must* at the time. And this is the reason why children, especially males, rarely escape being seized with this disease, because it is impossible to prevent *must*, (whose nature it is to make a hissing noise and to ferment) from changing into the state which happens to it after its hissing noise and its fermentation.

TREATMENT OF SMALLPOX

[From the *Treatise on the Smallpox and Measles*, Chapters v, vi, vii*]

It is necessary that blood should be taken from children, youths and young men who have never had the smallpox, or who have only had the chickenpox, (especially if the state of the air, and the season, and the temperaments of the individuals be such as we have mentioned above) before they are seized with a fever, and the symptoms of the smallpox appear in them. A vein may be opened in those who have reached the age of 14 years; and cupping glasses must be applied to those who are younger; and their bedrooms should be kept cool.

Let their food be such as extinguishes heat; soup of yellow lentils, broth seasoned with the juice of unripe grapes, and minced meat, kids foot jelly, veal broth, broth made of woodcocks, hens and pheasants, and the flesh of these birds minced and dressed with the juice of unripe grapes. Their drink should be water cooled with snow, or pure spring water cold, with which their dwellings may also be sprinkled. . . .

In the middle of the day let the patient wash himself in cold water and go into it and swim about in it. . . .

The following is the description of a medicine which restrains the ebullition of the blood, and is useful against heat and inflammation of the liver, and effervescence of the yellow bile:

> Take of Red Roses, ground fine, ten drachms
> Tabasheer, twenty drachms
> Sumach
> Broad leaves Dach seed
> Lentiles peeled
> Barberries
> Purslain seed
> White lettuce seed, of each five drachms

---

* These extracts have been selected to show the state of therapeutics at the time of Rhazes.

White Sanders, 2½ drachms
Common Camphor, 1 drachm

Let the patient take 3 drachms of this powder every morning in an ounce of the inspissated acid juice of citrons, or the inspissated juice of warted-leaved rhubarb, or the inspissated juice of pomegranates, or the juice of unripe grapes and the like. . . .

The eruption of the smallpox and measles is accelerated by well wrapping the patient up in clothes, and rubbing his body, by keeping him in a room not very cold, and by sipping cold water a little at a time. . . .

. . . As soon as the symptoms of smallpox appear, we must take especial care of the eyes, then of the throat, and afterwards of the nose, ears, and joints, in the way I am about to describe. . . .

As soon as the symptoms of the smallpox appear, drop rose water in the eyes from time to time. . . . The patient should gargle his mouth with acid pomegranate juice.

XII

THE MIDDLE AGES

Regimen sanitatis salernitanum WAS FIRST PRINTED IN 1484. DURING THE Middle Ages a great health resort and medical school flourished at Salerno, a town on the western Italian sea coast south of Naples. Its origin is legendary: possibly it was founded by Charlemagne. Later it was under the protection of the Norman dukes of Sicily. Its instruction attracted students from all over the world. The Emperor Frederick II instituted some ethical rules to govern Salernian medical students and practitioners. The school boasted many famous teachers—Gariopontus; Petrus Clericus; Benvenuto Grafeo, the oculist; Aegidius Carboliensis; Roger of Parma and Rolando Capelluti, surgeons. Salerno was situated in an ideal spot for a health resort, with warm sea bathing, mineral springs, and, not far away, mountain air.

Regimen sanitatis salernitanum is a Latin poem which sets forth the Salernian rules for hygiene and medical treatment. It was immensely popular, being translated often into French, English, German, Italian, Hebrew, Polish, Flemish, and Czech. The English translation by Sir John Harington (the inventor of the water closet), is given here.

REFERENCES

HARINGTON, *Sir* J. The School of Salernum (Regimen Sanitatis Salernitatum). The English Version. New York, Paul B. Hoeber, 1920.

RIESMAN, D. The Story of Medicine in the Middle Ages. New York, Paul B. Hoeber, 1936.

SCHACHNER, N.: The Mediaeval Universities. New York, Frederick A. Stokes, 1938.

THE SCHOOL OF SALERNO*

The Salerne Schoole doth by these lines impart
All health to Englands King, and doth aduise
From care his head to keepe, from wrath his heart,
Drinke not much wine, sup light, and soone arise,
When meate is gone, long sitting breedeth smart:

* Regimen Sanitatis Salernitatum. Translation by Sir John Harington, 1608.

76

And after-noone still waking keepe your eyes.
When mou'd you find your selfe to Natures Needs,
Forbeare them not, for that much danger breeds,
Vse three Physicians still; first Doctor Quiet,
Next Doctor Merry-man, and Doctor Dyet.

Rise early in the morne, and straight remember,
With water cold to wash your hands and eyes,
In gentle fashion retching euery member,
And to refresh your braine when as you rise,
In heat, in cold, in July and December.
Both comb your head, and rub your teeth likewise:
If bled you haue, keep coole, if bath' keepe warme:
If din'd to stand or walke will do no harme
Three things preserue the sight, Grasse, Glasse, & Fountains.
At Eve'n springs, at morning visit mountains.

Great harmes haue growne, & maladies exceeding,
By keeping in a little blast of wind:
So Cramps & Dropsies, Collickes haue their breeding,
And Mazed Braines for want of vent behind:
Besides we finde in stories worth the reading,
A certaine Romane Emperour was so kind,
Claudius by name, he made a Proclamation,
A Scape to be no losse of reputation.
Great suppers do the stomacke much offend,
Sup light if quiet you to sleepe intend.

To keepe good dyet, you should neuer feed
Until you finde your stomacke cleane and void
Of former eaten meate,

Sixe things, that here in order shall ensue,
Against all poysons haue a secret power,
Peare, Garlicke, Reddish-roots, Nuts, Rape, and Rue,
But Garlicke, chiefe; for they that it deuoure,
May drinke, & care not who their drinke do brew:
May walke in aires infected euery houre.
Sith Garlicke then hath powers to saue from death,

Beare with it though it makes vnsauory breath:
And scorne not Garlicke, like to some that thinke
It onely makes men winke, and drinke, and stinke.

Yet for your lodging roomes give this direction,
In houses where you mind to make your dwelling,
That neere the same there be no euill sents
Of puddle-waters, or of excrements,
Let aire be cleere and light, and free from faults,
That come of secret passages and vaults.

If to an vse you haue your selfe betaken,
Of any dyet, make no sudden change,
A custome is not easily forsaken,
Yea though it better were, yet seems it strange,
Long vse is as second nature taken,
With nature custome walkes in equall range.
Good dyet is a perfect way of curing:
And worthy much regard and health assuring.
A King that cannot rule him in his dyet,
Will hardly rule his Realme in peace and quiet.

If vnto Choller men be much inclin'd,
'Tis thought that Onyons are not good for those,
But if a man be flegmatique (by kind)
It does his stomack good, as some suppose:
For Oyntment iuyce of Onyons is assign'd,
To heads whose haire fals faster than it growes:
If Onyons cannot helpe in such mishap,
A man must get him a Gregorian cap.
And if your hound by hap should bite his master,
With Hony, Rew, and Onyons make a plaster.

Against these seuerall humors ouerflowing,
As seuerall kinds of Physicke may be good,
As diet, drinke, hot baths, whence sweat is growing,
With purging, vomiting, and letting blood:
Which taken in due time, not overflowing,
Each malladies infection is withstood.
The last of these is best, if skill and reason,

Respect age, strength, quantity, and season.
Of seuenty from seuenteene, if bloud abound,
The opening of a veine is healthfull found.

Three speciall Months (September, April, May)
There are, in which 'tis good to ope a veine;
In these 3 months the Moone beares greatest sway,
Then old or yong that store of bloud containe,
May bleed now, though some elder wizards say
Some dayse are ill in these, I hold it vaine:
September, April, May, haue dayes a peece,
That bleeding do forbid, and eating Geese,
And those are they forsooth of May the first,
Of other two, the last of each are worst.

Besides the former rules for such as pleases,
Of letting bloud to take more obseruation,
Know in beginning of all sharpe diseases,
'Tis counted best to make euacuation:
Too old, too yong, both letting bloud displeases.
By yeares and sicknesse make your computation.
First in the Spring for quantity you shall
Of bloud take twise as much as in the Fall:
In Spring and Summer let the right arme bloud,
The Fall and Winter for the left are good.

JOHN OF GADDESDEN

John of Gaddesden (1280-1361) was educated at Merton College, Cambridge, and at Montpellier, and became Prebendary of St. Paul's, London. His book *Rosa Anglica practica medicine a capite ad pedes* is a compendium of medicine in which diseases are taken up under headings— Cause, Signs, Prognosis, Cure. It begins, as can be seen in the title, at the head and ends at the feet. It was probably intended for the daily direction of those who had to treat the sick. In the Middle Ages most of the medical practice was done by the priest of the village or a priest in an abbey or monastery who had such duties assigned him by the abbot or father superior; he treated the other brothers and all the people living about the abbey, on the nearby feudal estate and so forth. The book was also undoubtedly used by the university doctor who could read Latin.

Chaucer (see page 93) mentions it as one of the books in the library of his typical physician. Undoubtedly also it was used by the lord of the manor for treating his tenants. The lord of the manor often was a man of quite considerable learning as may be seen by consulting Professor T. R. Lounsbury's essay on "The Learning of Chaucer," where is given a long list of works which Chaucer mentions with some knowledge of their contents.

The contents of the *Rosa Anglica* is sometimes rational, usually superstitious. Charms were of equal importance to recipes. The author recommends the royal touch for the King's Evil, and gives a list of the diseases which will bring the doctor the most money.

REFERENCES

CHOLMELEY, H. P. John of Gaddesden and the Rosa Medicinae. Oxford, Clarendon Press, 1912.

LOUNSBURY, T. R. Studies in Chaucer: His Life and Writings. London, 1892.

PREFACE

[From *The Rose of England, the Practice of Medicine from the Head to the Feet**]

Galen in the introduction to the seventh book of the *de Ingenio* says that it is impossible to become nearer to God by any other way than by the way of knowledge—therefore I have wished to write this book for the humble to read. Because since no book is without reproach, as Galen says in the second book of his *de Crise*, so neither will this one be. But all the same, I implore those who see it not gnaw it with an envious tooth, but to read it through humbly, for nothing is set down here but what has been proved by personal experience either of myself or others, and I, John of Gaddesden, have compiled the whole in the seventh year of my "lecture." And in regard to the whole book I intend to observe the following order of arrangement: first of all I try to investigate the name of any disease, secondly its definition, thirdly its incidence and cause. As Isaac says in the fourth book of his Fevers and in his section on Jaundice: "We can discuss everything which we wish to investigate in a triple fashion: we can consider either its name, which is a matter of arbitrary convention; or its definition, which indicates its nature; or its action, which indicates its effect, and in this use 'actio' is equivalent

* Rosa Anglica practica medicinae a capite a pedes. Based on translation by H. P. Cholmeley, 1912.

to incidence or cause." In the fourth place I give an account of the signs, both general and special, and what happenings to the patient are signs to the medical man, in accordance with Joanitius in his treatise on the signs of the official members. In the fifth place I give the prognosis and in the sixth place the cure, and here following Messue I give all things which are to be done for the cure of any dangerous disease which is capable of cure.

But before these matters are treated in the first chapter, I wish to give a name to the book, namely, the *Rosa Medicinae*, and I have so called it on account of five appendages which belong to the rose, as it were five fingers holding it, concerning which it is written:

That is to say, three of the parts surrounding the rose are hairy and two are smooth, and the same is the case with the five parts of my book. The first three are bearded with a long beard, for they treat of many things and about general diseases, and for a discussion of what constitutes a general or common disease look in the introduction the second book. The two following books treat of particular diseases, together with some matters omitted in the preceding books, and they are as without a beard (shorter).

And as the rose overtops all flowers, so this book overtops all treatises on the practice of medicine, and it is written for both poor and rich surgeons and physicians, so that there shall be no need for them to be always running to consult other books, for here they will find plenty about all curable disease both from the special and the general point of view. Pavia has attained to an eminence never before known. The editor therefore begs to dedicate the Rosa as "a learned and eminently instructive little present" to Ambrosius.

On Hydrops

[From *The Rose of England, the Practice of Medicine from the Head to the Feet*]

Idropisis is a watery disease inflating the body. The name "Idropisis" is derived from "idros," which is water, and "isis," which is inflation, that is to say a watery inflation. And so Haly in the third part of the commentary on the Tegni, 192, says, "Subtile and watery juices bring about watery sicknesses such as hydrops." And it is thus defined: Hydrops is a material sickness of which the cause is a cold matter, overflowing and entering into the limbs, and thence arise either all its manifestations, or empty spaces of those organs in which is carried on the government of the food and the humour. . . . Avicenna says "Hydrops is an error of the combining energy (virtutis unitivae) in the whole of the body, following on a change of the digestive energy in the liver." So much we may gather from Avicenna, Can. I, fen. I, doctrina 6, cap. 2, and also in book 6 de naturalibus virtutibus. For there he says that when the nutriment is combined in a limb, there it remains and swells it up. And Avicenna also says that when the nutriment does not cleave [to the members] thence arises hydrops. For the nutriment undergoes a triple dissolution: in the first it is digested and dispersed throughout the members; secondly, it is combined; and thirdly, it is assimilated. Others say that in the first place it is distributed; in the second it cleaves and in the third becomes fit for nourishment and is assimilated. When it is not distributed there arises "sinthesis" or widespread emaciation; when it does not combine, or cleave to the members, there arises hydrops; when it is not assimilated there arise leprosy or morphea, as will be shown further on.

The cure of hydrops is of two kinds, common and proper. The proper is by means of various appropriate medicines and by local measures. The common, as says Avicenna, is by extraction of the watery humidity and its drying up, and this extraction may be carried out in four ways, as Constantine lays down in the seventh book of his Practice. The first method is by diuretic medicines which provoke a flow of urine such as spica, cassia and the like. The second method is to purge out the yellow fluid by means of sweating and discharge from the bowels. For this latter effect use purging drugs such as the juice of mugwort and the juice of laurel. Emetics and clysters can also be used. Sweating can be brought about by sulphur bath or sea baths, or by suffumigations with water in which have been boiled such roots or herbs as pellitory or levisticus, together with bran; or with inunctions of hot oil, with laurel bark or a hot ointment such as arogon, agrippa, or martiaton. The third method is for the patient to drink his own urine. This remedy is good not only in hydrops, but also in jaundice and in the splenetic affection. The whey of goat's or cow's milk also purges. The fourth method is by means of an incision three fingers' breadth below the umbilicus, and a deep perforation made therein, or by a perforation made in the bursa testiculorum, or by intercutaneous scarifications between the joints of the feet, or above the feet or round the ankles. Incision, however, is dangerous, and must not be performed unless the patient is very strong. Avicenna says, "when the belly is full of water and the strength is well maintained, then make an incision and let out the water, but little by little and not all at once."

If the dropsy is not cured by any of these, and its energy seems still unabated, an incision must be made three fingers' breadth below the umbilicus. Care must be taken not to draw all the water off at once lest the patient suddenly die, as is laid down by Messue in the sixth book of his Particular Affections in the Aphorism "Of those suffering from empyema or who are hydropic."

Let the patient sit in a slightly elevated seat, and let the belly be forcibly compressed by the hands so that the watery matter may descend as far as possible. Then make an incision three fingers' breadth below the umbilicus with a sharp knife, the external skin being slightly elevated from the rest of the body, as far as Siphac, if the hydrops be from the intestines. If it come from the liver or the neighbouring parts make your incision to the right of and three fingers' breadth below the umbilicus. If from the spleen, make it on the left side. Lift up the skin lest Mirac be cut, then perforate Mirac, but make the hole in Mirac somewhat lower down than that in Siphac, so as to have them on different levels, that the water come not out continuously. Then put in a canula made of gold or silver or bronze. Then feel the patient's pulse, and if he be weak take out the canula and give medicine or dressing made of down dipped in wine or white of egg. Make the patient lie down and give him chicken broth with spicey medicines, or food of easy digestion such as partridge, kid or lamb. On the second day take off the dressing, replace the canula, and draw off some more water, and do this three or four times.

TREATMENT OF SMALLPOX

[From *The Rose of England, the Practice of Medicine from the Head to the Feet*]

Thus, I, in the case of the noble son of the English king, when he was infected with this disease, and I made everything around the bed to be red. [*Deinde capiatur scarletum rubeum, et involvatur variolosus totaliter, vel in panno alio rubeo, sic ego feci de filio no-*

bilissimi Regis Angliae quando patiebatur istos morbos, et feci omnia circa lectum esse rubea, et est bona cura, et curavi eum in sequenti sine vestigiis variolarum.]

TREATMENT OF PHTHISIS

[From *The Rose of England, the Practice of Medicine from the Head to the Feet*]

(1) Keep in check the catarrh and the rheumata; (2) cleanse the body; (3) divert and draw away the matter (of the disease) to a different part; (4) strengthen the chest and the head so that they do not take up the matter, and that it there multiply; (5) cleanse and dry up the ulcers and expel the matter from them; (6) consolidate them; (7) restrain and cure the cough by using demulcent drinks with ointments and stupes; (8) assist the patient to sleep; (9) strengthen and bring back the appetite; (10) keep in check the spitting of blood; (11) do what can be done to make the breathing more easy and to remove the asthma and the hoarseness; (12) regulate the way of life so far as the six non-naturals; (13) cure the putrid or hectic fever which goes with the disease.

As to food, the best is the milk of a young brunette with her first child, which should be a boy; the young woman should be well favored, [*bene complexionata et non utatur coitu*] and should eat and drink in moderation. Failing a wet nurse, the milk of other animals might be used in the following order of choice: the ass, the goat, and the cow. If the patient liked, he could take his milk straight from the udder; if not, it was to be boiled with a little salt and honey, so that it should not coagulate in the stomach, for in that case it was a very poison. Wine and milk should not be taken together, for wine coagulates milk in the stomach. If the patient has pain and colic after his milk, it does not agree with him. Therefore the dish should be washed with hot water and the milk milked into this, for then it is converted and changed quickly and becomes less harmful.

ON CHARMS

[From *The Rose of England, the Practice of Medicine from the Head to the Feet*]

Again, write these words on the jaw of the patient: In the name of the Father, the Son and the Holy Ghost, Amen. + Rex + Pax +

Nax + in Christo Filio, and the pain will cease at once as I have
often seen.

Again, whosoever shall say a prayer in honour of St. Apollonia,
Virgin, (Feb. 9) shall have no pain in his teeth on the day of the
prayer. The same thing is said of St. Nicasius the martyr (Oct. 11).
Again, draw characters on parchment or panel and let the patient
touch the aching tooth with his finger as long as he is drawing, and
he is cured. The characters are made in the shape of running water
by drawing a continuous line, not straight but up and down. Three
lines are to be drawn in the name of the Blessed Trinity and this is
to be done often.

Again, if the many-footed "worm" which rolls up into a ball
when you touch it, is pricked with a needle, and the aching tooth is
then touched with the needle, the pain will be eased.

Again, some say that the beak of a magpie hung from the neck
cures pain in the teeth and the uvula and the quinsy.

Again when the gospel for Sunday is read in the mass, let the
man hearing mass sign his tooth and his head with the sign of the
holy Cross and say a pater noster and an ave for the souls of the
father and mother of St. Philip, and this without stopping; it will
keep them from pain in the future and will cure that which may be
present, so say trustworthy authorities.

JOHN OF ARDERNE

John of Arderne (1307-1380) is the first English surgeon. Several of his
manuscripts have been preserved, and reprinted with a translation into
modern English. One of these is on fistula in ano. He saw war service
during the Hundred Years War. He practiced in London and in Newark.
His practice was among the nobility and his fees were enormous; tradition says that sometimes his fees consisted of ransoms for knights who
were held by the Turks after the Crusades. Since he took the ransom, the
knights never got home. Possibly he is the model for Chaucer's Doctor of
Physic. Note Chaucer's satirical reference to the Doctor's love of gold.

REFERENCES

POWER, Sir D'ARCY. *De Arte Phisicali et de Cirurgia of Master John
Arderne, Surgeon of Newark, Dated 1412.* Wellcome Historical Med-
ical Museum Research Studies in Medical History. New York, Wil-
liam Wood & Co., 1922.
POWER, Sir D'ARCY. *Treaties of Fistula in Ano, Haemorrhoids and
Clysters* by *John Arderne.* From an early fifteenth century manu-

script translation. London, Kegan Paul, Trench, Trubner & Co., Ltd., 1910.

A CASE OF TRAUMATIC TETANUS*

There was a gardener who, while he worked amongst the vines, cut his hand with a hook upon a Friday after the feast of (the translation of) Saint Thomas of Canterbury in summer (July 7), so that the thumb was wholly separated from the hand except at the joint where it was joined to the hand, and it could be bent backward to his arm and there streamed out much blood.

And as touching the cure. The thumb was first reduced to its proper position and sewn on and the bleeding was stopped with Lanfrank's red powder and with the hairs of a hare and the dressing was not removed until the third day. When it was removed there was no bleeding. Then there was put upon it those medicines which engender blood, redressing the wound once every day. The wound began to purge itself and to pour out matter. And on the fourth night after, the blood began to break out about midnight and he lost almost two pounds of it by weight. And when the bleeding was stopped the wound was redressed daily as before. Also on the eleventh night about the same time the bleeding broke out again in greater quantity than it did the first time. Nevertheless the blood was staunched, and by the morning, the patient was so taken with the cramp in the cheeks and in the arm that he was not able to take any meat into his mouth, nor could he open his mouth, and on the fifteenth day the bleeding broke out again, beyond all measure, and always the cramp continued and he died on the twentieth day.

BARTHOLOMEW ANGLICUS

Bartholomew Anglicus was an English Franciscan monk of the twelfth century, who wrote De proprietatibus rerum, a book of immense popularity, written to explain the allusions to natural objects in the scriptures. The seventh book is on medicine. It furnishes us with the most typical example of average practice in medieval times.

REFERENCE

STEEL, R. Medieval Lore: An Epitome of the Science, Geography, Animal and Plant Folk-Lore and Myth of the Middle Age: Being Classified

* From De Arte Phisicali et de Cirurgia of Master John Arderne, Surgeon of Newark. dated 1412. Translation by Sir D'Arcy Power, 1922.

Gleanings from the Encyclopedia of Bartholomew Anglicus on the Properties of Things. London, Elliot Stock, 1893.

FRENZY AND THE REMEDIES THEREOF

[From *On the Properties of Things,** Book VII]

These be the signs of frenzy, woodness and continual waking, moving and casting about the eyes, raging, stretching, and casting out of hands, moving and wagging of the head, grinding and gnashing together of the teeth; always they will arise out of their bed, now they sing, now they weep, and they bite gladly and rend their keeper and their leech: seldom be they still, but cry much. And these be most perilously sick, and yet they wot not then that they be sick. Then they must be soon holpen lest they perish, and that both in diet and in medicine. The diet shall be full scarce, as crumbs of bread, which must many times be wet in water. The medicine is, that in the beginning the patient's head be shaven, and washed in lukewarm vinegar, and that he be well kept or bound in a dark place. Diverse shapes of faces and semblance of painting shall not be shewed tofore him, lest he be tarred with woodness. All that be about him shall be commanded to be still and in silence; men shall not answer to his nice words. In the beginning of medicine he shall be let blood in a vein of the forehead, and bled as much as will fill an egg-shell. Afore all things (if virtue and age suffereth) he shall bleed in the head vein. Over all things, with ointment and balming men shall labour to bring him asleep. The head that is shaven shall be plastered with lungs of a swine, or of a wether, or of a sheep; the temples and forehead shall be anointed with the juice of lettuce, or of poppy. If after these medicines are laid thus to, the woodness dureth three days without sleep, there is no hope of recovery.

OF LEPROSY

[From *On the Properties of Things*, Book VII]

Universally this evil [leprosy] hath much tokens and signs. In them the flesh is notably corrupt, the shape is changed, the eyen become round, the eyelids are revelled, the sight sparkleth, the nostrils are straited and revelled and shrunk. The voice is hoarse, swelling groweth in the body, and many small botches and whelks hard and round, in the legs and in the utter parts; feeling is somedeal

* De Proprietatibus Rerum. *Circa* 1260. Translation by William Morris, 1880.

taken away. The nails are boystous and bunchy, the fingers shrink and crook, the breath is corrupt, and oft whole men are infected with the stench thereof. The flesh and skin is fatty, insomuch that they may throw water thereon, and it is not the more wet, but the water slides off, as it were off a wet hide. Also in the body be diverse specks, now red, now black, now wan, now pale. The tokens of leprosy be most seen in the utter parts, as in the feet, legs, and face; and namely in wasting and minishing of the brawns of the body.

To heal or to hide leprosy, best is a red adder with a white womb, if the venom be away, and the tail and the head smitten off, and the body sod with leeks, if it be oft taken and eaten. And this medicine helpeth in many evils; as appeareth by the blind man, to whom his wife gave an adder with garlick instead of an eel, that it might slay him, and he ate it, and after that by much sweat, he recovered his sight again.

The biting of a wood hound is deadly and venomous. And such venom is perilous. For it is long hidden and unknown, and increaseth and multiplieth itself, and is sometimes unknown to the year's end, and then the same day and hour of the biting, it cometh to the head, and breedeth frenzy. They that are bitten of a wood hound have in their sleep dreadful sights, and are fearful, astonished, and wroth without cause. And they dread to be seen of other men, and bark as hounds, and they dread water most of all things, and are afeared thereof full sore, and squeamous also. Against the biting of a wood hound wise men and ready used to make the wounds bleed with fire or with iron, that the venom may come out with blood, that cometh out of the wound.

GUY DE CHAULIAC

Guy de Chauliac (1300-1368) was the most eminent authority on surgery during the Middle Ages. His *Chirurgia magna* was written in 1363. Born in the countryside near Auvergne in France he took holy orders and was educated in medicine at Toulouse, Montpellier and Paris, with a special course in anatomy at Bologna. He settled in Avignon and was surgeon to the French popes. He operated for hernia and cataract, but hesitated to cut for the stone. He employed the cautery for cancer. He treated ulcers by investing them with a collar of steel. His discussion of fractures and dislocations is good. He used Theodoric's narcotic or soporific inhalant as an anesthetic. He did not believe in the power of nature in healing wounds, but in the surgeon's intervention with salves, plasters, etc.

REFERENCE

BRENNAN, W. A. Guy de Chauliac—on Wounds and Fractures. Chicago, 1923.

TREATMENT OF WOUNDS

[From *On Wounds and Fractures**]

The common object in every solution of continuity is union, as is said in the Third of the Techni. And this is the first indication learned from the essence of the malady itself, which rejects the contrary by its contrary. Which general and first intention is accomplished by two ways: first, by Nature as the principal worker, which operates by its own powers and by suitable nourishment; and secondly, by the physician as a servant working with the five objects which are subalternate one to the other.

The first object requires the removal of foreign substances, if there are any such among the divided parts.

The second is to approximate the separated parts to each other.

The third is to preserve the parts thus brought together in their proper form.

The fourth to conserve and preserve the substance of the organ.

The fifth teaches how to correct complications.

AIM OF THE FIRST INTENTION, WHICH IS TO REMOVE FOREIGN BODIES

[From *On Wounds and Fractures*]

The first aim is accomplished if the wound is not sufficiently open and that some foreign body (such as a sharp spicula of separated bone or some other infixed thing, as an arrow or other foreign substance, as a thorn), is between the parts, that the wound should be opened. And when it is sufficiently opened that the foreign substance should be drawn out or pulled out gently and without pain with the fingers or forceps or tenaculae or some other instrument which you invent.

* Chirurgia Magna, 1363. Translation by W. A. Brennan, 1923.

CONCERNING THE INSTRUMENTS FOR WITHDRAWING
ARROWS AND OTHER INFIXED BODIES

[From *On Wounds and Fractures*]

Because we extract infixed bodies by the invention of instruments. The indication of inventing them is taken from the considerations of the nature and differences of the infixed bodies and from the nature and consideration of the organs. From these two a third consideration is drawn, namely, the means of extracting and the invention of the instruments. And since the diversity of infixed bodies is infinite and could not certainly be described in words, hence for this reason it is advisable that the forms of the enemy's projectiles be examined. Nevertheless, Avicenna tries to comprise them under an octuple division, of which (to be very brief) I take the most common. Of infixed things some are of iron, others of thorns, others of bone, or of other nature. Besides, some are plain and others barbed. Moreover, some have a socket in which the shaft is set; others have a nail driven in the shaft. Besides this, some are poisonous, others not.

The diversity of the organs is learned from anatomy: thus some are principal and others not, and some are fleshy in which the infixed body scarcely takes hold; others are osseous in which the infixed body is firmly embedded. There are those which can be seen in which the infixed body has scarcely penetrated; others in which it has plunged deeply, so deeply, in fact, that it has reached to the opposite side.

The instruments that have been invented by reason of these considerations, although they are multitudinous, yet I have by me but eight of the most common: The first are the tenaculae of Avicenna, which are serrated and half moon shaped. . . .

The second, the tenaculae of Albucasis, which are serrated and like the beak of a bird. . . .

The third, cannulated tenaculae for barbed arrows. . . .

The fourth (Terebella), reversed augurs to seize the iron socket. . . .

The fifth, straight augurs, to widen the bone. . . .

The sixth, hollow and solid impulsors. . . .

The seventh, scissors to dilate the flesh in order that the arrows may be more easily extracted. . . .

The eighth is the arbalest (cross-bow). . . .

The method of operating which suits particular cases is such that if the infixed body cannot conveniently be extracted at the first attempt, it ought to be left alone until the flesh withers or corrupts and then by twisting it and moving it here and there the infixed body will be more easily drawn out, notwithstanding the dictum of Henric, who orders that they be extracted immediately because Avicenna, Albucasis and Brunus so wished. Then the wound should be cared for just like others, except that the blood altered by the infixed body must be expressed so that the wound may be assured against putrefaction; and warm oil must be poured into it, especially if there is question of pain. And if it is poisoned, let it be treated like a poisonous bite. But if it cannot be done easily by these means, the patient's armor being removed, and the things which must be prepared being prepared, and having made the prognosis, if there is need according to the formula already given, let the infixed body be seized with common tenaculae and let it be drawn out by twisting it. And if these should not prevail, let stronger ones be used. And if the arrows are barbed, let the arrows be seized with the cannulated tenaculae. If the shaft has loosened from the socket it can be drawn out with the reversed augur (terebella) inserted in the socket; and if the shaft is in the socket, let it be extracted with the straight augur. But if it cannot be otherwise extracted, if it be possible let the slit in the flesh be enlarged with a razor or that of the bone with straight augurs or with the trepans and let the infixed bodies be drawn out, as said. And if this is of no avail, the arbalest should be fixed with tenaculae and the patient, being well fortified, the arbalest is unloosed, and then one will extract the body. But if the arrow is reversed and cannot be extracted through the place where it entered, let it be pushed with the hollow or solid impulsors to the opposite side, and if it can conveniently be done, let it be extracted from that side; but if this is not possible, then let it remain until nature extracts it or demonstrates it. Albucasis tells of several in whom arrows have long remained hidden and who have lived with them without damage, and in some certain cases the infixed bodies were manifested by Nature and extracted and the patients recovered.

I do not care for the incantations and conjugations of Nicodemus which Theodorus and Gilbert relate.

Concerning the medicaments which extract affixed substances, I have operated upon thorns, corn ears, stones, glass, pieces of bone, and I have found one suitable medicament, which is taken from Avicenna:

℞ Fermenti (Leaven) .... ⎫ of each half a pound
    Honey .... ⎭
    Guy de chene ....a quarteron
    Ammoniae ....half a quarteron
    Oil ....a quarteron

Let a plaster be made of these and applied above the place.

Roger affirms that he has found that radix arundinis mixed with honey placed over an infixed substance draws it out without pain. Several other remedies are mentioned in the antidotary, and thus is the first object accomplished.

Concerning the second object, which is to bring together the separated parts:

The second object is accomplished by pulling with the hands and joining the separated parts and replacing the organ in its proper contour with the least pain which is possible, as will be mentioned in particular later on.

Concerning the third object, which is to hold the replaced parts together:

The third object is accomplished by a good and decent ligature and correct situation of the organ and suture if necessary.

OF THE METHOD AND QUALITY OF BINDING (LIGATURE)

[From *On Wounds and Fractures*]

With regard to the ligature, it is necessary to know that according to the intention of Avicenna in the Fourth, ligature is triple, namely, incarnative, expulsive and retentive.

The *incarnative ligature* is suitable for recent wounds and for fractures; it is made with a band folded from the two ends up to the middle, commencing from the part opposite the place wounded and conducting one of the ends toward the upper part of the limb and the other end towards the lower part, taking in as much of the neighboring parts as will seem expedient and tightening it more over the wounded place than in the adjoining parts. Sometimes too great stricture must be avoided, and also too much laxity, the limit being the good toleration of the patient. Let the ends of the bandage be sutured. And if there is need of several bandages, let them be put on the place and turned about by the same method. By this manner of ligature one lip of the wound is joined to the other and suppuration is prevented, as is proved in the Sixth of the Therapeutics. But some adapt a doubled cloth, tightening it and suturing it over the place of the solution.

The *expulsive ligature* is suitable for ulcers (wounds) and caverns in order to expel matter from their depth and prevent any other matter from coming into the place. This is made with a bandage folded from one end, commencing at the lower part of the limb, tightening it there very strongly, then by revolving, it is carried as far as the superior parts. I call the superior part, like Galen in the Fifth of the Therapeutics, that which is toward the heart or the liver, from which all the organs originate. With regard to myself, in unequal limbs, such as the legs, I adapt such a bandage by cutting it on one side from palm to palm and suturing, profiling it rigidly by the curved and incised part and loosely toward the back and nonincised part. In binding, I keep the wide dorsal part toward the gross part of the limb and the curved incised part toward the thin part of the limb. God knows what profit this ligature has been to me in ulcers and varices and in phlegmons of the legs.

The ligature for retention of medicaments is adapted to organs which one cannot stricture nor make any other ligature, as in the neck and in the abdomen and in all suppurative and painful dispositions. It is made with a bandage with one end or several ends or arms commencing over the wounded part and finishing in the opposite part.

CHAUCER

Chaucer's Doctor of Physic, is a picture of a medieval physician from the viewpoint of a layman.

REFERENCE

CURRY, W. C. Chaucer and the Mediaeval Sciences. New York, Oxford University Press, 1926.

With us ther was a Doctour of Phisik
In al this world ne was ther noon hym lyk
To speke of physick and of surgerye,
For he was grounded in astronomye;
He kepte his pacient a ful greet deel
In houres, by his magik naturel.
Wel coude he fortunen the ascendent
Of his images for his pacient.
He knew the cause of everich maladye
Were it of hoot or cold, or moiste or drye
And where engendered, and of what humour
He was a very parfit practisour.
The cause y known, and of his harm the rote
Anon he yaf (gave) the sick man his bote (remedy)
Ful redy hadde his apothecaries
To sende him drogges, and his letuaries.
For each of hem made other for to wynne
Her frendschipe n'as nat newe to bigynne.
Wel knew he the olde Esculapius
And Deiscorides, and eek Rufus
Old Ypocras, Haly and Galien
Serapion, Razis, and Avicen
Averrois, Damascien, and Constantyn
Bernard, and Gatesden, and Gilbertyn.
Of his diet mesurable was he
For it was of no superfuuitee
But of greet norissyng and digestible
His studie was but litel on the Bible.
In sangwyn and in pers he clad was al
Lyned with taffeta and with sendal
And yet he was but esy of dispence
He kept that he wan in pestilence.
For gold in physik is a cordial
Therfor he loved gold in special.

XIII

PARACELSUS

PARACELSUS (1493-1541) WAS AN ORIGINAL, ALTHOUGH ECCENTRIC, FIGURE, who is sometimes called the father of chemistry and the reformer of materia medica. He was born in Einsiedeln, near Zurich, Switzerland. He claimed to have traveled all over the world. He taught as professor of medicine in Basle and in other cities. His bombastic method of delivery, the license of his criticism of orthodox doctors, his theatrical tricks —such as publicly burning the works of Galen and Avicenna—led to his continual persecution, and he wandered from place to place in Germany, finally dying in Salzburg. Tradition says that he believed in, in fact owned, the philosopher's stone. His mind, his philosophy, looked back to the Middle Ages and forward to the Renaissance. He introduced mineral baths; he also introduced laudanum, mercury, lead, arsenic, iron, copper sulphate, tinctures and alcoholic extracts into the pharmacopeia. He discarded and ridiculed the doctrine of the four humours. His principal works are *Chirurgia magna* (1536), *De gradibus* (1568), *A Treatise on Diseases of Miners* (1567).

His mystical philosophy was first formulated in the youthful production "Paramirum." The nature of health and disease, states the "Paramirum," the determination of human destiny, depend upon five entia: First, the *"ens astrale,"* the influence of the stars; every man has his own constellation. Second, the *"ens venini,"* the influence of nourishment. Third, the *"ens naturale,"* the nature and functions of the physical body. Fourth, the *"ens spirituale,"* the nature of the spiritual side of man. Fifth, the *"ens Dei,"* the power of God to restore order, to bring health out of disease.

The flavor of his contribution to medicine is probably best sampled in "Sieben defensiones, antwort auf Etliche Verunglimpfungen seiner Misgönner" (Seven Arguments, Answering to Several of the Detractions of His Envious Critics). It was probably composed about 1537.

REFERENCES

HARTMANN, F. The Life of Philippus Theophrastus, Bombast of Hohenheim, Known by the Name of Paracelsus and the Substance of His Teachings. London, George Redway, 1887.

STODDART, A. M. The Life of Paracelsus. Philadelphia, David McKay, 1911.

SIGERIST, H. E. Paracelsus. In: The Great Doctors, A Biographical History of Medicine. New York, W. W. Norton, 1933. Page 109.

[From *Seven Arguments**]

FOREWORD TO READERS, by the learned Aureolus Theophrastus von Hohenheim, doctor of both schools.

Reader, give good heed so that I may inform you why I have written these arguments. In his time God laid well the foundation upon which the spirit of medicine had its beginnings, through Apollinus, Machaon, Podalirius and Hippocrates, allowing the light of nature to shine without any spirit of darkness. In marvelous manner he wrought great works, mighty wonders, great miracles from the mysteries, elixirs, arcana and essences of nature and in a wondrous way he implanted the arts of medicine in certain devout men as mentioned above. But just as the wicked foe with his corn cockle and weeds would let nothing grow in the clean wheat ground, so medicine was obscured from the first spirit of nature and fell prey to antimedical men. Accordingly it was jostled hither and yon by [persons and] sophistries, so much so that no one was able to come to the level of achievement reached by Machaon and Hippocrates; and in medicine that which is not tested in practice has lost its disputation and gains still less in debate. Now note, reader: When a valid doctrine is raised against sophistry's legion, should it not be easy for that work to overthrow idle prattle? Tell me, reader, to whom do I address myself? Surely to the saints, who do not give signs: The confluence and concourse might give to many a one such a fright that he would refrain from stopping the prattler's mouth. But the effluence and the recourse prove that no reliance is to be placed in the concourse. From it arose the erroneous belief that Hippocrates was an idle chatterer, and sophists would have it that the spirit of truth in medicine is but a clacking tongue. For wherein lie the excesses of a chatterer? Of this rabble some have allowed their mouths to be too hasty and have defended themselves by vituperation since, having once commenced to mouth the words of medicine, they must make their defense with mouths which can do

* From Sieben Defensiones. Edition of Prof. Karl Sudhoff, Klassiker der Medizin, No. 24, Leipzig, 1915. Translated by Julian F. Smith, Hooker Scientific Library, Central College, Fayette, Mo., 1941.

no other than scold and vituperate. I also have been a victim of
such a tongue of bitterness. But that is necessary because they do
not rest on the foundation stones of medicine but are established
on a scullery stone and, having forgotten the truths of medical art,
with tales of sophistry have aroused me and others not to leave such
talk unchallenged. But no such scurrilous language would emanate
from one who was devoted even to the first center of truth. Their
best trick is their rhetoric with all its brood and with the virtue
attributable to the pseudomedicos. Accordingly, reader, heed well
the answers so that you may know how to use them. Though there
is really no need to make any answer to such gentry, let them stand
as poetic physicians, rhetorical prescription scribblers and nebulous
pill compounders: In time the world will weary of these too. But to
let it be understood that a physician without works is nothing, and
that a physician's works do not reside in tongue wagging, I have
written these words to serve as instruction. For that cause, dear
reader, I have been harried in order that my writings might not see
the light of day: Nevertheless I have honored the archduchy of
Kärnten therewith, so that through these same worthy gentlemen
my work might reach you wherever in the world you might receive
it; for without this estate, reader, you would not see these words.
Accordingly esteem well the theory in this work, and indeed even
far more the achievements of art.

Done at Sanct Veit in Kärnten, August 18, [15]38.

FIRST ARGUMENT in discovery of the new medicine, by Doctor
Theophrastus.

That I introduce in this work a new theory and physic, together
with a new reasoning such as were never held before, nor understood, by philosophers, astronomers and physicians is due to the
causes of which I shall now inform you: That is, I report as one who
has proved to his own satisfaction that the doctrines and causes of
death in the old school of thought have been inaccurately and
ambiguously described and so much error has thereby been introduced and so long perpetuated that it is now held and regarded as
correct and incontrovertible. So it has become fast rooted and is so
held and retained that no other is sought, or any other is considered
to be in error. I wish to apprise you of this, for I must adjudge it a
great folly: When Heaven in the light of nature is constantly offer-

ing new ideas, new inventions, new arts, new cares, should not these be held valid? What profits the rain that fell a thousand years ago? Benefit comes from the rain that falls now. What gain comes to this present year from the cycle traversed by the sun a thousand years gone by? Did not Christ expound the manner in which we should judge this point, saying: Sufficient unto the day is the evil thereof: or as much as to say that it suffices if we perform the tasks which the day brings up. And He concludes by saying that the morrow should take thought for itself. So if cares must bear their own burden, and every day has twelve hours, and each hour has its own special duty to perform, then in what way does the twelfth hour hurt the first? Or how is the first hour any burden to the twelfth, so long as every matter is placed in its own realm in its own time? Our concern should be for the present, not for the past, and every realm of thought is concerned with all the light of nature. Hence God's wonders consist in changing the light of nature in many realms of thought, between the beginning and the end of the world, a circumstance which is often overlooked and not considered in accord with the body of thought in these realms. Accordingly from the power of the existing light of nature and from the predestined arrangement of present realms of thought I in my writings will go unchastised by many, and still less will I be harried and hindered by that sophistry which I have called a fallacy in medicine.

SECOND ARGUMENT relating to new diseases and names, by the aforesaid Doctor Theophrastus.

To defend, protect and shield myself in this, that I describe and propound a new disease never before described and bring forth new names, never before used but now given by me; to explain why this has happened by announcing myself because of the new diseases, now therefore give heed. I write of the senseless dance, called by common folk "St. Vitus dance," and of suicides, and of false diseases brought on by enchantments as in persons possessed. These diseases have never been described in medicine, but I deem it undesirable that they should be forgotten. But the immediate cause which impels me is that astronomy, which has never been taken up as yet by physicians, has taught me of such diseases. If other physicians had been so learned in astronomy they would certainly have made these discoveries and had this understanding long before me. But since the physicians have rejected astronomy they may not either recognize or understand these diseases, and other matters too, in their true nature. Now therefore since the medical science of other writers does not flow from the source from which medicine has its origin, that source and origin which I wish to extol, should I not then have power to write differently from other authors? For it is given to all to speak, to advise and to learn but not to every one is it given to speak and to teach that which has force. For you know that even the Evangelist bears witness how Christ taught, that he spoke as one having authority and not as the scribes and Pharisees. We should have respect for authority which proves itself by works, even though we do not wish to believe the words. To provide myself with such respect: As that man is rare who can stand his ground with respect to a situation which he has not seen with his own eyes, against one who has seen it with his own eyes, so we adjudge here the same distinction between those who speak without reason and those who speak with reason. It is no less true that all sickness belongs in the physician's province; it is also proper that he should know something of all diseases. Nevertheless that which he lacks in one he may make up in understanding of another. Then therefore the apostle's gifts are distributed and that which is given to a man, therein he has his honor, but that which is not granted him is unto him no shame. For whatever God wills each man to be, that he remains. Other writers cannot boast of such gifts; they rejoice in their fixed terminus, and that which they cannot effect beyond their terminus they claim to be incurable.

THIRD ARGUMENT by way of describing the new prescriptions.

But as for what has been announced, the outcry is still greater among the ignorant, opinionated quacks who claim that the prescription I write is poisonous, a corrosive and the essence of every evil poison in nature. My first reply to such sham pretense and clamor, if they were able to give any valid answer, would be to ask if they know what is poison and what is not poison? Or is not a poison one of the mysteries of nature? For on these very points they lack knowledge and understanding of nature's forces. For what thing is there, created by God, which is not endowed with some great gift for the good of mankind? Why then should poisons be thrown aside and despised when it is not the poison but nature that is being tried? I will give an example to illustrate my contention. Consider the toad, what an ugly and venomous creature it is; consider also the great mystery concerning the toad and the plague. Now if this mystery were to be despised because the toad is so ugly and venomous, what a mockery that would be! Who indeed compounded nature's prescription? Was it not God? Why should I despise His compound? even though He compounded something which I might deem insufficient! He it is in whose hand is all wisdom and He knows where He wishes to dispose each mystery. Why then should I be amazed or fearful, and because one component is poisonous should I despise the other component too? Every thing should be used for the purpose to which it is ordained and we should not stand in fear of it. For God is the physician and also the remedy itself. Every physician should take unto himself the power of God which Christ explained to us, saying: And though ye should drink poison ye shall not be harmed. If then the poison does not gain the mastery but is harmless so long as we use it as ordained by nature, why should poisons be despised? He who despises poison is ignorant of what resides therein. For the arcanum present in the poison is so endowed that it suffers no loss or harm from the poison. It is not, however, that I wish to content you with this chapter and verse, or that I feel I have adequately defended myself, but that I feel it essential to present to you an additional report so that I may explain poisons adequately.

FOURTH ARGUMENT concerning my travels.

It is necessary for me to make some reply concerning my travels and concerning the fact that I have been so unsettled. Now how can I oppose or overpower that which it is impossible for me to overpower? Or what can I add to or subtract from predestination? But in order that I may excuse myself to you in some fashion, and because I have had so many remonstrances, even in contumely and scorn, on the grounds that I have the wanderlust and that it detracts from my worth, let no man blame me that I trouble myself in this respect. My travels up to the present time have taught me that, for good reason, no disciple finds a master growing in his own back yard, or an instructor in the chimney corner. Again, the arts are not confined to one native land, but are scattered the world over. They are not to be found in one man or place but they must be sought,

collected and accepted where they are. The whole firmament bears
me witness that talents are not distributed by favor, not given to
every man in his village home, but in accordance with the content
of the uppermost spheres the radii go straight to the mark. Was it
not well and proper for me to search out this mark, to inquire and
examine and see what is effected in every person? If I were lacking
in this I would not be true to the Theophrastus who is myself. Is
it not therefore true that art seeks no man but must be sought?
Therefore it is with good reason and sense that I say I must seek
art, not art me. Take an example: If we would approach God we
must go to Him, for he says: Come unto me. That being the situation we must adapt ourselves to it, in what direction we will. It
follows therefore that if we wish to see a person, a country, a city
and to learn the location and character thereof, the climate and
weather, we must go there. For it is impossible that it will come to
us. Therefore it is the way of everyone who wishes to see and learn
something that he seeks it out and takes cognizance of it, and when
he has mastered it he goes on to learn more.

FIFTH ARGUMENT on banishing quacks and sham societies.
Since nothing is so clean as to be perfectly spotless it is necessary
to distinguish between the pure and the soiled. This is especially
true in medicine, which proves to be more bad than good. But as
Christ had twelve disciples and one was a traitor, how much more
shall it be true among men? It may be supposed that out of twelve
there will hardly be one true man. For there is good reason, since
we do all things for love but nothing comes of love, but only through
agreements and payments from which profits accrue; hence quacks
found their way into medicine, seeking money rather than following
the law of love. Now when any enterprise is directed toward selfish
gain the art becomes falsified, and the results as well; for art and
craftmanship must spring from love, without which there is no
complete whole. For in like manner we have two kinds of apostles,
one loving Christ for his own gain so that the purse of private gain
was given to him. Thus he had his reason for betraying Christ
himself for the sake of profit, and even delivering him up to death
for gain. Now if Christ had to endure being sold and betrayed for
personal profit how much more will quacks maim and cripple,
throttle and slay men in order that their own profits may increase

without let or hindrance. For as soon as love grows cold in our neighbor it ceases to bear good fruit for that neighbor and such fruit as is borne is taken for private gain. Therefore we should know that there are two kinds of physician, one acting from love and the other from greed, and both are known by their works because just men are known for their love while they do not violate love for their neighbor whereas the unrighteous, acting contrary to law, reap where they have not sowed and are like ravening wolves, reaping where there is anything to be reaped so that their own profit may increase regardless of the law of love.

SIXTH ARGUMENT, to excuse the author's odd manner and scornful mien.

As if it were not enough to attack me in certain articles, but to say that I am an eccentric, with an opposite answer to the effect that not everyone develops according to his own desires and not everyone responds in humility to his own plan of action: this they deem and consider as a great fault in me, whereas I myself esteem it as a great virtue and would not have the situation other than it is; my ways are almost entirely pleasing to me. But that I may reply by explaining my odd manner, note this: By nature I was not subtly spun, and indeed it is not the way in my country for men to achieve anything by spinning cocoons. Moreover, we were not brought up on figs or mead or wheat bread, but on cheese and milk and oaten cakes; he to whom there cling every day the attributes he received in youth cannot make a subtle companion. He can only counter broadly against subtleties, hairsplitting and fine-spun reasoning; for those who are dressed in soft raiment and were brought up among women cannot understand or be understood by us who grew up in the pine woods. Accordingly the coarse man must be adjudged coarse, though he may think himself a model of grace and charm. So it is with me; what I look upon as silk others call ticking or twill.

SEVENTH ARGUMENT, showing that although I am not omniscient and omnipotent I can do what may be required to meet every man's need.

I must confess that I cannot supply and fulfil every man's wish as he certainly and indubitably would like to have me do in matters which I cannot perform and which are not within my powers. But

God did not establish medical science according to the will of the many, that they might act every man according to the whim after which he runs. If it is not God's will to concede or give such people anything, what should I do about it? For I cannot master and overpower God, but He masters me and everyone else. Take, therefore, one common answer: If they were pleasing to God or acceptable to him for healing he would not have withheld nature from them. It is the same as though a man were to wish himself a fine, handsome specimen, outstanding among all his fellows, and were to wish all women and girls would be gracious to him, when actually he was born a cripple, has a hump on his back like a lute and in other respects too has no personal charm. How can women be gracious to him when his own nature was not gracious but ruined him in his mother's womb and from him produced nothing good? But simply that I may instruct you, know this: What good can nature grant to one to whom God has given nothing good? When favor is lacking with these two, what is the physician? or who can chide him? Now they say that when I visit a patient I do not know at once what ails him but that I require time in which to learn it. That is true; the fact that they make immediate diagnosis is the fault of their own folly, for the off-hand judgment is wrong from the start and as the days go by they know less and less the longer the time, and so they make themselves out as liars. But I seek day by day to arrive at the truth and the longer the time the more diligently I seek. For uncovering hidden diseases is not like recognizing colors. With colors the observer sees black, green, blue and so on; but if they had a curtain in front of them he would not know them; to see through a curtain would require spectacles such as never were. What the eyes tell may be diagnosed at once but it is useless to diagnose that which is hidden from sight, treating it as though it were plainly visible. Take a miner as an example; no matter how good, how honest, how clever, how skilled he may be, when he sees an ore for the first time he does not know what it contains and how he should treat it, roast, smelt, volatilize or burn it. First he must analyze it through a course of testing and examination before he can see his way. Then when he has sifted the ore he may treat it in a certain way and from then on, so must it be. So it is also with obscure chronic diseases, for which quick diagnosis is impossible, unless perhaps the humorists perform it. For it is impossible to find a dog so quickly, or a cat in a kitchen; how much less then in a dangerous mysterious matter? To consider the situation, to measure and make tests and to accord the tests their significance, that is not a matter for reproach; and having brought the proper art to bear, there is the adornment, the treasure which was sought. This is the way to proceed with such diseases. But those who rely on humors do not make their tests by experiment; instead they go to books for their tests and trials. For this reason many in Kirchhof get away before they learn the truth and in that way they never learn it. This, then, is their art, and should I be judged by such an art? I cannot do everything; what can they do? Those who think no one should get well unless it be by the "Summa" compendium have in it their Avicenna, their Moses. In brief, things are as they are: therein lie their aphorisms, nauseating among the least in Kirchhof.

CLOSING REMARKS

Therefore, reader, if you have comprehended my meaning to some degree in this reply and have recognized how I have gone about my attack with utmost mildness, you may now be able to measure for yourself the speech and actions of foolish idle people. You may also pause to think how all this has emanated only from physicians and to consider at the same time the kind of people medicine is burdened with, what a dissimilar pair were Podalirius and Apollo, and what people we have today. Might not nature itself be somewhat taken aback by this? For nature clearly recognizes her foe, as a dog recognizes the dog catcher. Holy Scripture sufficiently proves with what praise medicine should be extolled and what honor the physician should receive. It is sensible, however, that Hippocrates, Appollinus and Machaon are still quoted, for these healed through the true spirit of medicine, showing forth wonders, signs and works and appearing as luminaries of nature. Thus in my simple mind I can readily comprehend why Holy Scripture does not refer to those who are without works, to the claimants and mercenaries, but to those who followed in Machaon's footsteps. It is good to note in those Scriptures that toil and trouble exist in the world. But I consider that one with such a mastery of recognizing things, if well taught and having also lain sick in the selfsame hospital, would take on a friendlier appearance through love for others. The worthy County of Kärnten, finding the proper mood, represents Mecaenates and offers asylum to disciples of Hippocrates in our time, under protection and shield. In return may God grant them recompense, peace and unity, Amen.

XIV

FRACASTORIUS

GIROLAMO FRACASTORO (1478?-1553) OF VERONA, WAS A TYPICAL FIGURE OF the Italian Renaissance—poet, physician, mathematician, astronomer, and geologist. He is best known for the poem which gave the name to syphilis, a disease which was, as he states, in his day, new and of malignant epidemic proportions. *Syphilis, sive Morbus Gallicus* was published in 1530.

Translations: The "Syphilis" has several times been translated into English: (1) first in 1685, by Nahum Tate (1652-1692), poet laureate (partially reprinted in Major's "Classic Descriptions of Disease").

(2) Philmar Company, St. Louis, Mo., 1911, prose.

(3) Dr. William van Wych, "The Sinister Shepherd," Primavera Press, Los Angeles, Calif., 1934, in rhymed quatrains.

(4) Heneage Wynn-Finch, "Fracastor—Syphilis or the French Disease," London, William Heinemann, 1935 (with the Latin text on the opposite pages, prose rendering, good notes and splendid introduction by J. J. Abraham).

Reading Fracastorius should stimulate the student to investigate the history of syphilis. According to a widely popular theory the disease was brought from America by the men of the crew of Columbus, and after their return at the seige of Naples, spread through Europe. (Note first paragraph of Fracastorius.) The most formidable opponent of this theory was Professor Karl Sudhoff. Certain it is that the literature on the clinical recognition of syphilis is definite only about 1500.

Fracastorius' other great medical work *De Contagione* (1546), has a remarkable statement of the modern idea of the nature of infection. It has been translated by Professor Wilmer Cave Wright (G. P. Putnam's Sons, 1930).

REFERENCES

DENNIE, C. C., and SILVA, L. C. The Pestilence, a translation of the poem by Villalobos. *Bull. Soc. Med. Hist., Chicago,* January, 1935.

BAUMGARTNER, L., and FULTON, J. F. A Bibliography of the Poem Syphilis, Sive Morbus Gallicus by Girolamo Fracastoro of Verona. New Haven, Yale University Press, 1935.

HOLCOMB, R. C. Who Gave the World Syphilis? The Haitian Myth. New York. Froben Press, 1937.

Osler, W. Fracastorius. In: An Alabama Student and Other Biographical Essays. Oxford, Clarendon Press, 1908.

Pusey, W. A. The History and Epidemiology of Syphilis. Springfield, Ill., C. C. Thomas, 1933.

Major, R. H. Classic Descriptions of Disease (excerpts from Leoniceno, Villalobos, Almenar, de Vigo, von Hutton). Ed. 2. Springfield, Ill., C. C. Thomas, 1939. Pages 12-58.

The Different Types of Infection

[From *On Contagion, Contagious Diseases and Their Cure**]

The essential types of contagion are three in number:

(1) Infection by contact only.

(2) Infection by contact and by fomites as scabies, phthisis, leprosy (elephantiasis) and their kind. I call fomites such things as clothes, linen, etc., which although not themselves corrupt, can nevertheless foster the *essential seeds* of the contagion and thus cause infection.

(3) Finally there is another class of infection which acts not only by contact and by fomites but can also be transmitted to a distance. Such are the pestilential fevers, phthisis, certain ophthalmias, the exanthem that is called variola, and their like.

INFECTION BY CONTACT ALONE

The infection which passes between fruits is markedly of this kind, *e.g.,* as from one cluster of grapes to another and from one apple to another apple. . . . The putrefaction that thus passes from one fruit to another is really a dissolution of the combination innate heat and moisture by the process of evaporation.

The humidity (thus set free), softens and relaxes the parts and makes them separable, and the heat effects the separation. . . . I regard the particles of heat and of moisture separately, or in the case of moisture, perhaps in *combination* as the essential *germs* of the resulting putrefaction. I speak here of the particles of humidity in combination because in the evaporative process of putrefaction, it often happens that the very minute particles mingle themselves and thus generate new corruptions. This mingling is indeed especially favorable for the propagation of putrefactions and infections.

* De Contagionibus et Contagiosis Morbis et Eorum Curatione, 1546. Adapted from the French translation of Fournier by the editor.

INFECTION BY MEANS OF FOMITES

It may be questioned whether the infection by a fomes is of the same nature as infection that acts only by actual contact. The nature of infection by a fomes appears, indeed, to be different since having left its original focus and passed into a fomes it may there last for long unchanged. It is, indeed, wonderful how the infection of phthisis or pestilential fevers may cling to bedding clothes, wooden articles, and objects of that kind for two or three years, as we have ourselves observed.

On the other hand those minute particles given off by a body affected with putrefaction do not appear to preserve their virulence for long and on that account are not to be regarded as of identical essential nature either with those of fomites or with those that act by contact alone. . . . Not all substances are liable to become fomites, but only those that are porous and more and more or less calorific, for in their recesses the seeds of contagion can lurk hidden and unaltered either by the medium itself or by external causes, unless these are excessive, *e.g.,* they cannot withstand fire. Thus, iron, stone, and cold and impervious substances of this kind are hardly likely to act as fomites; on the other hand linen, cloth and wood are more apt to do so.

INFECTION AT A DISTANCE

It is well known that the pestilential fevers, phthisis and many other diseases are liable to seize on those who live with the infected, although they have come into no direct contact with them. It is no small mystery by what force the disease thus propogates itself. . . . For this type of contagion appears to be of quite a different nature and to act on a quite separate method from the others. . . . Thus a patient with ophthalmia may give his disease to another by merely looking at him. . . . This well illustrates the rapid and almost instantaneous penetrative power of this type of contagion . . . which may be compared to the poisonous glances of the catablepha.

THE AFFINITIES OF INFECTION

The affinities of infection are numerous and interesting. Thus there are plagues of trees which do not affect beasts and others of beasts which leave trees exempt. Again among animals there are diseases peculiar to men, oxen, horses, and so forth. Or, if separate kinds of living creatures are considered, there are diseases affecting children and young people from which the aged are exempt and vice versa. Some again only attack men, others women, and others again both sexes. There are some men that walk unharmed amid the pestilence while others fall. Again there are infections which have affinities for special organs. Thus ophthalmia affects only the eye. Phthisis has no effect upon that most delicate organ but acts especially upon the lungs. Alopeciae and Areae confine themselves to the head.

IS INFECTION A SORT OF PUTREFACTION?

We here consider whether all infection is a sort of putrefaction and also whether putrefaction is not itself infection. . . . Now with Rabies have we not infection without putrefaction? Again, when wine becomes vinegar have we not infection without putrefaction? For, if left to putrefy, it is later that it becomes fetid and undrinkable—the sure signs of putrefaction—and thus differs from vinegar which is pleasant to take and is indeed resistant to putrefaction.

But it must be remembered as regards putrefaction that sometimes there is but a simple dissolution of the combination of humidity and innate heat without any new *generation*—we then speak of it as *simple* putrefaction. Sometimes on the other hand, in the process of this dissolution, there is a true animal generation or generation of some substance definitely organized and arranged.

When there is simple putrefaction, there is no new generation but a fetor and a horrible taste arise . . . but when, on the other hand, there is neither the abominable smell nor taste but a definite redistribution of the qualities. As with wine . . . so also with milk and with phlegm, the first stage of putrescence is acidity. Similarly with Rabies, we must suppose a preliminary stage in which there is a certain amount (of the same preliminary type) of putrescence. It is, however, latent because putrefactions which take place in the living animal do not make themselves immediately apparent. It is an observed fact, however, that dogs which are becoming rabid are usually seized with febrile symptoms. If, therefore, we regard the matter inductively we shall consider that all infections may be reduced ultimately to putrefaction. . . . Furthermore, all putrefactions are liable to produce putrefactions like themselves, and, if all infection is putrefaction, infection in the ordinary sense of the word is nothing else than the passage of a putrefaction from one body to another either continuous with it or separated from it.

SYPHILIS

[From *Syphilis, or the French Disease**]

The poem is in three books. At the beginning of the first the author gives voice to the puzzlement that his generation feels about the mysterious malady that has suddenly afflicted them, and to the gossip that it had come from overseas (Columbus), breaking out when the French army (with Spanish mercenaries under Charles VIII) captured Naples (February 22, 1495). Fracastorius does not subscribe to this view, but gives his own opinion, after an acknowledgment to his patron Bembo, that the causes were very complicated, involving the politics of the Olympian gods—Jupiter, Saturn, Mars; resulting in a miasma that the sun's rays generated in the bosom of the earth. (Fracastorius probably did not himself believe in this appeal to the pagan Roman deities, but the miasma theory conforms to his ideas of contagion.) The peculiar nature of the methods of contagion of the disease, its signs and symptoms are then described in somewhat purple poesy.

The second book describes in a practical manner the treatment that had, in the experience of the author, been effective. The case history of a husbandman, named Ilceus, is recited; after several engaging mythological adventures, Ilceus finds his cure in a bath of the metal mercury, or quicksilver.

The third book tells fictionally of a visit of Spanish mariners to a land in the west where they are invited to a religious ceremony. The natives who attend are all afflicted with a loathsome disease, which manifests itself on the skin. The remedy which cures them is the juice of a tree—Hyacus (guaiac wood, recommended by Ulrich von Hutton). Then follows the immortal tale of Syphilus (*sic*) a shepherd, who left his indelible eponym on the disease.

BOOK I

I sing of that terrible disease, unknown to past centuries, which attacked all Europe in one day, and spread itself over apart of Africa and of Asia. I will explain what combination of events, which hidden germs have caused it, how it arose in Italy, at the time that

* Hieronymi Fracastorii Syphilis, Sivi Morbi Gallici, Libri Tres, Ad Petrum Bembum. Verona, 1530. Translation by the editor, based on the prose translation of the Philmar Company. By permission of Dr. Solomon Claiborne Martin, the Philmar Company and the *Urologic and Cutaneous Review.*

the French armies rendered desolate that unhappy country, which reason caused it to be called the French disease. I will tell how in these cruel circumstances the genius of man succeeded, with the help of the gods, in discovering the heroic remedy which abated the fury of the plague. . . .

My work is but a medical essay, but remember that Apollo himself did not look upon it as derogatory to his dignity divine to cultivate the healing art. This subject though poetical in form, takes on at times a serious interest, under a frivolous guise, will cause to appear before thee the great laws of nature, the decisions of destiny, and the mysterious origin of a frightful scourge.

Muse, what causes preside at the origin of this scourge? . . . Was it imported among us from those new worlds which were discovered by the brave mariners of Spain beyond the unknown seas of the Western world? Have we received from those far countries the germ where it is said it has reigned as supreme scourge from all eternity, numbering as many victims as there are inhabitants? Is it true that introduced, in that manner, among us it was spread throughout Europe by means of commercial relations? Is it true that it was born weak and obscure, to increase its force a hundredfold later on as it extended its ravages and invaded, little by little, the entire universe? Such as once, springing from a badly extinguished focus which an imprudent shepherd left in the country, a single spark sufficed to start a conflagration? . . .

No, it is not in this manner that this disease has developed itself. Incontestable testimony proves that it is not of a strange or foreign origin and that it was not necessary to cross the ocean to arrive in our midst.

Among the first victims who were attacked in our climate, I could mention a number of patients who were spontaneously attacked, without having exposed themselves to the least chance of contagion. Besides, how would it be possible to attribute to a contagious influence a disease which attacked so many people in such a short time? As a matter of fact it was on all sides at one time that the scourge was let loose upon us, in Italy, in the fertile fields of Sagra, in the forests of Ausonia, on the plains of Otrantes, on the banks of the Tiber, in the hundred cities that the Eriden enlarges by the hundred streams tributary to it, which laves them with its majestic waves. At the same time, in addition, it raged on foreign shores, and proud Spain, mother of the conquerors of the New World, did not suffer from the cruel attacks earlier than the people of whom the Pyrenees, the Rhine and the Alps are the boundaries.

All beings with which Nature has peopled land, air and the water, have not one mode of creation. The most simple, whose formation calls but for a few generating principles, are incessantly reproduced everywhere. Others, more complex, require for their being the help of many germs which are dispersed, succeeding but rarely and with difficulty in being born at certain times and in certain places. Others, finally come out of nothingness but after thousands of centuries, so many are the obstacles presented to the germs which are necessary to their genesis and to their union. Well, the same is true of diseases. All diseases do not have a common or identical origin. The ones, the majority even, have an easy development which accounts for their habitual frequency; but others are of a difficult sort to deliver and succeed but slowly to constitute themselves, after having long fought against the infinite difficulties which destiny opposed to their birth. Of this number is the French disease which, for a long time wrapped up in the darkness of the nothing, has suddenly freed itself from its bindings, after many centuries of waiting, to finally rise in the light and make an irruption among us.

The time has come when, after the completion of several centuries, the sovereign ruler of the world was, according to an eternal law, to future destinies of the earth and of the heavens, Jupiter prepared himself for this great work, to which he invited Mars and Saturn. On the appointed day, Cancer opened the doors of Olympus to the Immortals. Mars was the first to cross the threshold of the sacred place. From his impetuous gait, his glistening armor, the god of war was easily recognized, the cruel god who slakes his thirst with blood and carnage. Calm and majestic Jupiter next appeared, carried on a golden chariot, Jupiter always benevolent and good to mortals, unless contrary destinies claim his clemency.

Nevertheless, Jupiter sat on the throne where he alone had the right to sit. He consults the oracles and regulates the destinies of future periods. Despite this, it is settled, the voice of the gods has shaken Olympus and the fatal decree is pronounced.

A subtle poison at once spreads itself in the ether and disseminates its pernicious effluvia throughout the immensity of space.

What was the origin of this poison? Are we to believe that the sun's rays, associated with the malign influence of the stars, raised from the bosom of the earth and of the waters unhealthful vapors which spread in the air contagious miasms, the germs of a disease as yet unknown? Or on the other hand, were these miasms engendered in the upper regions of the atmosphere, from which as a consequence, they descended among us?

Let us now study the symptoms of this scourge which a celestial influx has caused and reproduced after centuries that it was forgotten. This disease does not affect the dumb inhabitants of the wave, nor the wild beasts of the forest, nor the birds of the sky, nor the horses or cattle. It only has to do with man; man alone is prey.

In the human body, it is the blood that it attacks at first, and, feeding on naught but fat and viscid humors, it is on the fat and corrupted parts of this fluid that it preferably attaches itself.

Here especially, O Muse, I claim thy help to limn the picture of this execrable pestilence. Deign also to inspire me, Apollo, god of the day, god of poetry, and make matters such that my work may, thanks to thee, remain through coming centuries. A day, in fact, may perhaps come when our great nephews will take the pleasure of consulting the description of a forgotten disease. Forgotten, yes, for no one doubts at a given time this disease will return into the clouds of nothingness. And no one doubts also but that, after another series of centuries, it will return to the light, to afflict anew the world and once more spread terror among the peoples of another age.

One of the most surprising facts is that, after having contracted the germ of the contagion, the victim attacked by the scourge does not often present any lesion that is well marked before the moon has four times accomplished its travels. The disease, in fact, does not show itself at once by accusing symptoms directly that it has penetrated the organism. For a certain time it broods in silence, as if it were gathering its forces for a more terrible explosion. During this period, at all events, a strange languor seizes the patient and depresses his whole being; his mind seems heavy, his limbs are soft, and weakening, fail to work; the eye loses its flash and the face is depressed in its expression and has become pale.

It is on the organs of generation that the virus first is transported, to irradiate from there to the neighboring parts and on the regions of the groin.

Soon after, more well defined symptoms show themselves. When the light of day disappears to give place to the shades of night, at the time when the inner heat of living bodies leaves the peripheral parts to concentrate upon the viscera, atrocious pains suddenly burst forth in the limbs charged with vitiated humors and torture the articulations, the arms, the shoulders, the calves. It is because at that moment, vigilant Nature, an enemy of all impurity, is at work to react against the putrid ferments which the disease has introduced into the veins and with which it has penetrated all the humors, all the nourishing juices of the organisms. She strains to drive them away; she energetically fights against them. But they resist; thick, viscid, not displacing themselves to the exsanguined framework of the tissue, and give rise to horrible sufferings wherever they adhere.

The most subtle of these humors, those which are the most easily evacuated, take refuge either in the skin or in the extremities of the limbs. They then produce hideous eruptions on those points and these exanthems soon spread over the whole body and over the face with a repulsive mask.

Unknown to our days, these eruptions consist of pustules and conical pimples, which, gorged with corrupted liquids, are not slow in opening to allow the escape of a mucous and virulent sanious liquid. Even, sometimes, the pimples that are similar develop in the depths of organs and noiselessly corrode the tissues. It is thus that horrible ulcers are seen covering the limbs, denuding the bones, eating the lips and penetrating the throat, from which there only issues a weak and plaintive voice.

At other times, again, there exhales from the skin thick humors which dry into fearsome crusts on the surface of the integument. Like these are seen the viscid juices which come from the cherry tree or the almond tree condensing in a gummy callus on the bark of these trees.

Ah! how many patients, sorrowful victims of this plague have contemplated with horror their faces and their bodies covered with the hideous taints, deploring their youth destroyed in its bloom, and have cursed the gods and threatened the sky! Unfortunate! Night which pours sweet repose upon all nature, has no more charms for

them, for sleep has fled from their eyes. For them, in the same manner, aurora comes without attractions, for day like night recalls their
pains. The pleasures of the table, joyous feasts, the intoxicating
gifts of Bacchus, the festivities of the city, the delights of the country, nothing smiles for them any more. Vainly do they search for a
respite to their sufferings on green banks made pleasant by the
purling of streams, in the shade of valleys, and in the solitude of
mountains. Desperate, lost, they return addressing ardent prayers to
the gods, burning expiatory incense in the temples, loading altars
with rich gifts. Useless trouble! The gods remain deaf to their voices
and disdain their sacrifices.

BOOK II

I will now, following my work, state here the diet and treatment
which are proper to oppose to the disease according to its phases
and different forms; I will reveal the marvelous agents which were
discovered to combat it.

At the beginning, amid the consternation produced by a disease
unknown up to that time, a thousand remedies were tried which
were all powerless. But stimulated by the darts of suffering, otherwise illuminated by his reverses, man learned how to find new arms
against the redoubted enemy; he fought against the scourge, he
threw it, he was enabled one day to proclaim himself victor.

Blood has not the same identical composition in all patients. If it
is pure, it is a good presage; if, on the contrary, it is thick, superabundant and charged with bile, the disease, under such conditions
will be more serious, more rebellious, and will only give way to the
use of more energetic measures, as well as more violent.

That which is the most essential to a cure, is to surprise the disease
at its inception, to strangle it in the form of a germ before it has time
to invade the viscera. For when it has penetrated into the organism,
when it has taken root and developed its ravages, it is alas! but at
the price of rough experience that one can succeed in expelling it.
Apply yourself then, before all, to combat it at its very inception,
and engrave in your mind the precepts which will follow.

Patients, the quality of the air that you breathe, is far from being
an indifferent matter. Learn how to avoid winds from the south; flee
from the fogs and from wet grounds with pernicious effluvia. Choose

for a stay a laughing country with uncovered horizon or some hill-
side bathed by the sun. Only there will you find pure air, continually
renewed by the winds and the friendly zephyrs.

Especially guard yourself against laziness and nonchalance. Go, go
drive in their dens the bear and the boar; hunt the deer from the
crests of the mountains to the foot of the valleys and into the depths
of the woods. As a fact, I have often seen the disease clear up by the
sweating and cure after long runs in the forests. This is not all.
Without false shame, take in hand the plough and turn its share in
the bosom of the earth; armed with a hoe tear up the underbrush,
strike with an axe the towering oak, uproot the sycamore.

Drive away far from you the anxieties, preoccupations, and re-
grets; far from you the trouble of passions and the assiduity of serious
study! What suits your state is the mild business of the muses, it is
joyous complete, and frolicsome dances. At all events do not succumb
to the attractions of love; nothing could be more harmful, and your
kisses would taint the tender daughters of Venus with a detestable
contagion.

Nothing is more important for you than the ordering of your diet
and the choice of foods. Upon this point redouble your attention
and vigilance. In the first place banish from the table all fish, no
matter which they are, fish from rivers or ponds, from fresh or salt
water. At most you may, in case of necessity, indulge in those that
are fished near bluffs or falls and whose meat is white, soft and delicate; such are, for example, the phycine, the dorade, the gudgeon,
the perch, friend of rocky shores, and the scare, sole ruminant of the
waves, the constant guest of the mouths of rivers. Also abstain from
aquatic birds which, living along the rivers or in swamps, feed on
nothing but fish. . . .

Do not let yourself be tempted either by the sparkling or frothy
wines of the shores of Corsica, of Falernum or of Puini, or those
which are produced in our small forms by the grape of Rhetia. Nothing will be less healthful for you than the light wines of the Sabine
or those of which the Naiads will have dulled the generous odors.

On the other hand, that of which you may freely use, is comprised
in all simple foods which are healthful and of which Nature is generous in gardens, and which are the delight of the gods; mint, cress,
chicory, hare's lettuce whose flower braves hoar frost, skirret, the

friend of small streams, sweet marlum, calamus with perfumed odor, the coquettish melissa, ox-tongue which thrives best on the edges of fountains, packet, spinach, sorrel, samphire with salty buds, hops which interlace with brush, byronia of which I advise you to gather the young before the adult branch has spoilt the shoots and spread its seeds that are turning green.

But it is vain, I suppose, that you have exhausted the entire series of these remedies; or, perhaps, impatient of such slowness, relying upon your strength and your health, you resolve to turn to more energetic agents, to end the matter with the hated enemy as soon as possible. Be it so! I will show you these violent and expeditious methods which can triumph, in a short time, over a disease that is usually long in duration, stubborn, subject to relapses and rebellious to mild medication. But also learn the price you will have to pay for your hasty deliverance.

First of all here is a treatment which consists in the use of fumigations composed of styrax, of cinnabar, of mibium, of antimony, and of incense. We have here, without a doubt, an active medication, which succeeds in cleaning the body of its awful taints; but it is excessively violent, irritating, and uncertain in its results. In addition to its bringing on respiratory difficulties and a true suffocation. Therefore, these fumigations, in my opinion, should never be used on the entire body; it is proper to limit their action to those parts which are the seat of eruptions or of ulcers. Another method of which mercury forms the basis is much preferable. As a fact, the action of mercury on the scourge is marvelous. . . .

In a valley of Syria, there formerly lived, it is said, a husbandman named Ilceus. He divided his tranquil life between the labors of his field, and the cultivation of a garden consecrated to the gods of the field. . . . Suddenly, O horrors, he was struck by the terrible scourge. The unfortunate man, in his distress, called the heavens to his aid: "Ye gods that I adore," he cried, "have pity on my torture! And thou, beneficent Callirhoe, thou who always curest our ills, do not forget that but a few days since I made an offering to thee, on the trunk of an oak, of the carcass of the deer that had fallen under my blows." . . .

At that time Callirhoe was bathing herself in a neighboring grotto. She heard his prayer and those vows. She at once answered Ilceus.

. . . She sent him sleep to assuage his pains; and, whilst he was resting in peace beneath the fresh shade of the willows, she appeared to him in a dream, arising from the bosom of the waters, and said: "Ilceus, at last the gods, in answer to my prayer, have taken pity on thee, but, alas! the remedy, the only remedy, that may cure thee of thy ills, thou shall hunt in vain in this part of the world that the Sun lights with his rays. Such in fact is the inexorable chastisement which has been visited upon thee by Diana and her brother Apollo, the very day that thou didst pierce with thy arrows the sacred deer of whose carcass thou didst make an offering to me. Diana saw thy victim panting upon the ground and bathed in blood; she saw thy fatal trophy suspended on one of the oaks of the neighboring forest; and in her grief she cursed thee! It is she and the son of Latona, excited against thee by the anger of his sister, who have afflicted thee with a horrible disease; and both have sworn that everywhere in which their empire extends thou shalt find no remedy for thy sufferings. There remains for thee, as the only resource, to seek thy safety in the bowels of the earth and the darkness of the infernal regions. Listen! Under a neighboring rock, a dark cavern opens and it reveals to the eyes of mortals a dense forest of oak trees. . . . Let the next sunrise see thee there. . . . Thy prayers will be heard, and a nymph will come offering herself to thee as a guide for thy steps in the dark roads which lead to the center of this earth. She herself, will also point out to thee the remedy that thou implorest." . . .

Ilceus awoke "intoxicated with joy." "Beneficent goddess," he cries, "I accept thy presage, I will obey thee; I will go, divine virgin, whither thy voice calls me!"

The next day, at the earliest streak of dawn, he proceeded to the cavern. He found its entrance under immense rocks which the tree of Jupiter has covered with branches. On the threshold of the chasm, he immolated a black sheep which he offered as a sacrifice to powerful Cybele. . . .

The nymphs of the earth who preside over the metals, were occupied at that moment in untying liquid sulphur with the silvered wave of mercury, a marvelous amalgam which hardened by the bath, transforms itself into pure gold. . . . When the voice of Cybele resounded, . . . the nymph Lipara responded and addressing Ilceus: "I know," she told him, "thy name and thy misfortunes; I know the design that brings thee here. Be without fear. . . . The remedy that thou seekest is here. Come, follow me in these dark paths, which lead to our domain; the nymph who is speaking to thee will guide thy footsteps." At these words she crossed the threshold of the cavern. Ilceus followed her without hesitation. What a picture then unrolled itself to his eyes! There were here gaping gulfs, there some subterranean rivers, at other places bottomless, abysms filled by eternal night. "We are here," says Lipara, "in the empire of Earth."

They then entered the avenue whose arches garnished with tutty are traversed by threads of gold and of sulphur of scintillating reflections. Then they arrive at the banks of a river with silvery waves. "Ilceus," says the nymph, "thou hast finally reached the end of thy troubles. When that sacred stream shall have passed over thy body three times, thou shalt be delivered of thy disease and its impure poison." At these words, she plunges her virgin hands into the river; three times she takes out of it the liquid metal and three times she spreads it on the limbs of Ilceus. O, prodigy! It is done! The disease at once disappears, and his hideous covering, on contact with this glowing flood, dissolves and disappears in a moment!

"Now leave," continued Lipara, "go find the day, the pure sky and the fortunate regions that the sun lights. But let thy care be to offer a sacrifice to Diana, to the gods of these gloomy places and to the goddess who has saved thy days."

At the beginning, mercury was employed associated with lard; later it was combined with the turpentine of Epirus and with the resin of the majestic birch. Certain physicians today combine it with horse fat or bear's grease, bellium and with the juice of cedar, others with myrrh, with male incense, with mibium and with burning sulphur. For my part, I prefer to alloy it with a mixture of black hellebore, orris root, galbanum, asafetida, oil of mastic, and oil of native sulphur.

Patients, a truce to the disgust which may be caused by this remedy! For if it is disgusting, the disease is still more so. Besides, your cure is at this price. So without hesitation, spread this mixture on your body and cover with it your entire skin, with the exception of the head and of the precordial region. Then, carefully wrap yourself in wool and tow; then get into bed, load yourself with bed covering and thus await until a sweat bathes your limbs with an impure dew.

Ten days in succession renew this treatment, for ten entire days you are to undergo this cruel trial whose beneficial effect will not cause you to wait. As a matter of fact, very soon an infallible presage will announce to you the hour of your freedom. Very soon you will feel the ferments of the disease dissolve themselves in your mouth in a disgusting flow of saliva, and you will see the virus, even the virus, evacuate itself at your feet in rivers of saliva.

If during the course of this treatment, small ulcers develop in your mouth, have a care to fight them with gargles of milk or by a decoction of pomegranate privet. This treatment being completed, you may then, without fear, recall Bacchus to your table and enjoy in full liberty the generous nectars of Phetia, of Falernum and of Chios.

BOOK III

Mothers and fathers, peasants and rulers, children and greybeards, stood mingled together: all tortured in soul and foul in body, with scabby skin from which matter oozed. Among them the high priest strode in white robes, purifying them with spring water and the leaves of the tree Hyacus.

The men of Europe were astonished to find this dreaded pestilence so easily overcome by the natives of Hesperis, and after each race could understand each other's language, he asked why a shepherd should stand in the sacred precincts covered with the dead bull's blood. The King replied that this was a yearly custom, the tradition of which was:

Once upon a time when Alcithous was King, one Syphilus, a shepherd, drove his oxen and sheep to pasture beside the river of this lovely country. It chanced that this was a very hot summer and the Dog-Star was burning up the parched fields and trees: so that the trees proffered no cool shade to the shepherds, nor were there any refreshing breezes. Syphilus, pitying the suffering of his flock, and himself maddened by the unending heat, raised his face and eyes towards the Almighty Sun and thus addressed the Sovereign: "Why, O Sun, do we call you Father? Why do we common folk raise sacred altars to you and worship you with sacrifices of oxen and rich incense,—if you do not in turn consider us, answer our prayers, and even the flocks of the King do not concern you? Or must I believe that You are consumed with envy? My flock holds a thousand snow-white heifers and a thousand sheep. I shall offer worship to my own King. *He* shall grant us soft breezes and bring the coolness of green woods to our cattle, and relieve us of this torrid heat."

After he had thus spoken, he forthwith raised altars to King Alcithous in the mountains and made sacrifice to him: and the rustic crowd and all the band of his fellow shepherds did likewise, offering incense kindled at their hearths, and seeking favorable omens with the blood of bulls and burnt offerings of smoking entrails. . . .

But when the Sun saw all this, he was very angry, and shot forth deadly rays of disease towards Father Earth. . . . Forthwith a pestilence unknown before sprung up on all lands. And first among them all, Syphilus, who had established the worship of the King with blood-sacrifices, and raised altars to him among the mountains, manifested the foul sores in his own body; first he knew sleepless nights, his bones ached mercilessly. And from him, the first to suffer it, the disease took its name and was called Syphilis by the native race. Before long the deadly infection spread itself among folk in all the cities, nor in its virulence did it spare the King's person.

XV

ANATOMY

GROSS ANATOMY, BOTH OF THE HUMAN AND THE ANIMAL BODY, WAS cultivated from the earliest times, but any systematization of knowledge was prevented by many causes—social and religious prohibitions, the inherent difficulty of the subject, and the encrustation of legend and prejudice which grew up around many important details. As illustrations, natural repugnance against dissection of the human body will easily be suggested; again, any medical student can tell you that actual practical dissection is far different from the clean-cut diagrams in his textbooks; and let it be remembered that up to the time of Vesalius it was solemnly believed that a male had one less rib than a female, after Adam's loss described in the Book of Genesis.

REFERENCES

SINGER, C. The Evolution of Anatomy. London, Kegan Paul, Trench, Trubner, 1925.
CORNER, G. W. Anatomy. Clio Medica series. New York, Paul B. Hoeber, 1930.
BALL, J. M. Andreas Vesalius, the Reformer of Anatomy. St. Louis, Medical Science Press, 1910.
ROTH, M. Andreas Vesalius. Berlin, 1892.

MONDINO DI LUCCI

Mondino di Lucci (1275-1326), professor of anatomy at Bologna, introduced the study of human anatomy instead of the dissection of pigs, and wrote the first original textbook on practical anatomy in the Middle Ages. His little book, which was no more than a dissecting manual, remained popular for two centuries and an early printed edition (1513) contains 79 pages.

[From the *Anatomy**]

Because as Galen says in the seventh book of his Method of Healing, on the authority of Plato, a work in any science or art is pro-

* Anathomia. Completed in 1316; first published in 1478. Translation by John D. Comrie, 1880.

pounded for three reasons. Firstly to satisfy one's friends, secondly that one may obtain a most useful exercise, that is by the mind; thirdly that so one may be saved from forgetfulness, which comes with age. Hence it is that, moved by these three causes, I have proposed to put together a certain work in medicine for my scholars.

ON THE ANATOMY OF THE UTERUS

And for these four causes the woman whom I anatomized in the past year, or A.D. 1315, in the month of January had a uterus twice as large as one whom I anatomized in the month of March in the same year. . . . And because the uterus of a pig which I anatomized in A.D. 1306 was a hundred times larger than it can ever be seen in a human being, there may be another cause, i.e., because it was pregnant and had in the uterus 13 little pigs. In this I showed the anatomy of the foetus or of pregnancy as I shall tell you.

LEONARDO DA VINCI

Leonardo da Vinci (1452-1519) was a cosmic figure, myriad-minded, not only typical but the epitome of the Renaissance. Mommsen, the great historian of Rome, says of Caesar that he was "the complete and perfect man." There have been very few of these in the course of history—Confucius, Aristotle, Goethe. But certainly in that group belongs Leonardo. He left a number of anatomical drawings which show that before Vesalius he made original and first-hand observations which earned William Hunter's comment that he was "the greatest anatomist of his epoch." They did not, however, influence his own time because they remained buried for two hundred years, until discovered by Blumenbach and William Hunter. They were finally collected from the Royal Library at Windsor, the Ambrosian Library at Milan, and the Institut de France and published 1898-1916. Every medical student should study good copies of them. The explanatory notes are in mirror writing. Leonardo left many notes on anatomy, geology, natural history, flying machines, hydraulics, and other subjects which have been collected and translated by Edward MacCurdy, from which the following excerpts on anatomy are made to show the modern quality of Leonardo's thought.

REFERENCES

MacCurdy, E. The Notebooks of Leonardo da Vinci. New York, Reynal & Hitchcock, 1938.

McMurrich, J. P. Leonardo da Vinci, the Anatomist. Printed for the Carnegie Institution of Washington. Baltimore, Williams & Wilkins Co., 1930.

[From *The Notebooks of Leonardo da Vinci**]

(Method for the study of the arm and the forearm)

You will first have these bones sawn lengthwise and then across, so that one can see where the bones are thick or thin; then represent them whole and disjoined, as here above, but from four aspects in order that one can understand their true shape; then proceed to clothe them by degrees with their nerves, veins and muscles.

(Motor muscles of hands and wings)

No movement either of the hand or the fingers is produced by the muscles above the elbow; and so it is with birds and it is for this reason that they are so powerful because all the muscles which lower the wings spring from the breast and these have in themselves a greater weight than that of all the rest of the bird.

(Action of muscles in breathing)

These muscles have a voluntary and an involuntary movement seeing that they are those which open and shut the lung. When they open they suspend their function which is to contract, for the ribs which at first were drawn up and compressed by the contracting of these muscles then remain at liberty and resume their natural distance as the breast expands. And since there is no vacuum in nature the lung which touches the ribs from within must necessarily follow their expansion; and the lung therefore opening like a pair of bellows draws in the air in order to fill the space so formed.

The intestines. As to these you will understand their windings well if you inflate them. And remember that after you have made them from four aspects thus arranged you then make them from four other aspects expanded in such a way that from their spaces and openings you can understand the whole, that is, the variations of their thicknesses.

(Heart and vessels proceeding from the heart, and comparison with the roots and ramifications of plants)

The plant never springs from the ramification for at first the plant exists before this ramification, and the heart exists before the veins.

* Arranged, rendered into English and introduced by Edward MacCurdy. [2 vol.] New York, Reynal & Hitchcock, 1938. By permission of Reynal & Hitchcock, Inc.

All the veins and arteries proceed from the heart; and the reason is that the maximum thickness that is found in these veins and arteries is at the junction that they make with the heart; and the farther away they are from the heart the thinner they become and they are divided into more minute ramifications. And if you should say that the veins start in the protuberance of the liver because they have their ramifications in this protuberance, just as the roots of plants have in the earth, the reply to this comparison is that plants do not have their origin in their roots, but that the roots and the other ramifications have their origin in the lower part of these plants, which is between the air and the earth; and all the parts of the plant above and below are always less than this part which borders upon the earth; therefore it is evident that the whole plant has its origin from the thickness, and, in consequence, the veins have their origin in the heart where is their greatest thickness; never can any plant be found which has its origin in the points of its roots or other ramifications; and the example of this is seen in the growing of the peach which proceeds from its nut as is shown above.

(Passage of the urine from the kidneys into the bladder by means of the ureters)

The authorities say that the uretary ducts do not enter directly to carry the urine to the bladder; but that they enter between skin and skin by ways that do not meet each other; and that the more the bladder becomes filled the more they become contracted; and this they say nature has done merely in order that when the bladder is filled it should turn the urine backwards whence it came; in such a way that in finding the ways between membrane and membrane to penetrate into the interior by narrow ways and not opposite to that of the first membrane, the more the bladder is filled, the more it presses one membrane against the other, and consequently it has no cause to spread itself out and turn back. This proof however does not hold, seeing that if the urine were to rise higher in the bladder than its entrance which is near the middle of its height it would follow that this entrance would suddenly close and no more urine would be able to pass into the bladder and the quantity would never exceed the half of the capacity of this bladder; the remainder of the bladder therefore would be superfluous, and nature does not create anything superfluous. We may say therefore, by the fifth (section) of the sixth (book) concerning waters, that the urine enters the bladder by a long and winding way, and when the bladder is full the uretary ducts remain full of urine, and the urine that is in the bladder cannot rise higher than their surface when the man is upright; but if he remains lying down it can turn back through these ducts, and even more can it do this if he should put himself upside down which is not often done; but the recumbent position is very usual, in which if a man lies on his side one of the uretary ducts remains above the other below; and that above opens its entrance and discharges the urine into the bladder, and the other duct below closes because of the weight of the urine; consequently a single duct transmits the urine to the bladder, and it is sufficient moreover that one of the emulgent veins purify the blood of the chyle of the urine which is mixed with it because these emulgent veins are opposite to one another and do not proceed from the vein of the chyle. And if the man sets himself with his back to the sky both the two uretary ducts pour urine into the bladder, and enter through the upper part of the bladder, because these ducts are joined in the back part of the bladder, and this part remains above when the body is facing downwards, and consequently the entrances of the urine are able to stand open, and to supply so much urine to the bladder that it fills it.

ANDREAS VESALIUS

Andreas Vesalius (1514-1564) of Brussels, afterwards professor of anatomy at Padua, published the first modern anatomical treatise, De Fabrica Humani Corporis, in 1543. Vesalius had been a pupil of Sylvius in Paris, who taught the old Galenical anatomy—of a five-lobed liver, two-horned uterus, one less rib for a man, the seven-segmented sternum—arrived at from dissection of animals. Vesalius broke with all this. He obtained human anatomical material (tradition has it, by cutting down the bodies of hanged criminals left exposed on the gibbets), often falling afoul of the authorities, especially the ecclesiastical authorities, in doing so. He left the North, traveled through Basel (where he met Oporinus, subsequently his publisher) and came to Padua. There he gave such remarkable public dissections that all the town flocked to hear him. Afterwards, as the result of quarrels, he went to Madrid and became court physician to Charles V. The "Fabrica" may be said to be the foundation of modern medicine. It has never been completely translated into English. It is illustrated by magnificent drawings, the work of Jan van Calcar.

The "Fabrica" corrects a number of errors: Luz—Adam's missing rib is replaced, the five-lobed liver, the bicornuate uterus, the seven-segmented sternum, the double bile duct, the interventricular pores, the hypothetic sutures in the maxillary, are all replaced by modern ideas.

Vesalius makes some errors. Some of his muscle figures show the recti going clear up to the clavicle, and scaleni muscles of dogs are drawn.

The "Fabrica" is divided into seven books:

The *first* book is on the skeleton. Skulls are divided into long and broad. The sphenoid bone is depicted for the first time. The incus and malleolus are described for the first time; the stapes is omitted. Galen's pre-maxillary bone is gone. There are some errors about the ribs. The vertebrae are well described. The sternum is shown as a thin segmented bone. The clavicle and wrist are excellent. The long bones of the upper arm are too short. The pelvis is good.

He regularly shows a small round bone in the carpus and tarsus that has no existence.

The *second* book describes the muscles. Vesalius worked out the action of each muscle. He describes a seventh muscle of the eye, the choanoides, found actually only in animals.

The *third* book describes the vascular system. The diagrams of the veins contain some errors. There are good diagrams and descriptions of the pulmonary veins.

The *fourth* book deals with the nervous system. Singer criticises this book, says it is inferior to those on the skeletal and the muscular systems. The optic nerve is wrongly drawn. The roots of the trigeminal and auditory nerves are very much confused. There is a clear and excellent drawing of the recurrent laryngeal nerves. Vesalius proved that cutting this nerve caused the voice to disappear.

The *fifth* book describes the abdominal viscera. The description is cursory but excellent. The attachments of the intestines are clear. The vermiform appendix is not mentioned in the text but it is shown in one of the illustrations. The description of the male generative organs is fair, that of the female organs is full of errors. The uterus is slightly bifid. The account of the embryo is poor.

The *sixth* book describes the thoracic viscera. The right lung is divided into only two lobes. The description of the heart is good. Vesalius says he has tried to put a probe through the pits in the interventricular septum, but always failed to get through.

The *seventh* book describes the brain and "includes a series of very fine figures of an absolutely pioneer character" (Singer). There are a series of horizontal sections of the brain. Vesalius describes the pituitary gland. There is a bad account of internal anatomy of the eye. He places the crystalline lens in the center of the eyeball and makes it spherical. He thinks it performs the functions we now ascribe to the retina.

The "Fabrica" ends with a chapter on "The Dissection of Living Animals."

PREFACE*

[From *The Fabric of the Human Body*]

To the Divine Charles the Fifth, Greatest and Most Invincible Emperor. The preface of Andrew Vesalius to his books On the Fabric of the Human Body

Whenever various obstacles stand seriously in the way of the study of the arts and sciences and keep them from being learned accurately and applied advantageously in practice, Charles, Most Clement Caesar, I think a great deal of damage is done. I think also that great harm is caused by too wide a separation of the disciplines which work toward the perfection of each individual art, and much more by the meticulous distribution of the practices of this art to different workers. The result is that men who have set the art before themselves as a goal, take up one part only. They leave aside things which point toward, and are inseparable from, that end; and as a result, they never accomplish anything outstanding. They never attain their proposed goal, but constantly fall short of the true essence of the art.

The other arts I shall pass over in silence, and I shall speak for a little while of that art which concerns the health of men.[1] To this, by far the most serviceable, basically necessary, difficult and laborious of all the arts which the genius of man has devised, nothing more unfortunate could have happened than the fact that some time ago —especially after the influx of the Goths and after the time of Mansor,[2] the King of Bochara in Persia (under whom the Arabs flourished, as deservedly familiar to us today as the Greeks)—medicine began to be torn apart. Its primary instrument, namely, the application of the work of the hand to curing, was so neglected that

* From *De Fabrica Humani Corporis*, 1543. There is no printed translation of Vesalius in any modern language. This is probably because he wrote in extremely corrupt Latin. This translation of the Preface was made by W. P. Hotchkiss, Associate Professor of History and Political Science, University of Kansas City. The Prefaces of the editions of 1543, 1555, and the Venetian edition of 15—, have all been collated to make this translation. Many passages are obscure and it has been thought best by Professor Hotchkiss and the editor to allow quite a loose translation in order to bring out any kind of sensible English meaning.

[1] M. Roth. *Andreas Vesalius Bruxellensis*, Berlin, Georg Reimer, 1892, pp. 197-8.

[2] During his study at Louvain, Vesalius reworked the ninth book of Rhazes' *Almansorem* and published it in February, 1537, at Louvain under title *Paraphrasis in novum librum Rhazes medici Arabis clariss. ad Regem Almansorem de singularuum corporis partium affectuum curatione.* Louvain, Restges Rescius, 1537. Roth, *op. cit.*, p. 76.

it seems to have been entrusted to laymen and to men who had no instruction in the sciences underlying the art of medicine.

Although once upon a time three schools of doctors existed, namely, the Logical, the Empirical and the Methodic, nevertheless the founders of these schools directed the aim of the art as a whole to the preservation of health and the elimination of diseases. In short, they referred to this end all the things which the individual men in their schools deemed necessary to their art, and they were accustomed to make use of three aids. Of these the first was a rational plan of diet; the second, all the uses of drugs; the third, surgery. The last shows with particular aptness that medicine consists in the supplying of deficiencies and the removal of superfluities; and it ever stands ready for the cure of affections. Time and use have shown that, as often as we engage in medical work, surgery is very helpful to the human race in its benefit to these affections.

This three-fold scheme of doctoring was equally familiar to the doctors of each sect. The doctors themselves accommodated their own hands to curing in accordance with the nature of the affections; and they expended no less energy in training their hands than in the business of arranging the diet or of knowing and compounding drugs. For instance, over and above his other books, the volumes which the divine Hippocrates wrote on the Rôle of the Doctor, on the Fractures of Bones, on the Dislocations of Joints, and evils of this type—the best written of all his works—show this clearly. Indeed, Galen, that prince of medicine after Hippocrates, in addition to boasting frequently that the care of the Pergamene gladiators had been entrusted to him alone, and to being unwilling, although his years were heavy upon him, that the apes which were to be dissected by himself should be skinned by the labor of his servants, frequently impresses upon [his readers] how much he delighted in the craft of the hand, and how zealously he and the other doctors of Asia practiced it.

But no one of the ancients seems to have handed down to posterity with equal care the curing which is wrought by the hand and that which is accomplished by diet and medicines. After the devastation of the Goths particularly, when all the sciences, which had previously been so flourishing and had been properly practiced, went to the dogs, the more elegant doctors at first in Italy in imitation of the ancient Romans began to be ashamed of working with their hands, and began to prescribe to their servants what operations they should perform upon the sick, and they merely stood alongside after the fashion of architects. When, soon, others also began to refuse the inconveniences of those practicing true medicine, meanwhile subtracting nothing from their profit and honor, they promptly fell away from the standards of the early doctors. They left the manner of cooking, and in fact the whole preparation, of the diet for the sick, to their attendants, and they left the composition of drugs to the vendors of medicines, and surgery to the barbers. And so in the course of time the technique of curing was so wretchedly torn apart that the doctors, prostituting themselves under the names of "Physicians," appropriated to themselves simply the prescription of drugs and diets for unusual affections; but the rest of medicine they relegated to those whom they call "Chirugians" and deem as if they were servants. They shamefully reject that which is the principal and the most ancient branch of medicine, the one which rests primarily upon the observation of nature (if indeed it is anything else!). Yet this branch of medicine even the kings in India practice today; and by the law of heredity in Persia they pass it all on to their children, as once the families of the Asclepiades[3] did. The Thracians, along with many other peoples cultivate and venerate it. Although this art accomplishes absolutely nothing without the help of nature, but rather desires to aid her as she works to free herself from disease; when a part of the art, which the Romans in times past have proscribed from the state as if designed to deceive and destroy men, has been almost wholly neglected, the result is that the utility of the art as a whole is removed and destroyed. To this primarily we owe the fact that this most sacred art is ridiculed, although many censures are normally cast at doctors anyway, since that part of the art which those educated in the liberal arts have shamefully allowed to be torn away is the part which permanently illuminates medicine with its especial glory.

When Homer, the fount of talents, affirms that the medical man is more pre-eminent than many, and when he and the other Greek poets celebrate Podalirius and Machaon, these sons of the divine Aesculapius are not lauded because they did away with a little fever —which Nature alone cures more easily without the aid of a doctor

---

[3] Cf. Roth, *op. cit.*, p. 108, where he mentions the Asclepiad family tradition and refers to Galen, *Anat. Adm.*, II, 1.

than when the aid is applied—or because they humored the palate
of men in peculiar and lamentable affections. They were celebrated
because they were especially pre-eminent in the cure of haemorrhages, dislocations, fractures, contusions, wounds, and the other
breaks in the continuity of the body. They freed the most noble soldiers of Agamemnon from arrow points, javelins, and other evils of
this sort which are principally caused by wars and which demand the
careful attention of the doctor.

But, Most August Caesar, Charles, I have in no manner proposed
to exalt any one instrument of medicine above the others, since the
aforesaid three-fold method of help absolutely can not be disjoined
and taken apart. The whole method belongs to one workman. To
effect this synthesis properly, all the parts of medicine should be constituted and prepared equally so that all the individual elements
can be put to use more advantageously, and each element in turn
unites all together more perfectly. Now and then an extremely rare
disease does turn up which does not immediately require the three-
fold instrument of the safeguards. And so a fitting plan of diet should
be instituted, and finally something must be attempted by medicines
and then by surgery. Therefore tyros in the art should be encouraged in all the methods, and, if it please the gods, scorning the
whisperings of the "physicians," they should apply their hands likewise to curing in whatever manner the nature of the art and reason
really demand, as the Greeks did. This they should do lest they turn
mutilated medicine to the destruction of the common life of man.
And they must be encouraged in this more diligently in proportion
as we see today that the men who are more thoroughly grounded
in the art abstain from surgery as from the plague. They are afraid
that they will be traduced by the fanatics of the medical profession
before the unlettered populace as "barbers." They also fear that
afterwards they may not get half the profit, honor, or reputation in
the eyes of either the unlearned mob or the leaders. This detestable
opinion of many people in the first place keeps us from taking up
the *whole* craft of healing; and when we arrogate to ourselves only
the cure of internal affections, we desire to be doctors only in a small
way (to tell the truth for once!). The ensuing damage to mortals
is great. Indeed, when all the compounding of drugs was relegated
to the pharmacists, the doctors in turn soon lost completely the
knowledge of the simple drugs that were necessary to them. The

shops were filled with barbarous labels and crooked pharmacists. Many of the finest preparations of the ancients were not available to us, and even today a great number lie hidden. They were the authors of this situation. They, in short, set an endless task not only for the most erudite men of our own age, but also for those who lived in the years immediately preceding. These scholars applied themselves to knowledge of simple medicines with such untiring zeal that, in their attempts to restore their learning to its pristine splendor, they seem to have contributed a great deal. Proof of this is furnished by a rare contemporary example among many celebrated men, Gerhard Vueldbik, Your Majesty's Secretary—a man remarkably adorned with varied erudition in sciences and languages, and very well grounded in the history of the families of our people.

Furthermore, this most perverse surrender of the instruments of healing to various artificers, has brought a much more execrable disaster and far more frightful calamity upon an outstanding part of natural philosophy. To anatomical study, Hippocrates and Plato gave a high rank, since it embraces the study of man and since it correctly must be considered the solidest foundation of the medical art and the beginning of the constitution. They did not doubt that it should be included in the first parts of medicine. When this subject used to be practiced exclusively by the doctors, they stretched every nerve to master it. But when they surrendered the surgical work to others and forgot their anatomical knowledge, it ultimately began to collapse.

As long as the doctors thought that only the curing of internal affections belonged to them, they considered that the mere knowledge of the viscera was abundantly sufficient. They neglected the fabric of bones, muscles, nerves, veins and of the arteries which creep through the bones and muscles, as being of no concern of theirs. When the whole business was committed to the barbers, not only did the true knowledge of the viscera disappear from among the doctors, but also their activity in dissecting straightway died. This went so far that the doctors did not even attempt cutting; but those barbers, to whom the craft of surgery was delegated, were too unlearned to understand the writings of the professors of dissection. It is far from the truth that this group of men preserved for us this most difficult art, transmitted manually to them; but it is true that this deplorable dispersion of the curative rôle brought a detestable procedure into our Gymnasiums, wherein some were accustomed to administer the cutting of the human body while others narrated the history of the parts. The latter, indeed, from a lofty chair arrogantly cackle like jackdaws about things which they never have tried, but which they commit to memory from the books of others or which they place in written form before their eyes. The former, however, are so unskilled in languages that they cannot explain the dissections to the spectators. They merely chop up the things which are to be shown on the instructions of the physician, who, having never put his hand to cutting, simply steers the boat from the commentary— and not without arrogance. And thus all things are taught wrongly, and days go by in silly disputations. Fewer facts are placed before the spectators in that tumult than a butcher could teach a doctor in his meat market.[4] I shall not mention those schools where they hardly ever think of dissecting the structure of the human body, with the result that ancient medicine declined from its pristine glory years ago.

When at length in the great happiness of this age, which the gods have willed to be ruled by Your power, medicine had begun to revive along with all the studies, and had begun to lift up its head from the profoundest darkness so that it almost seems to have recovered its old splendor in some Academies; and since medicine now needs nothing more acutely than the dead knowledge of the parts of the human body, I decided to go to work on this book with whatever strength and brains I had, and with the encouragement of the example of so many distinguished men. For fear that I alone might go slack at a time when all men are, with great success, essaying something for the sake of the common studies, or even for fear that I might fall away from the standards set by my progenitors, who were by no means obscure doctors, I thought that this branch of natural philosophy should be called back from the depths, so that, even if it should not be more complete among us than among the early doctors of dissection, nevertheless it should some day reach a point where one would not be ashamed to state that our method of dissection compares favorably with the ancient. And one might say that nothing had been so broken down by time, and then so quickly restored, as anatomy.

[4] Paraphrased by Roth, p. 103.

But this ambition of mine would never have succeeded if, when I was studying medicine at Paris, I myself had not applied my hand to this business, and incidentally had the pleasure of being present at several public dissections put on by certain barbers for my colleagues and me when some viscera were superficially shown. At that time, when we first saw the prosperous re-birth of medicine, anatomy was given rather perfunctory treatment there.[5] When some dissections of animals were being performed under the direction of that celebrated and most praiseworthy gentleman, Jacobus Sylvius, I was encouraged by colleagues and preceptors, although I had been trained only by my own efforts, to perform in public the third dissection[6] at which I ever happened to be present—a dissection which dealt purely and simply with the viscera as was the custom there—and I did it more thoroughly than was usual. Moreover, when I next attacked a dissection, I attempted to show the muscles of the hand along with the more accurate dissection of the viscera. For, aside from the eight muscles of the abdomen, badly mangled and in the wrong order, no one had ever shown a muscle to me, nor any bone, much less the succession of nerves, veins, and arteries.

Soon I had to go to Louvain because of the tumult of war,[7] and, since the professors of medicine there had not even dreamed about anatomy in eighteen years,[8] I delivered lectures on the human fabric accompanied by dissections which I did a little more accurately than at Paris, my object being to win the good will of the students at the University and at the same time to gain more experience for myself in material that was really obscure and yet basically important for me in general medicine.[9] The result was that the younger professors of that University now seem to expend a good deal of serious study in distinguishing the parts of man, properly understanding how re-

[5] Roth, *op. cit.*, p. 67. Public dissections filled not quite three full days from the entire school year. This section dealing with Paris is roughly paraphrased by Roth *ad loc. cit.*

[6] Roth, p. 69, dates these public demonstrations in 1535 and 1536. He mentions that Vesalius dissected countless dogs and other animals, studied bones in the cemetery of the Innocents, and frequented the place of public executions on Montfaucon. He had, from the latter, plenty of opportunity to study severed hands.

[7] Charles V's invasion of France in the second half of 1536.

[8] I.e., no public dissections had been performed in the past eighteen years. Roth, p. 74.

[9] The Bürgermeister was much interested in Vesalius' work and supplied him with plenty of material for his dissections, more than he was able to get at Padua. Roth, p. 74; cf. p. 99.

markable a stock of philosophizing the knowledge of these parts supplies to them.[10]

Next I worked on anatomy in investigating the construction of the human body at Padua in the most distinguished University of the entire world, for the treatment of this subject belongs to the professorship of surgical medicine which I have held for five years, supported by stipends from the illustrious Senate of Venice, well known for its exceptional liberality to the pursuits of learning.[11] Here and at Bologna I performed dissections rather more often, and, having exploded the ridiculous custom of the schools, I taught in such a way that in anatomy we might want nothing which has been handed down to us by the ancients.

But indeed it should be noted that the sluggishness of the doctors has taken too little care that the writings of Eudemus, Herophilus, Marinus, Andrew, Lycus and the other leaders in dissection be preserved for us; not even a fragment of any page survives from the many illustrious authors, more than twenty of whom Galen mentions in his second commentary on Hippocrates' book *De natura humana*. Why, hardly half of his own anatomical books have been saved from destruction! But those who have followed Galen, in which class I consider Oribasius, Theophilus, the Arabs, and all of our men however many I have chanced to read thus far (with your permission I would have written of these), if they handed down anything worth reading, they took it straight from Galen. And, by Heaven, to the man who is diligently dissecting they seem to have done nothing less than the dissection of the human body! And so, with their teeth set, the principal followers of Galen put their trust in some kind of talking, and relying upon the inertia of others in dissecting, they shamelessly abridge Galen into elaborate compendia. They do not depart from him a hair's breadth while they are following his sense; but to the front of their books they add writings of their own, stitched together completely from the opinions of Galen—and all of theirs is from him. The whole lot of them have placed their faith in him, with the result that you can not find a doctor who has thought

[10] But Vesalius was attacked by the theologians over his remarks on the location of the soul, and Louvain apparently got pretty hot for him.

[11] Vesalius went to Venice first (1537) to study the sick under clinical conditions, then went to Padua where he stood for his doctoral exams, finished them 5 December 1537 and started his lectures on anatomy the next day. Roth, p. 77.

that even the slightest slip has ever been detected in the anatomical volumes of Galen, much less *could* be found (now).[12]

Meanwhile (especially since Galen corrects himself frequently, and in later works written when he became better informed he points out his own slips perpetrated in certain books, and teaches the contrary) it now becomes obvious to us from the reborn art of dissection, from diligent reading of the books of Galen, and from impeccable restoration in numerous places of (the text of) these books,[13] that he himself never dissected the body of a man who had recently died.[14] Although the dried cadavers of men prepared, so to speak, for the inspections of the bones were available to him, he was misled by his apes, and he undeservedly censures the ancient doctors who had busied themselves with the dissection of men. Nay, you may even find a great many things in his writings which he has not followed correctly in the apes; not to mention the fact that in the manifold and infinite difference between the organs of the human body and the body of apes, Galen noticed almost none, except in the fingers and in the bending of the knee. This difference he doubtless would have omitted too, if it had not been obvious to him without the dissection of man.

But in the present work, I have in no wise set out to reprimand the false doctrines of Galen, easily the chief of the professors of dissection; and much less would I wish to be considered disloyal and too little respectful of authority toward that author of all good things right at the beginning of my work. For I am not unaware of how much disturbance the doctors—far less than the adherents of Aristotle—raise when they observe that Galen deviated more than two hundred times from the correct description of the harmony, use, and function of the parts of man in treatment of anatomy alone, as I now exhibit it in the schools, while they examine sharply the dissected particles with the greatest zeal in defending him.[15] Although these men, led by the love of truth, gradually grow milder and put

[12] Jacobus Sylvius, one of the professors under whom Vesalius studied at Paris, stated in his *Ordo et ordimis ratio in legendis Hippocratis et Galeni libris* (Paris, 1539) that Galen is infallible, that his *De usu partium* was divine and that any advance of knowledge past Galen was impossible. Cf. Roth, p. 65 and 125; cf. Roth, p. 144.

[13] Roth, p. 144, cf. p. 112, where Roth says that Vesalius had reached this conclusion by 1540.

[14] Vesalius was a close student of Galen and had edited part of the monumental edition of the Latin Galen published by Junta at Venice, 1541. See Roth, *op. cit.*, pp. 109-10.

[15] Roth freely paraphrases this passage on pp. 116-117.

a little more trust in their rational faculties and their eyes—by no means ineffectual eyes and brains—than to the writings of Galen, they are now writing hither and thither to their friends about these truly paradoxical things which have neither been borrowed from the attempts of others or buttressed by congeries of authorities so sedulously and they have been urging their friends to learn some true anatomy so eagerly and amicably, that there is hope of its being fostered in all our Universities as it once was practiced at Alexandria.[16]

In order that this may succeed under the happier auspices of the Muses, besides the works[17] on this subject which I published elsewhere and which certain plagiarists, thinking me far absent from Germany, sent forth as their own;[18] I have now prepared afresh, and to the best of my ability, the history of the parts of the human body in seven books, arranged in the order in which in this city, at Bologna,[19] and at Pisa I have been accustomed to treat it in the assembly of learned men. I have done this with the specific idea that those who have attended the dissector [at work] will have commentaries of the demonstrated facts, and will show anatomy to others with lighter work; although the [commentaries] will not be entirely useless to those to whom direct observation is denied, since of each particle of the human body the site, form, substance, connection to other parts, use, function, and very many things of this sort which we have been accustomed to turn up during dissections in the nature of the parts, together with the technique of dissecting both dead and live men, are pursued at adequate length. The books contain pictures of all the parts inserted into the context of the narrative, so that the dissected body is placed, so to speak, before the eyes of those studying the works of nature. And so, in the first book I have narrated the nature of all the bones and cartilages, which, because the other parts are supported by them and are marked off by them, come to be known first by students of anatomy. The second book dwells upon the ligaments, by the aid of which the bones and cartilages are in turn connected; and then the muscles, the workmen

[16] Cf. Roth, p. 148.

[17] Reference is to the *Tabulae anatomicae* published at Padua in 1538. Roth, p. 89.

[18] Pirated editions were published at Augsburg, Paris, Strassburg, Marburg, Frankfurt, Cologne, etc. Roth, p. 112.

[19] Vesalius performed public dissections at Bologna twice, in 1539 and 1540. Roth, p. 86.

of motion depending upon our will. The third includes the commonest series of the veins, which carry to the muscles and the bones and other particles the familiar blood to nourish them; and then of the arteries regulating the mixture of innate heat and vital spirit. The fourth teaches not only the layers of nerves which bear the animal spirit to the muscles, but all the rest of the nerves. The fifth treats the construction of the organs serving for nutrition, which is performed by food and drink; and furthermore, because of proximity of position, it contains also the instruments fashioned by the Supreme Creator of things for the succession of the species. The sixth is devoted to the heart, the tinder of the vital faculty, and to the particles aiding it. The seventh follows up the harmony of the brain and the senses of the organs, so that the series of nerves which trace their origin from the brain, as told in the fourth book, is not repeated. Indeed in the dividing up of the order of these books, I have followed the thought of Galen, who after the history of the muscles, veins, arteries, and nerves held that the anatomy of the viscera should be treated last.

Although, not improperly—and especially in the case of the beginner in this art—some[20] will contend that the knowledge of the viscera should be pursued along with the distribution of the vessels as I have set forth in the *Epitome*,[21] which I have prepared as a pathway to these books and an index of the thing demonstrated in them, and which was adorned by the splendor of the Most Serene Prince Philip, the son of your Majesty, and living example of the paternal virtues. But at this point I remember the judgment of certain men[22] who sharply condemned the fact that drawings were placed before the students of natural phenomena, drawings not only of plants but of the parts of the human body, on the ground that however exquisite the drawings might be, these things had to be learned not from pictures but from diligent dissection and from observation of the things themselves. Their criticism would be just if I had added to the context of my discourse the truest possible pictures of the parts—would that they never be distorted by the printers!—with the idea that students would rely upon these and thus let up on the dissection of bodies;

[20] This was a slap at Jacobus Sylvius, who had vigorously attacked the use of illustrations on the ground that Galen had decried the use of plant illustrations, and anyway, no one was smart enough to add anything to Galen's perfect descriptions. Roth, p. 125.
[21] The short compendium of the *Fabrica* published simultaneously with it.
[22] Cf. Roth, p. 150. A slap at Jacobus Silvius.

and the criticism would be just if I had not thought that by the pictures I was doing all in my power to encourage the candidates in medicine toward dissection with their own hands. Surely, if the custom of the ancients who kept their boys busy at home in doing their dissections, in painting the elements, and in reading had been brought down to this day and age, I would permit us to get along without the pictures and all the commentaries too, just as those ancients did. These men not only made the first attempts to write up the performance of dissections, but they thought it honorable, in the name of virtue, to communicate their art to their children and even to strangers whom they elevated.[23] When in the course of time the boys no longer were given practice in dissections, the result necessarily was that they learned their anatomy less thoroughly, when the practice which they were accustomed to begin in boyhood was abolished. The final result was that, when the family art of the Asclepiades had died out and had gone to the dogs for many centuries, men needed books which should preserve the entire observation of the art itself. Truly, there is no one who does not find out in geometry and in the other mathematical disciplines how much pictures help in the understanding of them, and place the matter before the eyes more clearly, even though the text itself is very explicit. But however that may be, in the whole of this work I have striven with the purpose that, in an exceedingly recondite and no less arduous business, I should help as many as possible and that I should treat as truly and completely as possible the history of the fabric of the human body, a fabric not built of ten or twelve parts,[24] as it appears to the casual observer, but of several thousand diverse parts; and finally that I might bring to the candidates in medicine a grist not to be scorned for the understanding of the books of Galen in this field which, among his other monuments, require especially the help of a preceptor.

But in the meanwhile, it is perfectly clear to me that my attempt will have all too little authority because I have not yet passed the twenty-eighth year of my life; it is equally clear that, because of the numerous indications of the false dogmas of Galen, it will be exceedingly unsafe from the attacks of the conservatives, who, as with

[23] Praef, p. 4, *alienis etiam viris quos virtutis nomine suspiciebant.*
[24] Cf. Roth, p. 131, esp. note 5.

us in the Italian schools, have sedulously avoided anatomy, and who, being old men, will be consumed with envy because of the correct discoveries of the young, and will be ashamed of having been blind thus far, along with the other followers of Galen. And they will be ashamed of not having noticed the things which we have just set forth even though they arrogate a great name to themselves in the art, unless my work should be published with the customary commendation of the patronage of some great power.

Now, because my work can not be protected more safely nor more splendidly adorned by any name greater than the imperishable name of the divine Charles, the greatest and most invincible emperor, I beg Your Majesty that this juvenile work of mine, defective in itself for many reasons and causes, be allowed to circulate in the hands of men under the splendid guidance and patronage of Your Majesty; until, through practical experience and the scholarly judgment which grows in time, I shall render this work worthy of the highest and best of princes; or until I shall present another book of different content derived from my art which will not be rejected.

Although I may be guessing, I think that nothing from the whole discipline of Apollo and, therefore, from the whole of natural philosophy could be printed which would be more acceptable to Your Majesty (or more pleasing) than a history in which the body and the mind and, in addition, that certain divine element derived from the symphony of the two are made known, in short, one in which we know ourselves, which is the true goal of man. And so, as I tie this up with numerous proofs, thus in the first place I contend that in the throng of books which were consecrated to Maximilian, the great Emperor of the Romans of happy memory, there never was a book more acceptable than a little volume on the present subject. Nor shall I ever forget with what pleasure you inspected my volume of plates, nor with what curiosity you lingered over the items which my father, Andreas, the most faithful chief pharmacist of your Majesty, once offered for your inspection.[25]

I shall pass over the incredible love of this man toward all the sciences, and especially that branch of mathematics which treats the knowledge of the earth and stars, and I shall not mention his prac-

[25] Andrew (Andreas) Vesalius, senior, was the herbapotheker of Charles V, traveled with him. Father Andreas presented the *Tabulae* to Charles V about the time of The Armistice of Nice, i.e., 1538. See Roth, p. 60.

tical knowledge and experience in this science, admirable as it is in so great a hero. Your scientific curiosity is so great that, just as you must be considered unique in the science of the world, so you must be likewise delighted at times by the contemplation of the fabric of the most perfect of all creatures; so also you must take pleasure in consideration of the dwelling place and implement of the immortal soul, since, due to its remarkable correspondence to the world in many of the names, it was called a "little world" by the ancients. Furthermore I have decided that I should not extol here the knowledge of the structure of the body as being the most worthy to man, although it is most highly commendable *per se*, and although the greatest men of Rome, distinguished alike in practical things as well as in the science of philosophy, were pleased to expend effort upon it. So too, although I am properly mindful that Alexander the Great did not wish to be painted except by Apelles, nor to be cast in bronze except by Lysippus, nor to be sculptured except by Pyrgotele, I have decided that none of your praises should be reviewed here. I am afraid that I should pour darkness instead of light upon them by my meagre and unpracticed oratory. It would be utterly reprehensible of me to follow the ritual which is all too general in prefaces and usually is unproductive of enjoyment and carried beyond just deserts, namely, the ritual in which they usually ascribe to everybody, as if from a pattern or a sort of formula, the following virtues:—admirable learning, singular prudence, marvelous clemency, sharp judgment, inexhaustible liberality, marvelous love toward studies and men of letters, mature speed in the execution of business and the whole chorus of virtues. Although I am not saying it here, there is simply no one who is not perfectly aware that Your Majesty excels all the remainder of mankind in all of these virtues as well as in dignity, happiness, and in triumphs for accomplished deeds. Whence it comes about that you are worshipped in the place of a supreme power, though you are yet alive, and I pray that the gods may not be hostile to your desires and to the whole world, but rather that they guard and keep you safe for mortals a long time, and happy forever. At Padua, August 1, in the year after the birth of Christ, 1542.[26]

[26] Vesalius took the mss. and the wood blocks to Venice and packed them for shipment. They were sent to Oporinus, c/o the Milanese merchants, the Danoni. Vesalius followed later and saw his book through the press. It was completed in June, 1543. Roth, 130.

ON DISSECTION OF THE LIVING*

[From *The Fabric of the Human Body*, Book VII, Chapter XIX]

WHAT MAY BE LEARNED BY DISSECTION OF THE DEAD AND WHAT
OF THE LIVING

Just as the dissection of the dead teaches well the number, position and shape of each part, and most accurately the nature and composition of its material substance, thus also the dissection of a living animal clearly demonstrates at once the function itself, at another time it shows very clearly the reasons for the existence of the parts. Therefore, even though students deservedly first come to be skilled in the study of dead animals, afterward when about to investigate the action and use of the parts of the body they must become acquainted with the living animal.

On the other hand since very many small parts of the body are endowed with different uses and functions, it is fitting that no one doubt that dissections of the living present also many contradictions.

THE USE OF LIGAMENTS

In the dead one sees the uses of the ligaments binding the bones together, and also that of the ligaments drawn across tendons. Thus, provided we divide the ligament placed transversely in the internal side of the foreleg and pull on the muscle flexing the second or third joints of the toes towards its origin, we discover that this tendon is so placed chiefly in order that it may hold the tendons together that they may not lift up from their bed. However it will be permitted to demonstrate this likewise in a living dog if thou shalt immediately free the skin from the elbow and paw, and shalt divide with a small knife the transverse ligament of the foreleg and the other parts which are held against the leg on the external side of the ulna and radius. Soon, forsooth, when the dog of its own volition flexes and extends its toes thou shalt see the tendons rise up from their sheaths.

THE USE AND FUNCTION OF THE MUSCLES

In a proper dissection thou shalt see the function of the muscles; notice during their own action they contract and become thick

* De Fabrica Corporis Humanis, 1543. Translated by Dr. Samuel W. Lambert. Reprinted from Proceedings of the Charaka Club, by permission of Columbia University Press and Dr. Samuel W. Lambert.

where they are most fleshy, and again they lengthen and become thin according as they in combination draw up a limb, either letting themselves go back and having been drawn out permit the limb to be pulled in an opposite direction by another muscle, or at other times indeed they do not put in action their own combination.

This is to be observed accurately at the elbow; indeed the skin of the same dog must be removed higher up in order that the whole leg and axilla shall be bared, and the nerves running forward into it (the leg) through the axilla may become visible. Thou shalt perceive that a chain of these nerves reaches to certain muscles and thou shalt intercept some one among them with a noose.

THE USE OF THE NERVE IN THE MUSCLES

In truth since the number and distribution of the nerves of the dog do not correspond exactly with those same in man I would advise when thou art about to perform this dissection in a living dog that thou hast at hand a dead one also in which thou shalt have separated the series of nerves running out through the foreleg and elbow, and thus thou wilt find some one of these nerves which is distributed to particular muscles. They will be arranged almost in this manner: In man one of them is the third, and is carried into the forearm along the anterior side of the elbow joint; another, in fact, the fifth, runs to the elbow next to the posterior portion of the internal tuberosity of the humerus. For in this manner the nerves are observed also in the dog. And these nerves having been tied somewhere before they reach the elbow joint, the motion of the muscles flexing the digits and arm will be abolished, and if thou wilt intercept with a band the nerve which in man is reckoned by me the fourth and is extended along the humerus to its external tuberosity, then the motion of the muscles extending the foreleg and digits will be abolished. . . .

When thou dividest the belly of a muscle straight through thou wilt observe that the muscle draws together and contracts in one part towards its insertion, in the other portion towards its origin.

But if on the other hand thou shalt cut the tendon of any muscle thou shalt see that the muscle contracts towards its origin. Likewise if thou shalt divide the origin it will contract towards its insertion. If in truth thou shalt have cut the insertion and the head at the same time the muscle will be bunched at its belly and the portion where it is most fleshy, and thus the functions of the muscles will be obvious to thee by the doing of these things.

EXAMINATION OF THE USES OF THE DORSAL MEDULLA

Even so if anyone may have considered examining the function of the dorsal medulla, it will be seen when the medulla has been injured how the parts below the injury lose sensation and motion. It will be permitted anyone to fasten a dog or to bind it to a block of wood in a way that one stretches out the back and neck.Therefore some of the spines of the vertebrae can be cut in front with a large knife and then the dorsal medulla can be laid bare in its bed, when anyone will get a view of the medulla about to be cut—for nothing is easier than thus to see that movement and sensation are abolished in the parts subjected to the section.

EXAMINATION OF THE USES OF THE VEINS AND ARTERIES

Also when inquiring into the use of the veins the work is scarcely one for the dissection of the living, since we shall become sufficiently acquainted in the case of the dead with the fact that these veins carry the blood through the whole body and that any part is not nourished in which a prominent vein has been severed in wounds.

Likewise concerning the arteries we scarcely require a dissection of the living although it will be allowable for anyone to lay bare the artery running into the groin and to obstruct it with a band, and to observe that the part of the artery cut off by the band pulsates no longer.

And thus it is observed by the easy experiment of opening an artery at any time in living animals that blood is contained in the arteries naturally.

In order that on the other hand we may be more certain that the force of pulsation does not belong to the artery or that the material contained in the arteries is not the producer of the pulsation, for in truth this force depends for its strength upon the heart. Besides, because we see that an artery bound by a cord no longer beats under the cord, it will be permitted to undertake an extensive dissection of the artery of the groin or of the thigh, and to take a small tube made of a reed of such a thickness as is the capacity of the artery and to insert it by cutting in such a way that the upper part of the tube reaches higher into the cavity of the artery than the upper part of the dissection, and in the same manner also that the lower portion of the tube is introduced downward farther than the lower part of the dissection, and thus the ligature of the artery which constricts its caliber above the canulla is passed by a circuit.

To be sure when this is done the blood and likewise the vital spirit run through the artery even as far as the foot; in fact the whole portion of the artery replaced by the cannula beats no longer. Moreover when the ligature has been cut, that part of the artery which is beyond the canulla shows no less pulsation than the portion above.

We shall see next how much force is actually carried to the brain from the heart by the arteries. Now in this demonstration thou shalt wonder greatly at a vivisection of Galen in which he advises that all things be cut off which are common to the brain and heart, always excepting the arteries which seek the head through the transverse processes of the cervical vertebrae and carry also besides a substantial portion of the vital spirit into the primary sinuses of the dura mater and also in like manner into the brain. So much so that it is not surprising that the brain performs its functions under these conditions for a long time, which Galen observed could easily be done, for the animal breathes for a long time during this dissection, and sometimes moves about. If indeed it runs, and therefore requires much breath, it falls not long afterwards although the brain will still afterwards receive the essence of the animal spirit from those arteries which I have closely observed seek the skull through the transverse processes of the cervical vertebrae. . . .

We see that the peritoneum is a wrapper for all the organs enclosed in it; that the omentum and likewise the mesentery serve in the best manner for the conduction and distribution of the blood vessels; that the stomach prepares the food and drink, and passes these through the stomach onward.

And however nothing may prevent our taking living dogs which have consumed food at less or greater intervals previously and examine them alive, and thus investigating the functions of the intestines. But on the contrary we are able to behold the functioning of the liver as also of the spleen or of the kidneys or of the bladder during the dissection of the living, scarcely better than in that of the dead; unless someone shall wish to excise the spleen in the living dog which I have once done, and have preserved the dog (alive) for some days.

And thus also have I once excised a kidney; in truth the management of this wound is more troublesome than the pleasant knowledge which is acquired in the doing of it; unless one undertakes these dissections not so much for the sake of knowledge of the organs as in order to train his hands, and to learn to sew up wounds of the abdomen in a fitting manner; which should also be diligently practiced upon the intestines, in order to become accustomed to sew them up when wounded, and to replace them in the abdomen when they shall have slipped out.

In truth these operations, just as the dislocations and fractures of bones, which we sometimes do on brute beasts, serve more for training the hands and for determining correct treatment rather than for investigating the functions of organs. . . .

EXAMINATION OF THE FETUS

Quite pleasing is it in the management of the fetus to see how when the fetus touches the surrounding air it tries to breathe. And this dissection is performed opportunely in a dog or pig when the sow will soon be ready to drop her young.

To be sure if thou dividest the abdomen of such an animal down to the cavity of the peritoneum, and then thou also openest the uterus at the site of a single fetus, and when the secundines have been separated from the uterus thou shalt place the fetus on a table, thou shalt see through its coverings and its transparent membrane how the fetus attempts in vain to breathe, and dies just as if suffocated. If in fact thou shalt perforate the covering of the fetus, and shalt free its head from the coverings, thou shalt soon see that the fetus as it were comes to life again and breathes finely.

And when thou shalt have investigated this in one fetus thou shalt turn to another which thou shalt not free from the uterus but thou shalt invert the uterus opened by the same management as that of the fetuses just described, and shalt turn the edges of the dissection already made backward until the lower part or site of the coverings of another and nearest fetus shall appear, and thou shalt free this site from the uterus even up to that place where the uterus is fused in the exterior covering of the fetus, and where its abundant flesh will possess a substance similar to the spleen, which interweaves vessels stretching out from the uterus into the external coverings of the fetus.

For that network of vessels must be preserved intact during this manipulation, and the remaining external covering must be removed from the uterus in order that thou shalt see through the transparent covering of the fetus that the arteries distributed by the covering and running to the umbilicus pulsate in the rhythm of the arteries running to the uterus, and that the naked fetus attempts and struggles for respiration and thereupon when the coverings are punctured and broken thou shalt see that then the fetus breathes and that pulsations of the arteries of the fetal membranes and of the umbilicus stop. Up to this moment the arteries of the uterus are beating in unison with the rest of the arteries outside of itself.

EXAMINATION OF THE FUNCTION OF THE HEART, AND LUNGS AND GREAT VESSELS

And we consider many things concerning the functions of the heart and lungs; of course the motion of the lungs, and whether their rough artery takes to itself a portion of those things which are inhaled; next, the dilation and contraction of the heart, and whether the pulsation of the heart and arteries act in the same rhythm; also in what manner the venous artery is dilated and contracted, and whereby an animal continues to live after its heart has been excised. For making these investigations we particularly require an animal endowed with a wide breastbone and possessing membranes dividing the thorax separated in such a way that when the sternum has been divided lengthwise we may be able to carry the dissection between these membranes even to the heart itself without causing a perforation of the cavities of the thorax in which the lungs are contained.

Indeed, when no such animal except man or a tailless ape is to be had, these dissections must be carried out in dogs and pigs and those animals of which an abundance is furnished us, as thou hast seen all this mentioned above.

Therefore and in like manner the lung follows the movement of the thorax. It is evident from this way, when a cut is made in an intercostal space penetrating the thoracic wall, that a portion of the lung on the injured side collapses, and distends with the thorax no longer. While until then the other lung following up to this time the movement of the chest, also soon collapses if thou shalt make a cut penetrating even into the cavity of the thorax on the other side. And then the animal, even if it moves the chest for some time, will die nevertheless just as if it were suffocated.

In fact it ought to be observed in this demonstration that thou shalt make the cut as close as possible to the upper edge of any rib lest thou mayest direct the incision by chance along the lower edge and perforate the vessels stretched along there. For the blood will flow forth thence at once, which rendered frothy in consequence of the air inhaled and exhaled through the wound, will appear to thee falsely as the lung, for this froth will appear to be the lung to those who dissect carelessly, and they will think that the lung is distended on such occasions by its own inherent force.

In order that thou mayest see the natural relationship of the lung to the thorax, thou shalt cut the cartilages of two or three median ribs from the opposite side, and when the incisions have been carried through the interspaces of these ribs thou shalt bend the separate ribs outward and break them where thou shalt decide to be the proper place through which thou seekest to observe the lung of the uninjured side, for since the membranes dividing the thorax in dogs are sufficiently translucent, it is very easy to inspect through them the portion of the lung which is following up to this time the movements of the thorax, and when these membranes have been perforated slightly, to consider in what manner that part of the lung collapses.

Before thou perforatest these membranes it will have been useful to grasp with the hands the branches of the venous artery in that part of the lung which now collapses, and in some other portion to free the substance of the lung from those vessels in order that thou mayest learn whether these vessels and the heart are moved in like manner.

For the movement of the heart is evident to thee even here especially if thou shalt divide the covering of the heart and shalt uncover the heart from it on that side where thou art carrying on this operation.

It will be fitting to try this on the left side as of course the right part of the lung may be lifted up to the point and at one time thou mayest conveniently grasp the main trunk of the venous artery easily in the hand. Also thou mayest in an operation of this kind carefully handle the base of the heart, and may cut off swiftly and at one time and with one ligature the vessels from their origins, and then thou mayest cut to excise the heart under the ligature, and when the bands have been loosened with which the animal has been tied, thou mayest allow it to run about.

Indeed we have at times seen dogs, but especially cats, treated in this manner, run around.

Besides thou wilt see the movement of the heart and also of the arteries more accurately if soon thou shalt bind the dog to a plank, and shalt carry the incision on both sides with a very sharp knife from the clavicle through the cartilages of the ribs where they are continuous with the bones, and thou shalt make a third incision even into the cavity of the peritoneum transversely from the end of one of the above incisions to the end of the other; and when the sternum has been lifted with its cartilages pressed to it, and when the transverse incision has been freed from them, thou shalt turn the sternum upward towards the head of the animal and soon when the covering of the heart has been opened thou shalt grasp the heart with one hand and with the other thou shalt take hold of the great artery extending into the back.

EXAMINATION OF THE RECURRENT NERVES AND THE LOSS OF VOICE FROM THE CUTTING OF THEM

And soon I begin in this wise an extended dissection in the neck with a rather sharp knife which divides the skin and muscles lying under it right to the trachea, avoiding this lest the incision may deviate to the side and wound especially the principal vein. Then by grasping the windpipe with the hands, and freeing it accurately from the overlying muscles by the use of the fingers as far as the arteries lying at its side; I seek out the nerves bordering upon the sixth pair of cerebral nerves; then I note the recurrent nerves lying on the sides of the rough artery (trachea) which I sometimes intercept with ligatures, at other times I cut. And first I do the same on the other side, in order that it may be clearly seen when one nerve has been tied or cut how half the voice disappears and is totally lost when both nerves are cut. And if I loosen the ligatures that the voice will return again. For this is carried out quickly and without an unusual loss of blood, and it is clearly proved how the animal struggles for deep breaths without its voice when the recurrent nerves have been divided with a sharp knife.

EXAMINATION OF THE FUNCTIONS OF THE DIAPHRAGM

In this wise I advise those standing near the dissection that they watch the movement of the diaphragm when it is held in the hand, and those at a distance that they observe the expansion and contraction of the stomach and liver similar to the movements in the cavity of the thorax.

Meanwhile I carry on an extensive dissection on the other side of the thorax to reach the bones of the ribs even nearly to that point where the ribs change into cartilages. And then I make transverse incisions along the bones of the ribs in order that I may free the bones in some part from the overlying muscles; and if it is seen that further painstaking work may follow I lift the intercostal muscles of two intercostal spaces from the tunic covering the ribs that I may thus tear away the intervening rib from the covering surrounding the rib with the help of my hands alone. And when that rib is broken from its cartilage and bent backward on the side, the great size of this tunic surrounding the ribs will be apparent. This transparent covering will demonstrate however the movements of the lung. After this covering has been punctured it will be seen how the lung collapses on this side although the thorax moves meanwhile just as before.

EXAMINATION OF THE MOVEMENTS OF THE LUNG

In fact in order that this may be made more evident I free many bones of the ribs from their cartilages, opening this side of the thorax as much as I am able in order that there may be presented through the membranes dividing the chest, another part of the lungs shows itself which being in the hitherto uninjured part of the thorax, moves well with it. And then when these membranes have been punctured this part also is soon seen to collapse from the perforation, and to fall together.

THE BONE OF LUZ
[From *The Fabric of the Human Body*, Book 1]

Vesalius breaks with tradition concerning the mythical bone of Luz—a sample of his vitriolic moods:

The dogma which asserts that man will be regenerated from this bone, of which we have just narrated the immense fiction, may be left for elucidation to those philosophers who reserve to themselves alone the right to free discussion and pronouncement on the resurrection and the immortality of the soul. And even on their account we should attach no importance whatever to the miraculous and occult powers ascribed to the internal ossicle of the right great toe, however much one may be concerned about it. At our public dissections and even as an amateur, we have often obtained a better supply of these bones than those truculent male strumpets of the Venetian horde, who to obtain the bone for purposes of comparison, as also the heart of an unpolluted male infant, lately killed a child, cut the heart from its living body, and were punished, as they richly deserved for the foulest of crimes. Moreover, this ossicle, called Albadaran by the Arabs and the truly occult and obscure philosophers alluded to. is less known to actual students of anatomy than to certain superstitious men who are capable of likening the fourth carpal bone to a chickpea.

XVI

WILLIAM HARVEY

WILLIAM HARVEY (1578-1657), THE GREATEST NAME IN ENGLISH MEDICINE and one of the greatest names in all experimental science, was born in Folkstone. He attended Granville and Caius Colleges, Cambridge, and afterwards studied at Padua where his professor was Fabricius, whose work on the valves in the veins and the development of the chick profoundly influenced Harvey's two great works. He returned to England, taught at the College of Physicians in London, saw military service in the Civil War on the Royalist side, died in London at an advanced age and was buried at Hempstead, Essex. *De motu cordis et sanguinis* (On the Motion of the Heart and Blood) was published in 1628 and went through many editions. *De generatione animalium* (On the Generation of Animals), published in 1651, gives him the right to the title of the father of embryology.

It has been claimed for others that they announced the circulation of the blood. Undoubtedly this is partially true. Fabricius described the valves in the veins, but he missed their significance. Servetus (Christianismi Restitutio, 1553), described the pulmonary circulation. Cesalpinus and Realdus Columbus both described the circulation, but from speculation. Harvey proved it by experiment.

Exercitatio anatomica de motu cordis et sanguinis in animalibus should be read in its entirety by every medical student.

REFERENCE

POWER, D'ARCY. William Harvey. London, T. Fisher Unwin, 1897.

*ANATOMICAL EXERCISES ON THE MOTION OF THE HEART AND BLOOD IN ANIMALS**

DEDICATION. To Charles I. It begins: "The heart of animals is the foundation of their life, the sovereign of everything within them, the sun of their microcosm, that upon which all growth depends, from which all power proceeds. The King in like manner is the foundation of his Kingdom."

* Exercitatio Anatomica de Motu Cordis et Sanguinis in Animalibus. Translated by Robert Willis. Sydenham Society, 1847.

INTRODUCTORY LETTER to Doctor Argent, President of the Royal College of Physicians, states that Harvey has repeatedly, for nine years, presented to the College his new views on the motion of the heart, and supported these views with arguments. Now in response to repeated requests he has committed them to writing.

INTRODUCTION. This gives an historical review of the theories of the circulation of Galen, Fabricius, Realdus Columbus, and Laurentius. His criticisms of all their statements are clear and refreshing.

CHAPTER I. The author's reasons for writing. Explains that he used dissection and experiment to study the motion of the heart and blood.

CHAPTER II. Motions of the heart as observed in animal experiments. The heart contracts in systole and forces blood out.

CHAPTER III. Movements of the arteries as seen in animal experiments. At the moment the heart contracts, the arteries dilate and receive the blood it forces out. Other experiments to prove that the blood goes out from the heart through the arteries to the body.

CHAPTER IV. The motion of the heart and its auricles as noted .in animal experiments. The auricles contract together first, followed by the contraction of the ventricles. Therefore the blood moves from the auricles to the ventricles.

CHAPTER V. Actions and functions of the heart. The veins empty blood into the auricles, and the semilunar valves at the opening of the pulmonary artery and the aorta prevent the blood from going back to the ventricles. Blood does not go from the right ventricle to the left through the septum of the heart, because there are no openings through the septum (as asserted by Galen and Vesalius).

CHAPTER VI. The way by which the blood passes from the vena cava to the arteries, or from the right ventricle of the heart to the left. Difference between animals with one chamber of the heart and man. Difference between the circulation in the embryo and in post-uterine life. "In the more perfect warm blooded animals as man, the blood passes from the right ventricle of the heart through the pulmonary artery to the lungs, from there through the pulmonary vein into the left auricle and then into the left ventricle."

CHAPTER VII. Passage of blood through the substance of the lungs from the right ventricle of the heart to the pulmonary vein and left ventricle. Blood filters through the lungs as sweat filters through the skin. In this chapter Harvey is not at his best, as he knew nothing of the capillaries.

CHAPTER VIII. Amount of blood passing through the heart from the veins to the arteries and the circular motion of the blood. First how much blood might be lost from cutting the arteries in animal experiments. The pulmonary artery and the aorta are the same size, so the same amount of blood must go out from each. This would not be true if the blood sent to the lungs was for the purpose of nourishing them, because the proportionate size of the lungs and the rest of the body presents too great a discrepancy. "The blood moves as in a circle. The arteries are the vessels carrying blood from the heart to the body and the veins returning blood from the body to the heart."

CHAPTER IX. The circulation of the blood is proved by a prime consideration. The left ventricle when filled contains two or three ounces. When it contracts it contains a fourth to an eighth less (this by reasonable conjecture). A single heart beat in man would force out half an ounce. The heart makes more than a thousand beats in half an hour, forcing out more than 500 ounces, a greater amount of blood than is present in the whole body. Therefore the blood must return to the heart over and over: there is no other explanation. This is one of the greatest demonstrations in the history of experimental physiology. For the first time quantitative measurements are introduced to prove a point.

CHAPTER X. The first proposition concerning the amount of blood passing from veins to arteries during the circulation of the blood is freed from objections and confirmed by experiment.

CHAPTER XI. The second proposition is proved. That the blood enters a limb through the arteries and returns through the veins. If a ligature is placed tightly on a limb the artery below stops pulsating and the extremity becomes bloodless. If the ligature is loosened somewhat but kept middling tight the extremity is suffused with blood and the veins are engaged.

CHAPTER XII. That there is a circulation of the blood follows from a proof of the second proposition.

CHAPTER XIII. The third proposition is proved, and the circulation of the blood is demonstrated from it. How does the blood from the extremities get back to the heart through the veins? The function of the valves in the veins. (The valves were discovered by Harvey's teacher Fabricius, and Harvey gives him full credit.) The discoverer of these valves did not appreciate their function. It is not to prevent blood from falling by its weight into areas lower down for there are some in the jugular vein which are directed downwards, and prevent blood from being carried upwards. It is impossible in dissecting veins to pass a probe from the main trunks very far into the smaller branches on account of the valvular obstructions. On the other hand, they yield to a probe introduced from without inwards. By tying off the arm of a subject as for blood-letting you observe at intervals in the veins knots or swellings. These are caused by the valves. If you clear the blood away from a nodule by pressing a thumb or finger below it you will see nothing can flow back being prevented by the valve. "With the arm bound and the veins swollen press on a vein a little below a swelling or valve and then squeeze the blood upwards beyond the valve, with another finger you will see that this part of the vein stays empty and that no back flow can occur through the valve. . . . So, as the veins are wide open passages for returning blood to the heart, they are adequately prevented from returning it to the heart. Above all, note this. With the arm of your subject bound, the veins distended, and the nodes or valves prominent, apply your thumb to a vein a little below a valve so as to stop the blood coming up from the hand, and then with your finger press the blood from that part of the vein up past the valve. Remove your thumb and the vein at once fills up from below."

CHAPTER XIV. Conclusion of the demonstration of the circulation of the blood.

CHAPTER XV. The circulation of the blood is confirmed by plausible methods. The blood circulates to keep the body warm. The distribution of food after digestion must be made by the blood.

CHAPTER XVI. The circulation of the blood is supported by its implications. How does it happen that in contagious conditions like poisoned wounds, bites of serpents or mad dogs the whole body may become diseased while the place of contact is often unharmed or healed? Without doubt the contagion first being deposited in a certain spot is carried by the returning blood to the heart, from which later it is spread to the whole body. This may also explain why certain medicinal agents when applied to the skin have almost as much effect as when taken by mouth.

CHAPTER XVII. The motion and circulation of the blood is established by what is displayed in the heart and elsewhere by anatomical investigation. Comparative anatomy and physiology of the heart. The function of the papillary muscles and chordae tendinal of the tricuspid and mitral valves.

THE AUTHOR'S MOTIVES FOR WRITING

[From Chapter 1]

When I first gave my mind to vivisections, as a means of discovering the motions and uses of the heart, and sought to discover these from actual inspection, and not from the writings of others, I found the task so truly arduous, so full of difficulties, that I was almost tempted to think, with Fracastorius, that the motion of the heart was only to be comprehended by God. For I could neither rightly perceive at first when the systole and when the diastole took place, nor when and where dilatation and contraction occurred, by reason of the rapidity of the motion, which in many animals is accomplished in the twinkling of an eye, coming and going like a flash of lightning; so that the systole presented itself to me now from this point, now from that; the diastole the same; and then everything was reversed, the motions occurring, as it seemed, variously and confusedly together. My mind was therefore greatly unsettled, nor did I know what I should myself conclude, nor what believe from others; I was not surprised that Andreas Laurentius should have said that the motion of the heart was as perplexing as the flux and reflux of Euripus had appeared to Aristotle.

At length, and by using greater and daily diligence, having frequent recourse to vivisections, employing a variety of animals for the purpose, and collating numerous observations, I thought that I had attained to the truth, that I should extricate myself and escape from this labyrinth, and that I had discovered what I so much desired, both the motion and the use of the heart and arteries; since which time I have not hesitated to expose my views upon these subjects, not only in private to my friends, but also in public, in my anatomical lectures, after the manner of the Academy of old. . . .

OF THE MOTIONS OF THE HEART, AS SEEN IN THE DISSECTION OF LIVING ANIMALS

[From Chapter II]

In the first place, then, when the chest of a living animal is laid open and the capsule that immediately surrounds the heart is slit up or removed, the organ is seen now to move, now to be at rest;—there is a time when it moves, and a time when it is motionless.

These things are more obvious in the colder animals, such as toads, frogs, serpents, small fishes, crabs, shrimps, snails, and shellfish. They also become more distinct in warm-blooded animals, such as the dog and hog, if they be attentively noted when the heart begins to flag, to move more slowly, and, as it were, to die: the movements then become slower and rarer, the pauses longer, by which it is made much more easy to perceive and unravel what the motions really are, and how they are performed. In the pause, as in death, the heart is soft, flaccid, exhausted, lying, as it were, at rest.

In the motion, and interval in which this is accomplished, three principal circumstances are to be noted:

1. That the heart is erected, and rises upwards to a point, so that at this time it strikes against the breast and the pulse is felt externally.

2. That it is everywhere contracted, but more especially towards the sides, so that it looks narrower, relatively longer, more drawn together. The heart of an eel taken out of the body of the animal and placed upon the table or the hand, shows these particulars; but

the same things are manifest in the heart of small fishes and of those colder animals where the organ is more conical or elongated.

3. The heart being grasped in the hand, is felt to become harder during its action. Now this hardness proceeds from tension, precisely as when the forearm is grasped, its tendons are perceived to become tense and resilient when the fingers are moved.

4. It may further be observed in fishes, and the colder blooded animals, such as frogs, serpents, etc., that the heart when it moves, becomes of a paler colour, when quiescent of a deeper blood-red colour.

From these particulars it appeared evident to me that the motion of the heart consists in a certain universal tension—both contraction in the line of its fibres, and constriction in every sense. It becomes erect, hard, and of diminished size during its action; the motion is plainly of the same nature as that of the muscles when they contract in the line of their sinews and fibres; for the muscles, when in action, acquire vigour and tenseness, and from soft become hard, prominent, and thickened: in the same manner the heart.

We are therefore authorized to conclude that the heart, at the moment of its action, is at once constricted on all sides, rendered thicker in its parietes and smaller in its ventricles, and so made apt to project or expel its charge of blood. This, indeed, is made sufficiently manifest by the fourth observation preceding, in which we have seen that the heart, by squeezing out the blood it contains becomes paler, and then when it sinks into repose and the ventricle is filled anew with blood, that the deeper crimson colour returns. But no one need remain in doubt of the fact, for if the ventricle be pierced the blood will be seen to be forcibly projected outwards upon each motion or pulsation when the heart is tense.

These things, therefore, happen together or at the same instant: the tension of the heart, the pulse of its apex, which is felt externally by its striking against the chest, the thickening of its parietes, and the forcible expulsion of the blood it contains by the constriction of its ventricles.

Hence the very opposite of the opinions commonly received, appears to be true; inasmuch as it is generally believed that when the heart strikes the breast and the pulse is felt without, the heart is dilated in its ventricles and is filled with blood; but the contrary of this is the fact, and the heart, when it contracts (and the shock is given), is emptied. Whence the motion which is generally regarded as the diastole of the heart, is in truth its systole. And in like manner the intrinsic motion of the heart is not the diastole but the systole; neither is it in the diastole that the heart grows firm and tense, but in the systole, for then only, when tense, is it moved and made vigorous. . . .

Of the Motions of Arteries, as Seen in the Dissection of Living Animals

[From Chapter III]

In connection with the motions of the heart these things are further to be observed having reference to the motions and pulses of the arteries:

1. At the moment the heart contracts, and when the breast is struck, when in short the organ is in its state of systole, the arteries are dilated, yield a pulse, and are in the state of diastole. In like manner, when the right ventricle contracts and propels its charge of blood, the arterial vein (the pulmonary artery) is distended at the same time with the other arteries of the body.

2. When the left ventricle ceases to act, to contract, to pulsate, the pulse in the arteries also ceases; further, when this ventricle contracts languidly, the pulse in the arteries is scarcely perceptible. In like manner, the pulse in the right ventricle failing, the pulse in the vena arteriosa (pulmonary artery) ceases also.

3. Further, when an artery is divided or punctured, the blood is seen to be forcibly propelled from the wound at the moment the left ventricle contracts; and, again, when the pulmonary artery is wounded, the blood will be seen spouting forth with violence at the instant when the right ventricle contracts.

So also in fishes, if the vessel which leads from the heart to the gills be divided, at the moment when the heart becomes tense and contracted, at the same moment does the blood flow with force from the divided vessel. . . .

I happened upon one occasion to have a particular case under my care, which plainly satisfied me of this truth: A certain person was affected with a large pulsating tumour on the right side of the neck, called an aneurism, just at that part where the artery descends into the axilla, produced by an erosion of the artery itself, and daily increasing in size; this tumour was visibly distended as it received the charge of blood brought to it by the artery, with each stroke of the heart: the connexion of parts was obvious when the body of the patient came to be opened after his death. The pulse in the corresponding arm was small, in consequence of the greater portion of the blood being divided into the tumour and so intercepted.

Whence it appears that wherever the motion of the blood through the arteries is impeded, whether it be by compression or infarction, or interception, there do the remote divisions of the arteries beat less forcibly, seeing that the pulse of the arteries is nothing more than the impulse or shock of the blood in these vessels.

Of the Motion of the Heart and Its Auricles, as Seen in the Bodies of Living Animals

[From Chapter iv]

Besides the motions already spoken of, we have still to consider those that appertain to the auricles.

Casper Bauhin and John Riolan,* most learned men and skilful anatomists, inform us from their observations, that if we carefully watch the movements of the heart in the vivisection of an animal, we shall perceive four motions distinct in time and in place, two of which are proper to the auricles, two to the ventricles. With all deference to such authority I say, that there are four motions distinct in point of place, but not of time; for the two auricles move together, and so also do the two ventricles, in such wise that though the places be four, the times are only two. And this occurs in the following manner:

There are, as it were, two motions going on together: one of the auricles, another of the ventricles; these by no means taking place simultaneously, but the motion of the auricles preceding, that of the heart itself following; the motion appearing to begin from the auricles and to extend to the ventricles. When all things are becoming languid, and the heart is dying, as also in fishes and the colder blooded animals, there is a short pause between these two motions, so that the heart aroused, as it were, appears to respond to the motion, now more quickly, now more tardily; and at length, and when

* Bauhin, lib. ii, cap. 21. Riolan, lib. viii, cap. 1.

near to death, it ceases to respond by its proper motion, but seems, as it were, to nod the head, and is so obscurely moved that it appears rather to give signs of motion to the pulsating auricle, than actually to move. The heart, therefore, ceases to pulsate sooner than the auricles, so that the auricles have been said to outlive it, the left ventricle ceasing to pulsate first of all; then its auricle, next the right ventricle; and, finally, all the other parts being at rest and dead, as Galen long since observed, the right auricle still continues to beat; life, therefore, appears to linger longest in the right auricle. Whilst the heart is gradually dying, it is sometimes seen to reply, after two or three contractions of the auricles, roused as it were to action, and making a single pulsation, slowly, unwillingly, and with an effort. . . .

But I think it right to describe what I have observed of an opposite character: the heart of an eel, of several fishes, and even of some (of the higher) animals taken out of the body, beats without auricles; nay, if it be cut in pieces the several parts may still be seen contracting and relaxing; so that in these creatures the body of the heart may be seen pulsating, palpitating, after the cessation of all motion in the auricle. But is not this perchance peculiar to animals more tenacious of life, whose radical moisture is more glutinous, or fat and sluggish, and less readily soluble? The same faculty indeed appears in the flesh of eels, generally, which even when skinned and embowelled, and cut into pieces, are still seen to move.

Experimenting with a pigeon upon one occasion, after the heart had wholly ceased to pulsate, and the auricles too had become motionless, I kept my finger wetted with saliva and warm for a short time upon the heart, and observed, that under the influence of this fomentation it recovered new strength and life, so that both ventricles and auricles pulsated, contracting and relaxing alternately, recalled as it were from death to life. . . .

I have also observed the first rudiments of the chick in the course of the fourth or fifth day of the incubation, in the guise of a little cloud, the shell having been removed and the egg immersed in clear tepid water. In the midst of the cloudlet in question there was a bloody point so small that it disappeared during the contraction and escaped the sight, but in the relaxation it reappeared again, red and like the point of a pin; so that betwixt the visible and invisible, betwixt being and not being, as it were, it gave by its pulses a kind of representation of the commencement of life.*

Of the Motion, Action, and Office of the Heart
[From Chapter v]

From these and other observations of the like kind, I am persuaded it will be found that the motion of the heart is as follows:

First of all, the auricle contracts, and in the course of its contraction throws the blood (which it contains in ample quantity as the head of the veins, the storehouse, and cistern of the blood), into the ventricle, which, being filled, the heart raises itself straightway, makes all its fibres tense, contracts the ventricles, and performs a beat, by which beat it immediately sends the blood supplied to it by the auricle into the arteries; the right ventricle sending its charge into the lungs by the vessel which is called vena arteriosa, but which, in structure and function, and all things else, is an artery; the left ventricle sending its charge into the aorta, and through this by the arteries to the body at large.

These two motions, one of the ventricles, another of the auricles, take place consecutively, but in such a manner that there is a kind of harmony or rhythm preserved between them, the two concurring in such wise that but one motion is apparent, especially in the warmer blooded animals, in which the movements in question are rapid. Nor is this for any other reason than it is in a piece of machinery, in which, though one wheel gives motion to another, yet all the wheels seem to move simultaneously; or in that mechanical contrivance which is adapted to firearms, where the trigger being touched, down comes the flint, strikes against the steel, elicits a spark, which falling among the powder, it is ignited, upon which the flame extends, enters the barrel, causes the explosion, propels the ball, and the mark is attained—all of which incidents, by reason of the celerity with which they happen, seem to take place in the twinkling of an eye. So also in deglutition: by the elevation of the root of the tongue, and the compression of the mouth, the food or drink is pushed into the fauces, the larynx is closed by its own

---

* At the period Harvey indicates, a rudimentary auricle and ventricle exist, but are so transparent that except with certain precautions their parietes cannot be seen. The filling and emptying of them, therefore, give the appearance of a speck of blood alternately appearing and disappearing.

muscles, and the epiglottis, whilst the pharynx, raised and opened
by its muscles no otherwise than is a sac that is to be filled, is lifted
up, and its mouth dilated; upon which, the mouthful being received, it is forced downwards by the transverse muscles, and then
carried farther by the longitudinal ones. Yet are all these motions,
though executed by different and distinct organs, performed har-
moniously, and in such order, that they seem to constitute but a
single motion and act, which we call deglutition. . . .

Even so does it come to pass with the motions and action of the
heart, which constitute a kind of deglutition, a transfusion of the
blood from the veins to the arteries. And if any one, bearing these
things in mind, will carefully watch the motions of the heart in
the body of a living animal, he will perceive not only all the particulars I have mentioned, *viz.* the heart becoming erect, and making
one continuous motion with its auricles; but farther, a certain obscure undulation and lateral inclination in the direction of the
axis of the right ventricle, (the organ) twisting itself slightly in performing its work. And indeed every one may see, when a horse
drinks, that the water is drawn in and transmitted to the stomach
at each movement of the throat, the motion being accompanied
with a sound, and yielding a pulse both to the ear and the touch;
in the same way it is with each motion of the heart, when there is
the delivery of a quantity of blood from the veins to the arteries,
that a pulse takes place, and can be heard within the chest.

OF THE COURSE BY WHICH THE BLOOD IS CARRIED FROM THE
VENA CAVA INTO THE ARTERIES, OR FROM THE RIGHT
INTO THE LEFT VENTRICLE OF THE HEART

[From Chapter VI]

. . . And now the discussion is brought to this point, that they
who inquire into the ways by which the blood reaches the left
ventricle of the heart and pulmonary veins from the vena cava, will
pursue the wisest course if they seek by dissection to discover the
causes why in the larger and more perfect animals of mature age,
nature has rather chosen to make the blood percolate the parenchyma of the lungs, than as in other instances chosen a direct and
obvious course—for I assume that no other path or mode of transit

can be entertained. It must be either because the larger and more perfect animals are warmer, and when adult their heat greater— ignited, as I might say, and requiring to be damped or mitigated; therefore it may be that the blood is sent through the lungs, that it may be tempered by the air that is inspired, and prevented from boiling up, and so becoming extinguished, or something else of the sort. But to determine these matters, and explain them satisfactorily, were to enter on a speculation in regard to the office of the lungs and the ends for which they exist; and upon such a subject, as well as upon what pertains to eventilation, to the necessity and use of the air, etc., as also to the variety and diversity of organs that exist in the bodies of animals in connexion with these matters, although I have made a vast number of observations, still, lest I should be held as wandering too wide of my present purpose, which is the use and motion of the heart, and be charged with speaking of things beside the question, and rather complicating and quitting than illustrating it, I shall leave such topics till I can more conveniently set them forth in a treatise apart. And, now, returning to my immediate subject, I go on with what yet remains for demonstration, *viz.* that in the more perfect and warmer adult animals, and man, the blood passes from the right ventricle of the heart by the vena arteriosa, or pulmonary artery, into the lungs, and thence by the arteriae venosae, or pulmonary veins, into the left auricle, and thence into the left ventricle of the heart. And, first, I shall show that this may be so, and then I shall prove that it is so in fact.

OF THE QUANTITY OF BLOOD PASSING THROUGH THE HEART FROM THE VEINS TO THE ARTERIES; AND OF THE CIRCULAR MOTION OF THE BLOOD

[From Chapter VIII]

Thus far I have spoken of the passage of the blood from the veins into the arteries, and of the manner in which it is transmitted and distributed by the action of the heart; points to which some, moved either by the authority of Galen or Columbus, or the reasonings of others, will give in their adhesion. But what remains to be said upon the quantity and source of the blood which thus passes, is of so novel and unheard-of character, that I not only fear injury to myself from the envy of a few, but I tremble lest I have mankind at large for my enemies, so much doth wont and custom, that become as another nature, and doctrine once sown and that hath struck deep root, and respect for antiquity influence all men: Still the die is cast, and my trust is in my love of truth, and the candour that inheres in cultivated minds. And sooth to say, when I surveyed my mass of evidence, whether derived from vivisections, and my various reflections on them, or from the ventricles of the heart and the vessels that enter into and issue from them, the symmetry and size of these conduits,—for nature doing nothing in vain, would never have given them so large a relative size without a purpose,—or from the arrangement and intimate structure of the valves in particular, and of the other parts of the heart in general, with many things besides, I frequently and seriously bethought me, and long revolved in my mind, what might be the quantity of blood which was transmitted, in how short a time its passage might be effected, and the like; and not finding it possible that this could be supplied by the juices of the ingested aliment without the veins on the one hand becoming drained, and the arteries on the other getting ruptured through the excessive charge of blood, unless the blood should somehow find its way from the arteries into the veins, and so return to the right side of the heart; I began to think whether there might not be A MOTION, AS IT WERE, IN A CIRCLE. Now this I afterwards found to be true; and I finally saw that the blood, forced by the action of the left ventricle into the arteries, was distributed to the body at large, and its several parts, in the same manner as it is sent through the lungs, impelled by the right ventricle into the pulmonary artery, and that it then passed through the veins and along the vena cava, and so round to the left ventricle in the manner already indicated. Which motion we may be allowed to call circular, in the same way as Aristotle says that the air and the rain emulate the circular motion of the superior bodies; for the moist earth, warmed by the sun, evaporates; the vapours drawn upwards are condensed, and descending in the form of rain, moisten the earth again; and by this arrangement are generations of living things produced; and in like manner too are tempests and meteors engendered by the circular motion, and by the approach and recession of the sun. . . .

That There Is a Circulation of the Blood Is Confirmed from the First Proposition

[From Chapter IX]

But lest any one should say that we give them words only, and make mere specious assertions without any foundation, and desire to innovate without sufficient cause, three points present themselves for confirmation, which being stated, I conceive that the truth I contend for will follow necessarily, and appear as a thing obvious to all. First,—the blood is incessantly transmitted by the action of the heart from the vena cava to the arteries in such quantity, that it cannot be supplied from the ingesta, and in such wise that the whole mass must very quickly pass through the organ; Second,—the blood under the influence of the arterial pulse enters and is impelled in a continuous, equable, and incessant stream through every part and member of the body, in much larger quantity than were sufficient for nutrition, or than the whole mass of fluids could supply; Third,—the veins in like manner return this blood incessantly to the heart from all parts and members of the body. These points proved, I conceive it will be manifest that the blood circulates, revolves, propelled and then returning, from the heart to the extremities, from the extremities to the heart, and thus that it performs a kind of circular motion.

Let us assume either arbitrarily or from experiment, the quantity of blood which the left ventricle of the heart will contain when distended to be, say two ounces, three ounces, one ounce and a half—in the dead body I have found it to hold upwards of two ounces. Let us assume further, how much less the heart will hold in the contracted than in the dilated state; and how much blood it will project into the aorta upon each contraction;—and all the world allows that with the systole something is always projected, a necessary consequence demonstrated in the third chapter, and obvious from the structure of the valves; and let us suppose as approaching the truth that the fourth, or fifth, or sixth, or even but the eighth part of its charge is thrown into the artery at each contraction; this would give either half an ounce, or three drachms, or one drachm of blood as propelled by the heart at each pulse into the aorta; which quantity, by reason of the valves at the root of the vessel, can by no means return into the ventricle. Now, in the course of half an hour, the heart will have made more than one thousand beats, in some as many as two, three, and even four thousand. Multiplying the number of drachms propelled by the number of pulses, we shall have either one thousand half-ounces, or one thousand times three drachms, or a like proportional quantity of blood, according to the amount which we assume as propelled with each stroke of the heart, sent from this organ into the artery; a larger quantity in every case than is contained in the whole body! In the same way, in the sheep or dog, say that but a single scruple of blood passes with each stroke of the heart, in one half-hour we should have one thousand scruples, or about three pounds and a half of blood injected into the aorta; but the body of neither animal contains above four pounds of blood, a fact which I have myself ascertained in the case of the sheep.

Upon this supposition, therefore, assumed merely as a ground for reasoning, we see the whole mass of blood passing through the heart, from the veins to the arteries, and in like manner through the lungs.

But let it be said that this does not take place in half an hour, but in an hour, or even in a day; any way it is still manifest that more blood passes through the heart in consequence of its action, than can either be supplied by the whole of the ingesta, or than can be contained in the veins at the same moment.

Nor can it be allowed that the heart in contracting sometimes propels and sometimes does not propel, or at the most propels but very little, a mere nothing, or an imaginary something: all this, indeed, has already been refuted; and is, besides, contrary both to sense and reason. For if it be a necessary effect of the dilatation of the heart that its ventricles become filled with blood, it is equally so that, contracting, these cavities should expel their contents; and this not in any trifling measure, seeing that neither are the conduits small, nor the contractions few in number, but frequent, and always in some certain proportion, whether it be a third, or a sixth, or an eighth, to the total capacity of the ventricles, so that a like proportion received with each stroke of the heart, the capacity of the ventricle contracted always bearing a certain relation to the capacity of the ventricle when dilated. And since in dilating, the ventricles cannot be supposed to get filled with nothing, or with an imaginary something; so in contracting they never expel nothing or aught imaginary, but always a certain something, *viz.* blood, in proportion to the amount of the contraction. Whence it is to be inferred, that if at one stroke the heart in man, the ox, or the sheep, ejects but a single drachm of blood, and there are one thousand strokes in half an hour, in this interval there will have been ten pounds five ounces expelled: were there with each stroke two drachms expelled, the quantity would of course amount to twenty pounds and ten ounces; were there half an ounce, the quantity would come to forty-one pounds and eight ounces; and were there one ounce, it would be as much as eighty-three pounds and four ounces; the whole of which, in the course of one half-hour, would have been transfused from the veins to the arteries. The actual quantity of blood expelled at each stroke of the heart, and the circumstances under which it is either greater or less than ordinary, I leave for particular determination afterwards, from numerous observations which I have made on the subject.

Meantime this much I know, and would here proclaim to all, that the blood is transfused at one time in larger, at another in smaller quantity; and that the circuit of the blood is accomplished now more rapidly, now more slowly, according to the temperament, age, etc., of the individual, to external and internal circumstances, to naturals and non-naturals,—sleep, rest, food, exercise, affections of the mind, and the like. But indeed, supposing even the smallest quantity of blood to be passed through the heart and the lungs with each pulsation, a vastly greater amount would still be thrown into the arteries and whole body, than could by any possibility be supplied by the food consumed; in short it could be furnished in no other way than by making a circuit and returning.

This truth, indeed, presents itself obviously before us when we consider what happens in the dissection of living animals; the great artery need not be divided, but a very small branch only, (as Galen even proves in regard to man,) to have the whole of the blood in the body, as well that of the veins as of the arteries, drained away in the course of no long time—some half-hour or less. Butchers are well aware of the fact and can bear witness to it; for, cutting the throat of an ox and so dividing the vessels of the neck, in less than a quarter of an hour they have all the vessels bloodless—the whole mass of blood has escaped. The same thing also occasionally occurs with great rapidity in performing amputations and removing tumours in the human subject.

Nor would this argument lose any of its force, did any one say that in killing animals in the shambles, and performing amputations, the blood escaped in equal, if not perchance in larger quantity by the veins than by the arteries. The contrary of this statement, indeed, is certainly the truth; the veins, in fact, collapsing, and being without any propelling power, and further, because of the impediment of the valves, as I shall show immediately, pour out but very little blood; whilst the arteries spout it forth with force abundantly, impetuously, and as if it were propelled by a syringe. And then the experiment is easily tried of leaving the vein untouched, and only dividing the artery in the neck of a sheep or dog, when it will be seen with what force, in what abundance, and how quickly, the whole blood in the body, of the veins as well as of the arteries, is emptied. But the arteries receive blood from the veins in no other way than by transmission through the heart, as we have already seen; so that if the aorta be tied at the base of the heart, and the carotid or any other artery be opened, no one will now be surprised to find it empty, and the veins only replete with blood.

And now the cause is manifest, wherefore in our dissections we usually find so large a quantity of blood in the veins, so little in the arteries; wherefore there is much in the right ventricle, little in the left; circumstances which probably led the ancients to believe that the arteries (as their name implies) contained nothing but spirits during the life of an animal. The true cause of the difference is this perhaps: that as there is no passage to the arteries, save through the lungs and heart, when an animal has ceased to breathe and the lungs to move, the blood in the pulmonary artery is prevented from passing into the pulmonary veins, and from thence into the left ventricle of the heart; just as we have already seen the same transit prevented in the embryo, by the want of movement in the lungs and the alternate opening and shutting of their minute orifices and invisible pores. But the heart not ceasing to act at the same precise moment as the lungs, but surviving them and continuing to pulsate for a time, the left ventricle and arteries go on distributing their blood to the body at large and sending it into the veins; receiving none from the lungs, however, they are soon exhausted and left, as it were, empty. But even this fact confirms our views, in no trifling manner, seeing that it can be ascribed to no other than the cause we have just assumed.

Moreover it appears from this that the more frequently or forcibly the arteries pulsate, the more speedily will the body be exhausted in an hemorrhage. Hence, also, it happens, that in fainting fits and in states of alarm, when the heart beats more languidly and with less force, hemorrhages are diminished or arrested.

Still further, it is from this that after death, when the heart has ceased to beat, it is impossible by dividing either the jugular or femoral veins and arteries, by any effort to force out more than one half of the whole mass of the blood. Neither could the butcher, did he neglect to cut the throat of the ox which he has knocked on the head and stunned, until the heart has ceased beating, ever bleed the carcass effectually.

Finally, we are now in a condition to suspect wherefore it is that no one has yet said anything to the purpose upon the anastomosis of the veins and arteries, either as to where or how it is effected, or for what purpose. I now enter upon the investigation of the subject.

XVII

OBSTETRICS

OBSTETRICS MUST HAVE BEEN PRACTICED IN SOME FORM FROM THE VERY earliest times, when man emerged from the brute stage and began to attempt to assist his fellows in distress. There are many references to obstetrical events in the Bible, in the Iliad, and in other early literature. It is evident from these that the midwife was the officiating figure. In other words, that obstetrics was in the hands of women. Their instructions were evidently verbal until a very late stage, just as they are among primitive tribes today. The first written accounts of antiquity have been lost or garbled.

Soranus of Ephesus, who was a physician in Rome, and is said to have lived in the reigns of Trajan and Hadrian, composed a book of obstetrics and gynecology. This, however, is first mentioned by Suidas, a Greek grammarian of the twelfth century, and we have nothing but fragments of it. It is possible, however, that the book was the original of Röslin's *Rosengarten* and Raynalde's *The Byrthe of Mankynde*. Our first actual literature on obstetrics, therefore, does not begin until the sixteenth century.

REFERENCES

ENGELMAN, G. J. Labor Among Primitive Peoples. St. Louis, J. H. Chambers, 1883.

FINDLAY, P. The Priests of Lucina, A History of Obstetrics. Boston, Little Brown, 1939.

JAMESON, E. M. Gynaecology and Obstetrics. Clio Medica series. New York, Paul B. Hoeber, 1936.

MILLER, J. L. Renaissance Midwifery. The Evolution of Modern Obstetrics, 1500-1700. Mayo Foundation Lectures on the History of Medicine. Philadelphia, W. B. Saunders, 1933.

THOMS, H. Classical Readings in Obstetrics and Gynaecology. Springfield, Illinois, C. C. Thomas, 1935.

EUCHARIUS RÖSLIN

Soon after the discovery of printing, Eucharius Röslin (d. 1526) published a little manual for midwives, the *Rosengarten* (1513). Imitations of it appeared in all languages. (*The Byrthe of Mankynde*, 1540, by Thomas Raynalde, is the English form.) Here, for the first time, was available a treatise on the signs of pregnancy, the course of labor, the ordinary practice of obstetrics. Röslin was a physician at Worms, afterwards at Frankfurt-on-Main; he is recorded as apothecary at Freiburg; and in 1502 he entered the service of the Duchess of Brunswick and Tuneberg, to whom his book is dedicated.

THE BYRTHE OF MANKYNDE*

The Byrthe of Mankynde first saw the light of day in the latter part of the reign of Henry VIII; it reappeared corrected, annotated, and illustrated several times during the stormy years of Bloody Mary, and of Edward VI; it was very popular during the long reign of Queen Elizabeth; it lived on during the Civil War, the Restoration, and by 1676, it was still virile enough for a final edition.

The first book is a store of general and preliminary information: there is a description of the genito-urinary system in the female (not always accurate nor always clear); an enumeration and a description of the skin layers, the abdominal muscles; a description of the peritoneum, the causes of menstruation and its utility; the origin of leucorrhea; the production of milk in the breast; and the foetal coverings.

The second book is concerned with obstetrics—the only one of the four in this small volume which is. The third book is allotted to infant feeding and care. The fourth book is a compilation of useful household remedies, and the methods of preparation thereof.

OF THE TYME OF BYRTH. AND WHICH IS CALLED NATURALL OR UNNATURALL

[From Book II, Chap. I]

In the fyrst booke we have sufficiently set foorth and described the maner, situation, and fourme of the Matrix wherein man is conceaved, with dyvers other matters appending and concernyng the better understandyng of the same. And nowe here in this second Booke, we wyll declare the maner of the quytyng and delvuerance of the Infant ourt of the mothers wombe, with other thinges thereto appertaynyng. And fyrst here in this Chapter we wil declare the tokens and signes whereby ye may perceive whether the tyme of labour be neare or not: for when the houre of labour approcheth neare, these signes following evermore proceede and come before.

First certayne dolours and paynes begin to growe about the futtes,

---

* A loose translation of Röslin's *Rosengarten*, by Thomas Raynalde. London, 1540.

the navyll, and in the raynes of the back, and lykewyse about the thyghes, and the other places beyng neare to the privie partes, which lykewyse then beginneth to swell and to burne, and to expell humours, so that it geveth a playne and evident token that the labour is neere.

But ye shall note, that there is two maner of byrthes, the one called naturall, the other not naturall. Naturall byrthe is when the chylde is borne in due season, and also in due fashion.

The due season is most commonly after the nynth moneth or about fourtie weekes after the conception, although some be delivered sometimes, in the seventh moneth, and the chylde proveth very well. But such as are borne in the eight moneth, other they be dead before the byrth, or els live not long after.

The due fashion of birth is this: first the head cometh forwarde, then followeth the necke and shoulders, the armes with the handes lying close to the body towarde the feete, the face and forepart of the chylde beyng towarde the face and forepart of the mother, as it appearth in fyrst of the byrth figures. For as hath been sayde in the fyrst Booke, before the tyme of delyveraunce, the chylde lyeth in the mothers wombe the head upward, and the feete downewarde, but when it should be delyvered, it is turned cleane contrary, the head downewarde, the feete upwarde, the face towardes the mothers belly, and that yf the byrth be naturall. Another thyng also is this, that yf the byrth be naturall, the one bodye and two heades, as appeareth in the XVIII. of the byrthe figures, such as of late was seene in the dominion of Werdenbergh.

Agayne, when it proceedeth, as when it cometh foorth with both feete or both knees together, or else with one foot onely, or with bothe feete downewards, and both handes upwardes, other els (the whiche is most perilous) sidelong, arcelong, or other els (having two at a byrth) both the other with his head, by those and dyvers other waynes the woman sustayneth great dolour, paine, and anguyshe.

Also there is great perill in labouring, when the secondine or latter birth is over firme or strong, and will not soone rise or breake asunder, so that the child may have his easy comming foorth. And contrarywyse, when it is overweake, slender or thin, so that it breaketh asunder before that the chylde be turned, or apt to issue foorth, for then the humours which are collect and geathered together about this secondine or seconde byrth, passe away sooner then they should do, and the birth shal lack his due humidities and moistures, which should cause it the easeyer to proceed, and with lesse payne.

Also yf the woman feele payne onlye in the backe and above the navell, and not under, it is a signe of harde labours; lykewyse if she were wont in times passed to be delivered with great payne, is an evidence lyklyhood of great labour alwayes in the birth.

Nowe signes and token of an expedite and easye deliveraunce, be suche as be contrary to all those that have be rehearsed before. As for example, when the woman hath been wont in tymes passed easyly to be delyvered, and that in her labour she feele but little strong or dolour, or though she have great paynes, yet they remayne not styll in the upper partes, but descende alwayes downwardes to the neather partes or bottome of the belly.

And to be short, in all paynefull and troublesome labours, these signes betoken and signifie good spede and lucke in the labour: unquietnes, muche styrring of the chylde in the mothers belly, all the thronges and paynes tombling in the forepart of the bottome of the belly, the woman strong and mightie of nature, such as can well and strongly helpe her selfe to the deliveraunce of the byrth. And agayne, evill signes be those, when she sweateth colde sweate, and that her pulces beate and labour over sore, and that she her selfe in the labouring faynt and swone, these be unluckie and mortall signes.

HOWE A WOMAN WITH CHYLDE SHALL USE HERSELFE, AND WHAT REMEDIES BE FOR THEM THAT HAVE HARD LABOUR

[From Book II, Chap. III]

To succour and helpe them that are in such difficult perill of labour, as we have spoken of before, yet muste observe, kepe and marke those things that we shal (by the grace of God) shewe you in this Chapter following.

First the woman with chylde must kepe two diets, the one a moneth before her labour, the other in the very labouring. And above al thinges she must issue suche occasions which may hynder the birth, to the uttermost of her power, the whiche occasions we rehearsed in the Chapter before. But if there be any such thyng whiche can not be avoyded, forsomuch as it commeth by nature, or by long continuance and custome in this case: yet ye that sue some suche remedies, the whiche may some what asswage it, molifie it, or make it more easy or tollerable, so that it hinder the byrth so much the lease.

But if it so be, that any infirmitie or disease, swelling, or other apostumation chaunce about the mother or the privie part, or about the vesicke or bladder, as the stone, the strangury, and such lyke, the whiche thynges maye cause suche straytnes and coarstation, that unneth without great and horrible payne, the partie can be deliverede or discharged. In these cases it behoveth such thynges to be loked unto and cured, before the tyme of labour commeth, by the advice of some experte surgion.

Also if the woman be overmuch constipat or bounds, most commonly she must use, the moneth before her labour, suche thynges the which may lenifie, mollifie, dissolve, and lose the belly, as apples fryed with suger, taken fastyng in the mornyng, and after that a draught of pure wyne along, or else temperd with the juyce of swete and very ripe apples.

Also to eat figges in the mornyng fasting and at night, loset well the belly. If these profite not, Caffia fistual taken 111 or 111.drams one halfe houre before dyner shall lose the belly without peril.

Agayne, in this case she must refrayne from all such thynges as do harden, restrayne, and constipat, as meates, broyled or rosted, and rice, hard egges, biefe, chestnuttes, and all sowre fruites, and such lyke.

Also yf father necessitie require, she may receave a clyster, but it must be very gently and easye made of a pynt of the broth of a chicken or other tender fleshe, thereto putting so muche course sugar or bony, as they make it reasonably sweete, and halfe a spoonefull of whyte salt. Or for the poore woman maybe be made a clyster of a pynt of water, wherein hath ben sod mallowes or holyoke, with bony sale, as before. She may use also some other easy and temperate purgation, to molifie and lose her with all, as Mercury sodden with flesh in potage, and divers suche other, or els suppositer tempered with sope, larde, or the yolkes of egges.

Now when the woman perceiveth the Matrix, or Mother to ware or loose, and to be dissolved, and the humours issue foorth in great plentie, then shal it be meete for her to sit downe, leanyng backwarede in maner upright. For the whiche purpose in some regions (as in Fraunce and Germanie) the Midwifes have stooles for the nonce, which beyng but lowe, and not bye far from the grounde, be made so compassewyse and cave or hollowe in the middes, that that may be recyved from underneth whiche is looked for, and the backe of the stoole leaning backward, receyveth the back of the woman: the fashion of the whiche stoole, is set in the beginning of the byrth figures hereafter.

And when the time of labour is come, in the same stoole, ought to be put many clothes or cloutes in the backe of it, the whiche the midwyfe may remove from one syde to another, accordyng as necessitie shal require. The midwyfe her selfe shal sit before the labouryng woman, and shal diligently onserve and waite, how muche, and after what means the child styrreth it selfe: also shal with her handes, first annointed with the oyle of Almondes, or the oyle of white lillies, rule and dyrecte every thing as shall seeme best.

Also the mydwyfe muste struct and comfort the partie, not onely refreshing her with good meate and drinke, but also with sweete woordes, gevyng her good hope of a speedfull deliveraunce, encouraging and so making her patience and tolleraunce, byddyng her to holde her breath in so much as shee may, also strekyng gentilly with her handes her belly about the Navell, for helpeth to depresse the byrth downewarde.

When the byrth commeth not naturally, then must the Mydwyfe do all her diligence and payne (yf it may be possible) to turne the byrth tenderlye with her annoynted handes, so that it may be reduced agayne to a naturall byrth. As for example: Sometyme it chaunceth the chylde to come the legges and both armes and handes downward, close to the sydes fyrst foorth, as appeareth in the seconde of the byrth figures. In this case the midwyfe must doo all her payne with tender handlyng and annoyntyng to receyve foorth the chylde, the legges being still close together, and the handes likewyse remayning, as appearth in the sayd second figure.

Also sometymes the byrth commeth foorth with one foote onely, the other being left upward, as appeareth in the fourth figure. And in this case it behoveth the labouring woman to lay her upryght upon her backe, holdyng up her thighes and belly, so that her head be the lower part of her body, then let the Mydwyfe with her hande returne in agayne the foote that commeth out first, in as tender

maner as she may be, and warne the woman that laboureth to styrring, the byrth may be turned the head downewarde, and so to make
a naturall byrth of it, and then to set the woman in the stoole
agayne, and to do as ye dyd in the fyrst figure. But yf it be so, that
notwithstanding the mothers styrryng and mouvng, the byrth do
not turne, then must the Midwyfe with her hande softly fetche out
the other lefte whiche remayned behynde, evermore takyng heede
of this that by handlyng of the chylde she do not remove ne set out
of theyr place the two handes hangyng downewarde towarde the
feete.

Remedies and Medicines by the Whiche the Labour may be made more Tolerable, easie, and without great payne

[From Book ii, Chap. v]

The thynges which helpe the byrth and make it more easie, are
these. First the woman that laboureth must eyther sytte grovelyng,
or els upright, leaning backwarde, according as it shall seeme com-
modius and necessary to the partie, or as she is accustomed. And in
winter or colde weather, the chamber wherein she laboureth must
be warmed, but in sommer, or hot weather, let in the ayre to refresh
her withall, lest betwene extreme heate and labour the woman faint
and fowne. And furthermore, she must be provoked to sneesyng, and
that eyther with the powder of Eleborus, or els of Pepper. Also the
sydes of the woman must be stroken downeward with the handes,
which thyng helpeth greatly and furthereth. And let the Mydwyfe
always be very diligent, providyng and seeyng what shal be necessary for the woman, annoynting the privities with oyl, or other such
grease as I spake of before, in this fashion.

How the Secondine or Second Byrth Shall be forced to issue foorth, if it come not freely of his own kinde

[From Book ii, Chap. v]

Here also sometime it commeth to passe, that the secondine which
is wont to come together with the byrth, remayne and tarry behynde,
and follow not, and that for divers cases. One is, for because perad-
venture the woman hath ben so fore weakened and feeblyshed with
travayle, dolour, and payne of that fyrst byrth, that she hath no
strength remayning to helpe her selfe, or the expectyng of the second

byrth. Another may be, that it be entangled, tred, or let within the
matirx (whiche unseth many tymes) or that it be destitute of
humours, so that the water be flowen from it sooner then tyme is,
whiche shoulde make the places more slipperie and more easie to
passe thorowe: Or els that the places over weeryed with long and
fore labour, for payne contract or geather together, and enclose them
selfe agayne, or that the places be swolen for anguyshe and payne,
and so let the commyng foorth of the second byrth.

IN THIS FIRST CHAPTER OF THE THIRDE BOOKE, IS FIRST DECLARED
THE MATTERS THEREIN CONTEYNED, AND THEN HOW THE INFANT
NEWLY BORNE MUST BE HANDLED, NOURYISHED, AND LOOKED TO

[From Book III, Chap. I]

In the Seconde Booke we have sufficientlye and at length declared
the maners and diversities of byrthes, with the daungers and per-
rylles often chauncyng to the woman at theyr labours, and after the
same. And now here in this third booke shal be entreated what is
to be done to the Infant borne. And howe to choose a Nurse, and
of her office: with manyfolde medicines and remedies agaynst sundry
infirmities, which eftsones happen to Infantes in their infancie.
Then after that the Infant is once come to lyght by and by the
Navell must be cutte three fyngers breadeth from the belly, and so
knyt up, and let be strued on the had of that that remayneth, of
the powder of Bose Armentaske, and Sanguis draconis, Sarcocola,
Myrrhe, and Cummin, of eache lyke muche beaten to powder: then
upon that binde a peece of wool, dypped in oyle Olive that the powder fall not of. Some use fyrst to knyt the Navell, and after to cutte
it so much, as is before rehearsed.
And furthermore some day, that of what length the rest of the
Navell is left, of the same length shall the chyldes tong be, yf it be
a man chyld. Item, Avicenna sayth, that divers thinges may be
knowen by markyng of the chyldes Navell: For (as he saith) when
the woman is delivered of her fyrst chylde, then beholde the Navell
of the chylde: which yf in that part of it whiche is next unto the
body it have never a wrincle, it protendeth and doth signifie perpetuall from thencefoorth sterilitie or barenesse: and yf it have any
wrincles in it, then so many wrinkles, so many chyldren shall the
woman have in time to come. Also some ad to this, and say, that if

there be lytle space betweene these wrincles in the Navell, then shall there be also little space betweene the bearing of the chyldren: if much, it signifieth long time betweene the bearyng of them: but these sayinges be neyther in the Gospell of the day, ne of the nyght.

Nowe to returne to our purpose, when that the Navell is cut of, and the rest knyt up: annoynt all the chyldes body with the oyle of Acornes, for that is singulerly good to confyrme thynges, whiche may chaunce from without, as smoke, colde, and such other thynges: which yf the infant be greeved withall strayght after the byrth, beyng yet very tender, it shoulde hurt it greatly.

After this annoyntyng, washe the infante with warm water, and with your finger (the navyle beyng pared) open the chyldes nose-sryls, and purge them of the fylthenesse. And also that the Nurse handle so the chyldes syttyng place, that it may be provoked to purge the belly. And chiefly it must be defended from overmuche cold, or overmuch heate.

After that the parte extant or the knot of the Navell is fallen (the which commonly chaunceth after the thyrde or fourth day) then on the rest remaynyng, strewe the powder or Ashes of a Calfes hoofe burnt, or of Snayle shelles, or of the powder of lead, called read lead, tempered with wine.

OF THE NURSE AND HER MYLKE, AND HOW LONG THE CHYLDE SHOULD SUCKE

[From Book III, Chap. II]

As concerning the brynging up, nourishment and gevyng of sucke to the chylde, it shal be beste that the mother geve her chylde sucke her selfe, for the mothers milke is more conveniente and agreeable to the Infant, then any other woman's, and more doth it nouryshe it, for because that in the mothers belly it was wont to the same. and fed with it, and therefore also it does more desyrously covet the same, as that with which it is best acquaynted. And to be short. the mothers milke is most holmsommest for the chylde, as Avicenna wryt-eth, it shal be sufficient to geve it sucke twice or thryce in a day. And always beare ye geve not the chylde to muche sucke at once in this tender age of it, for cloyying of it, and least also it loth it: but rather let it have often of it, and lytle at once, then fewe tymes, and movermuche at once, for suche is be over clyde with the mothers mylke, causeth theyr body to swell and inflate, and in theyr brine shall it appeare, that it is not overcome ne concoted or digested in the chylde: which thyng yet if it chaunce, let the Infant be kept fastyng untyll such time as that which it hath receyved already be completely digested.

Item, yf the mothers mylke be some what sharpe or choleryke, let her never geve the chylde her breast fastyng. If it be so that the mother can not geve the Infant sucke her selfe, eyther for because of sycknesse, or that her brestes be sore, and her mylk corrupted: then let her chose a holsome Nurse, with these conditions folowyng.

Fyrst, that she be of a good colour and complexion, and that her milke and brest be of good largenesse. Secondly, that it be two monethes after her labour at the least, and that (yf it may be) suche one whiche hath a man chylde. Thyrdly, that she be of meane and measurable lyking, neyther to fatte ne to leane. Fourthly, that she be good and honest of conversation, neyther over hastie or yrefull ne to sadde or forlorne, neyther to fearefull or tymorous: for these affections and qualities be pernitious and hurtfull to the mylke, corruptyng it, and passe foorth through the mylke into the chylde, makyng the chylde of lyke condition and maners. Also that they be not over lyght and wanton of behaviour. Fyfthly, that her brestes be full, and have sufficient plentie of mylke, and that they be neyther to great, soft, hangyng, and flaggyng ne to lytle, hard or contracte, but of a measurable quantitie.

Also looke upon her mylke, that it be not blackyshe, blewishe, gray, or reddysh, neyther sower, sharpe, saltyshe, or brackysh, neyther thinne and fluy, neyther over grosse or thycke, but temperately whyte, and pleasaunt in taste.

HUGH CHAMBERLEN

Hugh Chamberlen, in the preface to his translation (1696) of Mariceau's work on obstetrics, announced the obstetric forceps. He did not, it will be noticed, describe the instrument which he claimed to be so useful, nor give any hint which would allow others to construct a similar instrument. This reticence was in the tradition of the Chamberlen family. For several generations they practiced midwifery in London and let it be known that they had an instrument which would assist in difficult labor. Many attempts were made to wrest the secret from them. According to tradition, a patriotic group of citizens got up a purse which was accepted by Hugh Chamberlen but all he gave in exchange was one blade of the obstetric forceps. The true nature of the forceps, however, gradually leaked out. In 1733, Edmund Chapman said "there are different sorts of forceps which are well known."

William Giffard (-1731) is, according to Partridge, "the altruistic and honorable physician who should receive full credit for introducing the forceps into common use in England."

REFERENCE

DAS, KEDARNATH. Obstetric Forceps, Its History and Evolution. St. Louis, C. V. Mosby, 1929.

FIRST ANNOUNCEMENT OF THE OBSTETRIC FORCEPS

[From *The Diseases of Women with Child, and in Child-Bed**]

THE TRANSLATOR TO THE READER

. . . In the 17th Chapter of the Second Book, my Author justifies the fastning Hooks in the Head of a Child which comes right, and yet by reason of some difficulty or disproportion cannot pass; which I confess hath been, and is yet the practice of the most expert Artists in Midwifery, not only in England, but throughout Europe; and hath very much caused the report, That where a Man comes, one or both must necessarily die; and is the reason why many forbear sending, till the Child is dead, or the Mother dying. But I can neither approve of that practice, nor those delays; because my Father, Brothers, and my Self (tho none else in Europe that I know) have, by God's Blessing, and our Industry, attained to, and long practised a way to deliver Women in this case, without any prejudice to them or their Infants: tho all others (being obliged, for want of such an Expedient, to use the common way) do, and must endanger, if not destroy one or both, with Hooks. By this manual Operation may be dispatched, (when there is the least difficulty) with fewer pains, and in less time, to the great advantage, and without danger, both of Woman and Child. If therefore, the use of Hooks by Physicians and Chirurgeons, be condemned, (without thereto necessitated through some monstrous Birth) we can much less approve of a Mid-

* Traité des Maladies des Femmes Grosses, et de celles qui font accouchées. Paris, 1681. Translation by Hugh Chamberlen, M.D., 1696.

wife's using them, as some here in England boast they do; which rash presumption, in France, would call them in question for their Lives.

FIELDING OULD

Sir Fielding Ould (1710-1789) was born in Galway, Ireland, and studied in Paris, probably under Gregoire. He began practice in Dublin in 1736. He was appointed Master of the Rotunda Hospital, the great center of obstetric teaching, in 1759. His *Treatise on Midwifery* is valuable for its description of the mechanism of labor. Thoms states: "A study of the *Treatise* reveals many practices that are thought to belong to more modern obstetrics. Ould used opium in prolonged labor, recommended immediate delivery of the second twin, allowed the placenta to be born by the expulsive efforts of the uterus, used episiotomy, employed the forceps and performed version."

THE MECHANISM OF LABOR
[From *A Treatise on Midwifery**]

This every practitioner in midwifery has in his power to be certain of: yet it may not be amiss to prove it to the reader, who has not as yet practiced, by plain reasoning: first, it is evident that the head from the os frontis to the occipitus, is of an oblong figure, being very flat on each side; secondly, that the body, taking in the shoulders, makes still a more oblong figure, crossing that of the head; so that supposing the woman on her back, the head coming into the world, is a kind of elipsis in a vertical position; and the shoulders of the same form in an horizontal position; thirdly, that the pelvis is of an eliptical form, from one to the other hip. Now if the child presented with the face to the sacrum, the oblong figure of the head must cross that of the pelvis; and if it were possible that the head and pelvis could be formed to each other, so as to admit of its exit, it must of necessity, from what has been said above, acquire another form for the admission of the shoulders; which is very different from the constant uniformity in all the works of providence.

From what has been said, it is evident that when the child is turned, so as to have the chin on one shoulder, all the above objections are removed; for the head and shoulders are on a parallel line,

* Dublin, 1742.

in respect of their shape, and at the same time, both answer the form of the passage from the pelvis.

For want of this knowledge, many labours prove dangerous and tedious, that might have been very successful, had they been committed to nature. For it is too common for Midwives, immediately on the eruption of the waters, to move the child's head to and fro, in order to facilitate its exit; and this jogging, may very easily alter the position of the head, so as to make it what is generally esteemed natural; hence the crown of the head, near the joining of the coronal and sagital sutures, are by the efforts of the mother, forced against the os pubis; when this happens, the women tell you, the head is fixed on the share bone, which in reality, is the intersection of two elipses, namely, the head and the passage from the pelvis; but the misery does not end here; for the repeated throws of the mothers, forcing the head against the publis, at the point above-mentioned, pushes it so as to make the lambdoidal bone lie on the back, whereupon the face presents itself; the consequence whereof shall be mentioned in its proper place, when we treat of preternatural deliveries.

It is hoped that this opinion, being founded on theory, and confirmed by experience, will meet with few opponents; and without doubt, the due application of it will be of infinite use.

THE DELIVERY OF THE PLACENTA

[From *A Treatise on Midwifery*]

The child, as was said before, must be laid on the operator's lap, or on the bed, as far from the mother as the length of the funis will admit; which he must take in the right hand, about six finger's breadth from the pudendum, and roul it twice or thrice about his finger; then the first and second fingers of the left hand must be thrust into the vagina, by its direction; and the patient stopping her breath and forcing as if she were at stool, the naval-string must be gently pulled forward as she forces, the operator rather waiting for her expulsion of it, than being too desirous to extract it; for pulling the funis so as to extract the placenta forcibly, may probably cause a flooding; or perhaps break the naval-string whereby the placenta would be very difficultly brought forth; therefore let him just pull it sufficiently to make it incline forward still insisting on the patient's forcing down, which if she be not able to do of herself she must be compelled to it, by putting a finger into her throat, which will cause a pressure of the diaphragm, and the muscles of the belly, by her efforts to vomit, by these means it is commonly brought forth in about five minutes. When it comes away by expulsion, it always is whole, but it is subject to be broke, and part of it left in the womb, if any violence be used for its extraction.

Most authors give a strict charge to lose no time in the extraction of this extraneous body lest the orifice of the wombe should contract and obstruct its passage; and for this reason they advise the introduction of the operator's hand into the matrix; and by insinuating the fingers between it and the placenta, to cause their separation, the manner of doing which, shall be presently described. This fear of the womb closing makes many operators too hasty, which often produces fatal accidents.

Notwithstanding what has been said on this subject, it must be allowed that Mr. Deventer, whose authority has universal approbation, strenuously advises the constant extraction of this burthen by the introduction of the hand; and very much condemns the pulling it forth by the funis; therefore we must endeavour to remove his objections to this practice, which he allows to be the most general in all parts of the world.

First he says that immediately after the child comes forth, you may thrust not only the hand but the arm also into the womb, without giving any pain, the orifice being at this time sufficiently dilated; whereas if you try different means, as described by other authors, it will in the mean time, be so much contracted, that the hand cannot pass it without great pain. Here I allow, that were there a necessity for putting the hand into the matrix, the orifice is at this time more dilated, than it would be in some time after; but our author is certainly mistaken in the most material part of his argument; for it is not the passing of the hand through the orifice of the womb, that gives the patient such great pain; but it is its passing through the bones that make the opening into the pelvis, which I may venture to say, never alters as to its size: this indeed does give very extraordinary pain, which is the chief reason why the operation should be avoided when there is not a necessity for it.

CARL SIEGMUND FRANZ CREDÉ

Carl Siegmund Franz Credé (1819-1892), professor of obstetrics and gynecology at Leipzig, published *Die Verhütung der Augenentzündung der Neugeborenen* (The Prevention of Ophthalmia Neonatorum), in 1884. Thoms states that the procedure "has unquestionably prevented blindness in many thousands of newborn."

REFERENCE

THOMS, H. Classical Contributions to Obstetrics and Gynaecology. Springfield, Ill., C. C. Thomas, 1935.

[From *The Prophylactic Treatment of Ophthalmia Neonatorum**]

In other words within a period of almost three years, there occurred in 1160 children only one or at the most two, cases of blennorrhea. Thus I had reached a percentage-figure which can be accepted as attainable, because individual illnesses, in which the blame must be placed on omission or a false execution of the prophylactic process, will never be able to be completely checked.

The same percentage has continued to hold good from the conclusion of my third report up to now (end of March 1884), thus for one more whole year.

Here again I expressly point out, that all the other usual eyesicknesses of children, which formerly could be observed rather frequently along with blennorrhea in the first days after birth, and which are slight and not dangerous but annoying nevertheless: such as: inflammation of the conjunctiva (Bindehautcatarrh), slight inflammation of the conjunctiva (Bindehautenzündung), chafing (Wundsein), of the outer skin, etc., have all as good as disappeared since the introduction of my prophylaxis.

So as to prevent any misunderstandings, I shall proceed to describe exactly how the process was used in the Leipzig Institute:

After the ligature and division of the umbilical cord we first removed from the children in the usual manner the sebaceous matter and the blood, mucus, etc., which clung to them; then they were

* Die Verhütung der Augenentzündung der Neugeborenen. Berlin, 1884. Translation by Herbert Thoms. By permission of Dr. Thoms and the publisher, Charles C. Thomas.

brought to the bath and there, by means of a clean piece of cloth or better, by means of a clean Bruns' wadding for dressing not with the bath-water, but with other clean, ordinary, water, their eyes were cleansed on the outside: that is: all the sebaceous matter clinging to the eye-lids was removed. Then on the table where the child is swathed before clothes are put on the child, each eye is opened by means of two fingers, a single drop of a 2 per cent solution of silver nitrate hanging on a little glass rod is brought close to the cornea until it touches it, and is dropped on the middle of it. There is no further care given to the eyes. Especially in the next 24 to 36 hours, in case a slight reddening or swelling of the lids with secretion of mucus should follow, the instillation should not be repeated.

The solution of silver nitrate is contained in a little bottle of dark glass with a glass stopper. The neck of the bottle is 1 centimeter in diameter. The little glass rod used is 15 centimeters long, 3 millimeters thick, and smooth and rounded at both ends. The little bottle and rod are kept locked up in a small drawer in the swaddling table.

The solution is renewed about every 6 weeks, but can even be used much longer, without damaging its effects. To introduce the process into the widest circle of activity possible, the apparatus used for it must be as simple, cheap, and safe as possible. That is why I have again given up the syringes, brushes, pipettes, drop-glasses, which I used to use very often in treating already existing ophthalmic blennorrheas, because they did not achieve the desired end as well as the little glass rod I have described.

Rinsing of the female genitalia is performed for the sake of cleanliness, but it can also be omitted, because it has no influence on the treatment of the infection, even if sterilized water or antiseptic solutions are chosen.

Naturally it is not a question of the series of operations described above. In any case, the instillation does not have to be used before the ligature and division of the umbilical cord of the child takes place. In order not to lose control over those who are in labor during the third stage of labor; it is recommended for private practice, in case there is a lack of assistants who are familiar with the process, that first, immediately after the ligature and division of the umbilical cord, on the swaddling table, the eyes be cleaned on the outside, then the instillation made immediately—a thing which takes only a few minutes—then the placenta be completely cared for, and finally the child bathed.

In the Leipzig Lying-in Hospital, the instillations were made by the head-midwife alone, mostly without the supervision of a doctor; only one student-midwife can be useful to the extent that she delicately draws apart a little the child's eye-lids with one finger of her hand. By means of this assistance all the students are trained, and soon can carry out the process all by themselves.

JOHN STEARNS

The Introduction of Ergot*

John Stearns (1770-1848) first president of the New York Academy of Medicine, published a letter to Dr. S. Ackerly of Waterford in the *Medical Repository*, 1807, as follows:

In compliance with your request I herewith transmit you a sample of the pulvis parturiens, which I have been in the habit of using for several years, with the most complete success. It expedites lingering parturition, and saves to the accoucheur a considerable portion of time, without producing any bad effects on the patient. The cases in which I have generally found this power to be useful, are when the pains are lingering, have wholly subsided, or are in any way incompetent to exclude the fetus. Previously to its exhibition it is of the utmost consequence to ascertain the presentation, and whether any preternatural obstruction prevents delivery; as the violent and almost incessant action which it induces in the uterus precludes the possibility of turning. The pains induced by it are peculiarly forcing; though not accompanied with that distress and agony, of which the patients frequently complain when the action is much less. My method of administering it is either in decoction or powder. Boil half a drachm of the powder in half a pint of water, and give one third every twenty minutes until the pains commence. In powder I give from five to ten grains; some patients require larger doses, though I have generally found these sufficient.

If the dose is large it will produce nausea and vomiting. In most cases you will be surprised with the suddenness of its operation; it

* From a letter to Dr. S. Ackerly, published in the *New York Medical Repository*, Vol. XI, 1807.

is, therefore, necessary to be completely ready before you give the medicine, as the urgency of the pains will allow you but a short time afterward. Since I have adopted the use of this powder I have seldom found a case detained me more than three hours. Other physicians who have administered it concur with me in the success of its operation.

The *modus operandi* I feel incompetent to explain. At the same time that it augments the action of the uterus, it appears to relax the rigidity of the contracted muscular fibers. May it not produce the beneficial effects of bleeding without inducing that extreme debility, which is always consequent upon copious depletion? This appears to be corroborated by its nauseating effects on the stomach, and the known sympathy between this viscus and the uterus.

It is a vegetable, and appears to be a spurious growth of rye. On examining a granary where rye is stored, you will be able to procure a sufficient quantity from among that grain. Rye which grows in low, wet ground, yields it in greatest abundance. I have no objections to your giving this any publicity you may think proper.

The Use of Ergot

In 1822, fifteen years later, Dr. Stearns published his mature conclusions in a paper called "Observation on the Secale Cornutum, or Ergot, with directions for its use in Parturition."*

It was not till the year 1807 that the ergot ever appeared before the public in a form to arrest the attention of medical men. Some years previous to this, I was informed of the powerful effects produced by this article, in the hands of some ignorant Scotch woman, in the county of Washington. Determined to try its efficacy, I produced a quantity from a field of rye. My information was such as to impress upon my mind the necessity of extreme caution in my first experiments. The continued influence of this impression upon my subsequent practice has been a source of much consoling reflection. It has tended to prevent those fatal errors which have so often occurred, and which, I trust, will be satisfactorily explained in the ensuing remarks. . . .

The publication of my letter to Dr. Ackerly, in 1807, produced an immense number of applications from remote practitioners. I

* *J. Med. Rec.*, Vol. V, 1822.

immediately forwarded to each samples of the ergot, with directions for its use. . . . The success of the ergot is in no case more evident than in the selection of a suitable time of its exhibition. Although often given to procure abortion, it does not appear to have succeeded. It also generally fails to complete success when given in the early stages of labour, and before the os uteri is sufficiently dilated and relaxed. . . . I will now proceed to consider those indications which render its exhibition necessary and important.

The ergot is indicated, and may be administered.

I. When, in lingering labours, the child has descended into the pelvis, the parts dilated and relaxed, the pains having ceased, or being too ineffectual to advance the labour, there is danger to be apprehended from delay, by exhaustion of strength and vital energy from haemorrhage or other alarming symptoms.

II. When the pains are transferred from the uterus to other parts of the body, or to the whole muscular system, producing general puerperal convulsion.

III. When in the early stages of pregnancy, abortion becomes inevitable, accompanied with profuse haemorrhage and feeble uterine contractions.

IV. When the placenta is retained from a deficiency of contraction.

V. In patients liable to haemorrhage immediately after delivery. In such cases the ergot may be given as a preventive, a few minutes before the termination of the labour.

VI. When haemorrhage or lochial discharges are too profuse immediately after delivery, and the uterus continues dilated and relaxed without any ability to contract.

XVIII

AMBROÏSE PARÉ

AMBROÏSE PARÉ (1510-1590) WAS A GREAT FIGURE IN SURGERY AND OBSTETRICS.
He is notable for having solved the first great surgical problem—the control of hemorrhage. He first gained fame as an army surgeon. The practice of military surgeons, which he was taught, was to treat wounds by pouring in boiling oil. This he abandoned for simple dressings: he found the wounds healed quicker, and with less pain. The cautery was used to seal the bleeding arteries: he abandoned this for the ligature. After the wars he lived a long and useful life as a practitioner in Paris under five French sovereigns—François I, Henri II, Charles IX, Henri III, Henri IV. He wrote many works on surgery, copying the Vesalian anatomical figures to illustrate them. His *Apologie and Treatise of Voyages* is one of the most delightful of personal books. The first of the English translations of his works was made by Thomas Johnson (d. 1644).

REFERENCES

PAGET, S. Ambroïse Paré and His Times. New York, G. P. Putnam's Sons, 1897. (One of the great medical biographies.)
PACKARD, F. R. Life and Times of Ambroïse Paré. With a Translation of his Journeys in Divers Places. New York, Paul B. Hoeber, 1921.
SINGER, D. W. Selections from the Works of Ambroïse Paré, With Short Biography. Oxford, John Bale Sons and Danielsson, 1924.

OF VVOUNDS MADE BY GVNSHOT, OTHER FIERIE ENGENINES, AND ALL
SORTS OF VVEAPONS

[From *The Works of M. Ambroïse Paré**]

I have thought good here to premise my opinion of the originall, encrease, and hurt of fiery Engines, for that I hope it will be an ornament and grace to this my whole treatise; as also to intice my Reader, as it were with these junckets, to our following Banquet so much savouring of Gunpouder. For thus it shall bee knowne to all

* Les Oevers de M. Ambroïse Paré, 1575. Translation by Thomas Johnson, "The Works of that famous Chirurgion Ambrose Parey, translated out of the Latine and compared with the French," 1634.

whence Guns had their originall, and how many habits and shapes they have acquired from poore and obscure beginnings; and lastly how hurtfull to mankind the use of them is.

Polydore Virgill writes that a Germane of obscure birth and condition was the inventor of this new engine which we terme a Gun, being induced thereto by this occasion. He kept in a mortar covered with a tyle, or slate, for some other certaine uses a pouder (which since that time for its chiefe and new knowne faculty, is named Gunpouder.) Now it chanced as hee strucke fire with a steele and flint, a sparke thereof by accident fell into the mortar, whereupon the pouder suddainly catching fire, casts the stone or tyle which covered the mortar, up on high; he stood amazed at the novelty and strange effect of the thing, and withall observed the formerly unknowne faculty of the pouder; so that he thought good to make experiment thereof in a small Iron trunke framed for the purpose according to the intention of his minde.

When all things were correspondent to his expectation, he first shewed the use of his engine to the Venetians, when they warred with the Genovese at Fosse Clodiana, now called Chiozzia, in the yeare of our Lord 1380. Yet in the opinion of Peter Messias, their invention must have beene of greater antiquity; for it is read in the Chronicles of Alphonsus the eleaventh King of Castile, who subdued the isles Argezires, that when he beseiged the cheefe Towne in the yeare of our Lord 1343, the beseiged Moores shot as it were thunder against the assailants, out of iron mortars. But we have read in the Chronicles written by Peter Bishop, of Leons, of that Alphonsus who conquered Toledo, that in a certain sea fight fought by the King of Tunis, against the Moorish King of Sivill, whose part King Alphonsus favoured, the Tunetans cast lightning out of certaine hollow Engines or Trunkes with much noise, which could be no other, than our Guns, though not attained to that perfection of art and execution which they now have.

I think the deviser of this deadly Engine hath this for his recompence, that his name should be hidden by the darkenesse of perpetuall ignorance, as not meriting for this his most pernicious invention, any mention from posterity. Yet Andrew Thevet in his Cosmography published some few years agone, when hee comes to treate of the Suevi, the inhabitants of Germany, brings upon the authority and credite of a certaine old Manuscript, that the Germane the inventor of this warlike Engine was by profession a monke and Philosopher or Alchymist, borne at Friburge, and named Constantine Ancken. Howsoever it was, this kind of Engine was called Bombarda, i.e. a Gun, from that noise it makes, which the Greekes and Latines according to the sound call Bombus; then in the following ages, time, art, and mans maliciousnesse added much to this rude and unpolisht invention. For first for the matter, Brasse and Copper, mettalls farre more tractable, fusible and lesse subject to rust, came as supplies to Iron. Then for the forme, that rude and undigested barrell, or mortar-like-basse, hath undergone many formes and fashions, even so farre as it is gotten upon wheeles, that so it might run not onely from the higher ground, but also with more rapide violence to the ruine of mankinde; when as the first and rude mortars seemed not to be so nimbly traversed, nor sufficiently cruell for our destruction by the onely casting forth of Iron and fire.

Hence sprung these horrible monsters of Canons, double Canons, Bastards, Musquits, feild peices; hence these cruell and furious beasts, Culverines, Serpentines, Basilisques, Sackers, Falcons, Falconets, and divers other names not only drawne from their figure and making, but also from the effects of their cruelty. Wherefore certainly I cannot sufficiently admire the wisedome of our Ancestors, who have so rightly accommodated them with names agreeable to their natures; as those who have not onely taken them from the swiftest birds of prey, as Falcons; but also from things most harmefull and hatefull to mankinde, such as Serpents, Snakes, and Basilisks. That so wee might clearly discerne, that these engines were made for no other purpose, nor with other intent, but onely to be imployed for the speedy and cruell slaughter of men; and that by onely hearing them named we might detest and abhorre them, as pernicious enemies of our lives. I let passe other engines of this ofspring, being for their quantitie small, but so much the more pernicious and harmfull, for that they nearer assaile our lives, and may trayterously and forthwith seaze upon us not thinking nor fearing any such thing, so that we can scarse have any means of escape; such are Pistolls and other small hand guns, which for shortnesse you may carry in your pocket, and so privily and suddainly taking them forth oppresse the carelesse and secure.

THE FIRST DISCOVRSE

VVHEREIN VVOUNDS MADE BY GVNSHOT ARE FREED FROM BEING BVRNT,
OR CAVTERIZED ACCORDING TO VIGOES METHODE

[From *The Works of M. Ambroïse Paré**]

Fowling pieces which men usually carry upon their shoulders, are of the middle ranke of these engines, as also Muskets and Caleevers, which you cannot well discharge unlesse lying upon a Rest, which therefore may be called Breast-guns for that they are not laid to the cheeke, but against the Breast by reason of their weight and shortness; All which have beene invented for the commodity of footemen, and light horsemen. This middle sort of engine we call in Latine by a generall name Sclopus, in imitation of the sound, and the Italians who terme it Sclopetere; the French call it Harquebuse, a word. likewise borrowed from the Italians, by reason of the touch-hole by which you give fire to the peice, for the Italians call a hole Buzio.

In the year of our Lord 1536, Francis the French King, for his acts in warre and peace stiled the Great, sent a puissant Army beyond the Alpes, under the Government and leading of Annas of Mommorancie high Constable of France, both that he might releeve Turin with victualls, souldiers, and all things needefull, as also to recover the Cities of that Province taken by the Marquis du Guast, Generall of the Emperours forces. I was in the Kings Army the Chirurgion of Monsieur of Montejan, Generall of the foote. The Imperialists had taken the straits of Suze, the Castle of Villane, and all the other passages, so that the Kings army was not able to drive them from their fortifications but by fight. In this conflict there were many wounded on both sides with all sorts of weapons but cheefely with bullets. I will tell the truth, I was not very expert at that time in matters of Chirurgery; neither was I used to dresse wounds made by gunshot. Now. I had read in Iohn de Vigo that wounds made by Gunshot were venerate or poisoned, and that by reason of the Gunpouder; Wherefore for their cure, it was expedient to burne or

---

* Les Oevers de M. Ambroïse Paré, 1575. Translation by Thomas Johnson, "The Works of that famous Chirurgion Ambrose Parey, translated out of the Latine and compared with the French," 1634.

cauterize them with oyle of Elders scalding hot, with a little Treacle mixed therewith.

But for that I gave no great credite neither to the author, nor remedy, because I knew that caustickes could not be powred into wounds, without excessive paine; I, before I would runne a hazard, determined to see whether the Chirurgions, who went with me in the army, used any other manner of dressing to these wounds. I observed and saw that all of them used that Method of dressing which Vigo prescribes; and that they filled as full as they could, the wounds made by Gunshot with Tents and pledgets dipped in the scalding Oyle, at the first dressings; which encouraged me to doe the like to those, who came to be dressed of me.

It chanced on a time, that by reason of the multitude that were hurt, I wanted this Oyle. Now because there were some few left to be dressed, I was forced, that I might seeme to want nothing, and that I might not leave them undrest, to apply a digestive made of the yolke of an egg, oyle of Roses, and Turpentine. I could not sleep all that night, for I was troubled in minde, and the dressing of the precedent day, (which I judged unfit) troubled my thoughts; and I feared that the next day I should finde them dead, or at the point of death by the poyson of the wound, whom I had not dressed with the scalding oyle. Therefore I rose early in the morning. I visited my patients and beyond expectation, I found such as I had dressed with a digestive onely, free from vehemencie of paine to have had good rest, and that their wounds were not inflamed, nor tumifyed; but on the contrary the others that were burnt with the scalding oyle were feaverish, tormented with much paine, and the parts about their wounds were swolne. When I had many times tryed this in divers others I thought this much, that neither I nor any other should ever cauterize any wounded with Gun-shot.

XIX

THOMAS SYDENHAM

THOMAS SYDENHAM (1624-1689), KNOWN AS THE ENGLISH HIPPOCRATES, was the best clinical observer of the seventeenth century. Educated at Oxford and Montpellier he fought in the Civil War on the Puritan side. He enjoyed a large practice in London. His first-hand descriptions of gout, measles, scarlet fever, bronchopneumonia, dysentery, chorea, and hysteria are classics. His *Processus integri* (1692) contains his general philosophy of disease and its treatment. Garrison states that it was "the *vade mecum* of the English practitioner for more than a century, and an Oxford enthusiast is said to have committed it to memory."

REFERENCES

PAYNE, J. F. Thomas Sydenham. Masters of Medicine series. London, T. Fisher Unwin, 1900.
Selections from the Writings of Thomas Sydenham. *Medical Classics,* Vol. 4, No. 4, December, 1939.

ON THE MEASLES

[From *Processus integri,** Chapter XIV]

The measles generally attack children. On the first day they have chills and shivers, and are hot and cold in turns. On the second they have the fever in full—disquietude, thirst, want of appetite, a white (but not a dry) tongue, slight cough, heaviness of the head and eyes, and somnolence. The nose and eyes run continually; and this is the surest sign of measles. To this may be added sneezing, a swelling of the eyelids a little before the eruption, vomiting and diarrhoea with green stools. These appear more especially during teething time. The symptoms increase till the fourth day. Then—or sometimes on the fifth—there appear on the face and forehead small red spots, very like the bites of fleas. These increase in number, and cluster together, so as to mark the face with large red blotches. They are

---

* Translated from the Latin by R. G. Latham, M.D. Sydenham Society, 1848.

formed by small papulae, so slightly elevated above the skin, that their prominence can hardly be detected by the eye, but can just be felt by passing the fingers lightly along the skin.

2. The spots take hold of the face first; from which they spread to the chest and belly, and afterwards to the legs and ankles. On these parts may be seen broad, red maculae, on, but above, the level of the skin. In measles the eruption does not so thoroughly allay the other symptoms as in small-pox. There is, however, no vomiting after its appearance; nevertheless there is slight cough instead, which, with the fever and the difficulty of breathing, increases. There is also a running from the eyes, somnolence, and want of appetite. On the sixth day, or thereabouts, the forehead and face begin to grow rough, as the pustules die off, and as the skin breaks. Over the rest of the body the blotches are both very broad and very red. About the eighth day they disappear from the face, and scarcely show on the rest of the body. On the ninth, there are none anywhere. On the face, however, and on the extremities—sometimes over the trunk—they peel off in thin, mealy squamulae; at which time the fever, the difficulty of breathing, and the cough are aggravated. In adults and patients who have been under a hot regimen, they grow livid, and afterwards black.

> ℞   Pectoral decoction, Oiss;
>   Syrup of violets,
>   Syrup of maidenhair, aa ℥iss.

Mix, and make into an apozem. Of this take three or four ounces three or four times a day.

> ℞   Oil of sweet almonds, ℥ij;
>   Syrup of violets,
>   Syrup of maidenhair, aa ℥j;
>   Finest white sugar, q.s.

Mix, and make into a linctus; to be taken often, especially when the cough is troublesome.

> ℞   Black-cherry water, ℥ij;
>   Syrup of poppies, ℥j.

Mix, and make into a draught; to be taken every night, from the first onset of the disease, until the patient recovers: the dose being increased or diminished according to his age.

3. The patient must keep his bed for two days after the first eruption.

4. If, after the departure of the measles, fever, difficulty of breathing, and other symptoms like those of peripneumony supervene, blood is to be taken from the arm freely, once, twice or thrice, as the case may require, with due intervals between. The pectoral decoction and the linctus must also be continued; or, instead of the latter, the oil of sweet almonds alone. About the twelfth day from the invasion the patient may be moderately purged.

5. The diarrhoea which follows measles is cured by bleeding.

ON ST. VITUS DANCE

[From *Processus integri*, Chapter XVI]

This is a kind of convulsion, which attacks boys and girls from the tenth year to the time of puberty. It first shows itself by limping or unsteadiness in one of the legs, which the patient drags. The hand cannot be steady for a moment. It passes from one position to another by a convulsive movement, however much the patient may strive to the contrary. Before he can raise a cup to his lips, he makes as many gesticulations as a mountebank; since he does not move it in a straight line, but has his hand drawn aside by spasms, until by some good fortune he brings it at last to his mouth. He then gulps it off at once, so suddenly and so greedily as to look as if he were trying to amuse the lookers-on.

2. Bleed from the arm to eight ounces, more or less according to age.

3. The next day give half (more or less as the age of the patient requires it) of the common potion. At evening the following should be taken:

> ℞   Black-cherry water, Zj;
>   Aqua epileptica Langii, ziij;
>   Venice treacle, 3j;
>   Liquid laudanum, mviij.
>   Make into a draught.

Repeat the cathartic every other day three times, and the paregoric on the same nights.

4. Blood must again be drawn the next day, and the catharsis re-

peated; and so, bleeding and purging must alternate, until the third or fourth time, provided only that there be sufficient time between the alternate evacuations to ensure the patient against danger.

5. On the days when there is no purging—

℞   Conserve of Roman wormwood,
     Conserve of orange-peel, āā ʒj;
     Conserve of rosemary, ʒss;
     Venice treacle (old),
     Candied nutmeg, a a ziij;
     Candied ginger, ʒj;
     Syrup of lemon-juice, q.s.

Make into an electuary, of which a portion the size of a nutmeg is to be taken every morning and at five p.m. Wash down with five spoonsfuls of the following wine:

℞   Peony-root,
     Elecampane,
     Masterwort,
     Angelica, aa ʒj;
     Rue-leaves,
     Sage,
     Betony,
     Germander,
     White horehound,
     Tops of lesser centaury, of each a handful;
     Juniper-berries, ʒvj;
     Peel of two oranges.

Slice, and steep in six pints of cold Canary wine. Strain, and lay by for use.

℞   Rue-water ʒiv;
     Aqua epileptica Langii,
     Compoind bryony-water, aa ʒj;
     Syrup of peony, ʒvj.

Mix, and make into a julep; of which four spoonsful may be taken every night at bedtime, with the addition of eight drops of spirits of hartshorn.

6. Apply to the sole of the foot the emplastrum e caranna.

7. To guard against a relapse, bleed and purge for a few days that time next year, or a little earlier.

8. It is probable that this treatment may also cure the epilepsy of adults, but I have not tried. In adults, however, the bleeding and purging should be freer, since St. Vitus's dance is a disease of tender years.

ON SCARLET FEVER

[From *Observationes medicae circa Morborum acutorum historiam et curationem,* 1676]

1. Scarlet Fever (Scarlatina) may appear at any season. Nevertheless, it oftenest breaks out towards the end of summer, when it attacks whole families at once, and more especially the infant part of them. The patients feel rigors and shiverings, just as they do in other fevers. The symptoms, however, are moderate. Afterwards, however, the whole skin becomes covered with small red maculae, thicker than those of measles, as well as broader, redder, and less uniform. These last for two or three days, and then disappear. The cuticle peels off; and branny scales, remain, lying upon the surface like meal. They appear and disappear two or three times.

2. As the disease is, in my mind, neither more nor less than a moderate effervescence of the blood, arising from the heat of the preceding summer, or from some other exciting cause, I leave the blood as much as possible to its own despumation, and to the elimination of the peccant materials through the pores of the skin. With this view, I am chary both of bloodletting and of clysters. By such remedies, I hold that a revulsion is created, that the particles inimical to the blood become more intimately mixed therewith, and, finally, that the proper movement of Nature is checked. On the other hand, I am cautious in the use of cordials. By them, the blood may be over-agitated, and so unfitted for the regular and equable separation in which it is engrossed. Besides which, they may act as fuel to fever.

I hold it, then, sufficient for the patient to abstain wholly from animal food and from fermented liquors; to keep always indoors, and not to keep always in his bed. When the desquamation is complete, and when the symptoms are departing, I consider it proper to purge the patient with some mild laxative, accommodated to his age

and strength. By treatment thus simple and natural, this ailment—
we can hardly call it more—is dispelled without either trouble or
danger: whereas, if, on the other hand, we overtreat the patient by
confining him to his bed, or by throwing in cordials, and other
superfluous and over-learned medicines, the disease is aggravated,
and the sick man dies of his doctor.

3. This, however, must be borne in mind. If there occur at the
beginning of the eruption either epileptic fits, or coma—as they
often do occur with children or young patients—a large blister must
be placed at the back of the neck, and a paregoric draught of syrup
of poppies must be administered at once. This last must be repeated
every night until he recover. The ordinary drink must be warm
milk with three parts water, and animal food must be abstained
from.

ON BASTARD PERIPNEUMONY

[From *Observationes medicae circa Morborum acutorum historiam et
curationem*, 1676]

1. As the winter comes on, and oftener still as it is going off,
and as spring is approaching, there comes to light, every year, a fever
marked with numerous peripneumonic symptoms. It attacks by
preference, the stout and fat, those who have reached, or passed,
the heyday of life, and those who are over-addicted to spirituous
liquors, brandy more especially. The blood of these men becomes
loaded, during the winter, with an accumulation of phlegmatic
humours; whilst, as spring approaches, it is excited to a new motion.
Then cough takes occasion to set in, and administers to these same
phlegmatic humours, and determines them to fall upon the lungs.
And now, if the patient shall have lived carelessly, and if he still
keep on drinking freely, the matter which has excited the cough
grows gross, blocks up the passages of the lungs, and preys upon the
whole mass of the blood in the shape of fever.

2. At the first attack, the patient is hot and cold by turns, is
giddy, and complains of a shooting pain in the head, as often as the
coughing becomes importunate. He vomits up what he drinks,
sometimes coughing, sometimes not. The urine is turbid and intensely red, the blood the blood of pleurisy. He pants for wind, and
draws his breath frequently and by jerks. If he be inclined to cough,

his head feels as if it would split, and so he describes the feeling. The whole chest is in pain, and the wheezing of the lungs may be heard by the bystanders as often as the sick man coughs, since the lung is unable to dilate itself sufficiently, and its intumescences shut up the vital passages. This intercepts the circulation; and the blood being, as it were, smothered, there shall be (as there often is with stout people), an absence of the signs of fever. This same absence of the signs of fever may also arise from the excess of phlegmatic matter, which must so clog the blood as to disable it from rising to a full and sufficient ebullition.

3. In treating this fever, I make it my business to divert from the lungs, by means of venesection, the blood which creates the suffocation, and which lights up the inflammation. The lungs themselves I clear and cool with pectoral remedies; and by the help of a cooling diet I moderate the heat of the body at large. Now when it happens, on the one hand, that this sink of phlegm is lodged in the veins, is day by day supplying fuel to the fire of inflammation, and is, in consequence, appearing to indicate a frequent repetition of venesections, whilst, on the other hand, the most careful observations that I have been able to make, have taught me that such repetitions, with patients of gross habits, and with patients who have passed the prime of life, are the origin of much mischief, and when this latter fact dissuades me from bloodletting no less than the former conditions may indicate it, I say that in such cases I purge freely, and make such purging supplementary to the venesection; a substitute which is rightly applied in those cases that will not bear a large and repeated loss of blood.

4. Hence, I proceeded as follows: I kept my patient to his bed, bled him as he lay in it, and forbid him to get up for two or three hours. This I insist upon, because, whilst all losses of blood to a certain extent, shake and weaken the frame at large, this is the method for making them most tolerable. The patient who lies in bed, will suffer less from a ten ounce bleeding; than one who is up, from a bleeding of only six or seven ounces.

5. The next day I bleed again, then miss a day, and repeat the aforesaid purging drink. And so I bleed and purge, and purge and bleed, till the patient gets well. Such days as I do not purge, I recommend the pectoral decoction, sweet oil of almonds, and the like.

Meat, and meat-broths I forbid, and, still more strongly, spirituous liquors; in the place of which, I allow the patient, as his usual drink, a ptisam of barley-water and liquorice, and, if he particularly request it, a little thin small beer.

6. By this method can we overcome that bastard peripneumony, which originates in the over-abundant collection of phlegm, accumulated during the winter, and breaking out upon the lungs; a disease wherein we must purge as well as bleed. In the true peripneumony we must not do this. The true peripneumony is of the same nature as a pleurisy, except that it affects the lungs more universally. The true peripneumony and the true pleurisy are treated alike, that is, by bloodletting; by bloodletting in preference to cooling medicines, and to medicines of any other sort whatsoever.

7. This bastard peripneumony, although it somewhat approaches a dry asthma, and that in regard to the difficulty of breathing, as well as other symptoms, is still easily to be distinguished. The bastard peripneumony has fever, the asthma none. Yet the fever of the bastard peripneumony is far less than the fever of the true.

8. Now this must be carefully noticed, *viz.* that when the patients who are struggling with the disease have been addicted to brandy and such like liquors, it will be unsafe to deprive them of the same too suddenly. It must be done by degrees. By this means there is less likelihood of the abrupt change paving the way to dropsy. The same applies to all diseases thus originating. And now, as I am speaking of brandy, I will make a remark by the way. It were well if that spirit were either wholly banished, or limited to the restoration (not the extinction) of the vital spirits. Some may go further, and propose the entire abandonment of it for internal uses, confining it to surgical cases; i.e. as a dressing for the digestion of sores or as an application in burns. For these, it bears the bell from all other remedies. It defends the underlying skin from putrefaction, and, so doing, effects a quick cure, so quick as not to wait for the naturally slow and leisurely process of digestion, and the stages of digestion. Lint dipped in brandy, and applied, immediately after the injury, to any part of the body that shall have been scalded with hot water, or singed by gunpowder, will do this, provided that, as long as the pain last, the spirit be renewed. After it has fairly ceased, once or twice a day will be sufficient.

ON PERUVIAN BARK

[From *Epidemic Diseases*, Epistle I]

17. The Peruvian bark, commonly called Jesuit's bark, has, if I rightly remember, been famous in London for the cure of intermittent fevers for upwards of five and twenty years, and that rightly. The disease in question was seldom or never cured by any remedy before it. Hence agues were justly called the opprobria medicorum. A short time back, however, it went out of use, being condemned on two grounds, and those not light ones. Firstly, when given a few hours before the fit, as was the usual practice, it would sometimes kill the patient at once. This happened to an alderman of London, named Underwood, and also to a Captain Potter. Now this terrible effect of the powder, although rare, frightened the more prudent physicians, and that rightly. Secondly, the patient who by the help of the bark had been freed from an impending fit, would, at the end of a fortnight, generally have a relapse, as if the disease was still fresh, and had not abated in violence by running its course. All this shook the generality in their good opinion of the bark, since they considered it no great gain to put off the fit by endangering the life of the patient.

18. Now for many years I have been reflecting on the remarkable powers of this bark, considering, that with care and diligence, it was really the great remedy for intermittents. Hence I looked at two things, the danger to life, and the chance of a relapse. Guard against these, and I could cure the patient perfectly.

19. In respect to the danger to life, I laid it less to the bark than to its unseasonable administration. During the days when there is no fit, a vast mass of febrile matter accumulates in the body. Now, if in this case, we give the powder just before the fit, we check the method by which nature would get rid of it; so that being kept in, it endangers life. Now this I thought I could remedy by checking the generation of any new febrile matter. Hence I gave the powder immediately after the fit. This allayed the succeeding one. Then on the days of intermission I repeated it at regular intervals, until a fresh fit impended. Thus, by degrees, I brought the blood under the healing influence of the bark.

20. The relapse, which generally happens at the end of a fortnight, seemed to me to arise from the blood not being sufficiently
saturated with the febrifuge, which, efficient as it was, could not
exterminate the disease at once. From whence I concluded, that to
guard against this I must repeat the powder, even where the disease
was overcome for the present, at regular intervals, and before the
effects of the preceding dose had gone off.

21. On these principles my method was and is as follows: If I
visit a patient on (say) a Monday, and the ague be a quartan, and
it be expected that day, I do nothing, I only hope that he will escape
the fit next after. Then on the two days of intermission, the Tuesday
and Wednesday, I exhibit the bark thus:

> Rx Peruvian bark, very finely powdered, ounces one;
>   Syrup of cloves, or
>   Syrup of dried rose-leaves, q.s.
> Make into an electuary; to be divided into twelve parts,
>   of which one is to be taken every fourth hour,
>   beginning immediately after the paroxysm, and
>   washing down with a draught of wine.

If form of pill be preferred—

> Rx Bark, finely powdered, ounces one;
>   Syrup of cloves, q.s.
> Make into moderately-sized pills. Take six every four hours.

With less trouble and equal success you may mix an ounce of
bark with two pints of claret, and give it as before, in doses of eight
or nine spoonsful. On Thursday the fit is expected. I do nothing;
generally the fit keeps off. The remnants of the febrile matter have
been cleared away, and thrown off from the blood, by the sweats of
Monday's fit, and new accumulations have been checked, by the use
of the bark in the interval.

22. To prevent the disease from returning on the eighth day,
exactly, after the last dose, I give another exactly as before. Now
though this often puts an end to the ague, the patient is all the safer
for repeating the process three or even four times, especially if the
blood be weakened by the previous evacuation, or the patient have
exposed himself to the cold air.

On Gout

[From *Tractatus de Podagra et de Hydrope*, 1683]*

1. Either men will think that the nature of gout is wholly mysterious and incomprehensible, or that a man like myself who has suffered from it thirty-four years, must be of a slow and sluggish disposition not to have discovered something respecting the nature and treatment of a disease so peculiarly his own. Be this as it may, I will give a bona fide account of what I know. The difficulties and refinements relating to the disease itself, and the method of its cure, I will leave for Time, the guide to truth, to clear up and explain.

2. Gout attacks such old men as, after passing the best part of their life in ease and comfort, indulging freely in high living, wine, and other generous drinks, at length, from inactivity, the usual attendant of advanced life, have left off altogether the bodily exercises of their youth. Such men have generally large heads, are of a full, humid, and lax habit, and possess a luxurious and vigorous constitution, with excellent vital stamina.

3. Not that gout attacks these only. Sometimes it invades the spare and thin. Sometimes it will not wait for the advance of age. Sometimes even the prime of life is liable to it. This happens most where there is an unhappy hereditary tendency; or, even where (without such being the case) the patient has indulged in premature venery. The omission, too of any customary violent exercise brings it on. So, also, does the sudden change from over-hearty diet in the way of meats and drink, to a low regimen and thin potations.

5. Concerning this disease, in its most regular and typical state, I will first discourse; afterwards I will note its more irregular and uncertain phenomena. These occur when the unseasonable use of preposterous medicines has thrown it down from its original status. Also when the weakness and languor of the patient prevent it from rising to its proper and genuine symptoms. As often as gout is regular, it comes on thus. Towards the end of January or the beginning of February, suddenly and without any premonitory feelings, the disease breaks out. Its only forerunner is indigestion and crudity of the stomach, of which the patient labours some weeks before. His body feels swollen, heavy, and windy—symptoms which increase

* Translated by R. G. Latham, Sydenham Society, 1848.

until the fit breaks out. This is preceded a few days by torpor and a feeling of flatus along the legs and thighs. Besides this, there is a spasmodic affection, whilst the day before the fit the appetite is unnaturally hearty. The victim goes to bed and sleeps in good health. About two o'clock in the morning he is awakened by a severe pain in the great toe; more rarely in the heel, ankle, or instep. This pain is like that of a dislocation, and yet the parts feel as if cold water were poured over them. Then follows chills and shivers, and a little fever. The pain, which was at first moderate, becomes more intense. With its intensity the chills and shivers increase. After a time this comes to its height, accommodating itself to the bones and ligaments of the tarsus and metatarsus. Now it is a violent stretching and tearing of the ligaments—now it is a gnawing pain, and now a pressure and tightening. So exquisite and lively meanwhile is the feeling of the part affected, that it cannot bear the weight of the bedclothes nor the jar of a person walking in the room. The night is passed in torture, sleeplessness, turning of the part affected, and perpetual change of posture; the tossing about of the body being as incessant as the pain of the tortured joint, and being worse as the fit comes on. Hence the vain efforts, by change of posture, both in the body and the limb affected, to obtain an abatement of the pain. This comes only towards the morning of the next day, such time being necessary for the moderate digestion of the peccant matter. The patient has a sudden and slight respite, which he falsely attributes to the last change of position. A gentle perspiration is succeeded by sleep. He wakes freer from pain, and finds the part recently swollen. Up to this time, the only visible swelling had been that of the veins of the affected joint. Next day (perhaps for the next two or three days), if the generation of the gouty matter have been abundant, the part affected is painful, getting worse towards evening and better towards morning. A few days after, the other foot swells, and suffers the same pains. The pain in the foot second attacked regulates the state of the one first attacked. The more it is violent in the one, the more perfect is the abatement of suffering, and the return of strength in the other. Nevertheless, it brings on the same affliction here as it had brought on in the other foot, and that the same in duration and intensity. Sometimes, during the first days of the disease, the peccant matter is so exuberant, that one foot is insufficient for its discharge. It then attacks both, and that with equal violence. Generally, however, it takes the feet in succession. After it has attacked each foot, the fits become irregular, both as to the time of their accession and duration. One thing, however, is constant—the pain increases at night and remits in the morning. Now a series of lesser fits like these constitute a true attack of gout—long or short, according to the age of the patient. To suppose that an attack two or three months in length is all one fit is erroneous. It is rather a series of minor fits. Of these the latter is milder than the former, so that the peccant matter is discharged by degrees, and recovery follows. In strong constitutions, where the previous attacks have been few, a fortnight is the length of an attack. With age and impaired habits gout may last two months. With very advanced age, and in constitutions very much broken down by previous gout, the disease will hang on till the summer is far advanced. For the first fourteen days the urine is high-coloured, has a red sediment, and is loaded with gravel. Its amount is less than a third of what the patient drinks. During the same period the bowels are confined. Want of appetite, general chills towards evening, heaviness, and a troublesome feeling at the parts affected, attend the fit throughout. As the fit goes off, the foot itches intolerably, most between the toes; the cuticle scales off, and the feet desquamate, as if venomed. The disease being disposed of, the vigour and appetite of the patient return, and this in proportion to the violence of the last fits. In the same proportion the next fit either comes on or keeps off. Where one attack has been sharp, the next will take place that time next year—not earlier.

6. This is gout with its true and regular phenomena. When, however, either undue treatment or the prolonged delay of the disease has converted the whole body into a focus for the peccant matter, and when Nature is incompetent to its elimination, its course is different. The true seat of the disease is the foot—so much so, that when it appears elsewhere its character is changed, or else the constitution is weak. Then, however, it attacks the hands, wrists, elbows, knees, and other parts, the pains being as the pains of the feet. Sometimes it distorts the fingers, then they look like a bunch of parsnips, and become stiffened and immovable. This is from the deposit of chalkstone concretions about the ligaments of the knuckles. The effect of these is to destroy the skin and cuticle. Then you have chalk-stones like crabs' eyes exposed to view, and you may

turn them out with a needle. Sometimes the morbific matter fixes on the elbows, and raises a whitish tumour almost as large as an egg, which gradually grows red and inflamed. Sometimes the thigh feels as if a weight were attached to it, without however any notable pain. It descends however to the knee, and then the pain is intense. It checks all motion, nails the patient down to his bed, and will hardly allow him to change his posture a hair's breadth. Whenever, on account of the restlessness so usual in the disease, or from any urgent necessity, the patient has to be moved, the greatest caution is necessary. The least contrary movement causes pain, which is tolerable only in proportion as it is momentary. This movement is one of great troubles in gout, since, with perfect quiet, the agony is just tolerable.

9. Other symptoms arise—piles amongst others. Also indigestion, with rancid tastes in the mouth, whenever anything indigestible has been swallowed. The appetite fails, so does the whole system. The patient has no enjoyment of life. The urine, no longer high-coloured, is pale and copious, like the urine of diabetes. The back and other parts itch,—most at bedtime.

13. The body is not the only sufferer, and the dependent condition of the patient is not his worst misfortune. The mind suffers with the body; and which suffers most it is hard to say. So much do the mind and reason lose energy, as energy is lost by the body, so susceptible and vacillating is the temper, such a trouble is the patient to others as well as to himself, that a fit of gout is a fit of bad temper. To fear, to anxiety, and to other passions, the gouty patient is the continual victim, whilst as the disease departs the mind regains tranquillity.

15. For humble individuals like myself, there is one poor comfort, which is this, viz. that gout, unlike any other disease, kills more rich men than poor, more wise men than simple. Great kings, emperors, generals, admirals, and philosophers have all died of gout. Hereby Nature shows her impartiality: since those whom she favors in one way she afflicts in another—a mixture of good and evil pre-eminently adapted to our frail mortality.

20. In respect to the treatment, if we look to the humours themselves, and the indigestion from which they arise, it seems at first that we have to evacuate the aforesaid humours, and to guard against their increase by strengthening the concoctions; which is only what

is to be done in all humoural complaints. In gout, however, it seems as if it were the prerogative of Nature to exterminate the peccant matter after her own fashion, to deposit it in the joints, and afterwards to void it by insensible perspiration. In gout, too, but three methods have been proposed for the ejection of the causa continens —bleeding, purging, sweating. Now none of these succeed.

21. In the first place bleeding, however much it may promise great things, both in the evacuation of those humours which already have attacked the joints, and those which are ready to attack them, is still clearly contrary to that indication which is required by the antecedent cause. This is indigestion arising from the deprivation and defect of the spirits,—a deprivation and defect which blood-letting increases. Hence it is not to be applied either to ease a fit, or to guard against one, especially with old people, not even though the blood be that of pleurisy and rheumatism, diseases wherein bleeding does so much good; inasmuch as, if blood be taken during an intermission, however long after a fit, there is danger lest the agitation of the blood and humours bring on a fresh one, worse than the one that went before it. This is because the strength and vigour of the blood, that might serve to get rid of the peccant matter that supplies the disease, are weakened. If, on the other hand, a vein be opened soon after a fit, there is great risk lest Nature, whilst the blood is still weak, be so broken down as to open the door for a dropsy. Nevertheless, if the patient be young and have drunk hard, blood may be drawn at the beginning of the fit. If, however, it be continued during the following fits, gout will take up its quarters even in a young subject, and its empire will be no government, but a tyranny.

XX

MARCELLO MALPIGHI

MARCELLO MALPIGHI (1628-1694), PROFESSOR OF ANATOMY AT BOLOGNA, made most important discoveries in histological anatomy. His work in embryology, *De formatione pulli in ovo* (1673), described the aortic arches, the neural groove, the cerebral and optic vesicles. He described the red blood corpuscles, the rete mucosum or Malpighian layer of the skin, the structure of the liver, the kidneys and the spleen. His greatest contribution to science was the description of the capillaries in the lungs (*De pulmonibus*, 1661). This completed Harvey's work showing how the blood got from the arteries to the veins. According to Fraser Harris, "Harvey made their existence (i.e., the capillaries) a logical necessity; Malpighi made it an histological certainty."

ON THE LUNGS*

[From *De pulmonibus, Observationes anatomicae*, Epistle II]

And now, most famous man, I will handle the matter more closely. There were two things which, in my epistle about observations on the lungs, I left as doubtful and to be investigated with more exact study.

(1) The first was what may be the network described therein, where certain bladders and sinuses are bound together in a certain way in the lungs.

(2) The other was whether the vessels of the lungs are connected by mutual anastomosis, or gape into the common substance of the lungs and sinuses.

The solution of these problems may prepare the way for greater things and will place the operations of nature more clearly before the eyes. For the unloosing of these knots I have destroyed almost the whole race of frogs, which does not happen in that savage Batrachomyomachia of Homer. For in the anatomy of frogs, which, by favour of my very excellent colleague D. Carolo Fraccasato, I

---

* From De pulmonibus, Observationes anatomicae, Bologna, 1661. Translation by James Young, *Proc. Roy. Soc. Med.*, 1929. By permission.

had set on foot in order to become more certain about the membranous substance of the lungs, it happened to me to see such things that not undeservedly I can better make use of that [saying] of Homer for the present matter ——

"I see with my eyes a work trusty and great!"

Observations by means of the microscope will reveal more wonderful things than those viewed in regard to mere structure and connexion: for while the heart is still beating the contrary (i.e., in opposite directions in the different vessels) movement of the blood is observed in the vessels—though with difficulty—so that the circulation of the blood is clearly exposed. This is more clearly recognized in the mesentery and in the other great veins contained in the abdomen.

Thus by this impulse the blood is driven in very small [streams] through the arteries like a flood into the several cells, one or other branch clearly passing through or ending there. Thus blood, much divided, cuts off its red colour, and, carried round in a winding way is poured out on all sides till at length it may reach the walls, the angles, and the absorbing branches of the veins.

The power of the eye could not be extended further in the opened living animal, hence I had believed that this body of the blood breaks into the empty space, and is collected again by a gaping vessel and by the structure of the walls. The tortuous and diffused motion of the blood in divers directions, and its union at a determinate place offered a handle to this. But the dried lung of the frog made my belief dubious. This lung had, by chance, preserved the redness of the blood in (what afterwards proved to be) the smallest vessels, where by means of a more perfect lens, no more there met the eye the points forming the skin called Sagrino, but vessels mingled annularly. And, so great is the divarication of these vessels as they go out, here from a vein, there from an artery, that order is no longer preserved, but a network appears made up of the prolongations of both vessels. This network occupies not only the whole floor, but extends also to the walls, and is attached to the outgoing vessel, as I could see with greater difficulty but more abundantly in the oblong lung of a tortoise, which is similarly membranous and transparent. Here it was clear to sense that the blood flows away through the tortuous vessels, that it is not poured into spaces but always works through tubules, and is dispersed by the multiplex winding of the vessels. Nor is it a new practice of Nature to join together the extremities of vessels, since the same holds in the intestines and other parts; nay, what seems more wonderful, she joins the upper and the lower ends of veins to one another by visible anastomosis, as the most learned Fallopius has very well observed.

But in order that you may more easily get hold of what I have said, and follow it with your own sight, tie with a thread, just where it joins the heart, the projecting swollen lung of an opened frog while it is bathed on every side with abundant blood. This, when dried, will preserve the vessels turgid with blood. You will see this very well if you examine it by the microscope of one lens against the horizontal sun. Or you may institute another method of seeing these things. Place the lung on a crystal plate illuminated below through a tube by a lighted candle. To it bring a microscope of two lenses, and thus the vessels distributed in a ring-like fashion will be disclosed to you. By the same arrangement of the instruments and light, you will observe the movement of the blood through the vessels in question. You will yourself be able to contrive it by different degrees of light, which escape description by the pen. About the movement of the blood, however, one thing shows itself, worthy of your speculation. The auricle and the heart being ligatured, and thus deprived of motion and the impulse which might be derived from the heart into the connected vessels, the blood is still moved by the veins towards the heart so that it distends the vessels by its effort and copious flow. This lasts several hours. At the end, however, especially if it is exposed to the solar rays, it is agitated, not by the same continued motion, but, as if impelled by changing impulses, it advances and recedes fluctuating along the same way. This takes place when the heart and auricle are removed from the body.

From these things, therefore, as to the first problems to be solved, from analogy and the simplicity which Nature uses in all her operations, it can be inferred that that network which formerly I believed to be nervous in nature, mingled in the bladder and sinuses, is [really] a vessel carrying the body of blood thither or carrying it away. Also that, although in the lungs of perfect animals the vessels seem sometimes to gape and end in the midst of the network of rings, nevertheless, it is likely that, as in the cells of frogs and tortoise, that vessel is prolonged further into small vessels in the form of a network, and these escape the senses on account of their exquisite smallness.

Also from these things can be solved with the greatest probability the question of the mutual union and anastomosis of the vessels. For if Nature turns the blood about in vessels, and combines the ends of the vessels in a network, it is likely that in other cases an anastomosis joins them; this is clearly recognized in the bladder of frogs swollen with urine, in which the above described motion of the blood is observed through the transparent vessels joined together by anastomosis, and not that those vessels have received that connexion and course which the veins or fibres mark out in the leaves of nearly all trees.

To what purpose all these things may be made, beyond those which I dealt with in the last letter concerning the pulmonary mixing of the blood, you yourself seemed to recognize readily, nor is the opinion to be lessened by your very famous device, because by your kindness you have entrusted me with elaborate letters in which you philosophised subtly by observing the strange portents of Nature in vegetables, when we wonder that apples hang from trunks not their own, and that by grafting of plants the processes have produced bastards in happy association with legitimates. We see that one and the same tree has assumed diverse fashions in its branches,— while here the hanging fruits please the taste by a grateful acidity, there they fulfil every desire by their nectar-like sweetness, and you furnish credibility to the truth at which you wondered when in Rome, that the vine and the jasmine had come forth from the bole of the Massilian apple. He who cultivated the gardens with a light inserted fork made these clever things with bigger branches, and he taught the unreluctant trees the bringing forth of divers things. About this matter Virgil in the Georgics fitly sang:—

> "They ingraft the sprout from the alien tree
> And teach it to grow from the moist inner bark."

I have put these few little observations into a letter that I might increase the things found out about the lungs. You will bring out the truth and dignity of these matters by your authority and contrivance. Meantime, apply yourself happily to philosophy, and may you go on to render me altogether happy by increasing a little my very unimportant thoughts of your writings "De Animalium Motu." Farewell!

Concerning the Structure of the Kidneys*

[From *De Viscerum Structura Exercitatio Anatomica*]

In *De Viscerum Structura Exercitatio Anatomica,* published at Bonn in 1666, Malpighi described the structure and functions of the kidneys. Aristotle thought the bladder secreted the urine. Vesalius stated that the ureters carried the urine from the kidneys to the bladder. Observe the modern note of Malpighi's description: he used stains, and injected the arteries and veins.

For a long time the kidneys have been the subject of varying opinions, some even having regarded them as superfluous and unnecessary, a thought which is certainly not a tribute to Nature. More recently, however, because of their wonderful structure, and because of the very necessary function attributed to them, they have attained a place among the important parts of the body. So many different views regarding their composition are held by anatomists that there is little agreement.

The ancients conceived of a sieve which provided a means for separating the urine. Many have been satisfied simply with the name "parenchyma." In the meantime, the idea of fibers for drawing out the fluid pleased some, and this idea was strengthened by the similar structure of the heart. Among subsequent writers the existence of fibers in the kidney appeared doubtful and unlikely, whereupon they announced that when the substance of the kidney was cut, certain little canals were to be seen. Later some have contended that the substance of the kidneys is complex, parenchyma certainly, and fibers. And still more recently it has been stated in a very elaborate work that the substance of the kidney consists of a single fibrous substance, permeated by little canals. This was determined from cut sections in which, everything but vessels having been excluded, it was evident that the body of the kidneys consists of nothing but a collection of little canals or channels which increase in size uninterruptedly from the external surface toward the center.

The fact that the human mind has pondered these and similar

---

* Translation by Dr. J. M. Hayman, Jr., *Ann. Med. Hist.,* Vol. VII, No. 3, Sept., 1925. By permission of Dr. Hayman.

ideas about the kidneys through the ages stimulated me to further investigation, or at least to the confirmation of the statements of others. Study with me, then, a few things in the spirit of truth alone, so that we may establish the manner of Nature's operations in the individual viscera as I have revealed it in the liver and other organs. For this essay which I plan will perhaps shed light upon the structure of the kidneys. Do not stop to question whether these ideas are new or old, but ask, more properly, whether they harmonize with Nature. And be assured of this one thing, that I never reached my idea of the structure of the kidneys by the aid of books, but by the long, patient, and varied use of the microscope. I have gotten the rest by the deductions of reason, slowly, and with an open mind, as is my custom.

In man and other similar animals, the kidney is universally described as everywhere smooth; nevertheless, occasionally in the adult and often in the fetus its true formation is made clear; how it is pressed together from small portions which are separated by deep furrows. And in the adult indications of these divisions remain forever in the interior. By diligent research I have observed a rough and even a subdivided surface of the adult kidney. But to see this more plainly, the kidneys of new-born animals must be examined. In these, many-sided lobules stand out, surrounded by little furrows, which seem generally to be obliterated with the increasing age of the animal. Vestiges of these furrows, however, persist, and may be distinguished by a different color if liquid is injected into the artery or vein. Under these conditions they appear red, and the stretched-out lobules, which retain the milder flesh color, are surrounded by them.

These subdivisions of the kidney are found not only on the surface, but they extend inward and produce some division of flesh. In cattle and the higher domestic animals, and even in man himself, these remarkable little bodies (surrounded in some animals by conspicuous furrows) have a definite shape and exact delineation in the interior of the kidneys. They seem to form many-sided pyramids which are seen better in the turtle's kidney than in others, and less clearly in the human. Thus in these little aggregations one finds altogether the same subdivisions, which creep through the deeper substances of the kidney, described in man, the dog and the cat. The branches of the blood vessels which run through the outer portion of the kidney, and which arise from the renal vessels at their entrance, as is clearly seen in the cat, help and favor this division to some extent. These vessels follow along the cracks with remarkable little twigs, and more deeply embrace the subdivisions. The vessels supplying the deeper parts of the kidney, when they reach the outer surface are reflected and bound these same spaces.

Nor does this discontinuous structure of the kidney stop with the external surface, for when the capsule has been recently torn off, while the kidney substance is still soft, certain very small round bodies, like a coil of small worms, are observed, not unlike those which are found in the substance of the tests when the capsule is stripped off, or when it is .cut open through the middle. If ink is poured over the surface and immediately wiped off gently, these coils are brought out beautifully by the microscope, and, in addition, the marvelous branches of the vessels with the "globules" attached to them (of which more below) are now and then seen lying beneath the outer surface, and occasionally in compressed spaces. These worm-like prolongations of the vessels are clearly seen in the curved part of the kidneys of dogs and similar animals where they form furrows and so accommodate the blood vessels which extend there. After short, sharp bendings and convolutions, close to the outer surface, these worm-like vessels run in a straight course towards the pelvis. I have sometimes seen this when the outer surface of the kidneys has been broken with the hands only. These worm-like bodies, however, are seen better in the convexity of the bird's kidney, where they are surrounded by blood vessels. From these considerations it is clear that the extreme ends of these worm-like vessels, which form the outer part of the kidney, are continuous with the descending vessels which go to the pelvis. I confess that I have never been able to demonstrate this continuity, either on account of my clumsiness, perchance, or on account of the crudity of my instruments. Of course both reason and sense, aided by research, confirm the fact that certain continuous, tortuous spaces and little furrows (this established by very recent observations) run through the whole outer part of the kidneys, and these are outlined when ink is injected through the renal vessels. For by the injection of ink, or by some similar means, many irregular meshes appear, since the liquid makes paths of itself through the blood vessels. Sometimes these meshes are seen to be hexagonal and to surround the smaller lobules of the kidney. And, as has often been described, the different branches of the blood vessels, bifurcating here and there, are also seen. A different color is produced by the injected fluid and the propelled blood. On this account those who are experienced maintain that in a section of the human cadaver the color of the kidneys will often vary. For a red color, very frequently clear and bright, has been noticed by them; sometimes it is dark and black, and not seldom grey; all of these variations are caused by different quantities of blood, or its absence.

Among the vessels of the kidney the ureters seem to hold not the least important place. But about their origin and final insertion many may disagree. I will add this one thing to their arguments, to show how far indeed some were deceived by the idea of the pelvis running through the body of the kidney with the other vessels: namely, that the pelvis is an extension of the ureter, formed by the same membranes and sinew-like fibers of which the ureter is composed. In animals in which the urinary vessels end in a papillary body, a peculiar, nearly oval expansion of the ureter is observed in the middle part of the kidney, which in the region of the hilus forms certain tube-like appendages to the papillae. These, prolonged further towards the convexity, are bent back to the sides, as if forming an infundibulum, which is accepted by many as a second cavity or sieve, and surrounded closely one or another papilla projecting from them. From these divisions of the pelvis further branches go out, which follow the ramifications of the blood vessels and form exactly the same networks of these little tubes which we have said are made by the arteries and veins. From the terminations of the pelvis, which are formed into an arch, innumerable membranous, or if you prefer, sinuous fibers are prolonged almost to the convexity, whose terminations I have never been able to reach.

The constant trickle of urine from the kidneys through the ureters, thence carried to the bladder, and voided at a fixed interval, is sufficient indication of their function. But by what means this is accomplished is most obscure. It is reasonable to assume that this is wholly the result of the work of the glands: but since the minute and simple structure of the openings within the glands escapes us, we can only postulate some things in order to give a satisfactorily probable answer to this question. It is obvious that this mechanism accomplishes the work of separation of the urine by its internal arrangement. But whether this arrangement is similar to those devices which we make use of here and there for human needs, and in imitation of which we build rough contrivances, is doubtful. For although similar sponge-like bodies, structures with sieve-like fistulae, may be encountered, it is difficult to determine to which of these the structure of the kidneys is similar in all respects. And since the manifestation of Nature's working is most varied, we may discover mechanisms which are unknown to us and whose operations we cannot understand.

XXI

ANTONJ VAN LEEUWENHOEK

ANTONJ VAN LEEUWENHOEK (1632-1723), OF DELFT, HOLLAND, WAS AN enthusiastic and picturesque microscope maker and student of histology. He described bacteria in the mouth, red blood corpuscles, spermatozoa, and the capillary circulation. Most of his discoveries were published in the *Philosophical Transactions of the Royal Society of London*. The letter on the capillary circulation was sent to the Royal Society, but not published in the *Transactions*. It appeared first in the collected edition of Leeuwenhoek's works. It shares with Malpighi's work the distinction of first pointing out the capillary circulation, which was unknown to Harvey. Knowledge of its existence naturally made the mechanism of the circulation clear.

REFERENCE

DOBELL, C. Antony van Leeuwenhoek and his "Little animals." New York, Harcourt, Brace, 1932.

ON THE CAPILLARY CIRCULATION

[65th Missive]

Most Honourable Gentlemen, [etc.]

My last most respectful letter to you was of the 24th. of last month, in which I treated of the sting of the gnat, namely, that this sting, taken out of its case, consists of four distinct stings.

Herewith I again send you some of my trifling observations.

In our country we have two species of frogs. The first, which we used to have in great abundance round about our town, is commonly called frog. Of late years, however, these frogs have been very rare. I suppose because our sluggish water-courses have lately been filled, as it were, with a kind of noxious little fish (hitherto unknown, so far as I am aware), called stickle-backs, which devoured the frogs when they were still tadpoles.

I was greatly pleased to see very distinctly the circulation of the blood [in the tadpole], which was driven on from the parts that were nearest to the body to those on the outside, thus causing an uninterrupted, very rapid circulation. This circulation was not regular in its movement, but at very short intervals it was continually brought about anew with sudden impulses and before there was another sudden impulse we might (in case we had not observed a continual increase in the rapidity) have thought that a stoppage in the circulation would follow. But scarcely had the blood begun to move more slowly, when there was again a sudden impulse of the blood, so that there was an uninterrupted current; and trying accurately to measure the very short time in which each impulse took place, I found that in the time wanted to count rapidly to a hundred, there were as many as a hundred sudden impulses. From this I concluded that as often as these sudden impulses occurred, the blood was driven from the heart.

In another place I saw that three of the thinnest arteries, each running in a curve, all met together in one point and there formed a blood-vessel or vein, and consequently this blood-vessel was as wide as the three arteries mentioned. These three distinct vessels with their somewhat circular course, in which the circulation took place, were so small that a grain of sand could have covered them.

Such blood vessels running across each other I often noticed before when I tried to discover the junction of arteries and veins in other animals, but I was quite certain that the return circuit of the blood does not take place in the large vessels, but in the smallest or thinnest, for if it were otherwise, I concluded that all the parts of the body could not be fed. And as these discoveries seemed inscrutable to me, I gave up my investigations on this head for some years. If now we see clearly with our eyes that the passing of the blood from the arteries into the veins, in the tadpoles, only takes place in such blood-vessels as are so thin that only one corpuscle can be driven through at one time, we may conclude that the same thing takes place in the same way in our bodies as well as in that of all animals. And this being so, it is impossible for us to discover the passing of the blood from the arteries into the veins in our bodies or that of other animals, first because a single globule of blood being in a vein, has no colour; secondly because the blood does not move in the blood-vessels when we make this investigation.

I have said before that the corpuscles or globules that make the blood red, are so small that ten hundred thousand of them are not so big as a grain of coarse sand, and so we easily imagine how very small the blood-vessels are in which the circulation of the blood takes place.

The observations told here have not been made once, but they have been resumed repeatedly, giving me much pleasure, and every time on different tadpoles, and the result has almost always been the same. But it is remarkable that in the very small vessels mentioned above and placed furthest from the heart, as here at the end of the tail, the impulse was not by far so sudden and strong as in the vessels nearest to the heart. And though the uninterrupted current could be clearly observed, it could be distinctly seen that at each impulse from the heart the current was a little quicker.

When I looked along the length of the tail and at the thickest part of it, I could clearly see that on either side of the bone there was a large artery, through which the blood was carried to the extremity of the tail, and which on its way sent out several small branches.

When I looked at the part of the tail beside these arteries on the outside, I discovered there two large veins, which carried the blood back again to the heart, and moreover I saw that blood was driven into this large vein from several small veins. In short, I saw here the circulation of the blood to my perfect satisfaction, because there was nothing, though ever so slight, that caused me any doubt.

Also I observed the young frogs when they had changed from tadpoles into frogs and I also discovered in them a very large number of small blood-vessels which, continually running in curves, formed the vessels called arteries and veins, from which it was perfectly clear to me that the arteries and veins are one and the same continuous blood-vessels. But I saw them clearest of all and most of all at the end of the projecting parts of the legs, which we may call fingers, and of which the frog has four on each fore-leg and five on each hind-leg.

These blood-vessels (i.e. capillaries), called "arteries and veins" (being nevertheless identical) were exceedingly numerous at the ends of these fingers, and each ran in a curve, which made it impossible to follow the particular course of each vessel. All these vessels were so small or thin that no more than one corpuscle could pass through it at a time. But when I examined these fingers about the first or second joint, I found the blood-vessels there, which we call arteries and veins, bigger, so big even that the blood in these vessels had a red colour.

MEDICAL LIFE OF THE SEVENTEENTH CENTURY

MOLIÈRE (1622-1673) WROTE A NUMBER OF FAMOUS SCENES WHICH DEPICT the foibles of the doctors of his time. Some of them, such as the consultation in *L'Amour Médecin*, are eternal in their application, and their counterparts may be seen today. The medical profession in Molière's time was fair game for satire. The university doctors—the doctors of the long robe—were insufferable prigs and pedants, and all practice was formalized and bound round with customs and tradition. The real work of clinical medicine was done by the barbers, *inciseurs*, midwives, and apothecaries. The professors of medicine at the university of Paris took the following oath: "We swear and promise solemnly to give our lessons in a long black robe with wide sleeves, wearing the square cap on our head, and the scarlet shoulder-knot on our shoulder." Guy Patin's letters (Guy Patin, by Francis R. Packard, New York, Paul B. Hoeber) give a picture of the time from the viewpoint of a physician contemporary.

LOVE'S THE BEST DOCTOR*

L'Amour Médecin is said to have been suggested to Molière by a personal experience—a quarrel between his wife and the wife of a physician with whom she lodged. The date of the play's first performance (1665) is the same year that saw the beginning of Molière's long illness (tuberculosis) and his acquaintance with doctors. The plot is simple: Sganarelle has a daughter, Lucinde, whom he proposes to marry to a suitor of his choice, but she already has a lover, and objects to being bartered, so feigns a melancholy, which so alarms her parent that he calls in a consultation of physicians. Eventually Lucinde's lover, Clitandre, dresses up as a physician, is introduced into the household as a member of the faculty and declares that the only thing that will cure the patient is to make her believe she is being married. He offers to play the part of the bridegroom himself, and arranges for a real instead of a faked marriage, which is accomplished before Sganarelle finds he is duped.

We give the consultation scene. The four doctors were caricatured

* L'Amour Médecin. Translation by H. Baker and J. Miller, 1739.

from real characters well known in Paris at the time—Guy Patin says they were Guénaut, Brayer, Des Fougerais, and Valot. They attended the fatal illness of Cardinal Mazarin in 1661, wrangled and did not agree as to the cause of his trouble; Brayer said the spleen was affected, Guénaut the liver, Valot contended there was water in the chest cavity, and Des Fougerais maintained that there was an abscess in the mesentery. At a later time when Guénaut was one day entangled in a crowd of vehicles in the street, a cart driver shouted, "Let the Doctor go ahead. He's the one who did us the service to rid us of the Cardinal."

Act i

Scene vii

Sganarel. Here quick, let physicians be got, and in abundance; one can't have too many upon such an accident. Ah, my girl! My poor girl!

Act ii

Scene i

Sganarel, Lysetta

Lysetta. What will you do, sir, with four physicians? Is not one enough to kill any one body?

Sganarel. Hold your tongue. Four advices are better than one.

Lysetta. Why, can't your daughter die well enough without the assistance of these gentlemen?

Sganarel. Do the physicians kill people?

Lysetta. Undoubtedly; and I knew a man who proved by good reasons that we should never say, such a one is dead of a fever, or a catarrh, but she is dead of four doctors and two apothecaries.

Sganarel. Hush! Don't offend these gentlemen.

Lysetta. Faith, sir, our cat is lately recovered of a fall she had from the top of the house into the street, and was three days without either eating or moving foot or paw; but 'tis very lucky for her that there are no cat-doctors, for 'twould have been over with her, and they would not have failed purging her and bleeding her.

Sganarel. Will you hold your tongue, I say? What impertinence is this! Here they come.

Lysetta. Take care. You are going to be greatly edified; they'll tell you in Latin that your daughter is sick.

Scene II

Messrs. Thomès, Fonandrès, Macroton, Bahys, Sganarel, Lysetta

Sganarel. Well, gentlemen.

Mr. Thomès. We have sufficiently viewed the patient, and there are certainly a great many impurities in her.

Sganarel. Is my daughter much impure?

Mr. Thomès. I mean that there is much impurity in her body, an abundance of corrupt humours.

Sganarel. Oh! I understand you.

Mr. Thomès. But . . . We are going to consult together.

Sganarel. Come, let chairs be given.

Lysetta. (*To Mr. Thomès.*) Oh! sir, are you there?

Sganarel. (*To Lysetta.*) How do you know the gentleman?

Lysetta. By having seen him the other day at a friend of your niece's.

Mr. Thomès. How does her coachman do?

Lysetta. Very well. He's dead.

Mr. Thomès. Dead!

Lysetta. Yes.

Mr. Thomès. That can't be.

Lysetta. I don't know whether it can be or not; but I know well enough that so it is.

Mr. Thomès. He can't be dead, I tell you.

Lysetta. And I tell you that he is dead and buried.

Mr. Thomès. You are deceived.

Lysetta. I saw it.

Mr. Thomès. 'Tis impossible. Hippocrates says that these sort of distempers don't terminate till the fourteenth or twenty-first, and he fell sick but six days ago.

Lysetta. Hippocrates may say what he please; but the coachman is dead.

Sganarel. Silence, prate-apace, and let us go from hence. Gentlemen, I beg you to consult in the best manner. Though 'tis not the custom to pay beforehand, yet for fear I should forget it, and that the thing may be over, here ——

(*He gives them money, and each in receiving it makes a different gesture.*)

Scene III

Messrs. Fonandrès, Thomès, Macroton, and Bahys
(*They sit down and cough.*)

Mr. Fonandrès. Paris is wonderfully large, and one must make long jaunts when practice comes on a little.

Mr. Thomès. I must own that I have an admirable mule for that, and the way I make him go every day is scarce to be believed.

Mr. Fonandrès. I have a wonderful horse, and 'tis an indefatigable animal.

Mr. Thomès. Do you know the way my mule has gone to-day? I was first over against the arsenal, from the arsenal to the end of the suburb St. Germain, from the suburb St. Germain to the very end of the marshes, from the end of the marshes to the gate St. Honorius, from the gate St. Honorius to the suburb St. James's, from the suburb St. James's to the gate of Richelieu, from the gate of Richelieu hither, and from hence I must go yet to the Palace-Royal.

Mr. Fonandrès. My horse has done all that to-day, and besides I have been at Ruel to see a patient.

Mr. Thomès. But well thought on, what side do you take in the dispute betwixt the two physicians, Theophrastus and Artemius? for 'tis an affair which divides all our body.

Mr. Fonandrès. I am for Artemius.

Mr. Thomès. And I likewise; not but that his advice killed the patient and that of Theophrastus was certainly much the better; but he was wrong in the circumstances, and he ought not to have been of a different opinion to his senior. What say you of it?

Mr. Fonandrès. Without doubt. The formalities should be always preserved whatever may happen.

Mr. Thomès. For my part I am as severe as a devil in that respect, unless it's amongst friends. And three of us were called in t'other day to a consultation with a strange physician, where I stopped the whole affair, and would not suffer 'em to go on unless things went in order. The people of the house did what they could, and the distemper increased; but I would not bate an inch, and the patient died bravely during this dispute.

Mr. Fonandrès. 'Twas well done to teach people how to behave, and to show 'em their mistake.

Mr. Thomès. A dead man is but a dead man, and of no consequence: but one formality neglected does a great prejudice to the whole body of physicians.

Scene IV

Sganarel, Messrs. Thomès, Fonandrès, Macroton, and Bahys

Sganarel. Gentlemen, my daughter's oppression increases, pray tell me quickly what you have resolved on.

Mr. Thomès. (To Mr. Fonandrès.) Come, sir.

Mr. Fonandrès. No, sir, do you be pleased to speak.

Mr. Thomès. You jest sure, sir.

Mr. Fonandrès. I'll not speak the first.

Mr. Thomès. Sir.

Mr. Fonandrès. Sir.

Sganarel. Nay, pray gentlemen, leave all these ceremonies, and consider that things are pressing.

Mr. Thomès. Your daughter's illness ——

Mr. Fonandrès. The opinion of all these gentlemen together ——

Mr. Macroton. Af-ter ha-ving well con-sult-ed ——

Mr. Bahys. In order to reason ——

(They all four speak together.)

Sganarel. Nay, gentlemen, speak one after another, pray now.

Mr. Thomès. Sir, we have reasoned upon your daughter's distemper; and my opinion, as for my part, is that it proceeds from a great heat of blood: so I'd have you bleed her as soon as you can.

Mr. Fonandrès. And I say that her distemper is a putrefaction of humours, occasioned by too great a repletion, therefore I'd have you give her an emetic.

Mr. Thomès. I maintain that an emetic will kill her.

Mr. Fonandrès. And I, that bleeding will be the death of her.

Mr. Thomès. It belongs to you indeed to set up for a skilful man!

Mr. Fonandrès. Yes, it does belong to me; and I'll cope with you in all kinds of learning.

Mr. Thomès. Do you remember the man you killed a few days ago?

Mr. Fonandrès. Do you remember the lady you sent into the other world three days since?

Mr. Thomès. (To Sganarel.) I have told you my opinion.

Mr. Fonandrès. (To Sganarel.) I have told you my thoughts.

Mr. Thomès. If you don't bleed your daughter out of hand, she's a dead woman.   *(Goes out.)*

Mr. Fonandrès. If you do bleed her, she'll not be alive in a quarter of an hour hence.   *(Goes out.)*

Scene v

Sganarel, Messrs. Macroton and Bahys

Sganarel. Which of the two am I to believe, and what resolution shall I take upon such opposite advices? Gentlemen, I conjure you to determine me, and to tell me without passion, what you think the most proper to give my daughter relief.

Mr. Macroton. (Drawling out his words.) Sir, in these mat-ters, we must pro-ceed with cir-cum-spec-ti-on, and do no-thing in-con-si-de-rate-ly, as they say; for-as-much as the faults which may be com-mit-ted in this case are, ac-cor-ding to our ma-ster Hip-po-cra-tes, of a dan-ge-rous con-se-quence.

Mr. Bahys. (Sputtering out his words hastily.) 'Tis true. We must really take care what we do; for this is not child's play; and when we have once faltered 'tis not easy to repair the slip, and to re-estab-lish what we have spoilt. *Experimentum periculosum.* Wherefore we should reason first as we ought to do, weigh things, seriously consider the constitutions of people, examine the causes of the distemper, and see what remedies one ought to apply to it.

Sganarel. (Aside.) One creeps like a tortoise, and t'other rides post.

Mr. Macroton. For, sir, to come to fact, I find your daugh-ter has a chro-ni-cal dis-ease, and that she may be in jeo-par-dy if you don't give her some assis-tance; for-as-much as the symptoms which she has are in-di-ca-tive of a fu-li-gi-nous and mor-di-cant va-pour, which pricks the mem-branes of the brain; for this va-pour, which we call in Greek *at-mos,* is caus-ed by pu-trid, te-na-ci-ous, and con-glu-ti-nous humours, which are con-tain-ed in the abdomen.

Mr. Bahys. And as these humours were engendered there by a long succession of time; they are over-baked there, and have acquired this malignity, which fumes towards the region of the brain.

Mr. Macroton. So that to draw a-way, loos-en, ex-pel, e-va-cu-ate the said hu-mours, there must be a vi-go-rous pur-ga-tion. But first of all, I think it proper, and it would not be in-con-ve-ni-ent to make use of some lit-tle a-no-dyne me-de-cines; that is to say, lit-tle e-mol-li-ent and de-ter-sive cly-sters, and re-fresh-ing ju-leps and sy-rups, which may be mix-ed in her bar-ley wa-ter.

Mr. Bahys. Afterwards we'll come to purgation and bleeding, which we'll reiterate if there be need of it.

Mr. Macroton. Not but for all this your daughter may die; but at least you'll have done some-thing, and you'll have the con-so-la-ti-on that she di-ed ac-cord-ing to form.

Mr. Bahys. It is better to die according to the rules than to recover contrary to 'em.

Mr. Macroton. We tell you our thoughts sin-cere-ly.

Mr. Bahys. And have spoken to you as we would speak to our own brother.

Sganarel. (To Mr. Macroton, drawling out his words.) I ren-der you most hum-ble thanks. *(To Mr. Bahys, sputtering out his words.)* And am infinitely obliged to you for the pains you have taken.

Scene VI

Sganarel. (Alone.) So I'm just a little more uncertain than I was before. S'death, there's a fancy comes into my head, I'll go buy some orvietan, and make her take some of it. Orvietan is a remedy which many people have found good by. Soho!

Scene VII

Sganarel, The Operator

Sganarel. Sir, pray give me a box of your orvietan, which I'll pay you for.

The Operator. (Sings.)

> The gold in all lands which the sea doth surround,
> Can ne'er pay the worth of my secret profound:
> My remedy cures, by its excellence rare,
> More maladies than you can count in a year.

> > The scab,
> > The itch,
> > The scurf,
> > The plague,

> The fever,
> The gout,
> The pox,
> The flux,
> And measles ever,

Of orvietan such is the excellence rare.

Sganarel. Sir, I believe all the gold in the world is not sufficient to pay for your medicine, but however here's a half crown-piece which you may take if you please.

The Operator. (*Sings.*)

> Admire then my bounty, who for thirty poor pence,
> Such a marvellous treasure do so freely dispense.
> With this you may brave, quite devoid of all fear,
> All the ills which poor mortals are subject to here.

> The scab,
> The itch,
> The scurf,
> The plague,
> The fever,
> The gout,
> The pox,
> The flux,
> And measles ever,

Of orvietan* such is the excellence rare.

LE MALADE IMAGINAIRE

Le Malade Imaginaire is concerned with the character of Argan, a hypochondriac. He is discovered in the first scene going over his bills from apothecaries. He has an only daughter, Angelique, and a second wife, Beline, whose object it is to see him driven to his grave by the continued absorption of drugs. Argan finally determines to have his daughter marry a doctor, so he can have one in the house. The fun turns on the ridiculous sayings and doings of the doctors, and the efforts of Argan's sensible brother, Beralde, to induce the patient to throw the doctor's physic to the dogs. Beralde finally persuades Argan

* Orvietan was a popular sedative mixture.

to enter the medical profession himself and there is an exquisitely funny scene at the examination of a candidate for the doctor's degree. That this scene still has point is attested by the fact that on the fifteenth of January, Molière's birthday, every year the Theatre Français gives this scene. Professor Clarke describes it: "A bust of Molière occupies the center of the stage which represents the amphitheatre of a medical school. To the sound of music there enter two by two, the members of the Society of the Comedie Française, all robed in red. They bow to Molière and to each other, and take seats in a circle to represent the Faculty of Medicine. It is a sort of family reunion for the members of the company and an opportunity for the audience to recognize and applaud their favorite actors."

We give the first scene, and the last scene of the examination of the candidate, in their entirety.

REFERENCE

CLARKE. Molière and the Doctors. *Bull. Soc. Med. Hist., Chicago,* Vol. III, No. 2, 1923.

ACT I

Scene I

Scene: Argan's Chamber

Argan. (Sitting with a table before him, casting up his apothecary's bills with counters.) Three and two make five, and five makes ten, and ten makes twenty. Three and two make five. Item, the twenty-fourth, a little insinuative, preparative and emollient clyster to mollify, moisten, and refresh his worship's bowels. What pleases me in Mr. Fleurant, my apothecary, is, that his bills are always extremely civil. His worship's bowels, thirty sous. Ay, but Mr. Fleurant being civil isn't all, you ought to be reasonable too, and not fleece your patients. Thirty sous for a clyster! Your servant, I have told you of this already. You have charged me in your other bills but twenty sous, and twenty sous in the language of an apothecary is as much as to say ten sous; there they are, ten sous. Item, the said day, a good detersive clyster composed of double catholicum, rhubarb, *mel rosatum,* etc., according to prescription, to scour, wash and cleanse his honour's abdomen, thirty sous; with your leave ten sous. Item, the said day at night, an hepatic, soporific, and somniferous julep, composed to make his honour sleep, thirty-five sous; I don't complain of that, for it made me sleep well. Ten, fifteen, sixteen, seventeen sous, six deniers. Item, the twenty-fifth, a good purgative and corroborative medicine composed of *cassia recens* with *senna levantina,* etc., according to the prescription of Mr. Purgon to expel and evacuate his honour's choler, four livres. How! Mr. Fleurant, you jest sure, you should treat your patients with some humanity. Mr. Purgon did not prescribe you to set down four livres; put down, put down three livres if you please—fifty sous. Item, the said day, an anodyne and astringent potion to make his honour sleep, thirty sous. Good—fifteen sous. Item, the twenty-sixth, a carminative clyster to expel his honour's wind, thirty sous. Ten sous, Mr. Fleurant. Item, his honour's clyster repeated at night as before, thirty sous. Ten sous, Mr. Fleurant. Item, the twenty-seventh, a good medicine composed to dissipate and drive out his honour's ill humours, three livres. Good, fifty sous; I'm glad you are reasonable. Item, the twenty-eighth, a dose of clarified, dulcified milk, to sweeten, lenify, temper and refresh his honour's blood, twenty sous. Good, ten sous. Item, a cordial preservative potion, composed of twelve grains of bezoar, syrup of lemons, pomegranates, etc., according to prescription, five livres. Oh! Mr. Fleurant, softly, if you please, if you use people in this manner, one would be sick no longer, content yourself with four livres; sixty sous. Three and two make five, and five makes ten, and ten makes twenty. Sixty-three livres, four sous, and six deniers. So then in this month I have taken one, two, three, four, five, six, seven, eight purges; and one, two, three, four, five, six, seven, eight, nine, ten, eleven, twelve clysters; and the last month there were twelve purges and twenty clysters. I don't wonder if I am not so well this month as the last. I shall tell Mr. Purgon of it, that he may set this matter to rights. Here, take me away all these things. There's nobody there, 'tis in vain to speak, I'm always left alone; there's no way to keep 'em here. (*He rings a bell.*) They don't hear; my bell's not loud enough. (*Rings.*) No. (*Rings again.*) They are deaf. Toinet! (*Making as much noise with his bell as possible.*) Just as if I did not ring at all. Jade! Slut! (*Finding he still rings in vain.*) I'm mad. Drelin, drelin, drelin, the deuce take the carrion. Is it possible they should leave a poor sick creature in this manner! Drelin, drelin, drelin, oh! lamentable! Drelin, drelin, drelin. Oh! Heavens, they'll let me die here. Drelin, drelin, drelin.

The last part of *Le Malade Imaginaire* represents a burlesque ceremony of examination of a candidate for the doctorate of medicine. It is written in a hodge-podge of pig Latin, slang Spanish, Italian, English, and French. In no English edition is it ever translated, but it is easy enough to make out if assistance is given by a few translations as in the following rendition by Professor Clarke. Note that in answering the questions Argan gives the same remedies for every disease. "Clyster [enema] first, then bleeding, then purging," and if these do not work "Re-clyster, re-bleed, repurge."

INTERLUDE

First Entry

Upholsterers come in dancing to prepare the hall, and. place the benches to music.

Second Entry

A cavalcade of physicians to the sound of instruments.

Persons bearing clyster-pipes which represent maces, enter first. After them come the apothecaries with their mortars; surgeons and doctors two by two place themselves on each side of the stage, whilst the president ascends a chair, which is placed in the middle, and Argan who is to be admitted a doctor of physic, places himself on a low stool at the foot of the president's chair.

Praeses.   Scavantissimi doctores,
     Medicinae professores,
     Who here assemblati estis;
     Et vos other messiores,
     Of the opinions of the faculty
     Fideles executores;
     Chirurgiani and apothecari,
     Atque tota compania aussi,
     Health, honor and might,
     Atque bonum appetite.
     Non possum, docti confreri,
     Sufficiently admirari,
     What a bona inventio,

     Est medici professio;
     How rare and choice a thing is ista
     Medicina benedicta,
     Which by you nomine solo

Surprising miraculo
Since si longo tempore;
Has made in clover vivere
So many people omni genere.

Per totam terram now we see
The Grandam vogam in which we be;
Et quod grandes as well as small
Sunt de nobis infatuated all:
Totus mundus currens ad nostros remedios,
Nos regardat sicut deos,
Et nostris praescriptionibus
Principes and reges subjectos videtis.

'Tis therefore nostra sapientia,
Bonus sensus atque prudentia,
Strongly for ro travaillare,
A nos bene conservare
In tali credito, voga and honore;
And take care not to receive
Into our doctor's company
All persons unworthy
Of occupying positions
As doctores medicinae.

For this examination
I now turn over this
Miserable creature that
You see before you
So that you may ask him
All things that are proper
Et à bottom examinandum
Vestris capacitatibus.

First Doctor.   If to me leave is given by Dominus Praeses,
Et tanti docti doctores,
Et assistantes illustres,
Quem estimo and honoro,
Learnidissimo bacheliere

I will ask the cause and reason why
Opium causes sleep.

Argan. Mihi à docto doctore
Demandatur causam and rationem, quare
Opium facit dormire.
To which respondeo,
Because it has a
Soporific virtue
And is specific
In soothing our senses.

Chorus. Bene, bene, bene, bene respondere,
Dignus, dignus est intrare
In nostro docto corpore.
Bene, bene respondere.

Second Doctor. Cum permissione domini praesidis,
Doctissimae facultatis,
Et totius his nostris actis
Companiae assistantis,
Demandabo tibi, docte bacheliere,
Quae sunt remedia,
Quae in maladia
Called hydropisia
Convenit facere?

Argan. Clisterium donare,
Postea bleedare,
Afterwards purgare.

Chorus. Bene, bene, bene, bene respondere,
Dignus, dignus est intrare
In nostro docto corpore.

Third Doctor. Si bonum semblatur domine praesidi,
Doctissimae facultati
Et companiae praesenti,
Demandabo tibi, docte bacheliere,

>   Quae remedia eticis,
>   Pulmonicis atque asmaticis
>   Do you think à propos facere.

Argan.   Clisterium donare,
              Postea bleedare,
              Afterwards purgare.

Chorus.   Bene, bene, bene, bene respondere:
              Dignus, dignus est intrare
              In nostro docto corpore.

Fourth Doctor. Concerning illas maladias,
              Doctus bachelierus dixit maravillas:
              But if I do not tease and fret dominum praesidem,
              Doctissimam facultatem,
              Et totam honorabilem
              Companiam hearkennantem;
              Faciam illi unam quaestionem.
              Last night patientus unus
              Chanced to fall in meas manus:
              Habet grandam fiévram cum redoublamentis
              Grandum dolorem capitis,
              Et grandum malum in his si-de,
              Cum granda difficultate
              Et pena respirare.
              Be pleased then to tell me,
              Docte bacheliere,
              Quid illi facere.

Argan.   Clisterium donare,
              Postae bleedare,
              Afterwards purgare.

Fifth Doctor. But if maladia
              Opiniatria
              Non vult se curire,
              Quid illi facere?

Argan.	Clisterium donare,
	Postae bleedare,
	Afterwards purgare.
	Rebleedare, repurgare, and reclysterisare.
Chorus.	Bene, bene, bene, bene respondere:
	Dignus, dignus est intrare
	In nostro docto corpore.
The President.	(to Argan)—
	Swear to keep the statutes
	Per facultatem praescripta,
	Cum sense and jugeamento?
Argan.	Juro.
The President.	To be in omnibus
	Consultationibus
	Ancieni aviso;
	Aut bono,
	Aut Baddo?
Argan.	Juro.
The President.	That thou'lt never te servire
	De remediis aucunis,
	Than only those doctae facultatis;
	Should the patient burst-O
	Et mori de suo malo?
Argan.	Juro.
The President.	Ego cum isto boneto
	Venerablili and docto,
	Dono tibi and concedo
	Virtutem and powerantiam,
	Medicandi,
	Purgandi,
	Bleedandi,

Prickandi,
Cuttandi,
Slashandi,
Et occidendi
Impune per totam terram.

Third Entry

The surgeons and apothecaries do reverence with music to Argan.

Argan.   Grandes doctores doctrinae,
Of rhubarbe and of séné:
'Twou'd be in me without doubt one thinga folla,
Inepta and ridicula,
If I should m'engageare
Vobis loüangeas donare,
Et pretendebam addare
Des lumieras au soleillo,
Et des étoilas au cielo,
Des ondas à l'oceano,
Et des rosa to the springo.
Agree that in one wordo
Pro toto remercimento
Rendam gratiam corpori tam docto.
Vobis, vobis debeo
More than to nature, and than to patri meo;
Natura and pater meus
Hominem me habent factum:
But vos me, that which is plus,
Avetis factum medicum.
Honor, favor, and gratia,
Qui in hoc corde que voila,
Imprimant ressentimenta
Qui dureront in saecula.

Chorus.   Vivat, vivat, vivat, vivat, for ever vivat
Novus doctor, qui tam bene speakat,
Mille, mille annis, and manget and bibat,
Et bleedet and killat.

Fourth Entry

All the surgeons and apothecaries dance to the sound of the instruments and voices, and clapping of hands, and apothecaries' mortars.

First Surgeon.	May he see doctas
	Suas praescriptionas
	Omnium chirurgorum,
	Et apotiquarum
	Fillire shopas.

Chorus.	Vivat, vivat, vivat, vivat for ever vivat
	Novus doctor, qui tam bene speakat,
	Mille, mille annis, and manget and bibat.
	Et bleedet and killat.

Second Surgeon.	May all his anni
	Be to him boni
	Et favorabiles,
	Et n'habere jamais
	Quàm plaguas, poxas,
	Fiévras, pluresias
	Bloody fluxies and dissenterias.

Chorus.	Vivat, vivat, vivat, vivat, for ever vivat
	Novus doctor, qui tam bene speakat,
	Mille, mille annis, and manget and bibat,
	Et bleedet and killat.

Fifth and Last Entry

While the chorus is singing, the doctors, surgeons, and apothecaries go out all according to their several ranks, with the same ceremony they entered.

XXIII
EARLY GROSS PATHOLOGY

ISOLATED AUTOPSIES WERE FREQUENTLY DONE DURING THE MIDDLE AGES. Benivieni (?-1502) left in his posthumous *De Abditis Causis Morborum* (On the Secret Causes of Disease, 1507) some clear accounts that showed his grasp of the importance of autopsies to clinical medicine.

Théophile Bonet was the author of an enormous compendium, *Sepulchretum* (1679). This is a collection of postmortem observations arranged logically in so far as the first concerns a disease of the head and the last a disease of the feet.

Giovanni Battista Morgagni (1682-1771), professor of anatomy at Padua, published, in 1761, *De Sedibus et Causis Morborum* (On the Seats and Causes of Disease). This work marks the beginning of pathology as a systematized science. It consists of a number of letters, written by the professor to a young friend, which contain early accounts of aneurism, acute yellow atrophy of the liver, heart block, valvular disease of the heart. Morgagni also described the sequence of changes in pneumonia, showed that brain abscess follows middle ear suppuration, and noted the decussation of the pyramids.

Matthew Baillie (1761-1823) was a nephew of John and William Hunter, and succeeded to their practice. He published *Morbid Anatomy of Some of the Most Important Parts of the Human Body*, in 1793. It was the first book to treat pathology as an entity. It was illustrated by accurate copper plate engravings made by William Clift. The diseases of the organs are taken up systematically.

REFERENCES

LONG, E. R. A History of Pathology. Baltimore, William and Wilkins, 1928.
LONG, E. R. Selected Readings in Pathology. Springfield, Ill., C. C. Thomas, 1929.

ANTONIO BENIVIENI

[From *On the Secret Causes of Disease**]

ON STONES FOUND IN THE TUNIC OF THE LIVER

A noble woman suffered for a long time from a pain in the region of the liver, and on account of it consulted a great many physicians, but was unable to escape the malady by the use of any remedy, wherefore she resolved to try our skill along with others. We came together, many physicians, and discussed from this time on in many a conversation what might be the hidden causes of this disease. In truth we did not agree so that a decision could be reached in the midst of our many doubts. While some judged that an abscess of the liver might be present, and others an ill habit, we ourselves believed the fault lay in the covering membrane of the organ. And when, after a few days, with increasing severity of the disease she passed away, even as we had by common consent predicted from certain signs, we took care that the body of the deceased should be opened. And we found, packed in the sloping portion of the covering of the liver, stones of varying size and color. Some were round, some angular, and some quadrate, according as situation and chance determined. These by their own weight, made from the tunic of the organ a small sac as long as the palm of the hand and broad as two fingers. After observing this cause of death we judged it idle and without profit to dispute further on the obscurities.

CALLOUS GROWTH IN THE STOMACH

Antonius Brunus, a relative of mine by marriage, could retain the food he had eaten but a short time, and then vomited it back, undigested, quite as it was. He was assiduously supported by all the medicaments that can be applied for ailments of the stomach, but these helped in no way. From lack of nourishment of the body he wasted away until scarcely skin and bone remained, and finally little by little was brought to the end. In the interest of the public good the defunct body was then opened. The stomach was found grown into a callus at the lowest part of its orifice, so that nothing could be passed through to the parts below. Inevitably death resulted.

* De Abditis Causis Morborum. Translation by Dr. Esmond R. Long. By permission of Dr. Long and the publisher, Charles C. Thomas, Springfield, Illinois.

CALLUS IN THE MESARAIC VEINS

The son of Peter Aldimarius in his eighth year developed a perverted appetite, eating stones and plaster until he finally reached a state of lingering mild fever. With the passage of time the stomach began to fail to retain food and the bowels became loose. With these two ailments the boy was in such distress that within a short time he died. With the consent of the father we opened the corpse and found a callous substance among those veins which are called the mesaraic. As a result of this, with all the veins obstructed so that the blood could not be carried further, inevitably the boy departed from this life.

DEATH FROM DIFFICULTY IN BREATHING

We have known many who suffered from difficulty in breathing. Among others was one who supported existence only by holding his neck continually erect, while gradually, with failing strength, becoming pale and cyanotic throughout his body. Yet there was no pain in the chest or praecordium. On the other hand a whistling sound was to be observed, the result of the difficulty with which the breath was exhaled. From this one could conclude that he suffered from an ailment of the chest. After some days, with increasing severity of the affliction, remedies failing, the sick man passed away. We decided then to open the body, so that we might apprehend the hidden and unknown causes of his illness. When this was done we found enough black bile and blacker blood had come together in his heart, so that it is no wonder that the mixture on diffusing through the veins corrupted the breath itself and brought the man to a hurried end. As we know, such a fact was touched upon elsewhere by the author Galen.

GIOVANNI BATTISTA MORGAGNI

PREFACE

[From *On the Seats and Causes of Diseases**]

There are two sayings of C. Lucilius, as you have it in Cicero: I mean "That he neither wished to have his writings fall into the

* De Sedibus et Causis Morborum. The extracts given here are from the translation by Benjamin Alexander, M.A. London, 1769.

hands of the most learned, nor of the most unlearned readers," which I should equally make use of on the present occasion, if it were not my desire to be useful to the unlearned, as well as to be assisted by the learned reader. For I have had two views in publishing these writings; the first that I might assist the studies of such as are intended for the practice of medicine; the second, and this the principle view, that I might be universally useful though this cannot happen without the concurrence and assistance of the learned in every quarter.

2. Theophilus Bonetus, was a man who deserved the esteem of the faculty of medicine in particular, and of mankind in general, in an equal degree with any other, on account of his publishing those books which are entitled the Sepulchretum. For by collecting in as great a number as possible and digesting into order, the dissections of bodies, which had been carried off by disease, he formed them into one compact body: and thereby caused those observations, which when scattered up and down through the writings of almost innumerable authors, were but of little advantage, to become extremely useful when collected together and methodically disposed.

Now then; in an affair wherin everyone is concerned, and not only in the present but in the future ages: in order to judge more easily what may be expected from me alone, and how far it is just to expect it, I must by no means conceal the circumstance which first gave occasion to my writing these books.

The anatomical writings of Valsalva being already published, and my epistles upon them, it accidently happened that being retired from Padua, as in those early years, I was wont frequently to do in the summer time, I fell into company with a young gentleman, of strict morals and an excellent disposition, who was much given to the study of the sciences and particularly to that of medicine. This young gentleman, having read those writings and those letters likewise, every now and then engaged me in a discourse, than which nothing could be more agreeable to me; I mean a discourse in respect to my preceptors, and in particular Valsalva and Albertini, whose methods in the art of healing, even the most trifling he was desirous to know; and he even sometimes enquired after my own observations and thoroughly as well as theirs.

And having among other things, as frequently happens in conversations, opened my thoughts in regard to the Sepulchretum, he never ceased to entreat me by every kind of solicitation, that I would apply to this subject in particular: and as I had promised in my little memoire upon the Life of Valsalva, to endeavor that a great number of his observations, which were made with the same view, should be brought to public light, he begged that I would join mine together with them, and would show in both his and mine, by example as it were, what I should think wanting to complete a new edition of the Sepulchretum, which he, perhaps, if he could engage his friends to assist him would at some time or another undertake.

With this view then I began upon returning to Padua to make a trial of that Nature, by sending some letters to my friend. And that he was pleased with them appears from two circumstances; the first that he was continually soliciting me to send him more and more after that till he drew me on so far as to the seventieth; the second, that when I begged them of him in order to revise their contents, he did not return them, till he had made me solemnly promise, that I would not abridge any part thereof.

ON THE PALSY
[From *On the Seats and Causes of Diseases*, Letter XI]

What convulsion, of which I wrote in the last letter, is to the epilepsy, the same is palsy, of which I shall write at present, to the apoplexy. For as to the section which succeeds next in the Sepulchretum, de Stupore, Torpore, Tremore, Horrore, Rigorem, Anxietate, doubtless you perceive that some of these disorders belong to the palsy, some to convulsion, and some should be placed under other heads: and this the observations also in that section, when compar'd demonstrate, inasmuch as, a few of the principal only accepted, some of them are taken from one section, and some from another, as you will see. And indeed that fifteenth section, which is entitled de Paralysis, as many in like manner, which are transferred from other places. But I, however, shall preserve my custom, and shall bring over again in this place, under the head of apoplexy; nor shall I take any from those that relate to blows and wounds, which you will receive hereafter. Wherefore I shall give you only three from Valsalva's papers, and four from mine.

An old man of sixty, being troubled with a flux of the belly, joined with gripings, and continual watchings besides, anointed his belly with oil of quinces. And although the flux continued quite to the time of his death, yet on the following night, without any previous symptom of his head, he was suddenly seized with hemiplegia, so that the whole right side of his body remained immoveable. On the first day, however, after bleeding, and irritating medicines being applied to the soles of his feet, he could move both his hand and foot a little; yet the day after could not move them at all. As to the other parts, his right eye was half-shut, his cheeks were red, he scarcely spoke at all, and when he did, he stammered; but he answered in such a manner, by nods and signs, to those who asked him questions, that you might perceive his internal senses to be strong and perfect. In the beginning of the disorder his respiration was easy; but it became difficult in a day or two before his death, which happened on the beginning of the fourth day.

While the brain was taken out from the cranium, and especially while the infundibulum was divided from the pituitary gland, a limpid serum, and a fluid blood came forth. On the left, and by the sides of the sanguiferous vessels of the meninges, a little matter was observed which had the appearance of a jelly. And on the same side, in like manner, under the pia mater, the very substance of the brain seemed to be a little eroded in two places: which was more manifest in the ventricle of the same side. For the corpus striatum was found to be entirely separated from the remainder of the cerebrum, by reason of an erosion, perhaps brought on by serum which stagnated in the ventricles.

Another old man, of the same age, fell down suddenly, at the same time lost the power of moving, and feeling, in the right side of his body. When he was asked questions, he scarcely answer'd at all; yet what he did say was stammering. In all the time that he survived this stroke, he made but little water, and never went to stool without the assistance of clysters. Finally, in the beginning of the 21st day, a difficult respiration coming on, he died. The thorax being opened, the lungs, but especially the right lobe, were found to have been seized with a phlegmon, in their posterior parts. The ventricles of the heart contain'd small polypous concretions, which extended themselves into the neighboring vessels. In sawing through the cranium, the dura mater being wounded, a limpid water flow'd out. The same kind of water was found in the right ventricle of the brain;

but in the left, it was tinged with aeruginous colour, and had formed an ulcerous cavity in the basis thereof.

An old man of seventy, who had been very voracious in his diet, being seized with an apoplexy long before, and after that with a palsy of the whole right side of the body, was frequently agitated on the other side with convulsions. His senses were affected; and he sometimes discharged calculi with his urine. The abdomen being open'd after death, the omentum was seen to be so far drawn upward, as to cover the whole anterior part of the stomach. But the left lobe of the liver, which is us'd to lie over a part of the stomach, scarcely touched it at all, in consequence of being drawn up by the diaphragm, to which it was firmly attached. Moreover, the stomach, although it was corrugated, was, however, when extended, much bigger than it generally is. And the spleen was evidently twice as big as it ought to have been, and of a very dark colour. In the left kidney were found four stones; one of the bigness of a chestnut, the others less. The thorax was not at all open'd. While the brain was taken out of the cranium, some serum, which was contained betwixt the dura and pia mater, flow'd out. In the left ventricle, the plexus choroides had in it a body of the bigness of a horse-bean, made up of several hydatids: and under the same ventricle was a sinus, the sides of which consisted of the substance of the cerebrum, that was yellow and flaccid, and seemed also to be corrupted. . . .

However the doctrine of Valsalva, which I have before spoken in commendation of, is confirm'd by this observation of Wepfer, as well as by the three foregoing dissections of the old men. For the hemiplegia had been in the left side of the body, whereas the injury, as we have seen, was on the right side of the brain: and this is what I had never attended to, till I look'd over, very accurately, this section of the Sepulchretum; where the paralysis was found to be in the left side, but the imposthume in the right side of the brain. Yet surely it is not so much to be wonder'd at, that these things should have escap'd me, as that Wepfer, who had observed the circumstances once, and again, should have taken so little notice of it. For he says "I do not indeed deny, that those tumors of the right ventricle may, in some measure, have conspired to the production of a hemiplegia in the left side; for I myself, with many others, have observ'd, that one side of the brain being affected, the opposite side of the body had been seized with the palsy: but I believe that the concomitant, and perhaps primary cause, of the hemiplegia, in these cases, was serum"; without doubt, that with which he thought the small pores of the brain were afterwards obstructed. If he had not said, that he had observed it, "with many others," it might perhaps be suspected, that he had observed it from a long series of dissections, where it happen'd from an internal cause, as Valsalva has done since him. But now we naturally understand, that Wepfer had seen the same thing that "many others" had seen before him; I mean, that this contrast betwixt the injury of the brain, and the palsy of the body, had been frequently the consequence of blows and of wounds. Moreover, he does not only attribute but little to these tumors, which had compress'd the right side of the brain so long, and consequently prevent'd, or at least diminish'd, the influx of animal spirits, into the left part of the spinal marrow, for a long time; but he even did not think, that there was much more to be attributed to this part of the spinal marrow, which for that reason, perhaps, "seem'd less than the right." . . .

ON ANEURISM

[From *The Seats and Causes of Diseases*]

A man who had been too much given to the exercise of tennis and the abuse of wine, was, in consequence of both these irregularities, seized with a pain of the right arm, and soon after of the left, joined with a fever. After these there appeared a tumour on the upper part of the sternum, like a large boil: by which appearance some vulgar surgeons being deceived, and either not having at all observed, or having neglected, the pulsation, applied such things as are generally used to bring these tumours to suppuration; and these applications were of the most violent kind. As the tumour still encreased, others applied emollient medicines, from which it seemed to them to be diminished; that is, from the fibres being rubbed with ointments and relaxed; whereas they had been before greatly irritated by the applications. But as this circumstance related rather to the common integuments, than to the tumour itself, or to the coats that were proper thereto, it not only soon recovered its former magnitude, but even was, plainly, seen to encrease every day. Wherefore, when the patient came into the Hospital of Incurables, at Bologna, which was,

I suppose in the year 1704, it was equal in size to a quince; and what was much worse, it began to exsude blood in one place; so that the man himself was very near having broken through the skin (this being reduced to the utmost thinness in that part, and he being quite ignorant of the danger which was at hand) when he began to pull off the bandages, for the sake of showing his disorder. But this circumstance being observed, he was prevented going on, and ordered to keep himself still, and to think seriously and piously of his departure from this mortal life, which was very near at hand, and inevitable. And this really happened on the day following, from the vast profusion of blood that had been foretold, though not so soon expected by the patient. Nevertheless, he had the presence of mind, immediately as he felt the blood gushing forth, not only to commend himself to God, but to take up with his own hands a basin that lay at his bed-side; and, as if he had been receiving the blood of another person, put it beneath the gaping tumour, while the attendants immediately ran to him as fast as possible, in whose arms he soon expired.

In examining the body before I dissected it, I saw that there was no longer any tumour, inasmuch as it had subsided after the blood, by which it had been raised up externally, and had been discharged. The skin was there broken through, and the parts that lie beneath it with an aperture, which admitted two fingers at once. The membrana adiposa of the thorax discharged a water during the time of dissection, with which some vessels were also turgid, that were prominent, here and there, upon the surface of the skin in the feet and the legs. In both the cavities of the thorax, also, was a great quantity of water, of a yellowish colour. And there was a large aneurism, into which the anterior part of the curvature of the aorta itself being expanded, had partly consumed the upper part of the sternum, the extremities of the clavicles which lie upon it, and the neighbouring ribs, and partly had made them diseased, by bringing on a caries. And where the bones had been consumed or affected with the caries, there not the least traces of the coats of the artery remained: to which, in other places, a thick substance every where adhered internally, resembling a dry and lurid kind of flesh, distinguished with some whitish points; and this substance you might easily divide into many membranes, as it were, one lying upon another, quite different in their nature from those coats to which they adhered, as they were evidently polypous. And these things being accurately attended to, nothing occurred besides that was worthy of remark.

The deplorable exit of this man teaches, in the first place, how much care ought to be taken in the beginning, that an internal aneurism may obtain no increase: and in the second place, if, either by the ignorance of the persons who attempt their cure, or the disobedience of the patient, or only by the force of the disorder itself, they do at length encrease so that they are only covered by the common integuments of the whole body; that then we ought to take care lest the bandages, especially when they are already dried to the part, be hastily taken off: and finally, if the case proceed to such an extremity, that the rupture of the skin is every day impending, and bleeding, either on account of the constitution or infirmity of the patient, or on the score of other things which I have hinted at already, is dangerous; that every thing is to be previously studied, by which, for some days at least, life may be prolonged. That is to say, besides the greatest tranquility of body and mind, and the greatest abstinence that can be consistently observed, so that no more food be taken than is barely necessary for the preservation of life, and that in small quantities, and of such a quality as is by no means stimulating; besides that situation of body, by which the weight of the blood being lessened, does not press upon the skin, and other things of the like kind; something ought to be thought of by the surgeon, by way of defence; as, for instance, if the bladder of an ox, four times doubled, were applied, or a bandage of soft leather; and the edges of this bandage were all daubed over with a medicine, by which they would be firmly glued down to the neighbouring skin that lay around the tumour, and was as yet sound and entire.

On Calculi

[From *The Seats and Causes of Diseases*]

But I have said that the passage of the bile is prevented from calculi, either by means of compression, or obstruction. For if any one should say that calculi are sometimes formed in the little glandular bodies of the liver themselves, and that to this class, without doubt, belonged those lesser calculi, which Riedlinus saw "on the external surface of the liver," I should not contest his opinion, although I believe they are more frequently generated in the very branches of the hepatic duct, as those who have very minutely traced them, have found. And as, certainly nothing had happened more frequently to Ruysch, in oxen and sheep, than to find calculi in the pori biliarii, so nothing happened "more rarely," than to find these concretions in the "parenchymatous substance of the liver itself"; so that, although he very attentively "dissected away all the fleshy part," in more than a hundred livers, yet he found in one only, a calculus "buried in the parenchymatous substance, and not at all affixed to the porus biliarius."

Nor can I suppose, that the ancient observations of Platerus, of hepatic calculi resembling "a tophaceous concretion, ramified in the manner of coral, and hollow internally," are to be referred to any other part, than to the same biliary branches, especially as I read Glisson expressly asserting, that similar observations "of tubuli of so great a length, that if they could but have been taken out in their perfect state, they would, like coral, have resembled a great number of the ramifications of the porus biliarius, in one continued stony series," were made by him on the livers of oxen, and even within the same pore or duct. The branches of which Reverhorst, also, found to be internally beset with a calculous crust, in the body of man.

Nor have I found calculi, in the human liver, in any other place than in these branches. Nor do I suppose that those stones, which by Columbus, and Camenicenus, were supposed to be found in the ven portarum, had any different situation: yet my reasons for thinking thus, although not sufficiently attended to by some authors of eminence, as I have already given them on a former occasion, I shall not repeat here. These calculi, therefore, when at length from tubular bodies, by continual and fresh accretions of similar matter, they are made perfectly solid, as happens in aquaeducts, must without any doubt whatever, occupy the whole passages whereof I have spoken, and prevent the transit of the bile.

I have also said this; that calculi of the liver, though large, do not bring on a jaundice, is not to be wondered at, unless they are in such situations as necessarily to obstruct these passages. And I believe that this disease was present, for I cannot now positively affirm it, in a certain man, whose liver had a stone in the center of the concave surface, of the form and magnitude of a pigeon's egg, as an anatomical friend of mine, who had dissected the body, informed me by letter, many years ago. But I do not wonder that this disorder had not been observed in three women, who, although they had a much larger stone, or a greater number of concretions, and more heavy ones, within the membrane of the liver, nevertheless, had them in such a situation, that they seemed to be rather on the outside of the liver, than within its substance: and this was the reason I did not make mention of them above. For that membrane being drawn away from this viscus, by the included weight, and being extended downwards, had formed a sacculus in two of them of the length of a span; for in the third it was described only as a follicle, pendulous downwards.

This last observation is from Benivenius, and is totally different, as you will easily perceive by comparing them, from the second, which is given in the Sepulchretum, from the third chapter of his book. And a similar observation to his; except that in the sacculus not many calculi were contained, but one large calculus, only, was included together with a great quantity of glutinous humour, and that the woman never complained of any thing but of a heat in her liver; the observation of Georgius Greselius, is subjoined. And it was in consequence of bearing these examples in my mind, and observing therefrom, that besides the gall-bladder itself being enlarged, another kind of cyst, distended likewise with a fluid, might sometimes hang below the liver, which, although it was entirely preternatural, would, nevertheless resemble this natural cyst; it was in consequence, I say, of reasoning from these examples, that in the case of Laurence Bacchetti, formerly a physician at Padua, the history of whose disease, and dissection, two other learned men have published, since Dominic Militia, I carried myself with so much caution, as not to affirm any thing for certain, though I made no scruple to declare my opinion.

This gentleman had a tumour hanging below the liver, which you immediately felt by applying your hand to the abdomen: it was globular, and moveable, so that you could easily bring it towards the right side, or towards the left, by means of the hand with which you laid hold of it. When different physicians seemed to have different opinions, as you will read in Militia, who declares the several opinions of all; to me, who saw him once after others, this tumour seemed to be the gall-bladder, enlarged by an immoderate distension of fluid, and produced downwards which I declared to Dominic Sephanelli, a physician and friend of the patient, who with great politeness attended me home, and very earnestly desired my opinion;

yet I made this declaration in such a manner, as to affirm nothing for certain. What I had thus declared was so evidently confirmed by the dissection, that although the declaration might be passed over by some, yet the appearance itself could be concealed by nobody.

CASE HISTORIES

[From *The Seats and Causes of Diseases*]

To begin with a great man; whose history is for that reason more accurately described by Valsalva: The Cardinal Antonio Francesco Sanvitalis was of a moderate stature, or somewhat taller, of a full fleshy habit, and a florid colour: he had been much given to study and close application: was also subject to the gout, and had some years before been attacked with a certain ineffectual irritation of the fauces, to spit: and beside this, he was also troubled at intervals, with convulsive motions in his feet and hands. Finally, when he was five and fifty years of age, having lived for two months together in a mountainous country, on which the south winds generally blew, and the air of which he had at other times found extremely inimical to him; and being also troubled with cares and anxieties of mind, and the winter solstice of the year 1714 being at hand, he fell into a vertiginous disorder; after which, although he was freed therefrom, he shewed a constant sadness, and propensity to sleep. Within about twenty days the vertiginous disorder returned, and brought a vomiting with it. Yet both these were in a short space removed, and after that a violent pain of the head, which had succeeded them. But the day following, at the same hour on which the vertigo had seized him, all sense of feeling and power of motion was lost in the left part of his body, and he lay as if overcome with a profound sleep. His respiration, however, was natural; but his pulse frequent, large, and vehement; and though it was in vain to irritate the left limbs, yet the same irritations being applied to the sole of the right foot, and the usual ones to the nostrils, he was somewhat roused, so as to say many things by signs, and some even by proper words. But these irritations had a happier effect after blood being taken away; more especially on the sixth day from the apoplexy, when the right jugular vein was opened by Valsalva's order; for about four hours after, his internal senses were awakened; and his speech, for more than an hour, was restored. The same change happened, about the same hour, on the following night, and was more evident, and of longer duration. But this rousing was his last: for from that time he gradually declined; and was seized with convulsive motions on his right side, especially in his hand and foot: his whole face was likewise convulsed, but especially about his eyes, and perhaps the heart itself; for he frequently at the same time lay entirely without pulse. In fine, these symptoms recurring about the beginning of the tenth day, he died.

In the belly and chest every thing was found in a natural state. The brain, however, was flaccid; and in the left ventricle was a little serum; but the right contained more than two ounces of coagulated blood. The plexus choroides was here torn through, and the parietes of the ventricle, even on the external side, toward the back part, were corroded into the form of a deep ulcer.

Many things concurred to dispose this great man to an apoplexy: studies, close application to important business, anxieties of mind, and even the gout itself, which often draws after it a calculus, and at other times an apoplectic affection. In reading histories of this kind, please to observe, among the rest, those of a prince, and a count, both of whom were gouty, both apoplectic: and besides cystic calculi in each, the lateral ventricles of one were full of serum; but those of the other, which is more to our present purpose, were full of extravasated blood. Many of these common signs which Caelius Aurelianus formerly collected, foretold the apoplexy of the cardinal, to wit, the convulsive motions of the hands and face; and even, as I think, the convulsive motion of the fauces too: next to these, the repeated vertigoes, which were followed by a proneness to sleep, sadness, and a violent pain of the head; which so far proclaimed the approach of an apoplexy, that the last vertigo may be in some measure taken for a kind of slight apoplectic paroxysm, inasmuch as a more heavy one succeeded it on the following day, at the same hour. That this apoplexy was sanguineous, the quantity of blood, demonstrated by the florid colour, might have shewn as also the rarefaction brought on by the south winds; the vessels, now growing rigid with age, being presently streightened by the winter, and on both these accounts made easily liable to rupture. Nor yet was the apoplexy violent in its beginning or progress; as the natural respiration, and the power of feeling and moving, not being wholly taken away, even on the left side, concurred to shew. This was also testified from the speech, together with the internal senses, being once, and again, and even a third time restored; till at length, a laceration in the brain being encreased, and the blood more effused, the disorder became fatal. Nor was the febrile pulse of the least advantage, though it attended from the very beginning of the disorder: nor yet the fever itself, if you will allow me to suppose what, I think, the remission and exacerbation, sometimes observed at the same hour, did in some measure shew. I will even venture to say, that it was of great disadvantage, by strongly agitating and impelling the blood; so that among the many and various things given out by the interpreters of Hippocrates, and other ancient as well as modern physicians, of a fever succeeding an apoplexy, this seems to claim the first place here, that although, in the serous apoplexy, it may sometimes be useful, yet it is rather hurtful in the sanguineous; and the very experienced Werlhof has affirmed, that an apoplexy is rarely solved by a succeeding fever. But on the other hand, blood letting had all the advantage it could possibly have, especially from the jugular vein; and that the right too, as Valsalva, who flew from Bologna to the cardinal, has laid down as a maxim, taken from his observations on patients afflicted with a hemiplegia, and as dissection also at that time confirmed. For the mischief was in the right side of the brain, whereas the left side of the body was resolved; which you will find was also the case in the dissections that follow.

A strumpet, of about twenty years of age, had laboured many months under a slow fever, a cough, an ill-conditioned expectoration, and a wasting of the whole body. She complained of a pain in the left part of the thorax, so that she could scarcely bear to lie down upon it. She was troubled with a difficulty of breathing. To which was added a copious spitting of blood: but this being checked, and two days after a south wind blowing hard, in which state of air those who labour under a similar disorder, for the most part, perish, death put an end to her disease.

The right lobe of the lungs adhered very little to the ribs. Both of them abounded with hard tubercles, which inclined to a white colour, and resembled glandular bodies. Besides, the upper lobules of the lungs, on both sides, had other disorders in that upper part. For the right lobe, towards the sternum, contained a large hollow ulcer, and in this a purulent matter; but the left, towards the side, contained a hard substance, equal to the bigness of a large pear, which, in some measure, resembled the substance of the pancreas, when indurated; and in the middle of this substance was a small ulcer, full of pus. In the pericardium was a small quantity of serum: in the left ventricle of the heart was a small polypous concretion, in the right was one of a moderate size, the greater production of which was inserted into the neighbouring auricle.

MATTHEW BAILLIE

DISEASED APPEARANCES OF THE LUNGS

[From *The Morbid Anatomy of Some of the Most Important Parts of the Human Body,* Chapter IV]

INFLAMMATION

When a portion of the lungs is inflamed, its spungy structure appears much redder than usual, the colour being partly florid and partly of a darker hue. This arises from a much greater number of small vessels than usual, being distributed upon the cells of the lungs which are capable of admitting the red globules of the blood. There is also an extravasation of the coagulable lymph into the substance of the lungs, and sometimes of the blood. The extravasated blood has been said upon some occasions to be in very large quantity; but this has never fallen under my own observation.

In consequence of the greater quantity of blood being accumulated in the inflamed portion of the lungs, they become considerably heavier, and will frequently sink in water. The pleura covering the inflamed portion of the lungs has generally a similar affection; it is crowded with fine red vessels, and has generally lying upon it a layer of coagulable lymph.

This inflamed state of the lungs is to be distinguished from blood accumulated in some part of them after death, in consequence of gravitation. From the body lying in the horizontal posture after death, blood is often accumulated at the posterior part of the lungs, giving them there a deeper colour, and rendering them heavier. In this case there will be found no crowd of fine vessels filled with blood, nor any other mark of inflammation of the pleura. Where blood too is accumulated in any part of a lung after death from gravitation, it is always of a dark colour; but where blood is accumulated from inflammation, portions of the inflamed part will appear florid.

It is very common to find abscesses formed in the lungs. These sometimes consist of small cavities containing pus, and at other times the cavities are very large, so that the greater part of the substance of the lungs has been destroyed. These cavities sometimes communicate only with branches of the trachea, which are destroyed in the progress of the ulceration; at other times they open into the cavity of the chest, emptying their contents there, and forming the disease which is called empyema. Where abscesses are deeply seated in the substance of the lungs, the pleura is commonly not affected; but where abscesses are formed near the surface, it is almost constantly inflamed. The lungs round the boundaries of an abscess, when it has arisen from common inflammation, are more solid in their texture, in consequence of coagulable lymph being thrown out during the progress of the inflammation. When the abscesses are scrofulous, the texture of the lungs in the neighborhood is sometimes not firmer than usual, but presents the common natural appearance. This I believe to be principally the case when the abscesses are small, and placed at a considerable distance from each other. When a portion of the lungs is crowded with tubercules, and some of these are converted into abscesses, the intermediate substance of the lungs is often of a very solid texture. When blood vessels are traced into an abscess of the lungs, I have found them, upon examination, very much contracted, just before they reach the abscess, so that the opening of their extremities has been closed up entirely. On such occasions it will require a probe to be pushed with a good deal of force, in order to open again their extremities. The late ingenious Dr. Stark has found in some of these vessels the blood coagulated. This change in the blood vessels is no doubt with a view to prevent large haemorrhages from taking place, which would certainly be almost immediately fatal.

When abscesses of the lungs are the consequence of common inflammation, they are comparatively under the most favorable circumstances for recovery; but they are much more frequently the consequence of an inflammation depending on a particular constitution, viz. what is called scrofulous; and in this case they are almost always fatal.

TUBERCLES

There is no morbid appearance so common in the lungs as that of tubercles. These consist of rounded firm white bodies, interspersed through their substance. They are I believe formed in the cellular structure, which connects the air cells of the lungs together, and are not a morbid affection of glands, as has been frequently imagined. There is no glandular structure in the cellular connecting membrane of the lungs; and on the inside of the branches of the trachea, where there are follicles, tubercles have never been seen. They are first very small, being not larger than the heads of very small pins, and in this case are frequently accumulated in small clusters. The smaller tubercles of a cluster probably grow together, and form one large tubercle. The most ordinary size of tubercles is about that of a garden pea, but they are subject in this respect to much variety. They adhere pretty closely to the substance of the lungs, and have no peculiar covering or capsule. When cut into, they are found to consist of a white smooth substance, having great firmness, and often contain in part a thick curdly pus. When a tubercle is almost entirely changed into pus, it appears like a white capsule in which the pus is lodged. When several tubercles of considerable size are grown together, so as to form a pretty large tuberculated mass, pus is very generally found upon cutting into it. The pus is frequently thick and curdly, but when in considerable quantity it is thinner, and resembles very much the pus from a common sore. In cutting into the substance of the lungs, a number of abscesses is sometimes found from pretty large tubercles having advanced to a state of suppuration. In the interstices between these tubercles, the lungs are frequently of a harder, firmer texture, with the cells in a great measure obliterated. The texture of the lungs on many occasions, however, round the boundaries of an abscess, is perfectly natural.

I have sometimes seen a number of small abscesses interspersed through the lungs, each of which was not larger than a pea. The pus in these is rather thicker than what arises from common inflammation, and resembles scrofulous pus. It is probable that these abscesses have been produced by a number of small scattered tubercles taking on the process of suppuration. The lungs immediately surrounding these abscesses are often of a perfectly healthy structure, none of the cells being closed up by adhesions.

When tubercles are converted into abscesses, it forms one of the most distinctive diseases in this island, viz. pthisis pulmonalis. Tubercles are sometimes found in the lungs of children at a very early age, viz. two or three years old, but they most frequently occur before the completion of the growth. They are apt likewise to be formed at rather an advanced age.

DISEASED APPEARANCES OF THE STOMACH

[From *The Morbid Anatomy of Some of the Most Important Parts of the Human Body*]

INFLAMMATION

It sometimes happens, although not very frequently (unless poisons have been swallowed) that inflammation takes place in the stomach, so as to spread over a very considerable portion of its inner membrane, or perhaps the whole of it. It is much more common, however, for inflammation to occupy a smaller portion of the stomach. In such cases too, the inflammation is generally not very violent. The stomach upon the outside, at the inflamed part, shows a greater number of small vessels than usual, but frequently not much crowded. In opening into the stomach, it is found to be a little thicker at the inflamed part, the inner membrane is very red from the number of small florid vessels, and there are frequently spots of extravasated blood. It does not often occur that a common inflammation of the stomach proceeds to form pus, or to terminate in gangrene.

When arsenic has been swallowed (which is the poison most frequently taken) the stomach is affected with a most intense degree of inflammation. Its substance becomes thicker, and in opening into its cavity there is a very great degree of redness in the inner membrane, arising partly from the very great number of minute vessels, and partly from extravasated blood. Portions of the inner membrane are sometimes destroyed, from the violent action that has taken place in consequence of the immediate application of the poison. I have also seen a thin layer of the coagulable lymph thrown out upon a portion of the inner surface of the stomach. Most commonly too some part of the arsenic is to be seen in the form of a white powder, lying upon different portions of the inner membrane.

In opening the bodies of persons who have died from hydrophobia, the inner membrane of the stomach is frequently found inflamed at the cardia, and its great end. This inflammation, however, is generally inconsiderable, is, I believe, never very great, and in some instances has been said to be wanting.

ULCERS OF THE STOMACH

Opportunities occasionally offer themselves of observing ulcers of the stomach. These sometimes resemble common ulcers in any other part of the body, but frequently they have a peculiar appearance. Many of them are hardly surrounded with any inflammation, nor have they irregular eroded edges as ulcers have generally, nor is there any particular diseased alteration in the structure of the stomach in the neighbourhood. They appear very much as if some little time before a part had been cut out from the stomach with a knife, and the edges had healed, so as to present an uniform smooth boundary round the excavation which had been made. These ulcers sometimes destroy only a portion of the coats of the stomach at some one part, and at times destroy them entirely. When a portion of the coats is destroyed entirely, there is sometimes a thin appearance of the stomach surrounding the hole, which has a smooth surface, and depends on the progress of the ulceration. At other times the stomach is a little thickened surrounding the hole; and at other times still, it seems to have the common natural structure.

SCIRRHUS AND CANCER OF THE STOMACH

This affection of the stomach is not very uncommon towards an advanced period of life, and, I think, is more frequently met with in men than in women. This, perhaps, arises from the greater intemperance in the one sex, than in the other. It cannot, however, be produced entirely by intemperance; there must be added a considerable predisposition of the parts towards this disease. Hence, when there is no previous disposition the stomach does not become affected with this disease, whatever be the intemperance. When, however, there is the previous disposition, there is reason to think that it is encouraged and brought forwards by this mode of living.

Scirrhus sometimes extends over almost every part of the stomach, but most commonly it attacks one part. The part which is affected with scirrhus has sometimes no very distinct limit between it and the sound structure of the stomach, but most commonly the limit is very well marked. When scirrhus attacks a portion of the stomach only, it is generally towards the pylorus. The principal reason of this probably is, that there is more of glandular structure in that part of the stomach than any other; and it would appear that glandular parts of the body are more liable to be affected with scirrhus, than parts of the body generally.

When the whole stomach, or a portion of it, is scirrhus, it is much thicker than usual, as well as much harder. When the diseased part is cut into, the original structure of the stomach is frequently marked with sufficient distinctness, but very much altered from the natural appearance. The peritoneal covering of the stomach is many times thicker than it ought to be, and has almost a gristly hardness. The muscular part is also very much thickened, and is intersected by frequent pretty strong membranous septa. These membranous septa are, probably, nothing else than the cellular membrane intervening between the fasciculi of the muscular fibres, thickened from disease. The inner membrane is also extremely thick and hard, and not unfrequently somewhat tuberculated towards the cavity of the stomach.

It very frequently happens that this thickened mass is ulcerated upon its surface, and then a stomach is said to be cancerous. Sometimes the inner membrane of the stomach throws out a process which terminates in a great many smaller processes, and produces what has been commonly called a fungous appearance.

It also happens that the stomach at some part loses entirely all vestige of the natural structure, and is changed into a very hard mass, of a whitish or brownish colour, with some appearance of membrane intersecting it; or it is converted into a gristly substance, like cartilage somewhat softened. The absorbent glands in the neighbourhood are at the same time commonly enlarged, and have a very hard white structure.

I have seen several instances of a scirrhus tumour being formed in the stomach about the size of a walnut, while every other part of it was healthy. This tumour has most frequently a small depression near the middle of its surface, and appears a little radiated in its structure. While this tumour remains free from irritation, the functions of the stomach are probably very little affected by it; when, however, it is irritated, it must occasion very considerable disorder in the functions of the stomach, and perhaps lay the foundation of a fatal disease.

A part of the stomach is occasionally formed into a pouch by mechanical means, although very rarely. I have seen one instance of a pouch being so formed, in which five halfpence had been lodged. The coats of the stomach were thinner at that part, but were not inflamed or ulcerated. The halfpence had remained there for some considerable time, forming a pouch by their pressure, but had not irritated the stomach in such a manner as to produce inflammation or ulceration.

The orifice of the stomach may be almost, or perhaps entirely, shut up by a permanent contraction of its muscular fibres, either at the cardia or the pylorus. It is much more likely, however, to occur most frequently at the pylorus, from natural circumstances: There is both less opposition to prevent the contraction at this orifice of the stomach than at the cardia, and there is also belonging to it a stronger and more direct circular muscular power. A less contraction too, at the pylorus, will produce an obstruction in the canal, than at the cardia. I have seen one instance of this contraction at the pylorus, which, even there, is a very rare disease. The contraction was so great as hardly to admit a common goose quill to pass from the stomach into the duodenum, and it had prevented a number of plum stones from passing, which were therefore detained in the stomach.

The stomach is sometimes found so contracted through the whole of its extent as not to be larger than a portion of the small intestine; at other times, it is enlarged to much more than its ordinary size. Neither of these appearances are to be considered as arising from disease. They depend entirely on the muscular fibres of the stomach being in a state of contraction or relaxation at the time of death. It happens, I think, more frequently that the stomach is dilated than contracted.

The stomach is very commonly found, in a dead body, flaccid and almost empty, but not unfrequently it is found more or less distended with air: this air may have been formed after death, but it is often formed during life. When this is the case, we may suppose it to be produced by a new chemical arrangement in the contents of the stomach; but, I believe, it more frequently happens that air is separated from the blood in the blood vessels of the stomach, and poured by the small exhalants into its cavity. This has been more particularly taken notice of by Mr. Hunter, in his Essay upon Digestion, and by myself, in a paper which is published in the Medical and Chirurgical Transactions.

In looking upon the coats of the stomach at its great end, a small portion of them there appears frequently to be thinner, more transparent, and feels somewhat more pulpy than is usual, but those appearances are seldom very strongly marked. They arise from the action of the gastric juice resting on that part of the stomach in greater quantity than anywhere else, and dissolving a small portion of its coats. This is therefore not to be considered as the consequence of a disease, but as a natural effect, arising from the action of the gastric juice, and the state of the stomach after death. When the gastric juice has been in considerable quantity, and of an active nature, the stomach has been dissolved quite through its substance at the great end, and its contents have been effused into the general cavity of the abdomen. In such cases the neighbouring viscera are also partially dissolved. The instances, however, of so powerful a solution are rare, and have almost only occurred in persons who, while in good health, had died suddenly from accident. The true explanation of these appearances was first given by Mr. Hunter, and published, at the request of Sir John Pringle, in the Philosophical Transactions.

Tumours consisting of a suetty substance, have been sometimes found in the stomach, but they are to be considered as a very rare appearance of disease. Ruysch relates, that he has seen a tumour from the stomach of a man which contained hair, together with some dentes molares; and this he has preserved in his collection.

Calculi with different appearances have been described as being occasionally found in the stomach. They never come under my own observation, and are to be reckoned very uncommon.

Papillae and pustules, somewhat resembling the small-pox, have also been described as being formed on the inner membrane of the stomach, but these are exceedingly rare.

Even true small-pox pustules have been said to be formed in the stomachs of persons who have died from the disease. In later dissections, however, this appearance has not been observed, so that, at least, it must be considered as very uncommon.

XXIV

PEDIATRICS

CHILDREN'S DISEASES ARE REFERRED TO, IN PASSING, BY SOME OF THE ANCIENT authors—Soranus of Ephesus, Paul of Aegina, and Rhazes, for example. There was a short manuscript, *Liber de passionibus puero um Galeni* (Book on the Diseases of Children of Galen), which was circulated during the Middle Ages; it was probably not by Galen. Sudhoff thought that it originated about the sixth century, and that it was probably founded on an ancient text. The "Rosengarten" books contained instructions for the care of infants during the puerperium.

The first printed book to be devoted especially to children's diseases was that of Paulus Bagellardus (?-1492), *Libellus de ægritudinibus infantum* (The Little Book of the Diseases of Children), which was printed in 1472. Bagellardus was an Italian. The text is translated in extenso in Ruhräh's *Pediatrics of the Past*.

Similar volumes are those of Metlinger (1473) in German; Roelants (1484), translated with embellishments by Austrius (1538); the *The Boke of Children* of Phayre, the first to be printed in English (1544); the books on children's diseases of Würtz (circa 1570), and of Mercuriali (1583).

Francis Glisson (1597-1677), Regius Professor of Physic, at Cambridge, a founder of the Royal Society and Royal College of Physicians, published in 1650, a splendid account of infantile rickets and scurvy—*De rachitide sive morbo puerili qui vulgo the rickets dicitur, tractatus*. The quotation given here is from the translation of Nicholas Culpepper.

Nils Rosén von Rosenstein (1706-1773) wrote, in 1752, a systematic treatise on the treatment of infants and children. It has been said that he laid the foundation of pediatrics as a specialty.

William Cadogan (1711-1797), of London, published, in 1748, a treatise on the care and feeding of infants, the first formal presentation of the subject from a scientific standpoint.

Eduard Heinrich Henoch (1820-1910), of Berlin, in his *Lectures on Diseases of Children* (1881), may be said to have initiated the modern concept of pediatrics, which has since been elaborated by Escherich, Baginsky, Czerny, Huebner, Still, Finkelstein, Jacobi, Holt, and Abt.

More extensive selections and discussion of pediatrics may be found in *Pediatrics of the Past*, by Ruhräh (New York, 1925), and in *History of Pediatrics* by Still (Oxford Press, 1931).

PAULUS BAGELLARDUS

ON THE CARE OF INFANTS DURING THE FIRST MONTH

[From *The Little Book of the Diseases of Children**]

When the infant at the command of God emerges from the womb, then the midwife with eager and gentle hand should wrap it up in a linen cloth which is not rough, but rather smooth and old, and place it on her lap, noting whether the infant be alive or not or spotted, i.e., whether black or white or of bluish color and whether it is breathing or not. If she find it warm, not black, she should blow into its mouth, if it has no respiration, or into its anus; but if, as sometimes happens, the anus is closed by a little skin, she should cut it with a sharp knife or hot gold thread or some similar instrument. If the infant is alive and of bluish color, then she should cut the umbilicus or umbilical vein, letting it out to four fingers in length and tying it with the twisted cord itself or with twisted wool or silk, yet with a loose knot, lest the infant suffer pain. And thus you allow it to stay until the fall or consolidation of the umbilicus. But if the umbilicus does not consolidate, then she should cover it with powdered myrrh or aloe, or what is better, with powdered myrtle.

Then having tied the umbilicus, the midwife should lay the infant in a basin or mostellum [pot?] or some similar vessel filled with sweet water, comfortably warm, not stinging nor cold, or salty, according to the custom of the Greeks. And she should introduce the infant into this water or bath, its head elevated with her left hand, while with her right hand she should shape its head, its sightless eyes, cleanse its nostrils, open its mouth, rub its jaws, shape its arms and its hands and everything. Next she should wrap it up in a linen cloth made comfortably warm and rub the infant's body.

After this, she should cover the infant's head with a fine linen cloth after the manner of a hood. Then secure a soft linen cloth and with the infant placed on the midwife's lap in such a way that its head is toward her feet and its feet rest upon her body, the midwife should roll it in the linen cloth, after it has been bathed, wrapping its feet. First with its arms raised above, she should wrap its breast

* Libellus de egritudinibus infantium, 1472. Translation by Ruhräh in Pediatrics of the Past. New York, 1925.

and bind its body with a band, by three or four windings. Next the midwife takes another piece of linen or little cloth and draws the hands of the infant straight forward towards the knees and hips, shaping them evenly, so that the infant acquires no humpiness. She then, with the same assisting band, binds and wraps the infant's arms and hands, all of which will be correctly shaped.

On Convulsions of Children

[From *The Little Book of the Diseases of Children,* Chapter III]

Convulsions happen to children from repletion or inanition; from repletion especially in the fleshy, from inanition in infants either because of fevers which give rise to convulsions or because of prolonged crying, whence it happens sometimes that infants stiffen so that they cannot be bent either upward or downward, which condition according to some is called alcuhes or alcuses. If, therefore, they suffer convulsions from repletion, the nurse should be watched, lest her milk decline to excessive humidity, i.e., lest the milk be watery, but she should be nourished more on foods which tend to fatten the milk, and the infant must abstain from excessive nursing and excessive sleep, likewise its members should be smeared with the oil of iris, i.e., of white lilies, or with the oil of alkirum.

If, however, it happens that the infant is dried up, the nurse should be fed with foods which are moist in substance as well as quality, such as are praiseworthy meats boiled with cold and moist things, and especially if such a convulsion exist after continued fevers. Moreover, the joints should be smeared with violet oil and a little wax.

Some, however, put the infant into a bath. From boiling down the heads of young porkers or wether sheep or young goats until their melting, they make a decoction of these meats, and put the infant in such a bath. Then they make an ointment of the above-mentioned oils. Yet I know from experience that I have seen many infants so stiff that they could not be bent upward or downward, who, by the mere application on the spondyles of the neck of oil of white lilies or wet hyssop, are relieved and cured by the favor of the Lord from such a contraction. But among the remedies especially tried, particularly when a humid convulsion persists, is theriaca magna or andromachi.

NICHOLAS ROSEN VON ROSENSTEIN

CHILDREN'S DISEASES

[From *The Diseases of Children and their Remedies**]

DENTITION

[Chapter VIII]

As soon as a child arrives at the age of four months, and becomes then indisposed, we generally suspect its indisposition is occasioned by its teeth, though this oftentimes has quite a different appearance, which not being taken notice of, increases, and gets the better of the child: we ought therefore not to cure it inconsiderately, but make a diligent enquiry into the real cause of the disease, whether it is or is not occasioned by dentition. We may easily know this by the following signs.

1. By the child's frequently putting its finger, or any thing it lays hold of, into the mouth, or by biting hard upon the nipples of the breast; the reason of its doing this, is, to allay the itching which the pressure of the growing teeth occasions in the gums.

2. If the child spits much, or swallows its spittle. In the latter case, it will for the most part have reachings or a diarrhoea, which is of great relief in the cutting of teeth.

3. If the gums are tender, swelled, or already inflamed; the first of these we may know by the nurse's feeling in the child's mouth, and the latter may be seen by looking at them. The child will also feel some kind of pain or smart by sucking.

4. If the tonsils, eyes, or cheeks, swell and become red.

Should all these signs be observed, the child is beyond a doubt affected by its teeth, and every thing will happen in the above-mentioned order. When the teeth are emitted one by one, the child's pain will be trifling, but if several pierce the gums at the same time, dentition will be accompanied with fever, anxiety, startings, convulsions in the muscles of the face and whole body, and sometimes much worse, ending either in a lethargy or death itself. Dentition is both early and easy, when the child is born the full time after con-

---

* Underrattelser on bornasjukdomar och deras botemedel, Stockholm, 1765. Translated from the Swedish by Andrew Sparrmann. London, 1776.

ception, and of healthy parents; and its mother during her pregnancy has not been subject to violent passions, but on the contrary been of a good and cheerful disposition, has eaten good food, and given the child wholesome milk. The more we neglect the above-mentioned rules, the more difficult dentition will be, and the child consequently sooner lose its life. Those which we call eye-teeth (*Dentes Canini*) and those opposite them, in the under jaw, are the most troublesome of all; and more especially if they, as generally happens, shoot out later than the foremost grinders; as they then must be squeezed between them and the fore-teeth, which are very often situated very close to one another.

A tooth before it can shoot out must first pierce the gums; in case the teeth should not be hard enough, they cannot work their way through, or if the gums are too thick, more time will be required for piercing them, as a greater number of fibres are then to be torn asunder, this occasions irritation and pain, both of which cause a heat in the mouth, a greater flow of humours to the part affected, swellings, inflammations, and restlessness; from this we learn that dentition becomes easy if we gain by giving the child wholesome mother's or nurse's milk; and the former may be done by the nurse's rubbing her finger gently on the gums in the manner above mentioned. By this means the gums will become so thin, by the time the child is at the age of three months, that the teeth will shoot out almost without occasioning the least sensation. I do not write this merely as a theoretical reasoning, but from repeated experience. It will equally be of service, to let the child bite upon a wolf's tooth, or any other hard thing; the only inconvenience this may be attended with, is, that the child by playing with it, may easily hurt itself in the face, and especially in the eyes; if by neglecting part or the whole of these precautions, dentition already be very difficult, with the above-mentioned bad symptoms, we must (1) relax the gums, and (2) prevent as much as possible the sensation of pain occasioned by the pressure of the tooth. As to the first, the gums may be softened and relaxed by frequently touching the tumid and pained part with warm honey, syrup of violets, althaea root, or some good oil of olives, or of almonds cold drawn; fresh marrow or butter, fresh brains of a hare, or a calf, or any other animal; the more relaxing and aperient to the gums, the better they are.

ON WORMS

[Chapter XXII]

The *ascarides* are often expelled by eating raw carrots, or drinking birch-juice, or by sucking the juice of the young bark of fir, till one gets a looseness; also by tying a string to a piece of fresh pork, introducing it into the *intestinum rectum*, and pulling it out again after a little time; for a number of these worms will then always follow. This must be done repeatedly, changing the pork at each time, in order to evacuate them all. One may likewise eradicate them with clysters of tepid milk and a little salt, or with our common mineral waters and salt; likewise with a clyster of a drachm of fine sugar and an equal portion of rats-dung, well rubbed together, and mixed with tepid milk (not boiled), to be injected five or six nights running.

The following clysters will likewise prove a good remedy: Take one pint of water and an ounce of quicksilver, boil it gently in a covered earthen pot, and add a little honey to it. This, being injected repeatedly, will certainly deliver the patient from these guests.

But the most efficacious remedy is a clyster of tobacco-smoke.

ON HOOPING COUGH

[Chapter XIX]

The hooping-cough, like the small-pox, measles, and the venereal disease, never appeared in Europe originally, but was transported thither from other parts of the world by means of merchandise, seamen, and animals: it was a new disease to our ancestors in Europe, and probably was conveyed to them either from Africa or the East-Indies, where it was rooted before.

Its first appearance in *Sweden* cannot be determined with any certainty; but in *France* it began in the year 1414.

It is likewise observable that the hooping-cough always appears as an epidemical disease. I think its nature is easily to be understood, since I have many times plainly perceived it to be contagious, and that it infects only such children who have not yet had it. Therefore it infects in the same manner as the measles or small-pox. I knew the hooping-cough conveyed from a patient to two other children in a different house by means of an emissary. I have even myself carried it from one house to another undesignedly.

A person who has once had the hooping-cough is as secure from the danger of catching that disorder again as those who have had the small-pox and measles are with regard to those respective diseases. During my practice I never found or heard of any one who has been infected with the hooping-cough more than once.

It comes on only by degrees, and is at first dry, but when it has continued ten or twelve days, it turns humid, and the matter which is then coughed up looks ripe; nevertheless it increases more and more, leaving long intervals; the fits return at certain hours, but continue at each time with such violence and for so long a time, that the child grows blue in the face, its eyes look as if they were forced out, and they run besides, and a bleeding of the nose is sometimes brought on; it coughs till it is quite out of breath, that one is in apprehension of its being choaked; for if the patient now and then is capable of drawing some breath, it is with a sounding which very much indicates with what difficulty the lungs can admit the air. The coughing continues, and does not leave off for that time, till the child vomits up a quantity of slime. If at any time the coughing should intermit without the paroxysm being ended with a vomiting, it will immediately return again, and will not cease but after a vomiting. If the paroxysm happens to come on immediately after the child has taken nourishment, it will grow blue in the face, stumble, and be stifled if one does not quickly excite vomiting by irritating its throat with a finger. Therefore such patients should not be left alone by themselves, but be attended by a sensible person who has a presence of mind, and who will besides be of service in preventing them from hurting themselves, for they will generally tumble down during the fit. However, they commonly lay hold of something when the coughing seizes them, for instance a chair or table, keeping it fast with all their strength, whilst they during that time are stamping with their feet. The chincough is called *Coqueluche* in *France*, because they formerly thought it to arise from a running of the head, and that it was to be cured by keeping the head warm by a cap. We have not received any particular name for the chincough from the ancient Romans and Greeks, as it was not then known to them.

It is worth our while to take this into consideration; for the disease is both tedious and severe. When it is left to the course of nature alone to be worked out, it commonly will last eleven or twelve weeks, nay frequently half a year. What is still worse, the disease is very dangerous and often fatal. A number of patients are stifled by it, getting convulsions and apoplexies, others pine away entirely, others again are puffed up by it and die. Besides a great number contract ruptures hereby, or become deformed.

FRANCIS GLISSON

THE SIGNS OF THE RACHITIS
[From *De Rachitide**]

. . . We divide the Diagnostical Signs into Pathognomanical and Synedremental. . . .

First a certain laxity and softness, if not a flaccidity of all the first affected parts is usually observed in this affect. The skin is also soft and smooth to the touch, the musculous flesh is less rigid and firm, the joints are easily flexible and many times unable to sustain the body. Whereupon the body being erected it is bent forward or backwards, or to the right side or to the left.

Secondly a certain debility, weakness and enervation befalleth all the parts subsequent to motion. This weakness dependeth much upon the laxity, softness and weakness of the parts aforesaid: for which reason we have placed those signs before this, as also this before the slothfulness and stupification in the next place to be enumerated, which owe much both to the looseness and softness. Moreover, this debility beginneth from the very first rudiments of the disease. For if children be infected within the first year of their age or thereabouts, they go upon their feet later by reason of that weakness, and for the most part they speak before they walk, which amongst us English men, is vulgarly held to be a bad omen. But if they be afflicted with this disease, after they have begun to walk, by degrees they stand more and more feebly upon their legs, and they often stagger as they are going, and stumble upon every slight occasion: neither are they able to sustain themselves long upon their legs without sitting, or to move and play up and down with an usual alacrity, till they have rested. Lastly, upon a vehement increase of the Disease they totally lose the use of their feet; yea, they can scarce

* Batavia, 1650. Translation by Nicholas Culpepper: Glisson on the Rickets. London, 1668.

sit with an erected posture, and the weak and feeble neck doth scarcely, or not at all, sustain the burden of the head.

Thirdly: a kind of slothfulness and numbness doth invade the Joynts preventing after the beginning of the disease, and by little and little is increased, so that daily they are more and more averse from motion and are not delighted like other children with the agitation of their bodies.

The Signs which belong to the disproportioned nourishment of the parts.

First, there is an unusual Bigness of the Head.

Second, the fleshy parts are daily more and more worn away.

Third, certain swellings and knotty excrescences, are observed about some of the joynts.

These are chiefly conspicuous in the wrists, and somewhat less in the ankles. The like tumors also are in the tops of the Ribs, where they are enjoyed in the gristles in the Breast.

Fourthly, some bones wax crooked, especially the bone called the shank bone, and the Fibula or small bone of the leg.

Fifthly, the Teeth come forth both slowly and with trouble.

Sixthly, the Breast in the higher progression of the disease becomes narrow on the sides.

WILLIAM CADOGAN

[From *An Essay upon Nursing and the Management of Children, from their Birth to Three Years of Age*]

You perceive, Sir, by the hints I have already dropped, what I am going to complain of is, that Children in general are over-cloathed and over-fed; and fed and cloathed improperly. To these causes I impute almost all their diseases. But to be a little more explicit. The first great mistake is, that they think a new-born infant cannot be kept too warm: from this prejudice they load and bind it with flannels, wrappers, swathes, stays, &c. which altogether are almost equal to it's own weight; by which means a healthy child in a month's time is made so tender and chilly, it cannot bear the external air; and if, by any accident of a door or window left carelessly open too long, a refreshing breeze be admitted into the suffocating atmosphere of the lying-in bed-chamber, the child and Mother sometimes catches irrecoverable colds. But, what is worse than this, at the end of the month, if things go on apparently well, this hotbed plant is sent out into the country to be reared in a leaky house, that lets in wind and rain from every quarter. Is it any wonder the child never thrives afterwards? The truth is, a new-born Child cannot well be too cool and loose in its dress; it wants less cloathing than a grown person in proportion, because it is naturally warmer, as appears by the thermometer, and would therefore bear the cold of a winter's night much better than any adult person whatever. There are many instances, both antient and modern, of infants exposed and deserted, that have lived several days. As it was the practice of antient times, in many parts of the world, to expose all those whom the parents did not care to be incumbered with; that were deformed, or born under evil stars; not to mention the many Foundlings picked up in LONDON streets. These instances may serve to shew, that Nature has made Children able to bear even great hardships, before they are made weak and sickly by their mistaken Nurses. But, besides the mischief arising from the weight and heat of these swaddling-cloaths, they are put on so tight, and the Child is so cramped by them, that its bowels have not room, nor the limbs any liberty, to act and exert themselves in the free easy manner they ought. This is a very hurtful circumstance; for limbs that are not used will never be strong, and such tender bodies cannot bear much pressure: the circulation restrained by the compression of any one part, must produce unnatural swellings in some other, especially as the fibres of infants are so easily distended. To which doubtless are owing the many distortions and deformities we meet with everywhere; chiefly among Women, who suffer more in this particular than the Men.

I would recommend the following dress: A little flannel waistcoat, without sleeves, made to fit the body, and tie loosely behind; to which there should be a petticoat sewed, and over this a kind of gown of the same material, or any other that is light, thin, and flimsey. The petticoat should not be quite so long as the Child, the gown a few inches longer; with one cap only on the head, which may be made double if it be thought not warm enough. What I mean is, that the whole coiffure should be so contrived, that it might be put on at once, and neither bind nor press the head at all: the linen as usual. This I think would be abundantly sufficient for the day; laying aside all those swathes, bandages, stays, and contrivances.

that are most ridiculously used to close and keep the head in it's place, and support the body. As if Nature, exact Nature, had produced her chief work, a human creature, so carelessly unfinished as to want those idle aids to make it perfect. Shoes and stockings are very needless incumbrances, besides that they keep the legs wet and nasty, if they are not changed every hour, and often cramp and hurt the feet: a child would stand firmer, and learn to walk much sooner without them. I think they cannot be necessary till it runs out in the dirt. There should be a thin flannel shirt for the night, which ought to be every way quite loose. Children in this simple, pleasant dress, which may be readily put on and off without teazing them, would find themselves perfectly easy and happy, enjoying the free use of their limbs and faculties, which they would very soon begin to employ when they are thus left at liberty. I would have them put into it as soon as they are born, and continued in it till they are three years old; when it may be changed for any other more genteel and fashionable: tho' I could wish it was not the custom to wear stays at all; not because I see no beauty in the sugar-loaf shape, but that I am apprehensive it is often procured at the expence of the health and strength of the body. There is an odd notion enough entertained about change, and the keeping of children clean. Some imagine that clean linen and fresh cloaths draw, and rob them of their nourishing juices. I cannot see that they do any thing more than imbibe a little of that moisture which their bodies exhale. Were it, as is supposed, it would be of service to them; since they are always too abundantly supplied, and therefore I think they cannot be changed too often, and would have them clean every day; as it would free them from stinks and sournesses, which are not only offensive, but very prejudicial to the tender state of infancy.

The feeding of Children properly is of much greater importance to them than their cloathing. We ought to take great care to be right in this material article, and that nothing be given them but what is wholesome and good for them, and in such quantity as the body calls for towards it's support and growth; not a grain more. Let us consider what Nature directs in the case: if we follow Nature, instead of leading or driving it, we cannot err. In the business of Nursing, as well as Physick, Art is ever destructive, if it does not exactly copy this original. When a Child is first born, there seems to be no provisions at all made for it; for the Mother's milk, as it is now man-

aged, seldom comes till the third day; so that according to this appearance of Nature a Child would be left a day and a half, or two
days, without any food. Were this really the case, it would be a sufficient proof that it wanted none; as indeed it does not immediately;
for it is born full of blood, full of excrement, it's appetites not awake,
nor it's senses opened; and requires some intermediate time of abstinence and rest to compose and recover the struggle of the birth
and the change of circulation (the blood running into new channels), which always put it into a little fever. However extraordinary
this might appear, I am sure it would be better that the Child was
not fed even all that time, than as it generally is fed; for it would
sleep the greatest part of the time, and, when the milk was ready for
it, would be very hungry, and suck with more eagerness; which is
often necessary, for it seldom comes freely at first. But let me endeavour to reconcile this difficulty, that a Child should be born thus
apparently unprovided for. I say apparently, for in reality it is not
so. Nature neither intended that a Child should be kept so long
fasting, nor that we should feed it for her. Her design is broke in
upon, and a difficulty raised that is wholly owing to mistaken management. The Child, as soon as it is born, is taken from the Mother,
and not suffered to suck till the Milk comes of itself; but is either
fed with strange and improper things, or put to suck some other
Woman, whose Milk flowing in a full stream, overpowers the new-
born infant, that has not yet learnt to swallow, and sets it a coughing,
or gives it a hiccup; the Mother is left to struggle with the load of
her Milk, unassisted by the sucking of the Child. Thus two great
evils are produced, the one a prejudice to the Child's health, the
other, the danger of the Mother's life, at least the retarding her
recovery, by causing what is called a milk-fever; which has been
thought to be natural, but so far from it, that it is entirely owing to
this misconduct. I am confident, from experience, that there would
be no fever at all, were things managed rightly; were the Child kept
without food of any kind, till it was hungry, which it is impossible it
should be just after the birth, and then applied to the Mother's
breasts; it would suck with strength enough, after a few repeated
trials, to make the milk flow gradually, in due proportion to the
Child's unexercised faculty of swallowing, and the call of it's stomach.
Thus the Child would not only provide for itself the best of nourishment, but, by opening a free passage for it, would take off the

Mother's load, as it increased, before it could oppress or hurt her; and therefore effectually prevent the fever; which is caused only by the painful distension of the lacteal vessels of the breasts, when the milk is injudiciously suffered to accumulate. . . .

There is usually milk enough with the first Child; sometimes more than it can take: it is poured forth from an exuberant, overflowing urn, by a bountiful hand, that never provides sparingly. The call of Nature should be waited for to feed it with any thing more substantial, and the appetite ever precede the food; not only with regard to the daily meals, but those changes of diet, which opening, increasing life requires. But this is never done in either case, which is one of the greatest mistakes of all Nurses. Thus far Nature, if she be not interrupted, will do the whole business perfectly well; and there seems to be nothing left for a Nurse to do, but to keep the Child clean and sweet, and to tumble and toss it about a good deal, play with it, and keep it in good humour.

When the Child requires more solid sustenance, we are to enquire what, and how much, is most proper to give it. We may be well assured, there is a great mistake either in the quantity or quality of Children's food, or both, as it is usually given them; because they are made sick by it; for to this mistake I cannot help imputing nine in ten of all their diseases. . . .

It is not common for people to complain of ails they think hereditary, 'till they are grown up; that is, 'till they have contributed to them by their own irregularities and excesses, and then are glad to throw their own faults back upon their Parents, and lament a bad constitution, when they have spoiled a very good one.

EDUARD HEINRICH HENOCH

PURPURA

[From *Lectures on Diseases of Children**]

Under this term are included various morbid conditions, the real nature of which is unknown, which present the common peculiarity of producing hemorrhages into the integument, the mucous membranes, and even the parenchyma of organs. These hemorrhages

* Vorlesungen über Kinderkrankheiten, 1881. Translated from the German in Wood's Standard Library of Medical Authors. 1882.

usually occur spontaneously, without any exciting cause—not, as in the congenital hemorrhagic diathesis, from injuries to the skin or the mucous membranes.

I have previously said that we must be very careful, especially in hospital practice and among the poor, not to mistake the remains of flea-bites for true patches of purpura. Especially in infectious diseases (typhoid fever, scarlatina) I was often in doubt whether the specks of blood were due to flea-bites or were caused by the disease, since, as you will remember, true petechiae and larger hemorrhages into the skin may result from infectious processes or from endocarditis. You should therefore never fail to examine the heart in febrile purpura. In a case of endocarditis after scarlatina, which did not even present a distinct valvular murmur, but merely an indistinct first sound, I made this diagnosis, especially on account of a quite extensive purpura, and this was verified by the autopsy.

But I shall now simply discuss those hemorrhages which develope independently of a febrile, general disease, or of endocarditis, and are described as purpura simplex when they merely affect the skin, and as purpura hemorrhagica, or morbus maculosus, when associated with hemorrhages into the mucous membranes. Unfortunately, we know nothing of the nature of these morbid conditions, or even of the anatomical causes of the numerous hemorrhages. The old view that it is due to a vice in the composition of the blood can be proven neither by chemical nor microscopical examination. In several mild cases of morbus maculosus I found the red blood-globules large, full, and in nowise changed with respect to color and number. Small forms were visible only here and there, and the number of white globules was not greater than in the normal condition. Nor has the former theory of diminished coagulability of the fibrin been confirmed, and it was therefore natural that the small blood-vessels should be held responsible. As the hemorrhages could occur from rupture of the vessels, as well as from migration of the red globules through their walls, abnormal friability of the latter was thought of, and, in fact, microscopical changes of the small arteries and capillaries, which are calculated to produce such a result, have been described by various investigators (Hayem, Straganow, and others). Although the occurrence of these changes cannot be denied, I think that they can be taken into consideration in severe and fatal cases alone. If we remember how suddenly morbus maculosus sometimes develops, and how quickly it may disappear, the assumption of any considerable structural changes in the vessels is hardly allowable in such cases, and this very fact proves that we have to deal with various conditions in this disease. The severe form depends, perhaps, upon the changes in the small vessels, while in milder cases we may think of a vaso-motor neurosis, which gives rise to stasis of blood, rupture of the walls of the vessels, or migration of red blood-globules from paralytic dilatation of the smallest vessels. The complication with slight oedema in a series of cases also favors this hypothesis.

Simple purpura, in which hemorrhages into the mucous membranes are absent, occurs occasionally in poorly nourished children. It is more frequently associated with leukaemia and enlargement of the spleen. The specks of blood then are few in number, and, at the most, as large as a lentil. Purpura is observed most frequently, and at the same time more profusely, in children who also complain of pains in the limbs, especially in various joints, or the latter are perhaps swollen, or these symptoms may have been present a few days before. Numerous smaller, and larger, dark red, or bluish, round patches are especially noticed on the legs and feet, while the upper portions of the body are free, or present but few specks. They are not changed by pressure, and here and there present in the centre a papular or more diffuse hardness and prominence caused by coagulation of fibrin. Apart from the previously mentioned spontaneous pains, the tibia, small bones of the feet, and the soles are not infrequently tender on pressure, and movements of the joints are painful, so that walking may be rendered more or less difficult. Occasionally a wheal-like efflorescence (erythema nodosum) is also present, in the middle of which a bluish extravasation of blood can be seen and felt, and I have not infrequently noticed a slight oedema of the dorsum of the foot, though the urine did not contain any albumen. In one boy, in whom extravasations were present on the arms and face, the eyelids, cheeks, and wings of the nose were also oedematous. After a few days the specks usually grow pale, but again form as soon as the pains or swellings in the joints reappear, or perhaps without the latter as soon as the little patients begin to walk, so that a number of weeks may elapse before recovery is complete. In the majority of cases observed by me the affection ran an apyrevial course, was rarely associated with slight irregular elevations of temperature, and with slight or no disturbance of the general condition, and always terminated in complete recovery. In a girl, aged eleven, the pulse was only 68 per minute, and was not entirely regular, though examination of the heart revealed nothing abnormal. In another case the purpura was associated with pemphigoid vesicles as large as a pea, with sero-bloody contents.

There is at present no explanation for the undoubted connection of the purpura with the pains and swellings in the limbs and joints. It is questionable whether the term purpura rheumatica is justified, because the influence of cold and wet cannot always be demonstrated. This etiological factor is especially absent in a more complicated form, in which, in addition to previously mentioned symptoms, vomiting, intestinal hemorrhage, and colic, are also present. I have observed five cases belonging to this variety.

A boy, aged fifteen; gastro-duodenal catarrah, with slight jaundice in consequence of indigestion. A few days later, pains in the joints of the fingers; a few days afterward, purpura upon the thighs, with colic, vomiting, and black stools. At times the colic was extremely severe; region of transverse colon tender and distended. Moderate fever. Disappearance of the symptoms in five days, but a relapse at the end of three days; convalescence in a week. Three relapses in the next few weeks, always attended with bloody stools. Finally complete recovery.

A boy, aged four, suffering from colic, tenesmus, and scanty, bloody stools. At the same time large patches of purpura on the elbows and thighs. Improvement in three days, but new patches on scrotum and prepuce. A few days later another attack of diarrhoea, with streaks of blood and severe colic, then constipation and fresh exacerbation of purpura. Entire duration three weeks.

A healthy girl, aged twelve. For a week rheumatic pains in the limbs, followed by tenderness and swelling of the wrist and ankle joints, with a slight fever; heart intact. A few days later purpura on the abdomen and lower extremities; very severe colic, repeated vomiting, and bloody diarrhoea. Disappearance of the symptoms in five days. Four relapses within a month; finally, complete recovery.

The purpura in these cases was always combined with colic, tenderness of the colon, vomiting, intestinal hemorrhage, and, with one exception, with rheumatic pains, the swelling of the joints being less constant. There was also a characteristic development of the symptoms in exacerbations, with intervals of several days, or even a week, so that the disease was prolonged to three to seven weeks. Fever was not constant, and, when present, was always very moderate. That these symptoms are mutually connected cannot be denied, but how this connection can be explained I am unable to state. I must leave it undecided whether we have to deal with a process similar to that described by Zimmermann in an adult, viz, stenosis of the small intestinal arteries by cellular and nuclear proliferation of the tunica adventitia and media, and consequent multiple necrosis of the intestinal mucous membrane.

The application of an ice-bag to the abdomen, iced milk as nourishment, and an emulsion of almonds and oil, to which I added extr. opii (0.05 : 120.0) when the pains were severe, appeared to act best in treatment. Absolute quiet in bed is here requisite as in ordinary purpura rheumatica. In many of the latter cases I think I have obtained good results by the administration of iodide of potassium.

The forms of purpura previously considered are distinguished from the variety for which I would reserve the term purpura hemorrhagica or morbus maculosus, by the entire absence of pain, swelling of the joints and the previously mentioned symptoms. We merely observe purpura and hemorrhages which are confined, in the majority of cases, to the gums and nose. Only in exceptional cases could I detect the presence of blood or albumen in the urine. We often find small extravasations of blood on the mucous membrane of the lips and cheeks; they are not situated loosely upon the mucous membrane, but are infiltrated into its superficial layer, so that a flat loss of substance is visible after their removal. In almost all of my cases the disease began suddenly in the midst of perfect health. Hemorrhagic spots of a dark-red, occasionally brownish red or bluish color, from the size of a millet-seed to that of a five-cent piece, spread irregularly over the entire surface of the skin, so that it may be flecked like a leopard's skin within twenty-four to thirty-six hours. Here and there the hemorrhages are in streaks or spread out over the surface. These spots never disappear on pressure, but occasionally a red zone appears around a central spot of coagulation, and its extreme hyperaemic border may pale momentarily on pressure. If hemorrhage from the mouth occurs, the coagula situated between the teeth may impede mastication. Contact with the gums may produce hemorrhages as readily as contusion of the skin, and even scratching it with the finger-nail. Pricks of a pin bled freely, and the introduction of a hypodermic needle almost always produced quite a large infiltration of blood into the skin and subjacent connective tissue. The general condition was usually undisturbed. No enlargement of the spleen could be determined, nor abnormalities of the heart or hemorrhages into the fundus of the eye. Exacerbations of the hemorrhages rarely occurred and prolonged the duration of the affection, which usually lasted ten or fourteen days before the spots paled completely. Fever was never observed, and the temperature was not infrequently below the normal (36.9—37.2).

Serious accidents occurred rarely, as, for example, in one boy a profuse epistaxis occurred twice, so that a tampon was required, and in another child the extraction of a tooth gave rise to a hemorrhage lasting thirty-six hours. There is not much danger of exhaustion from profuse hemorrhages; this danger characterizes the severe form of morbus maculosus, which occurs much more rarely, and is perhaps dependent on permanent molecular changes of the small vessels. The gradual development, numerous exacerbations, chronic course, and continually increasing anaemia, distinguish this variety from the ordinary one, which usually runs an acute course; in addition, there are profuse, constantly renewed hemorrhages from various parts—the nose, stomach, intestinal canal, kidneys, external auditory canal, lungs. These rare cases may terminate fatally after lasting from a month to a year, in consequence of exhaustion complicated by anasarca and dropsy of the cavities, or suddenly by hemorrhage into a vital organ, especially the brain. During its long course there are long intervals of apparent good health, and may give rise to delusive hopes which are negatived by the sudden reappearance of the hemorrhages.

I could arrive at no definite conclusion with regard to the cause of morbus maculosus in any of my cases. The majority of the children were from eight to thirteen years old, and appeared to be otherwise perfectly healthy. It was sometimes preceded by scarlatina or measles, concerning whose relations to morbus maculosus I have previously spoken.

The treatment of the acute, milder variety may be purely expectant, and I have entirely discontinued the use of ergotin, which I formerly recommended. However, if you wish to use this remedy, it should be given internally alone, as subcutaneous injections almost always produced considerable infiltrations of blood, which even terminated in suppuration. In the chronic form I would recommend ferruginous preparations, especially liquor ferri sesquichlorati, and the country or mountain air, but only at a moderate elevation; cold-water treatment, which gave good results in at least two of my cases, may also be tried. Individual hemorrhages, if they prove serious, must be treated according to their locality; trial may always be made of ergotin in such cases.

XXV

EIGHTEENTH CENTURY CLINICAL MEDICINE

IT IS DIFFICULT FOR THE MODERN STUDENT TO UNDERSTAND WHY BOERHAAVE and Haller loomed so large in their own time. The explanation perhaps lies in the mysterious thing called personality—they suited their period and their influence as teachers was great. Both were systematists, and to read their writings today reminds the reader of the gibe that medicine is a "succession of forgotten theories."

Hermann Boerhaave (1668-1738), of Leyden, was the founder of the Eclectic School. He made the medical school of Leyden one of the great centers of clinical teaching. His pupils, van Swieten and de Haen, founded the Vienna school. Boerhaave taught chemistry, physics, botany, ophthalmology, and introduced bedside teaching. He was world famous as a consultant, and the story is told that a letter addressed simply "To the Greatest Physician in the World" was delivered to him.

Albrecht von Haller (1708-1777), of Bern, Switzerland was a botanist, anatomist, physiologist, bibliographer, and teacher. His concept of the nervous system made a differentiation between sensation and motor impulse. His system was founded on the conception of irritability as the basic function of life.

The selections that follow indicate the state of the average enlightened practice of medicine during this period.

REFERENCE

Sigerist, H. E. (1) Hermann Boerhaave. (2) Albrecht von Haller. In: The Great Doctors, A Biographical History of Medicine. New York, W. W. Norton, 1933. Pages 185 and 191.

HERMANN BOERHAAVE

Institutes of Medicine
[From *Institutes of Medicine**]

I imagine a man applying himself to study the first principles of medicine, taken up, as it were, with a geometrical consideration of

* Institutiones Medicae, 1708. Translated by an unknown author, 1720.

figures, bodies, weights, speed, construction of machines, and the forces which these produce in other bodies.

While exercising his mind thus, he learns by precept and example how to distinguish clearly the evident from the obscure, the false from the true, and by his very slowness in forming opinions to gain mental balance. . . .

With a vision now cleared by the light of geometry he silently takes in the lessons from dissected bodies and from living animals opened to his gaze. . . .

Indeed whilst adapting to his own use things accomplished and discovered only by the hardest labour of others, he builds for himself a clear idea of the human frame. To this he adds a knowledge of the vital fluids; and tests it in the living person and his excretions by the aids of anatomy, chemistry, hydrostatics, and even of the microscope. . . .

Now he opens and explores the bodies of those whose maladies he has noted; now he studies the disease he induces upon animals; again he groups together all the results of diseases and of remedies he has tried; again he learns the same things from the best authors; finally, arranging and pondering, he adjusts all things to one another, and, by the means which Theory has shown, he obtains at length a certain grasp of the history and cure in each disease.

See then before you the finished picture of the perfect doctor!

Aphorisms on the Diagnosis and Cure of Diseases

617. The symptoms of an acute special fever are particularly these: coldness, shivering, anxiety, thirst, nausea, eructation, vomiting, weakness, warmth, flushing, dryness, delirium, coma, sleeplessness, convulsion, sweating, diarrhoea and inflammatory rash.

986. Respiration that is deep and infrequent points to a blackage of the brain and diseases that have thus arisen or will arise, like coma, lethargy, delirium, etc.

999. Urine that is red but devoid of sediment, in acute diseases teaches (1) a strong movement of inflammation between the constituent parts of a body-fluid, or between vessels and fluids; (2) an intimate and firm union of oil, salt, earth and water with these fluids: (3) hence therefore great overloading in the disease; (4) a duration of long prospect; (5) and great danger therein.

1196. If an ulcer of the lungs has so advanced that the whole quality of the body is thence consumed, consumption of the lungs is said to affect the patient.

1197. The origin of this ulcer is traced from any cause which can so stop the blood in the lungs that it is forced to change into purulent material.

ALBRECHT VON HALLER

ANEURISMS

[From *On the Diseases of the Aorta and the Vena Cava**]

1. Aneurisms of the aorta, near the heart, are no longer of rare occurrence; nevertheless, I am persuaded that I shall not do anything that is displeasing to well-informed medical men, if I relate two cases that have lately occurred, which I saw when the bodies were dissected, and both of which are observations of interest.

2. The first was that of a woman, whose case Wincklerus has related. In her the aorta, where it is attached to the heart, had become so enlarged as to attain a circumference of five inches and two lines. In this dilated part, which was bounded by those vertebrae that were near that vessel, there was considerable ulceration, the internal membrane of the artery being everywhere changed into projecting floating tufts, and being torn and rugged. These tufts consisted for the most part of scales, that were either osseous or that resembled bone; but the muscular and internal coats were healthy. The whole of the thoracic and abdominal aorta was in a similar condition. In the smaller vessels, such as the hypogastric, iliac, uterine, and the other arteries of the pelvis (which we have lately described), there were incrustations, partly osseous, partly flexible and hard, which were so closely connected to the muscular fibres, that the transverse direction of these was imprinted upon them by so many transverse lines. The valves of the aorta were partly indurated, and partly studded with knobs of a stony hardness. The other valves situated in the heart were healthy and natural.

3. But that disease was a much more remarkable one which we saw in the month of January of the present year, in the body of a beggar-woman, who, being found dead at the door of an inn in the

---

* De Aortae Venaeque Cavae gravioribus quibusdam Morbis, Gottingae, 1749. Translation by John E. Erichsen. Sydenham Society, 1844.

neighboring village of Weenden, was, according to law, brought to our school. The body appeared sufficiently healthy, the omentum covering all the viscera of the abdomen. The stomach was very much contracted, being for the most part of smaller diameter than the rectum. The ovaries gave evidence of age, being scirrhous and dry, as they are about the fiftieth year.

4. When we opened the thorax a tumour of the aorta immediately appeared. After leaving the heart this vessel continued of its usual diameter for two inches; but at the lowest part, and in the whole of the arch, as far as the vertebrae, it was of larger size, and its diameter was three inches greater than it usually is. As soon as it reached the vertebrae, it returned to its natural diameter, and continued so for the remainder of its course.

5. On cutting into the tumour, the nature of which we did not well understand, there was found a large quantity of grumous blood about the centre of the artery. A great part of it was collected not so much into coagula as into broad laminae, scarcely a line in thickness; these were tough, pale, resembling membrane, but softer, and were free and floating, being indeed distinct polypi. Lastly, as the coats of the aorta appeared to be five or six lines in thickness, we found that a new accessory membrane growing from the tumour had adhered to its internal tunic, and might easily have been taken for a part of it. It was white, pulpy, and lamellated, being partly of a membranous character and partly composed of a kind of cruor, but it everywhere adhered equally to the internal coat of the artery. In the artery itself there were found many white hard scales, appearing as if full of pus, such as we have just described.

6. But we were much more surprised to find that this membrane was prolonged into the left carotid artery alone, and not into any of the other arterial trunks; so that the vessel throughout the whole of the neck was full of it, no cavity whatever being left: and this abnormal pulp, which was continued from the aorta into the carotid artery, was white, soft but tough, and separable with difficulty, from the true coats of the artery, filling up the vessel in the form of a continuous cylindrical polypus, as far as its division, where it also, being divided, passed into the different branches. In the external carotid, the polypus was continued as far as the origin of the labial artery, which was the first one that was pervious. In the internal carotid, the trunk of which was narrow and constricted, the polypus was prolonged up as far as the carotid canal.

7. But it appeared even more surprising to us that a similar polypous and fibrous pulp existed in the internal jugular vein of the same side, filling up its whole cavity, and terminating at the transverse branch which joins the internal to the external jugular under the parotid gland. At the lower part this pulp was inseparably attached to the parietes of the vein. The other vessels in the body were perfectly free, and the right carotid artery and jugular vein, being pervious, transmitted the blood without difficulty. We made preparations, and had drawings taken of the arteries of the thigh.

8. A little more than three years ago we saw a similar, if not more remarkable disease, in the vena cava. The vessel was blocked up in a woman not much more than forty years of age, between the renal and iliac veins, so that nothing was left in its canal except a fibrous, polypous, and hard fleshy mass. The right spermatic vein, which was enormously enlarged, being an inch in diameter, returned the blood from the lower part of the aorta, opening into the vein of the right ureter, which arose from the iliac of the same side.

9. On inquiring into the progress and formation of these very serious diseases, it is not at all improbable that the aneurism has arisen from the large number of osseous scales which did not admit of dilatation, and which were formed of the concretion of a yellow humour. These offer such an obstacle to the heart, that it being, during life, excited to overcome this resistence, gradually distends the artery very forcibly, thus giving rise to an aneurismal sac. Hence, as often as osseous laminae are found about the heart, so often will a dilatation of the trunk of the artery be met with. The blood moving more slowly in the sac, as is the case in all aneurisms, coagulates, and being converted into grumous masses, from these into membrane and polypi, and lastly into a diseased or abnormal membrane, is pressed by the pulsations of the heart against the parietes of the aorta, to which it adheres. The same polypi being forced into the carotid fill it, and are changed into a pulpy mass, such as is found in the umbilical arteries. I cannot explain the occurrence of these bloody polypi in the accompanying vein in any other way than that the artery being obstructed, the venous blood of the same side is moved on less freely, whence it stagnates, and is converted into a similar matter. I confess that I cannot assign any reason for the obstruction of the vena cava, as in the body no cause that could compress the vein, nor any disease in the arteries, was discovered.

XXVI

MEDICAL LIFE IN THE EIGHTEENTH CENTURY

ALAIN-RENÉ LE SAGE (1668-1747) WAS A BRETON, WHO KNEW HIS WORLD AND his age well by the time he published *Gil Blas of Santillane*, the first two volumes in 1715, the last two in 1735. This picaresque novel details many phases of existence, among them apprenticeship to a Doctor Sangrado, whose methods as described in the excerpt on page 287, may be taken as typical of the early eighteenth century barber-surgeon. The descriptions of the bleeding, purging, and vomiting methods of Sangrado are not unduly exaggerated, as may be seen by comparing the actual instructions from a barber-surgeon's manual given in the first of the following excerpts. This is taken from the *Prattica nova, di tutto quello, ch'al Barbiero s'appartiene* of Cintio D'Amato, Venice, 1669. It is one of those rare books describing and illustrating the barber-surgeon's job.

CINTIO D'AMATO

[From *On the Correct Method of Making Incisions in Veins of the Hand, and on Their Treatment,** Chapter XIX]

Two veins in the hand are used for bleeding, one of which descending from the cephalic vein passes to a point between the thumb and index finger, where it ends, sending out two branches one of which goes down to the thumb while the other goes up to the index finger, as has been observed in anatomy. The other is commonly called the salvatelle [vena cephalica pollicis] which ends between the little finger and the ring finger according to my observations, since it has its origin in the basilic vein, commonly called the liver vein.

The first vein, coming down from the head, should be opened only where the cephalic vein joins the trunk vein, or at the elbow; or in case of aversion such as I have observed in angina, before

*Prattica nova, di tutto quello, ch'al Barbiero s'appartiene, Venice, 1669. Translated by Julian F. Smith, Hooker Scientific Library, Central College, Fayette, Mo., 1940.

letting blood from the hand as already mentioned and then under the tongue both for drawing off matter from the head and for treating ailments of the mouth or face, or for the physician's other purposes. Whatever is required by the physician the vein should be opened above the thumb; nevertheless bleeding should not be effected at this point if the termination of a branch above this digit, as already noted, does not permit such copious bleeding as from the vein itself. If the digit has nerves here it will be more sensitive and the vein will be missed (which heaven forbid!) undoubtedly causing a spasm by piercing the nerve, but between one finger and the other there is less danger and the intended purpose is more easily accomplished in that blood is let more copiously and quickly. This incision can be made either in the right or left hand.

The second vein, as the salvatelle, is concerned in the left hand with affections of the spleen, and in the right hand with affections of the liver, whether inflammation or chronic disease or affections of the transverse membrane.

But to bleed this vein well it is essential first for the attending barber to prepare hot water in which the patient's hand is placed so that the heat swells it and the vein becomes more prominent. In a short time its location can be traced by gentle rubbing with the left thumb so that it appears more quickly, whereupon it is tied with tape or twine at the pulse. When it is tied the patient opens his hand two or three times to bring out the blood present therein. Also, if the vein is found in one place or another in the finger and its location is not positively certain two fingers may be bled well apart from the place.

Be it noted that in thin persons the ligature should not be too tight because of the pain it causes, and all the more because such persons are more sensitive and their vein is harder to find, though it can be detected by palpation. In fat persons the ligature should be very tight and located farther away so that the vein is not shut off. It is then not difficult to find, as may be ascertained in all cases by careful observation.

Palpation being satisfactory, and the vein to be bled having been located with the left hand, and taken in a linen napkin for a firmer hold, so that the patient's hand cannot free itself, his hand is held by the finger and stretched somewhat while the vein is pierced longitudinally. This procedure is safer and less painful (as appears in the drawing). After making the incision the hand is again dipped in hot water so that blood will flow more freely instead of slowly, heat having the property of opening the vein and liquefying the blood. When sufficient blood has been drawn off the wound is bound up as has been described for other bleedings; but note that after cutting the vein the tape or cord should be loosened somewhat so that the coarse, impure blood which is putrefied and infected can escape while coarse and impure blood retained at the opening will prevent escape of the fine healthy blood, which would debilitate the patient, aggravate his illness and corrupt his humors, his health returning unless he suffers a prolonged illness or dies.

ALAIN-RENÉ LE SAGE

THE CANON'S ILLNESS: HIS TREATMENT: THE CONSEQUENCE: THE LEGACY TO GIL BLAS

[From *The Adventures of Gil Blas of Santillane,* Book II, Chapter II]

I stayed three months with the Licentiate Sédillo, without complaining of bad nights. At the end of that time he fell sick. The distemper was a fever; and it inflamed the gout. For the first time in his life, which had been long, he called in a physician. Doctor Sangrado was sent for; the Hippocrates of Valladolid. Dame Jacintha was for sending for the lawyer first, and touched that string; but the patient thought it was time enough, and had a little will of his own upon some points. Away I went therefore for Doctor Sangrado; and brought him with me. A tall, withered, wan executioner of the sisters three, who had done all their justice for at least these forty years! This learned forerunner of the undertaker had an aspect suited to his office: his words were weighed to a scruple; and his jargon sounded grand in the ears of the uninitiated. His arguments were mathematical demonstrations: and his opinions had the merit of originality.

After studying my master's symptoms, he began with medical solemnity: The question here is, to remedy an obstructed perspiration. Ordinary practitioners, in this case, would follow the old routine of salines, diuretics, volatile salts, sulphur and mercury; but purges and sudorifics are a deadly practice! Chemical preparations are edged tools in the hands of the ignorant. My methods are more simple, and more efficacious. What is your usual diet? I live pretty much upon soups, replied the canon, and eat my meat with a good deal of gravy. Soups and gravy! exclaimed the petrified doctor. Upon my word, it is no wonder you are ill. High living is a poisoned bait; a trap set by sensuality, to cut short the days of wretched man. We must have done with pampering our appetites: the more insipid, the more wholesome. The human blood is not a gravy! Why then you must give it such a nourishment as will assimilate with the particles of which it is composed. You drink wine, I warrant you? Yes, said the licentiate, but diluted. Oh! finely diluted, I dare say, rejoined the physician. This is licentiousness with a vengeance! A frightful course of feeding! Why, you ought to have died years ago. How old are you? I am in my sixty-ninth year, replied the canon. So I thought, quoth the practitioner, a premature old age is always the consequence of intemperance. If you had only drank clear water all your life, and had been contented with plain food, boiled apples, for instance, you would not have been a martyr to the gout, and your limbs would have performed their functions with lubricity. But I do not despair of setting you on your legs again, provided you give yourself up to my management. The licentiate promised to be upon his good behaviour.

Sangrado then sent me for a surgeon of his own choosing, and took from him six good porringers of blood, by way of a beginning, to remedy this obstinate obstruction. He then said to the surgeon: Master Martin Onez, you will take as much more three hours hence, and to-morrow you will repeat the operation. It is a mere vulgar error, that the blood is of any use in the system; the faster you draw it off the better. A patient has nothing to do but to keep himself quiet, with him, to live is merely not to die; he has no more occasion for blood than a man in a trance; in both cases, life consists exclusively in pulsation and respiration.

When the doctor had ordered these frequent and copious bleedings, he added a drench of warm water at very short intervals, maintaining that water in sufficient quantities was the grand secret in the materia medica. He then took his leave, telling Dame Jacintha and me, with an air of confidence, that he would answer for the patient's life, if his system was fairly pursued. The housekeeper, though protesting secretly against this new practice, bowed to his superior authority. In fact, we set on the kettles in a hurry; and, as the physician had desired us above all things to give him enough, we began with pouring down two or three pints at as many gulps. An hour after we beset him again; then, returning to the attack time after time, we fairly poured a deluge into his poor stomach. The surgeon, on the other hand, taking out the blood as we put in the water, we reduced the old canon to death's door in less than two days.

This venerable ecclesiastic, able to hold it out no longer, as I pledged him in a large glass of his new cordial, said to me in a faint voice—Hold, Gil Blas, do not give me any more, my friend. It is plain death will come when he will come, in spite of water; and, though I have hardly a drop of blood in my veins, I am no better for getting rid of the enemy. The ablest physician in the world can do nothing for us, when our time is expired. Fetch a notary; I will make my will. At these last words, pleasing enough to my fancy, I affected to appear unhappy; and concealing my impatience to be gone: Sir, said I, you are not reduced so low, thank God, but you may yet recover. No, no, interrupted he, my good fellow, it is all over, I feel the gout shifting, and the hand of death is upon me. Make haste, and go where I told you. I saw, sure enough, that he changed every moment: and the case was so urgent, that I ran as fast as I could, leaving him in Dame Jacintha's care who was more afraid than myself of his dying without a will. I laid hold of the first notary I could find; Sir, said I, the Licentiate Sedillo, my master, is drawing near his end; he wants to settle his affairs; there is not a moment to be lost. The notary was a dapper little fellow, who loved his joke; and enquired who was our physician. At the name of Doctor Sangrado, hurrying on his cloak and hat: For mercy's sake! cried he, let us set off with all possible speed; for this doctor dispatches business so fast, that our fraternity cannot keep pace with him. That fellow spoils half my jobs.

With this sarcasm, he set forward in good earnest, and as we pushed on, to get the start of the grim tyrant, I said to him: Sir, you are aware that a dying testator's memory is sometimes a little short; should my master chance to forget me, be so good as to put in a word in my favour. That I will, my lad, replied the little proctor; you may rely on it. I will urge something handsome, if I have an opportunity. The licentiate, on our arrival, had still all his faculties about him. Dame Jacintha was by his bedside, laying in her tears by wholesale. She had played her game, and bespoken a hand-some remembrance. We left the notary alone with my master, and went together into the ante-chamber, where we met the surgeon, sent by the physician for another and last experiment. We laid hold of him. Stop, Master Martin, said the housekeeper, you cannot go into Signor Sédillo's room just now. He is giving his last orders; but you may bleed away when the will is made.

We were terribly afraid, this pious gentlewoman and I, lest the licentiate should go off with his will half finished; but by good luck, the important deed was executed. We saw the proctor come out, who, finding me on the watch, slapped me on the shoulder, and said with a simper; Gil Blas is not forgotten. At these words, I felt the most lively joy; and was so well pleased with my master for his kind notice that I promised myself the pleasure of praying for his soul after death, which event happened anon; for the surgeon having bled him once more, the poor old man, quite exhausted, gave up the ghost under the lancet. Just as he was breathing his last, the physician made his appearance, and looked a little foolish, notwith-standing the universality of his death-bed experience. Yet far from imputing the accident to the new practice, he walked off, affirming with intrepidity, that it was owing to their having been too lenient with the lancet, and too chary of their warm water. The medical executioner, I mean the surgeon, seeing that his functions also were at an end, followed Doctor Sangrado.

XXVII

SMALLPOX

THE DEVASTATING NATURE OF SMALLPOX BEFORE JENNER'S DAY CAN BE
gathered from Macaulay's description. Before vaccination came into
general use, inoculation was very commonly practiced. With inoculation
variolous matter taken from a human pustule is introduced under the
skin of another human being. This procedure was introduced into
Europe in 1718, by Lady Mary Wortley Montagu, the wife of the English ambassador to Turkey.

Edward Jenner (1749-1823) was born at Berkeley, Gloucestershire.
He became a pupil of John Hunter with whom in later life he carried on
a lively correspondence. After he entered upon practice in Gloucestershire he frequently heard the rumor that dairy maids who had been
inoculated with the cowpox were not capable of having smallpox. He
put this theory to the test, proved it to be true, and in 1798 published
his book, *An Inquiry into the Causes and Effects of the Variolae Vaccinae.* Jenner's discovery established not only the procedure of vaccination but also the science of preventive medicine.

Soon after vaccination was put into actual practice, it was plain to
everyone with experience that it was superior to inoculation (as pointed
out in the report to the House of Commons on the resolution to make
a grant to Edward Jenner) because an inoculated person could spread
smallpox whereas a vaccinated person could not. Furthermore, vaccination itself was a mild affection and inoculation sometimes caused serious
illness. As a matter of historical fact, vaccination rapidly supplanted
inoculation.

Benjamin Waterhouse (1754-1846), professor of physic at Harvard
Medical School, made the first vaccination in America, using his own
children as subjects. His *History of the Kinepox* (1800) is one of the
great American medical classics.

REFERENCES

BARON, J. The Life of Edward Jenner, M.D. London, Henry Colburn,
1838.

CROOKSHANK, E. M. History and Pathology of Vaccination. London,
H. K. Lewis, 1889.

DREWITT, F. D. The Life of Edward Jenner, M.D., F.R.S., Naturalist,
and Discoverer of Vaccination. London, Longmans, Green, 1931.

RODDIS, L. H. Edward Jenner and the Discovery of Smallpox Vaccination. *Military Surgeon,* Vol. 65, Nos. 5, 6, 1929; Vol. 66, No. 1, 1930.

THOMAS BABINGTON MACAULAY

DEATH OF MARY

[From *History of England,** Chapter XX]

He had but too good reason to be uneasy. His wife had, during two or three days, been poorly; and on the preceding evening grave symptoms had appeared. Sir Thomas Millington, who was physician in ordinary to the King, thought that she had the measles. But Radcliffe, who, with coarse manners and little book learning, had raised himself to the first practice in London chiefly by his rare skill in diagnostics, uttered the more alarming words, small pox. That disease, over which science has since achieved a succession of glorious and beneficent victories, was then the most terrible of all the ministers of death. The havoc of the plague had been far more rapid: but the plague had visited our shores only once or twice within living memory; and the small pox was always present, filling the church-yards with corpses, tormenting with constant fears all whom it had not yet stricken, leaving on those whose lives it spared the hideous traces of its power, turning the babe into a changeling at which the mother shuddered, and making the eyes and cheeks of the betrothed maiden objects of horror to the lover. Towards the end of the year 1694, this pestilence was more than usually severe. At length the infection spread to the palace, and reached the young and blooming Queen. She received the intimation of her danger with true greatness of soul. She gave orders that every lady of her bedchamber, every maid of honour, nay, every menial servant, who had not had the small pox, should instantly leave Kensington House. She locked herself up during a short time in her closet, burned some papers, arranged others, and then calmly awaited her fate.

* London, 1848 and 1861.

LADY MARY WORTLEY MONTAGU
[From the *Letters**]

I am going to tell you a thing that I am sure will make you wish yourself here. The smallpox, so fatal and so general amongst us, is here rendered entirely harmless by the invention of ingrafting, which is the term they give it. There is a set of old women who make it their business to perform the operation every autumn, in the month of September, when the great heat is abated. People send to one another to know if any of their family has a mind to have the smallpox; they make parties for this purpose, and when they are met (commonly fifteen or sixteen together), the old woman comes with a nutshell full of the matter of the best sort of smallpox, and asks what veins you please to have opened. She immediately rips open that you offer to her with a large needle (which gives you no more pain than a common scratch), and puts into the vein as much venom as can lie upon the head of her needle, and after binds up the little wound with a hollow bit of shell; and in this manner opens four or five veins. The Grecians have commonly the superstition of opening one in the middle of the forehead, in each arm, and on the breast to mark the sign of the cross; but this has a very ill effect, all these wounds leaving little scars, and is not done by those that are not superstitious, who choose to have them in the legs, or that part of the arm that is concealed. The children or young patients play together all the rest of the day, and are in perfect health to the eighth. Then the fever begins to seize them, and they keep their beds two days, very seldom three. Every year thousands undergo this operation; and the French ambassador says pleasantly, that they take the smallpox here by way of diversion, as they take the waters in other countries. There is no example of anyone that has died in it; and you may believe I am very well satisfied of the safety of this experiment, since I intend to try it on my dear little son.

* Letter 31, edition of 1779.

EDWARD JENNER

[From *An Inquiry into the Causes and Effects of the Variolae Vaccinae,
a Disease Discovered in some of the Western Counties of England,
Particularly Gloucestershire, and Known by the Name
of the Cow-Pox**]

. . . *quid nobis certius ipsis Sensibus esse potest, quo vera ac falsa
notemus.*—LUCRETIUS.

The deviation of man from the state in which he was originally
placed by nature seems to have proved to him a prolific source of
diseases. From the love of splendour, from the indulgence of lux-
ury, and from his fondness for amusement he has familiarized himself with a great number of animals, which may not originally have
been intended for his associates.

The wolf, disarmed of ferocity, is now pillowed in the lady's lap.
The cat, the little tiger of our island, whose natural home is the
forest, is equally domesticated and caressed. The cow, the hog, the
sheep, and the horse, are all, for a variety of purposes, brought under
his care and dominion.

There is a disease to which the horse, from his state of domestica-
tion, is frequently subject. The farriers call it the grease. It is an
inflammation and swelling in the heel, from which issues matter
possessing properties of a very peculiar kind, which seems capable of
generating a disease in the human body (after it has undergone the
modification which I shall presently speak of), which bears so strong
a resemblance to the smallpox that I think it highly probable it may
be the source of the disease.

In this dairy country a great number of cows are kept, and the
office of milking is performed indiscriminately by men and maid
servants. One of the former having been appointed to apply dressings to the heels of a horse affected with the grease, and not paying
due attention to cleanliness, incautiously bears his part in milking
the cows, with some particles of the infectious matter adhering to his
fingers. When this is the case, it commonly happens that a disease
is communicated to the cows, and from the cows to dairy maids,
which spreads through the farm until the most of the cattle and

* London, 1798.

domestics feel its unpleasant consequences. This disease has obtained the name of the cow-pox. It appears on the nipples of the cows in the form of irregular pustules. At their first appearance they are commonly of a palish blue, or rather of a colour somewhat approaching to livid, and are surrounded by an erysipelatous inflammation. These pustules, unless a timely remedy be applied, frequently degenerate into phagedenic ulcers, which prove extremely troublesome. The animals become indisposed, and the secretion of milk is much lessened. Inflamed spots now begin to appear on different parts of the hands of the domestics employed in milking, and sometimes on the wrists, which quickly run on to suppuration, first assuming the appearance of the small vesications produced by a burn. Most commonly they appear about the joints of the fingers and at their extremities; but whatever parts are affected, if the situation will admit, these superficial suppurations put on a circular form, with their edges more elevated than their centre, and of a colour distantly approaching to blue. Absorption takes place, and tumours appear in each axilla. The system becomes affected—the pulse is quickened; and shiverings, succeeded by heat, with general lassitude and pains about the loins and limbs, with vomiting, come on. The head is painful, and the patient is now and then even affected with delirium. These symptoms, varying in their degrees of violence, generally continue from one day to three or four, leaving ulcerated sores about the hands, which, from the sensibility of the parts, are very troublesome, and commonly heal slowly, frequently becoming phagedenic, like those from whence they sprung. The lips, nostrils, eyelids, and other parts of the body are sometimes affected with sores; but these evidently arise from their being heedlessly rubbed or scratched with the patient's infected fingers. No eruptions on the skin have followed the decline of the feverish symptoms in any instance that has come to my inspection, one only excepted, and in this case a very few appeared on the arms: they were very minute, of a vivid red colour, and soon died away without advancing to maturation; so that I cannot determine whether they had any connection with the preceding symptoms.

Thus the disease makes its progress from the horse to the nipple of the cow, and from the cow to the human subject.

Morbid matter of various kinds, when absorbed into the system, may produce effects in some degree similar; but what renders the cow-pox virus so extremely singular is that the person who has been thus affected is forever after secure from the infection of the small-pox; neither exposure to the variolous effluvia, nor the insertion of the matter into the skin, producing this distemper.

In support of so extraordinary a fact, I shall lay before my reader a great number of instances.

Case I. Joseph Merret, now as under gardener to the Earl of Berkeley, lived as a servant with a farmer near this place in the year 1770, and occasionally assisted in milking his master's cows. Several horses belonging to the farm began to have sore heels, which Merret frequently attended. The cows soon became affected with the cow-pox, and soon after several sores appeared on his hands. Swellings and stiffness in each axilla followed, and he was so much indisposed for several days as to be incapable of pursuing his ordinary employment. Previously to the appearance of the distemper among the cows there was no fresh cow brought into the farm, nor any servant employed who was affected with the cow-pox.

In April, 1795, a general inoculation taking place here, Merret was inoculated with his family; so that a period of twenty-five years had elapsed from his having the cow-pox to this time. However, though the variolous matter was repeatedly inserted into his arm, I found it impracticable to infect him with it; an efflorescence only, taking on an erysipelatous look about the centre, appearing on the skin near the punctured parts. During the whole time that his family had the smallpox, one of whom had it very full, he remained in the house with them, but received no injury from exposure to the contagion.

It is necessary to observe that the utmost care was taken to ascertain, with the most scrupulous precision, that no one whose case is here adduced had gone through the smallpox previous to these attempts to produce that disease.

Had these experiments been conducted in a large city, or in a populous neighborhood, some doubts might have been entertained; but here, where population is thin, and where such an event as a person's having had the smallpox is always faithfully recorded, no risk of inaccuracy in this particular can arise.

Case II. Sarah Portlock, of this place, was infected with the cow-pox when a servant at a farmer's in the neighborhood, twenty-seven years ago.

In the year 1792, conceiving herself, from this circumstance, secure from the infection of the smallpox, she nursed one of her own children who had accidentally caught the disease, but no indisposition ensued. During the time she remained in the infected room, variolous matter was inserted into both her arms, but without any further effect than in the preceding case.

Case XVII. The more accurately to observe the progress of the infection I selected a healthy boy, about eight years old, for the purpose of inoculating for the cow-pox. The matter was taken from a sore on the hand of a dairymaid, who was infected by her master's cows, and it was inserted on the 14th day of May, 1796, into the arm of the boy by means of two superficial incisions, barely penetrating the cutis, each about an inch long.

On the seventh day he complained of uneasiness in the axilla and on the ninth he became a little chilly, lost his appetite, and had a slight headache. During the whole of this day he was perceptibly indisposed, and spent the night with some degree of restlessness, but on the day following he was perfectly well.

The appearance of the incisions in their progress to a state of maturation were much the same as when produced in a similar manner by variolous matter. The difference which I perceived was in the state of the limpid fluid arising from the action of the virus, which assumed rather a darker hue, and in that of the efflorescence spreading round the incisions, which had more of an erysipelatous look than we commonly perceive when variolous matter has been made use of in the same manner; but the whole died away (leaving on the inoculated parts scabs and subsequent eschars) without giving me or my patient the least trouble.

In order to ascertain whether the boy, after feeling so slight an affection of the system from the cow-pox virus, was secure from the contagion of the smallpox, he was inoculated the 1st of July following with variolous matter, immediately taken from a pustule. Several slight punctures and incisions were made on both his arms, and the matter was carefully inserted, but no disease followed. The same appearances were observable on the arms as we commonly see when a patient has had variolous matter applied, after having either the cow-pox or smallpox. Several months afterwards he was again inoculated with variolous matter, but no sensible effect was produced on the constitution.

After the many fruitless attempts to give the smallpox to those who had had the cow-pox, it did not appear necessary, nor was it convenient to me, to inoculate the whole of those who had been the subjects of these late trials; yet I thought it right to see the effects of variolous matter on some of them, particularly William Summers, the first of these patients who had been infected with matter taken from the cow. He was, therefore, inoculated from a fresh pustule; but, as in the preceding cases, the system did not feel the effects of it in the smallest degree. I had an opportunity also of having this boy and William Pead inoculated by my nephew, Mr. Henry Jenner, whose report to me is as follows: "I have inoculated Pead and Barge, two of the boys whom you lately infected with the cow-pox. On the second day the incisions were inflamed and there was a pale inflammatory stain around them. On the third day these appearances, were still increasing and their arms itched considerably. On the fourth day the inflammation was evidently subsiding, and on the sixth day it was scarcely perceptible. No symptoms of indisposition followed.

"To convince myself that the variolous matter made use of was in a perfect state I at the same time inoculated a patient with some of it who never had gone through the cow-pox, and it produced the smallpox in the usual regular manner."

These experiments afforded me much satisfaction; they proved that the matter, in passing from one human subject to another, through five gradations, lost none of its original properties, J. Barge being the fifth who received the infection successively from William Summers, the boy to whom it was communicated from the cow.

I shall now conclude this inquiry with some general observations on the subject, and on some others which are interwoven with it.

Although I presume it may not be necessary to produce further testimony in support of my assertion "that the cow-pox protects the human constitution from the infection of the smallpox," yet it affords me considerable satisfaction to say that Lord Somerville, the President of the Board of Agriculture, to whom this paper was shown by Sir Joseph Banks, has found upon inquiry that the statements were confirmed by the concurring testimony of Mr. Dolland, a surgeon, who resides in a dairy country remote from this, in which these observations were made. With respect to the opinion adduced "that the source of the infection is a peculiar morbid matter arising in the horse," although I have not been able to prove it from actual experiments conducted immediately under my own eye, yet the evidence I have adduced appears sufficient to establish it.

They who are not in the habit of conducting experiments may not be aware of the coincidence of circumstances necessary for their being managed so as to prove perfectly decisive; nor how often men engaged in professional pursuits are liable to interruptions which disappoint them almost at the instant of their being accomplished: however, I feel no room for hesitation respecting the common origin of the disease, being well convinced that it never appears among the cows (except it can be traced to a cow introduced among the general herd which has been previously infected, or to an infected servant) unless they have been milked by some one who, at the same time, has the care of a horse affected with diseased heels.

The spring of the year 1797, which I intended particularly to have devoted to the completion of this investigation, proved, from its dryness, remarkably adverse to my wishes; for it frequently happens, while the farmers' horses are exposed to the cold rains which fall at that season, that their heels become diseased, and so cow-pox then appeared in the neighborhood.

The active quality of the virus from the horses' heels is greatly increased after it has acted on the nipples of the cow, as it rarely happens that the horse affects his dresser with sores, and as rarely that a milkmaid escapes the infection when she milks infected cows. It is most active at the commencement of the disease, even before it has acquired a pus-like appearance; indeed, I am not confident whether this property in the matter does not entirely cease as soon as it is secreted in the form of pus. I am induced to think it does cease, and that it is the thin, darkish-looking fluid only, oozing from the newly-formed cracks in the heels, similar to what sometimes appears from erysipelatous blisters, which gives the disease. Nor am I certain that the nipples of the cows are at all times in a state to receive the infection. The appearance of the disease in the spring and the early part of the summer, when they are disposed to be affected with spontaneous eruptions so much more frequently than at other seasons, induces me to think that the virus from the horse must be received upon them when they are in this state, in order to produce effects: experiments, however, must determine these points. But it is clear that when the cow-pox virus is once generated, that the cows cannot resist the contagion, in whatever state their nipples may chance to be, if they are milked with an infected hand.

Whether the matter, either from the cow or the horse, will affect the sound skin of the human body, I cannot positively determine; probably it will not, unless on those parts where the cuticle is extremely thin, as on the lips, for example. . . .

A medical gentleman (now no more), who for many years inoculated in this neighborhood, frequently preserved the variolous matter intended for his use on a piece of lint or cotton, which, in its fluid state, was put into a vial, corked, and conveyed into a warm pocket; a situation certainly favorable for speedily producing putrefaction in it. In this state (not infrequently after it had been taken several days from the pustules) it was inserted into the arms of his patients, and brought on inflammation of the incised parts, swellings of the axillary glands, fever, and sometimes eruptions. But what was this disease? Certainly not the smallpox; for the matter having from putrefaction lost or suffered a derangement in its specific properties, was no longer capable of producing that malady, those who had been inoculated in this manner being as much subject to the contagion of the smallpox as if they had never been under the influence of this artificial disease; and many, unfortunately, fell victims to it, who thought themselves in perfect security. The same unfortunate circumstance of giving a disease, supposed to be the smallpox, with inefficacious variolous matter, having occurred under the direction of some other practitioners within my knowledge, and probably from the same incautious method of securing the variolous matter, I avail myself of this opportunity of mentioning what I conceive to be of great importance; and, as a further cautionary hint, I shall again digress so far as to add another observation on the subject of inoculation. . . .

Thus far have I proceeded in an inquiry founded, as it must appear, on the basis of experiment; in which, however, conjecture has been occasionally admitted in order to present to persons well situated for such discussions objects for a more minute investigation. In the meantime I shall myself continue to prosecute this inquiry, encouraged by the hope of its becoming essentially beneficial to mankind.

BENJAMIN WATERHOUSE

THE HISTORY OF THE KINE-POX, COMMONLY CALLED THE COW-POX

[From *A Prospect of Exterminating the Smallpox—Being the History of the Variolae Vaccinae or Kine-pox commonly called the Cow-pox as it has appeared in England: With an account of a series of Inoculations performed for the Kine-pox in Massachusetts**]

CHAPTER I

In the beginning of the year 1799 I received from my friend Dr. Lettsom of London, a copy of Dr. Edward Jenner's "Inquiry into the causes and effects of the variolae vaccinae, or Cow-pox": a disease totally unknown in this quarter of the world. On perusing this work I was struck with the unspeakable advantages that might accrue to this, and indeed to the human race at large, from the discovery of a mild distemper that would ever after secure the constitution from that terrible scourge, the smallpox.

As the ordinary mode of communicating even medical discoveries in this country is by newspapers, I drew up the following account of the Cow-pox, which was printed in the Columbian Centinal March 12, 1799.

Something Curious in the Medical Line

Everybody has heard of these distempers accompanied by pocks and pustules, called the small-pox, and chickenpox and the swine-pox, but few have ever heard of the cow-pox, or if you like the term better, the cow small-pox; or to express it in technical language, the variolae vaccinae. There is however such a disease which has been noticed here and there in several parts of England, more particularly in Gloucestershire, for sixty or seventy years past, but has never been an object of medical inquiry until lately.

This variolae vaccinae is very readily communicated to those who milk cows infected with it. This malady appears on the teats of the cows. . . . Those who milk the cows thus affected, seldom or ever fail catching the distemper, *if there be cracks, wounds or abrasions of the hands*. . . . But what makes this newly discovered disease so

* Boston, 1800.

very curious, and so extremely important is that every person thus affected is EVER AFTER SECURED FROM THE ORDINARY SMALLPOX, *let him be ever so much exposed to the effluvian of it, or let ever so much ripe matter be inserted into the skin by inoculation.*

Dr. Edward Jenner is the physician in England who has collected and arranged a series of facts and experiments respecting the disease there called the Cow-pox.

CHAPTER II

Under the serious impression of effecting a public benefit, and conceiving it moreover a duty in my official situation in this University, I sent to England for some of the vaccine, or cow-pox matter for trial. After several fruitless attempts, I obtained some by a short passage from Bristol, and with it I inoculated all the younger part of my family.

The first of my children that I inoculated was a boy of five years old, named Daniel Oliver Waterhouse. I made a slight incision in the usual place for inoculation in the arm, inserted a small portion of the infected thread, and covered it with a sticking plaster. It exhibited no other appearances than what would have arisen from any other extraneous substance, until the sixth day when an increased redness called forth my attention. On the eighth day he complained of pain under the inoculated arm and on the ninth the inoculated part exhibited evident signs of virulency. By the tenth anyone much experienced in the inoculated small-pox would have pronounced the arm infected. The pain and swelling under his arm went on gradually encreasing and by the eleventh day from inoculation his febrile symptoms were pretty strongly marked. The sore in the arm proceeded exactly as Drs. Jenner and Woodville described, and appeared to the eye very like the second plate in Dr. Jenner's elegant publication.

The inoculated part in this boy was surrounded by an efflorescence which extended from his shoulder to his elbow, which made it necessary to apply some remedies to lessen it; but the "symptoms," as they are called, scarcely drew him from his play more than an hour or two; and he went through the disease in so light a manner as hardly even to express any marks of peevishness. A piece of true skin was fairly taken out of the arm by the virus, the part appearing as if eaten out by a caustick, a never failing sign of thorough section of the system by the inoculated small-pox.

Satisfied with the appearances and symptoms in this boy I inoculated another of three years of age with matter taken from his brother's arm, for he had no pustules on his body. He likewise went through the disease in a perfect and very satisfactory manner. The child pursued his amusements with as little interruption as his brother. Then I inoculated a servant boy of about 12 years of age, with some of the infected thread from England. His arm was pretty sore and his symptoms pretty severe. He treated himself rather harshly by exercising unnecessarily in the garden when the weather was extremely hot (Fahrt. Therm. 96 in the shade!) and then washing his head and upper parts of his body under the pump, and setting, in short, all rules at defiance in my absence. Nevertheless this boy went through the disorder without any other accident than a sore throat and a stiffness of the muscles of the neck. All which soon vanished by the help of a few remedies.

Being obliged to go from home a few days, I requested my colleague Dr. Warren to visit these children. Dr. Danforth as well as some other physicians, came to Boston out of curiosity, and so did several practitioners from the country. I mention this because it gave rise to a groundless report, that one of the children had so bad an arm that I thought it prudent to take the advice of some of my brethren upon it.

From a full matured pustule in my little boy three years old I inoculated his infant sister, already weaned, of one year. At the same time and from the same pustule, I inoculated its nursery maid. They both went through the disease with equal regularity. . . .

CHAPTER III

Having thus traced the most important facts respecting the causes and effects of the Kine-pox up to their source in England, and having confirmed most of them by actual experiment in America, one experiment only remained behind to complete the business. To effect this I wrote the following letter to Dr. Aspinwall, physician to the smallpox hospital in the neighborhood of Boston,

Cambridge, August 2, 1800.

Dear Doctor:

You have doubtless heard of the newly discovered disorder, known in England by the name of cow-pox, which so nearly resembles the smallpox, that it is now agreed in Great Britain, that the former will pass for the latter.

I have procured some of the vaccine matter, and therewith inoculated seven of my family. The inoculation has proceeded in six of them exactly as described by Woodville and Jenner; but my desire is to confirm the doctrine by having some of them inoculated by you.

I can obtain variolous matter and inoculate them privately, but I wish to do it in the most open and public way possible. As I have imported a new distemper, I conceive that the public have a right to know exactly every step I take in it. I write this, then to enquire whether you will on philanthropic principles try the experiment of inoculating some of my children who have already undergone the cow-pox. If you accede to my proposal, I shall consider it as an experiment in which we have co-operated for the good of our fellow citizens, and relate it as such in the pamphlet I mean to publish on the subject.

I am, etc.

B.W.

Hon. William Aspinwall, Esq.
Brookline.

To this letter the doctor returned a polite answer, assuring me of his readiness to give any assistance in his power, to ascertain whether the cow-pox would prevent the small-pox; observing that he had at that time fresh matter that he could depend on, and desiring me to send the children to the hospital for that purpose. Of the three which I offered, the doctor chose to try the experiment on the boy of 12 years of age, mentioned in page 20, whom he inoculated in my presence by two punctures, and with matter taken that moment from a patient who had it pretty full upon him. He at the same time inserted an infected thread and then put him into the hospital, where was one patient with it the natural way. On the fourth day, the doctor pronounced the arm to be infected. It became every hour sorer, but in a day or two it died off, and grew well, without producing the slightest trace of a disease; so that the boy was dismissed from the hospital and returned home the twelfth day after the experiment. One fact, in such cases, is worth a thousand arguments.

XXVIII

LEOPOLD AUENBRUGGER

LEOPOLD AUENBRUGGER (1722-1809), OF VIENNA, WAS BORN IN GRAZ, AUSTRIA. His *Inventum Novum* (1761) described percussion and pointed out the percussion signs and their significance as proved by postmortem examination. Though the book was twice reprinted in its original Latin form and translated into French by both Rozier and Corvisart, its reception was somewhat cool until Corvisart's translation appeared. The first sentence in the book has been called the most scientific statement ever made.

The influence of the work on the development of clinical medicine was far more profound than we can realize by reading it now—it furnished clinicians with an objective method of investigation, and swept aside all the tortuous theorizing of the eighteenth century: it initiated modern scientific medicine.

The whole work occupies only twenty-four pages of ordinary modern print—little more than an average magazine article. We give the first third in full: in the latter part we have deleted only the passages that are very old-fashioned. It is arranged in formal fashion—a series first of observations, stated very simply, then a scholium or dissertation enlarging on the simple statement of observation.

REFERENCE

FORBES, J. On Percussion of the Chest, Being a Translation of Auenbrugger's Original Treatise Entitled "Inventum Novum ex Percussione Thoracis Humani, ut Signo Abstrusos Interni Pectoris Morbos Detegendi. (Vienna, 1761.)" Baltimore, Johns Hopkins Press, 1936.

[From *On Percussion of the Chest**]

PREFACE

I here present the Reader with a new sign I have discovered for detecting diseases of the chest. This consists in the Percussion of the human thorax, whereby, according to the character of the par-

* Inventum novum ex percussione thoracis humani, ut signo abstrusos interni pectoris morbos detegendi, Vienna, 1761. Translated by John Forbes, M.D., London, 1824. Baltimore, The Johns Hopkins Press, 1936.

ticular sounds thence elicited, an opinion is formed of the internal state of that cavity. In making public my discoveries respecting this matter, I have been actuated neither by an itch for writing, nor a fondness for speculation, but by the desire of submitting to my brethren the fruits of seven years' observation and reflexion. In doing so, I have not been unconscious of the dangers I must encounter; since it has always been the fate of those who have illustrated or improved the arts and sciences by their discoveries, to be beset by envy, malice, hatred, detraction, and calumny.

This the common lot, I have chosen to undergo; but with the determination of refusing to everyone who is actuated by such motives as these, all explanation of my doctrines. What I have written I have proved again and again, by the testimony of my own senses, and amid laborious and tedious exertions;—still guarding, on all occasions, against the seductive influence of self-love.

And here, lest any one should imagine that this new sign has been thoroughly investigated, even as far as regards the diseases noticed in my Treatise, I think it necessary candidly to confess, that there still remain many defects to be remedied—and which I expect will be remedied—by careful observation and experience. Perhaps also, the same observation and experience may lead to the discovery of other truths, in these or other diseases, of like value in the diagnosis, prognosis and cure of thoracic affections. Owing to this acknowledged imperfection, it will be seen, that, in my difficulties, I have had recourse to the Commentaries of the most illustrious Baron Van Swieten, as containing every thing which can be desired by the faithful observer of the nature; by which means I have not only avoided the vice of tedious and prolix writing, but have, at the same time possessed myself of the firmest basis whereupon to raise, most securely and creditably, the rudiments of my discovery. In submitting this to the public, I doubt not that I shall be considered, by all those who can justly appreciate medical science, as having thereby rendered a grateful service to our art,—inasmuch as it must be allowed to throw no small degree of light upon the obscurer diseases of the chest, of which a more perfect knowledge has hitherto been much wanted.

In drawing up my little work, I have omitted many things that were doubtful, and not sufficiently digested; to the due perfection of which it will be my endeavour henceforth to apply myself. To

conclude, I have not been ambitious of ornament in my mode or style of writing, being contented if I shall be understood.
December 31, 1760.

FIRST OBSERVATION

Of the natural sound of the Chest, and its character in different parts

I

The thorax of a healthy person sounds, when struck.

Scholium. I deem it unnecessary to give in this place, any description of the thorax. I think it sufficient to say, that by this term I mean that cavity bounded above by the neck and clavicles, and below by the diaphragm: in the sound state, the viscera it contains are fitted for their respective uses.

II

The sound thus elicited (I) from the healthy chest, resembles the stifled sound of a drum covered with a thick woollen cloth or other envelope.

III

This sound is perceptible on different parts of the chest in the following manner:

1. On the right side anteriorly it is observed from the clavicle to the sixth true rib; laterally, from the axilla to the seventh rib; and posteriorly, from the scapula to the second and third false ribs.

2. The left side yields this sound from the clavicle to the fourth true rib, anteriorly: and on the back and laterally, in the same extent as the other side: over the space occupied by the heart the sound loses part of its usual clearness, and becomes dull.

3. The whole sternum yields as distinct a sound as the sides of the chest, except in the cardiac region where it is somewhat duller.

4. The same sound is perceptible over that part of the spinal column which contributes to form the chest.

Scholium. The sound is more distinct in the lean, and proportionably duller in the robust; in very fat persons it is almost lost. The most sonorous region is from the clavicle to the fourth rib anteriorly; lower down, the mammae and pectoral muscles deaden the sound.

Sometimes, owing to the presence of muscle, the sound is dull beneath the axilla. In the scapular regions on the back, owing to the obstacle afforded by the bones and thick muscles there, it is also less distinct. Sometimes, but rarely, it exists over the third false rib —owing, I conceive, to a very unwonted length of the thoracic cavity.

SECOND OBSERVATION

On the method of Percussion

IV

The thorax ought to be struck, slowly and gently, with the points of the fingers, brought close together and at the same time extended.

Scholium. Robust and fat subjects require a stronger percussion; such, indeed, as to elicit a degree of sound equal to that produced by a slight percussion, in a lean subject.

V

During percussion the shirt is to be drawn tight over the chest, or the hand of the operator covered with a glove made of unpolished leather.

Scholium. If the naked chest is struck by the naked hand, the contact of the polished surfaces produces a kind of noise which alters or obscures the natural character of the sound.

VI

During the application of percussion the patient is first to go on breathing in the natural manner, and then is to hold his breath after a full inspiration. The difference of sound during inspiration, expiration, and the retention of the breath, is important in fixing our diagnosis.

VII

While undergoing percussion on the fore parts of the chest, the patient is to hold his head erect, and the shoulders are to be thrown back; in order that the chest may protrude, and the skin and muscles be drawn tight over it: a clear sound is thus obtained.

VIII

While we are striking the lateral parts of the chest, the patient is to hold his arms across his head; as, thereby, the thoracic parietes are made more tense, and a clearer sound obtained.

IX

When operating on the back, you are to cause the patient to bend forwards, and draw his shoulders toward the anterior parts of the chest, so as to render the dorsal region rounded; and for the same reasons, as stated in VIII.

Scholium. Any healthy person may make experience of percussion in his own person or that of other sound subjects; and will thus be convinced, from the variety of the sounds obtained, that this sign is not to be despised in forming a diagnosis.

THIRD OBSERVATION

Of the preternatural or morbid sound of the Chest, and its general import

X and Scholium

To be able justly to appreciate the value of the various sounds elicited from the chest in cases of disease, it is necessary to have learned by experience on many subjects, the modifications of sound, general or partial, produced by the habit of body, natural conformation as to the scapulae, mammae, the heart, the capacity of the thorax, the degree of fleshiness, fatness, etc., inasmuch as these various circumstances modify the sound very considerably.

XI

If, then, a distinct sound, equal on both sides, and commensurate to the degree of percussion, is not obtained from the sonorous regions above mentioned, a morbid condition of some of the parts within the chest is indicated.

Scholium. On this truth a general rule is founded, and from this certain predictions can be deduced, as will be shown in order. For I have learned from much experience that diseases of the worst description may exist within the chest, unmarked by any symptoms, and undiscoverable by any other means than percussion alone.

A clear and equal sound elicited from both sides of the chest indicates that the air cells of the lungs are free, and uncompressed either by a solid or liquid body. (Exceptions to this rule will be mentioned in their place.)

XII and XIII

If a sonorous part of the chest, struck with the same intensity, yields a sound duller than natural, disease exists in that part.

XIV

If a sonorous region of the chest appears, on percussion, entirely destitute of the natural sound—that is, if it yields only a sound like that of a fleshy limb when struck,—disease exists in that region.

XVI

If a place, naturally sonorous, and now sounding only as a piece of flesh when struck, still retains the same sound (on percussion) when the breath is held after a deep inspiration,—we are to conclude that the disease extends deep into the cavity of the chest.

FOURTH OBSERVATION

Of the diseases in general in which the morbid sound of the Chest is observed

XVIII

The preternatural or morbid sound occurs in acute and chronic diseases; it always accompanies a copious effusion of fluid in the thoracic cavity.

Scholium. It must be admitted that whatever diminishes the volume of air within the chest, diminishes the natural sound of the cavity; but we know from the nature, the causes, and the effects, of acute and chronic diseases of the chest, that such a result is possible in these cases; and the fact is finally demonstrated by examinations after death. The effect of effused fluids in producing the morbid sound, is at once proved by the injection of water into the thorax of a dead body; in which case it will be found that the sound elicited by percussion, will be obscure over the portion of the cavity occupied by the injected liquid.

XXV·

The following corollaries are the result of my observation of inflammatory diseases of the chest, studied under the sign of morbid resonance:

1. The duller the sound, and the more nearly approaching that of a fleshy limb stricken, the more severe is the disease.

2. The more extensive the space over which the morbid sound is perceived, the more certain is the danger from the disease.

3. The disease is more dangerous on the left than on the right side.

4. The existence of the morbid sound on the superior and anterior part of the chest (i.e. from the clavicle to the fourth rib) indicates less danger, than on the inferior parts of the chest.

5. The want of the natural sound behind, indicates more danger than it does on the anterior and superior part of the chest.

6. The total destitution of sound over one whole side, is generally (passim) a fatal sign.

7. The absence of sound along the course of the sternum is a fatal sign.

8. The entire absence of the natural sound over a large space in the region of the heart, is a fatal sign.

Scholium. I have sometimes observed that the fatal prognostics given in the corollaries 6 and 7, were not verified when the matter made its way outwards, or abcesses formed in parts less essential to life. And this natural process has been often happily imitated by the ancients by cauterising, or otherwise incising, the affected parts.

RÉNÉ-THÉOPHILE-HYACINTHE LAËNNEC

RÉNÉ-THÉOPHILE-HYACINTHE LAËNNEC (1781-1826), BORN IN QUIMPER, Brittany, was the greatest French clinician of the nineteenth century. His most notable contribution to medicine was the stethoscope. He studied first at Nantes. To complete his medical studies he went to Paris and came under the influence of Corvisart and was undoubtedly impressed with the possibilities of objective methods of accumulating data as expounded by his master, who had translated Auenbrugger's treatise on percussion. At the Necker Hospital first, and afterwards at the Charité, he was associated with such brilliant men as Louis, Bayle, Bichat, and Andral. His studies on cirrhosis of the liver, mitral disease, and pulmonary disease were important, but pale into insignificance beside his work on auscultation. The stethoscope was apparently discovered largely through accident; it was at first no more than a hollow wooden cylinder, but with this crude instrument Laënnec described most of the signs of pulmonary and cardiac disease. His book, *De l'auscultation mediate*, was first published in 1819; another edition appeared in 1826, just before Laënnec's death.

An idea of the scope of his book can be obtained by the following outline of the contents:

PART FIRST

OF THE EXPLORATION OF THE CHEST

The tracheal rhonchus
The dry sonorous rhonchus
The dry sibilous rhonchus
Dry crepitous rhonchus
Of the metallic tinkling

PART SECOND

DISEASES OF THE BRONCHI, LUNGS, AND PLEURA

Book First

DISEASES OF THE BRONCHI

Book Second

DISEASES OF THE LUNGS

Chap. I. Of Hypertrophy of the Lungs
Chap. II. Of Atrophy of the Lungs
Chap. III. Of Emphysema of the Lungs
Chap. IV. Of Oedema of the Lungs
Chap. V. Of Pulmonary Apoplexy
Chap. VI. Of Pneumonia
Chap. VII. Of Phthisis Pulmonalis
Chap. VIII. Of Cysts in the Lungs
Chap. IX. Of Hydatids in the Lungs
Chap. X. Of Concretions in the Lungs
Chap. XI. Of Melanosis of the Lungs

Book Third

DISEASES OF THE PLEURA

Chap. I. Of Pleurisy
Chap. II. Of Hydrothorax
Chap. III. Of Haemathorax
Chap. IV. Of Pneumothorax

PART THIRD

DISEASES OF THE HEART AND ITS APPENDAGES

Book First

OF THE EXPLORATION OF THE ORGANS OF CIRCULATION

GENERAL REMARKS

Chap. I. Of the Extent of the Heart's Pulsations
Chap. II. Of the Shock or Impulse
Chap. III. Of the Sound
Chap. IV. Of the Rhythm
Chap. V. Of Anomalous Sounds

REFERENCE

WEBB, G. B. Réné Théophile Hyacinthe Laënnec, a Memoir. New York, Paul B. Hoeber, 1928.

MEDIATE AUSCULTATION

[From *On Mediate Auscultation**]

INTRODUCTION

In 1816 I was consulted by a young woman presenting general symptoms of disease of the heart. Owing to her stoutness little information could be gathered by application of the hand and percussion. The patient's age and sex did not permit me to resort to the kind of examination I have just described (i.e., direct application of the ear to the chest). I recalled a well-known acoustic phenomenon: namely, if you place your ear against one end of a wooden beam the scratch of a pin at the other extremity is most distinctly audible. It occurred to me that this physical property might serve a useful purpose in the case with which I was then dealing. Taking a sheaf of paper I rolled it into a very tight roll, one end of which I placed over the praecordial region, whilst I put my ear to the other. I was both surprised and gratified at being able to hear the beating of the heart with much greater clearness and distinctness than I had ever done before by direct application of my ear.

* De L'Auscultation Médiate, ou Traité du Diagnostic des Maladies des Poumons et du Coeur, Fondé Principalement sur ce Nouveau Moyen D'Exploration, Paris, 1819. Translation by John Forbes. London, 1834.

I at once saw that this means might become a useful method for studying, not only the beating of the heart, but likewise all movements capable of producing sound in the thoracic cavity, and that consequently it might serve for the investigation of respiration, the voice, râles and even possibly the movements of a liquid effused into the pleural cavity or pericardium.

With this conviction, I at once began and have continued to the present time, a series of observations at the Hospital Necker. As a result I have obtained many new and certain signs, most of which are striking, easy of recognition, and calculated perhaps to render the diagnosis of nearly all complaints of the lungs, pleurae and heart both more certain and more circumstantial, than the surgical diagnosis obtained by use of the sound or by introduction of the finger.

I shall divide my work into four parts. The first will deal with the signs that can be obtained from the voice when heard by means of the cylinder (i.e., stethoscope); the second with those furnished by the respiratory sounds; the third with those supplied by râles; and, by way of appendix, the results at which I have arrived in my investigations on the effusion of liquids into the various cavities of the thorax; the fourth will contain an analysis of the heart-beats in health and in sickness, and an account of the special signs characterizing diseases of the heart and aorta.

Before proceeding with my subject, I consider it my duty to record the various attempts that I have made to improve upon the exploring instrument I at present use; these attempts have proved almost entirely vain, and, if I mention them, it is in the hope that any other investigator seeking to perfect the instrument will strike out a fresh path.

The first instrument employed by me consisted of a cylinder or roll of paper, sixteen lines in diameter and one foot long, made of three quires of paper rolled very tightly round, and held in position with gummed paper and filed smooth at both ends. However tight the roll may be, there will always remain a tube three or four lines in diameter running up the centre, because the sheets of paper composing it can never be rolled completely on themselves. This fortuitous circumstance gave rise, as will be seen, to an important observation upon my part: I found that for listening to the voice the tube is an indispensable factor. An entirely solid body is the best instrument that can be used for listening to the heart; such an instrument would indeed suffice also for hearing respiratory sounds and râles; yet these last two phenomena yield greater intensity of sound if a perforated cylinder is used, hollowed out at one end into a kind of funnel 1½ in. in depth.

The densest bodies are not, as analogy would lead us to suppose, the best materials for constructing these instruments. Glass and metals, apart from their weight and the sensation of cold that they impart in winter, are not such good carriers of the heart-beats and the sounds produced by breathing and râles, as are bodies of lesser density. Having noticed this fact, which at first caused me some surprise, I attempted to use the least dense substances possible, and consequently I had a cylinder made of goldbeater's skin, which could be inflated with air by means of a tap, while the central tube space was held open by another tube made of card. This cylinder is the worst of all; it yields less intensity of sound and, as a further drawback, becomes deflated every few minutes, particularly when the air is cold; in addition, it is of all cylinders the one most liable to render audible a sound other than that which is being explored, owing to the crepitation of its own walls and to friction with the patient's clothing or the observer's hand.

Substances of medium density, such as paper, wood and cane, are those which have always appeared to me preferable to all others. This result may be in contradiction with an axiom of physics; none the less I consider it to be quite established.

I consequently employ at the present time a wooden cylinder with a tube three lines (a French line-0.0888 in. or 2.256 mm.) in diameter bored right down its axis; it is divisible into two parts by means of a screw and is thus more portable. One of the parts is hollowed out at its end into a wide funnel-shaped depression 1½ in. deep leading into the central tube. A cylinder made like this is the instrument most suitable for exploring breath sounds and râles. It is converted into a tube of uniform diameter with thick walls all the way, for exploring the voice and the heart beats, by introducing into the funnel or bell a kind of stopper made of the same wood, fitting it quite closely; this is made fast by means of a small brass tube running through it, entering a certain distance into the tubular space running through the length of the cylinder. This instrument is sufficient for all cases, although, as I have already said, a perfectly solid body might perhaps be better for listening to the beating of the heart.

The dimensions indicated above are not altogether unimportant; if the diameter is larger it is not always possible to apply the stethoscope closely against all points of the chest; if the instrument is longer, it becomes difficult to hold it exactly in place; if it were shorter, the physician would often be obliged to adopt an uncomfortable position, which is to be avoided above all things if he desires to carry accurate observations.

I shall be careful, when discussing each variety of exploration, to mention the positions which experience has taught me to be most favourable for observation and least tiring for both physician and patient.

Suffice it to say for the moment that in all cases the stethoscope should be held like a pen, and that the hand must be placed quite close to the patient's chest in order to make sure that the instrument is properly applied.

The end of the instrument intended to rest on the patient's chest, that is to say the end provided with the stopper, should be very slightly concave; it is then less liable to wobble, and as this concavity is easily filled by the skin it in no case leaves an empty space even when placed on the flattest part of the chest.

When excessive emaciation has destroyed the pectoral muscles, so as to leave depressions between the ribs so deep that the whole of the surface of the end of the stethoscope cannot rest at the same time on the chest, they may be filled up with lint or with cotton wrapped in linen or a sheet of paper. The same precaution must be taken when it is the heart that is being examined in those patients whose sternum is bent inwards at its lower end, as often happens with cobblers and some other handicraftsmen.

I have made other modifications in the stethoscope and I have tried various instruments of different shape, but as they cannot be of general use, I shall not speak of them till occasion arises.

Some of the signs obtained by mediate auscultation are very easy to distinguish, and if they have been heard once there will be no difficulty in again recognizing them; such are the signs indicative of cavities in the lungs, considerable hypertrophy of the heart, fistulous communication between the pleural cavity and the bronchi, etc. But there are others that require more study and practice, and for the very reason that this method of examination permits of much greater precision in diagnosis than any other, pains must be taken to use it to the very best advantage.

Mediate auscultation must not, however, lead us to neglect Auenbrugger's method; to which it gives, on the contrary, an entirely new importance, extending its use to numerous maladies in which percussion alone would be of no help or might positively lead to error. Thus it is, by a comparison of the results of both methods, that positive and evident signs can be obtained in emphysema of the lung, pneumothorax, and liquid effusions in the pleural cavity. Nor must we neglect other methods of examination that are more limited in their scope, especially Hippocratic succussion, mensuration of the thorax, and even auscultation with the ear directly on the chest or immediate auscultation. These methods which have lapsed into oblivion and are by themselves often as liable to mislead the practitioner as to enlighten him, become in the cases which will be set forth in the present work, useful means of confirming the diagnosis established by mediate auscultation and percussion and of imparting to it the highest degree of certainty and accuracy that can be obtained in a physical science.

For various reasons, it is only in hospitals that a familiarity and thorough skill in the practice of mediate auscultation can be acquired, because it is necessary, at least sometimes, to have the conclusions reached by means of the stethoscope verified by post-mortem examination, in order that we may acquire confidence in ourselves and in the instrument, and that we may have reliance upon our own powers of observation and be convinced by the eye of the accuracy of the signs perceived by the hearing. It is sufficient, however, to have observed a disease on two or three occasions to learn how to recognize it with certainty; most of the affections of the lungs and heart are so common, that after a week's experience of them in a hospital, all that will be left to be studied are a few rare cases, almost all of which will be encountered in the course of a year, if every patient is examined with care. It would be doubtless imposing too much upon a physician whose whole time is devoted to private practice to require him to attend a hospital for so long a period; but a hospital physician whose duty it is to carefully examine all his patients daily, can easily save his professional brethren this trouble by informing them whenever he comes upon a rare or specially instructive case.

TUBERCULOUS AFFECTION OF THE LUNG

[Book II, Chapter VII, Section 1]

The cavities producing pectoriloquy are those vulgarly known as ulcers of the lung. They are not, as was supposed, and is still believed by the majority of medical practitioners, an outcome of inflammation and suppuration of the pulmonary tissue. The recent advances in pathological anatomy have proved beyond all manner of doubt that these cavities are due to the softening and subsequent evacuation of a peculiar kind of accidental formation which modern anatomists have specially designated by the name tubercle, formerly applied without distinction to any sort of unnatural protuberance or tumour.

The presence of tubercles in the lung is the cause, and constitutes the peculiar anatomical characteristic of pulmonary phthisis.

The cavities produced by tubercle differ essentially from an ulcer, inasmuch as this spreads by eating into the tissue in which it is formed, whereas the first are produced by the spontaneous destruction of accidental formations which have pushed aside and compressed the pulmonary tissue, but not destroyed it, and have no tendency to increase at its expense.

The manner in which these morbific formations develop has been described by M. Bayle with much greater completeness and precision than had ever been achieved before. However, certain observations made since the publication of his Researches having enabled me to rectify or extend some of his remarks, I esteem it requisite, for the understanding of many things I shall have to say, to set forth in an abridged form the characteristics and mode of development of tubercles, points upon which I might otherwise have referred to the excellent work quoted above.

The tubercles develop in the form of small semi-transparent granules, grey in colour, though occasionally they are entirely transparent and almost colourless. Their size varies from that of millet-seed to that of hemp-seed; when in this state they may be termed *miliary tubercles*. These granules increase in size and become yellowish and opaque, first in the centre and then progressively throughout their substance. Those nearest together unite in the course of growth and then form more or less voluminous masses, pale yellow in colour, opaque, and comparable, as regards density, with the most compact kinds of cheese; they are called *crude tubercles*.

It is usually at about this stage in the growth of the tubercles that the pulmonary tissue, healthy up till now, begins to harden and to become greyish and semi-transparent around the tubercles owing to the fresh growth of tuberculous matter in its early or semi-transparent stage which infiltrates it.

Sometimes indeed tuberculous masses of great size are formed as a consequence of this impregnation or infiltration without any previous development of miliary tubercles. The pulmonary tissue thus congested is dense, moist and quite impregnable to air; when it is cut the incisions reveal a smooth polished surface. As these indurations gradually pass into the state of crude tubercles we may observe the appearance in them of a multitude of minute opaque yellow specks which multiplying and growing larger end by invading the whole of the hardened part.

In whatsoever manner these crude tubercles may be formed, they finish at the end of a longer or shorter time of very variable duration by becoming softened and liquefied. This softening starts near the middle of each mass, which daily becomes softer and moister until the softening has reached the periphery and become complete.

In this phase the tuberculous matter may present itself under two different forms: sometimes it resembles thick pus, but it is odourless and yellower than the crude tubercles; sometimes it separates into two parts, one of which is very liquid and more or less transparent and colourless unless it is contaminated with blood; the other part is opaque and has the consistency of soft, friable cheese. In this latter state, specially met with in scrofulous patients, it often bears a complete resemblance to whey in which small fragments of caseous matter are floating.

When the tuberculous matter has completely softened it breaks its way into the nearest bronchial tube. The resulting aperture being narrower than the cavity with which it communicates, both remain of necessity fistulous even after the complete evacuation of the tuberculous matter.

It is exceedingly rare to find only a single cavity in a lung thus affected. Usually these cavities are surrounded with crude tubercles and miliary tubercles which soften successively, burst into the main cavity and give rise to the anfractusities generally to be observed in it, and which in some cases spread step by step until they reach the surface of the lung.

Cords or columns of dense pulmonary tissue generally permeated with tuberculous matter often traverse these cavities and somewhat resemble the fleshy columns in the ventricles of the heart; they are thinner towards the middle than at their extremities.

As the cavity gradually empties, its walls become coated with a sort of false membrane, thin, smooth, of a white colour, almost opaque, of a somewhat soft almost friable consistency, and it comes away readily when scraped with a scapel. This membrane is generally complete, lining the whole inner surface of the cavity. Its place is occasionally taken, however, by a pseudo-membranous exudation, thinner, more transparent, less friable, and more closely adherent to the walls of the cavity which it usually covers only in patches. If it does happen to line the entire cavity, it presents at places much greater thickness, and this fact seems to point to its being the product of an exudation starting simultaneously at several different points.

Frequently this second membrane is found beneath the first which is then not completely adherent and is torn at various places.

Finally, sometimes we cannot find any perceptible trace of either variety of false membrane, and the walls of the cavity are formed by pulmonary tissue, which is generally hardened, red, and infil-trated with tuberculous matter in various stages of development. Judging from these facts I am of opinion that the second kind of false membrane is merely the early stage in the development of the first, that when this is completely formed it tends to peel off, and is then expectorated in fragments and replaced by new membrane, and that this matter forms one constituent of the sputum of con-sumptives.

M. Bayle believes that this false membrane secretes the pus which the patients expectorate. This opinion is based on the analogy which exists between it and that which forms on the surface of blisters and other ulcers. However that may be, it appears to me clear that the greater part of the sputum expectorated by sufferers from phthisis is formed by the bronchial secretion increased because of the condition which exists in the lungs. Although I do not intend to deny abso-lutely that it might be formed in the cavities, I would remark that when they are lined by the soft membrane, described above, they are often completely empty, or if they contain puriform matter, this resembles much less the sputum of the patient than does that which is contained in the bronchial tubes.

Should the disease come to a standstill for a length of time there will soon be formed here and there below the false membrane, greyish white semi-transparent patches in texture like cartilage, but a little softer, and very closely adherent to the lung tissue. In the course of growth these patches coalesce, and completely line the ulcerous cavity and end by continuity of substance with the inner membrane of the bronchial tubes opening into the cavity.

When this cartilaginous membrane is completely formed it is usually white or pearl grey in colour; should it assume an apparently reddish or purple tinge, this will be due to the thinness and semi-transparency of the membrane through which the colour of the lung-tissue itself is visible.

At times, however, even when the cartilaginous membrane is of considerable thickness, its inner surface is pink or red, and this colour cannot be removed by washing; it is probably due to the presence of a network of fine vessels, although no distinct vessels can be discerned doubtless because of their minuteness.

In a few rare cases tubercles are found completely softened, or very nearly so, in the midst of thoroughly crepitant pulmonary tissue; in such cases (I have encountered only two or three in the course of eighteen years) the walls of the cavity are quite smooth and seem to be formed only of the pulmonary tissue slightly com-pressed without any kind of accidental membrane.

INVESTIGATION OF RESPIRATION

[Part i, Chapter iv, Section 1]

Auscultation of respiration by means of the stethoscope furnishes us with signs easy to apprehend, and peculiarly fitted to reveal the existence and extent of most of the organic complaints affecting the thoracic viscera—more especially peripneumonia, pulmonary phthisis, oedema and emphysema of the lung, the various kinds of tumours or new growths which make their appearance in that organ, pleurisy, blood-spitting, pleural effusion, and pneumothorax, or the collection of air in the pleural cavities.

The description and physical signs of some of these complaints will be reserved for the third part of this work, because the most important and most easily recognizable evidence of their existence is obtained from the study of rales.

The stethoscope hollowed out at its extremity in the shape of a funnel is the one which should be employed for listening to respiration. If this form of stethoscope is placed against the chest of a man in good health, there will be heard, during both inspiration and expiration, a slight but most distinct murmur, which indicates the entrance of air into the pulmonary tissue and its expulsion. This murmur may be compared to the sound produced by a pair of bellows, the valve of which is noiseless, or, better still, to that heard by the unaided ear when a man, in deep and peaceful sleep, periodically draws a long inspiration. It is almost equally audible at all points of the chest, but especially at those spots where the lungs are nearest to the cutaneous surface—that is to say, in the anterior, superior, lateral, and lower posterior regions. The hollow of the armpit and the space between the clavicle and upper edge of the trapezius muscle are the points where it is heard with greatest intensity.

We can hear it equally well over the larynx and over the bare or cervical portion of the trachea—and even in many men over the whole length of the windpipe down to the base of the sternum; but over the trachea, and also in some degree at the roots of the bronchi, the respiratory sound has a special character, clearly showing that the air is passing through a roomier channel than the air cells; further, it often appears as if the patient, during inspiration, were drawing in the air contained in the tube of the stethoscope, again expelling it during expiration.

If the state of respiration is to be correctly judged by the stethoscope, we must not only rely upon the first few moments of examination. The placing of the ear in such a manner as to apply the instrument causes a sensation of buzzing. The timidity, constraint, and uneasiness experienced by the patient, causing him automatically to reduce the extent of his respiration; sometimes, too, the awkward position of the observer himself; the heart-beats, if unusually loud, and first heard by the ear, are all causes which may preclude at the outset a correct appreciation, or may, indeed, altogether prevent the hearing of inspiration and expiration. It is only after a lapse of a few seconds that a right judgment can be formed.

It is hardly requisite to add that no kind of noise must be going on in the neighborhood of the patient.

Clothes, even when their thickness is considerable, do not appreciably diminish the intensity of the sound produced by respiration; but care must be taken that no friction occurs between the clothes and the stethoscope; for this, especially when the garments are of silk, or thin dry wool, gives rise to a noise calculated to cause error, owing to its similarity to the respiratory sounds.

Excessive stoutness and infiltration of the walls of the chest are no serious obstacle to hearing the respiratory sounds.

An increase in the frequency of respiration causes the respiratory sound to be more sonorous. A very deep inspiration, if performed slowly, is sometimes scarcely audible, whereas an incomplete inspiration, in which the dilation of the walls of the thorax is almost imperceptible to the eye, may be very loud if performed rapidly; consequently, when using the stethoscope to examine respiration, it is advisable, especially for the unpractised observer, to request the patient to breathe somewhat quickly. This precaution becomes superfluous, however, in most diseases of the thoracic organs, which cause a fairly evident difficulty of breathing; for, as dyspnoea nearly always quickens respiration, it is, therefore, necessarily more perceptible. The same occurs in fever and nervous agitation.

Several other causes may lead to variations in the intensities of the respiratory sound; age especially plays an important part in this respect. In children, respiration is very sonorous, and even noisy; it is easily heard through several layers of thick clothing. It is not necessary in these cases to press the stethoscope firmly down in order to prevent friction between the garments; any sound which might arise from this cause being drowned by the intensity of that respiration.

It is not only by its intensity that the respiration of children differs from that of adults. There is also a very marked difference in the character of the sound which, like all elementary sensations, it is impossible to describe, but which can be easily recognized by comparison. It appears with all children as if we could distinctly feel the dilation of the air cells to their full capacity; whereas with adults we might believe them to be only half filled, or that their walls being harder are incapable of distending to the same degree. This difference of sound is chiefly noticeable in inspiration; in expiration it is much less marked. The expansion of the chest accompanying each inspiration is also much greater in the child than in the adult. The younger the child the more marked are these characteristics, which usually persist in varying degree till a little beyond puberty.

With the adult the respiratory sound varies much in intensity. There are men in flourishing health in whom it is scarcely audible, unless they draw a very long breath; and even then, though it is heard quite well and is quite pure, that is to say, unmixed with râles or other foreign sounds, it only has half the volume of sound and fremitus that is heard in the majority of men. Such persons will be generally found to be those whose respiration is not habitually frequent, and often they are not ordinarily subject to dyspnoea and loss of breath from any cause whatever.

The respiration of others is loud enough naturally to be quite easily heard even with ordinary inspiration, but they are not either more or less liable to shortness of breath than those just mentioned. Finally, a very few persons retain till extreme old age a respiration similar to that of childhood, to which for this reason I shall sometimes apply the name *puerile respiration* in the course of the present work. Such persons are nearly always women or men of nervous temperament. Something of the excitability, and especially the irritability of children is often to be noticed in their character. Some have not, strictly speaking, any disease of the respiratory organs, but they easily become out of breath after exertion, at the same time they are thin, and readily take cold. Others are afflicted with chronic catarrh coupled with dyspnoea, and this combination constitutes, as we shall see later, one of the maladies known by the name of *asthma*.

Apart from these exceptional cases, an adult, however strenuous his acts of inspiration may be, cannot give to his respiration the sonority and peculiar character possessed by it in childhood. In a few pathological cases, however, the respiration reverts to its puerile character spontaneously, without the patient appearing to inspire more forcibly than usual. This is observable especially when the whole of one lung or a considerable portion of both lungs has become impermeable to air as a result of some disease, and more particularly an acute disease. Over the parts of the lung which have remained healthy, the respiratory sound is exactly like that heard in children. The same phenomenon is observable throughout the whole extent of the lung in patients suffering from fairly serious fevers or certain nervous affections.

When, for the first time, a comparison is made between the respiration of a child and that of an adult, one might be tempted to believe that the greater intensity of sound observed in the case of the child depended upon the thinner muscular covering of the walls of the thorax, and the greater flexibility of the pulmonary tissue; but the first of these causes is only in a very small degree responsible for the difference, for the respiration of the fattest children is more forcibly heard even through thick clothing than is that of the thinnest adult examined without any clothing, and of the adults who exhibit the phenomenon of *puerile respiration* many are exceedingly stout; in women who are both grown up and fat, the respiration is often audible with great force, even through the breasts.

Nor is the less noisy respiration of the adult due to any hardening or lack of flexibility of the pulmonary tissue, since it may at times accidentally become again what it was in childhood. I am of opinion that the difference proves rather that children require a larger amount of air and that consequently inspiration is more complete with them than with adults, either on account of their more active circulation, or because of some differences in the chemical composition of their blood. It is at least quite likely that the blood of children is considerably more oxygenated than that of adults. The same may be said of asthmatical patients exhibiting puerile respiration when compared with those suffering from another kind of asthma which will be discussed in a subsequent chapter of Part II (*vide* Emphysema of the lung), and in whom respiration, already very weak, is entirely suspended for hours on end, now in one portion of the lung, now in another. The first often have a rosy tint which shows them to be in perfect health, whilst the second are always livid in the face and at the extremities. It is to the last only that we can with truth sometimes apply the popular proverb that asthma is a warrant of long life.

May this not be, because breathing but little, they live on a lower plane, just as a lamp whose very small wick sheds but a meagre light and threatens to go out at the faintest puff of air, may yet continue to burn for a long time because it consumes its oil very slowly?

Whatever the case may be, it appears to me certain that the constitution of the pulmonary organ most conducive to health and longevity is that found in men who habitually need only a moderate dilation of the lungs and whose respiration is much less loud than that of children. This state should therefore be considered the natural state: *id est maxime naturale, quod, fieri optime patitur.*

[From *On Mediate Auscultation*, Chapter xviii, Section 11]

INDURATION OF THE MITRAL VALVES

The symptoms of ossification of the mitral valve are little different from those attending the same affection of the sigmoid. According to M. Corvisart the principal sign of the mitral lesion is "a peculiar rustling sensation (*bruissement*) perceived on the application of the hand to the region of the heart." This peculiar sensation is nothing else than the *purring-thrill* already described. It is assuredly very frequently observed in the case of ossification of the mitral or sigmoid valves when this exists in a high degree; but, as I formerly stated, it may exist when these valves are perfectly sound, and it is almost always absent when the induration is not so extensive as materially to obstruct the orifices. The bellows-sound is a much more constant sign: it accompanies the contraction of the left auricle when the mitral valve is affected, and that of the ventricle, when the induration is in the sigmoid. But even this is wanting when the alteration is not extensive, and as it is, moreover, very common when the heart is perfectly sound, we must lay no stress on it as a sign, unless it be combined with other circumstances calculated to confirm the diagnosis. Accordingly, when the sound of the bellows, rasp, or file, persists in the left auricle, either continuously or interruptedly, for several months;—when it is found only there, and exists even in the greatest quietude; when it is scarcely lessened by venesection, or, when lessened, if it still leaves behind it a degree of roughness in the sound of the auricle,—or, yet more, when the purring-thrill co-exists with this;—we may be assured that the left auriculo-ventricular orifice is contracted. If the same phenomena occur, under similar circumstances, in the left ventricle, we may be equally certain that the aortal orifice is contracted.

Three or four times, during the last four years, I have discovered this lesion, by means of these signs. Three similar examples, equally verified by dissection, are recorded in M. Bertin's work (Obs. 49, 50, 51) and a fourth is given in the collection of cases published by Dr. Forbes. (Case VII.) But if these phenomena exist only for a time, although as much as two or three months,—if they accompany the increase of any other nervous or organic disease of the heart, we must not depend upon them as indications of the lesions now in question; since all the facts formerly recounted prove that these sounds are not produced (as might be imagined at first) by the passage of the blood over a rough or rugged surface, but to the spasmodic energy requisite in the muscular contraction to overcome the obstacles opposed to it. It follows, therefore, that any other cause besides diminution of the size of the orifices, which occasions con-traction of the heart, is equally capable of giving occasion to the bellows-sound and purring-thrill; and it is fair to admit that, in the first edition of this work, I laid too much stress on these two phe-nomena as signs of valvular disease.

A slight degree of cartilaginous or bony induration of the valves may exist for a long time without any visible alteration of the health, or even of the action of the heart; and even by proper measures of precaution and by seasonable bleedings, we may frequently preserve for a long time the life of individuals, who present every sign of considerable contraction of the orifices. The following case is a proof of this.

Case XLV.—A very muscular young man, aged sixteen, came into the Necker Hospital in February, 1819, complaining of oppression on the chest and palpitation; symptoms which had seized him sud-denly, together with haemoptysis and epistaxis, two years before. These symptoms were relieved at the time by rest; but returned as often as he made any considerable degree of exertion. He presented the following symptoms on coming into the hospital: respiration and resonance good over the whole chest; the hand applied to the region of the heart feels the pulsation strongly, and accompanied with the *purring* vibration. This vibratory sensation is not continuous, but returns at regular intervals. The stethoscope, applied between the cartilages of the fifth and seventh ribs, gives the following results:— contraction of the auricle extremely prolonged, accompanied with a dull but strong sound exactly like that produced by a file on wood. This sound is attended by a vibration sensible to the ear, and which is evidently the same as that felt by the hand. Succeeding this, a louder sound and a shock synchronous with the pulse point out the contraction of the ventricle, which occupies only one fourth the time, and has something harsh in its sound. Under the lower end of the sternum the contractions of the heart are quite different. Here the impulse of the right ventricle is very great, its contraction accompanied by a very distinct sound, and being of the ordinary duration; viz. twice as long as that of the auricle. The sound of the auricle is somewhat obtuse, but without any thing analogous to the vibratory character of the left. The action of the heart is audible below both clavicles, on both sides, but feebly, especially on the right. Over the whole sternum, on the right side and below the left clavicle, the contractions of the heart have the same rhythm as at the end of the sternum. On the left side, on the contrary, the whizzing sound of the left auricle already described is much feebler than in the left precordial region. From these signs the following diagnostic was given— *Ossification of the mitral valve, slight hypertrophy of the left ventricle; perhaps slight ossification of the sigmoid valves of the aorta; great hypertrophy of the right ventricle.* The pulse, in this case, was pretty strong and very regular, and all the functions natural, only the sleep was habitually disturbed by frightful dreams, and the lad could not use any severe exercise, nor even walk rather fast, without being attacked by strong palpitations and a feeling of suffocation. Four venesections, after intervals of a few days, gave much relief. After the first, the pulse became weak; and immediately after each bleeding the *purring vibration* became imperceptible to the hand, and the whizzing of the auricle changed from the sound of a *file* to that of a *bellows*, the valve of which we keep open by the hand; the shock of the right ventricle continued to be very strong. This patient left the hospital after a month, being, in his own opinion, pretty well. He came afterwards several times to consult me, and was bled occasionally. I saw him once more in 1822. I found that he had abandoned his laborious occupation of gardener, and had an easy place as the servant of a priest. Since his change of situation he has been much easier: but his former symptoms still exist, although in a slighter degree.

XXX

MEDICAL LIFE IN THE EARLY NINETEENTH CENTURY

WILLIAM MAKEPEACE THACKERAY

[From *Pendennis*, Chapter II]

Thackeray's *Pendennis* describes the life of an apothecary. It will be noticed that the apothecary, in addition to managing his shop, performed the duties that are today assigned to a practitioner of internal medicine, including pediatrics, dermatology, and to some extent obstetrics. The Society of Apothecaries is one of the ancient guilds in England; it held examinations and licensed apothecaries to practice their professions, as Mr. Pendennis did. This society still licenses all apothecaries in England, though apothecaries no longer practice medicine. Many practitioners, however, still find it an advantage to be an L.S.A. (Licentiate Society Apothecaries). The student apothecary formerly obtained his training in one of the great hospitals. In the early nineteenth century, the rise of the medical schools and the universities reduced the scope of the Society of Apothecaries.

Early in the Regency of George the Magnificent, there lived in a small town in the west of England, called Clavering, a gentleman whose name was Pendennis. There were those alive who remembered having seen his name painted on a board, which was surmounted by a gilt pestle and mortar over the door of a very humble little shop in the city of Bath, where Mr. Pendennis exercised the profession of apothecary and surgeon; and where he not only attended gentlemen in their sick-rooms, and ladies at the most interesting periods of their lives, but would condescend to sell a brown-paper plaster to a farmer's wife across the counter,—or to vend toothbrushes, hair-powder, and London perfumery.

And yet that little apothecary who sold a stray customer a penny-worth of salts, or a more fragrant cake of Windsor soap, was a gentleman of good education, and of as old a family as any in the whole county of Somerset. He had a Cornish pedigree which carried the

Pendennises up to the time of the Druids,—and who knows how much farther back? They had intermarried with the Normans at a very late period of their family existence, and they were related to all the great families of Wales and Brittany. Pendennis had had a piece of University education too, and might have pursued that career with honour, but in his second year at Oxbridge his father died insolvent, and poor Pen was obliged to betake himself to the pestle and apron. He always detested the trade, and it was only necessity, and the offer of his mother's brother, a London apothecary of low family, into which Pendennis's father had demeaned himself by marrying, that forced John Pendennis into so odious a calling.

He quickly after his apprenticeship parted from the coarseminded practitioner his relative, and set up for himself at Bath with his modest medical ensign. He had for some time a hard struggle with poverty; and it was all he could do to keep the shop in decent repair, and his bedridden mother in comfort: but Lady Ribstone happening to be passing to the Rooms with an intoxicated Irish chairman who bumped her Ladyship up against Pen's very door-post, and drove his chair-pole through the handsomest pink-bottle in the surgeon's window, alighted screaming from her vehicle, and was accommodated with a chair in Mr. Pendennis's shop, where she was brought round with cinnamon and sal-volatile.

Mr. Pendennis's manners were so uncommonly gentlemanlike and soothing, that her Ladyship, the wife of Sir Pepin Ribstone, of Codlingbury, in the county of Somerset, Bart., appointed her preserver, as she called him, apothecary to her person and family, which was very large. Master Ribstone coming home for the Christmas holidays from Eton, over-ate himself and had a fever, in which Mr. Pendennis treated him with the greatest skill and tenderness. In a word, he got the good graces of the Codlingbury family, and from that day began to prosper. The good company of Bath patronised him, and amongst the ladies especially he was beloved and admired.

CHARLES DICKENS

[From *Martin Chuzzlewit*]

In Sairy Gamp Dickens drew the picture of the early nineteenth century midwife and nurse; this was before Miss Nightingale put nursing on a professional basis. In 1857, the London *Times* described the nurses of the city hospitals thus:

"Lectured by Committees, preached at by chaplains, scowled on by treasurers and stewards, scolded by matrons, sworn at by surgeons, bullied by dressers, grumbled at and abused by patients, insulted if old and ill-favored, talked flippantly to if middle-aged and good humoured, seduced if young—they are what any woman would be under the same circumstances."

Mr. Pecksniff was in a hackney cabriolet, for Jonas Chuzzlewit had said "Spare no expense." Mankind is evil in its thoughts and in its base constructions, and Jonas was resolved it should not have an inch to stretch into an ell against him. It never should be charged upon his father's son that he had grudged the money for his father's funeral. Hence, until the obsequies should be concluded, Jonas had taken for his motto "Spend, and spare not!"

Mr. Pecksniff had been to the undertaker, and was now upon his way to another officer in the train of mourning: a female functionary, a nurse, and watcher, and performer of nameless offices about the persons of the dead: whom he had recommended. Her name, as Mr. Pecksniff gathered from a scrap of writing in his hand, was Gamp; her residence in Kingsgate Street, High Holborn. So Mr. Pecksniff in a hackney cab, was rattling over Holborn stones, in quest of Mrs. Gamp.

This lady lodged at a bird-fancier's, next door but one to the celebrated mutton-pie shop, and directly opposite to the original cat's-meat warehouse; the renown of which establishments was duly heralded on their respective fronts. It was a little house, and this was the more convenient; for Mrs. Gamp being, in her highest walk of life, a monthly nurse, or, as her sign-board boldly had it, "Midwife," and lodging in the first-floor front, was easily assailable at night by pebbles, walking-sticks, and fragments of tobacco-pipe: all much more efficacious than the street-door knocker, which was so constructed as to wake the street with ease, and even spread alarms of fire in Holborn, without making the smallest impression on the premises to which it was addressed.

It chanced on this particular occasion, that Mrs. Gamp had been up all the previous night, in attendance upon a ceremony to which the usage of gossips has given that name which expresses, in two syllables, the curse pronounced on Adam. It chanced that Mrs. Gamp had not been regularly engaged, but had been called in at a crisis, in consequence of her great repute, to assist another professional lady with her advice; and thus it happened that, all points of interest in the case being over, Mrs. Gamp had come home again to the bird-fancier's and gone to bed. So when Mr. Pecksniff drove up in the hackney cab, Mrs. Gamp's curtains were drawn close, and Mrs. Gamp was fast asleep behind them.

If the bird-fancier had been at home, as he ought to have been, there would have been no great harm in this; but he was out, and his shop was closed. The shutters were down certainly; and in every pane of glass there was at least one tiny bird in a tiny bird-cage, twittering and hopping his little ballet of despair, and knocking his head against the roof: while one unhappy goldfinch who lived outside a red villa with his name on the door, drew the water for his own drinking, and mutely appealed to some good man to drop a farthing's-worth of poison in it. Still, the door was shut.

Noting these circumstances, Mr. Pecksniff, in the innocence of his heart, applied himself to the knocker; but at the first double knock every window in the street became alive with female heads; and before he could repeat the performance whole troops of married ladies (some about to trouble Mrs. Gamp themselves very shortly) came flocking round the steps, all crying out with one accord, and with uncommon interest, "Knock at the winder, sir, knock at the winder. Lord bless you, don't lose no more time than you can help; knock at the winder."

Acting upon this suggestion, and borrowing the driver's whip for the purpose, Mr. Pecksniff soon made a commotion among the first-floor flower-pots, and roused Mrs. Gamp, whose voice—to the great satisfaction of the matrons—was heard to say, "I'm coming."

"He's as pale as a muffin," said one lady, in allusion to Mr. Pecksniff.

"So he ought to be, if he's the feelings of a man," observed another.

A third lady (with her arms folded) said she wished he had chosen any other time for fetching Mrs. Gamp, but it always happened so with her.

It gave Mr. Pecksniff much uneasiness to find, from these remarks, that he was supposed to have come to Mrs. Gamp upon an errand touching—not the close of life, but the other end. Mrs. Gamp herself was under the same impression, for, throwing open the window, she cried behind the curtains, as she hastily attired herself:

"Is it Mrs. Perkins?"

"No!" returned Mr. Pecksniff, sharply. "Nothing of the sort."

"What, Mr. Whilks!" cried Mrs. Gamp. "Don't say it's you, Mr. Whilks, and that poor creetur Mrs. Whilks with not even a pincushion ready. Don't say it's you, Mr. Whilks!"

"It isn't Mr. Whilks," said Pecksniff. "I don't know the man. Nothing of the kind. A gentleman is dead; and some person being wanted in the house, you have been recommended by Mr. Mould the undertaker."

As she was by this time in a condition to appear, Mrs. Gamp, who had a face for all occasions, looked out of the window with her mourning countenance, and said she would be down directly. But the matrons took it very ill that Mr. Pecksniff's mission was of so unimportant a kind; and the lady with her arms folded rated him in good round terms, signifying that she would be glad to know what he meant by terrifying delicate females "with his corpses," and giving it as her opinion that he was quite ugly enough to know better. The other ladies were not at all behind-hand in expressing similar sentiments; and the children, of whom some scores had now collected, hooted and defied Mr. Pecksniff quite savagely. So when Mrs. Gamp appeared, the unoffending gentleman was glad to hustle her with very little ceremony into the cabriolet, and drive off, overwhelmed with popular execration.

Mrs. Gamp had a large bundle with her, a pair of pattens, and a species of gig umbrella; the latter article in colour like a faded leaf, except where a circular patch of a lively blue had been dexterously let in at the top.

She was a fat old woman, this Mrs. Gamp, with a husky voice and a moist eye which she had a remarkable power of turning up, and only showing the white of it. Having very little neck, it cost her some trouble to look over herself, if one may say so, at those to whom she talked. She wore a very rusty black gown, rather the worse for snuff, and a shawl and bonnet to correspond. In these dilapidated articles of dress she had, on principle, arrayed herself, time out of mind, on such occasions as the present; for this at once expressed a decent amount of veneration for the deceased, and invited the next of kin to present her with a fresher suit of weeds: an appeal so frequently successful, that the very fetch and ghost of Mrs. Gamp, bonnet and all, might be seen hanging up, any hour in the day, in at least a dozen of the second-hand clothes shops about Holborn. The face of Mrs. Gamp—the nose in particular—was somewhat red and swollen, and it was difficult to enjoy her society without becoming conscious of a smell of spirits. Like most persons who have attained to great eminence in their profession, she took to hers very kindly; insomuch that, setting aside her natural predilections as a woman, she went to a lying-in or a laying-out with equal zest and relish.

"Ah," said Mrs. Gamp; for it was always a safe sentiment in cases of mourning. "Ah dear! When Gamp was summoned to his long home, and I see him a-lying in Guy's Hospital with a penny-piece on each eye, and his wooden leg under his left arm, I thought I should have fainted away. But I bore up."

If certain whispers current in the Kingsgate Street circles had any truth in them, she had indeed borne up surprisingly; and had exerted such uncommon fortitude as to dispose of Mr. Gamp's remains for the benefit of science. But it should be added, in fairness, that this had happened twenty years before; and that Mr. and Mrs. Gamp had long been separated on the ground of incompatibility of temper in their drink.

"You have become indifferent since then, I suppose?" said Mr. Pecksniff. "Use is second nature, Mrs. Gamp."

"You may well say second natur, sir," returned the lady. "One's first ways is to find sich things a trial to the feelings, and so is one's lasting custom. If it wasn't for the nerve a little sip of liquor gives me (I never was able to do more than taste it), I never could go through with what I sometimes has to do. 'Mrs. Harris,' I says, at the very last case as ever I acted in, which it was but a young person, 'Mrs. Harris,' I says, 'leave the bottle on the chimley-piece, and don't ask me to take none, but let me put my lips to it when I am so dispoged, and then I will do what I'm engaged to do, according to the best of my ability.' 'Mrs. Gamp,' she says, in answer, 'if ever there was a sober creetur to be got at eighteen pence a day for working people, and three and six for gentlefolks—night watching,'" said Mrs. Gamp, with emphasis, "'being a extra charge—you are that inwallable person.' 'Mrs. Harris,' I says to her, 'don't name the charge, for if I could afford to lay all my feller creeturs out for nothink, I would gladly do it, sich is the love I bears 'em. But what

I always says to them as has the management of matters, Mrs. Harris' ": here she kept her eye on Mr. Pecksniff: " 'be they gents or be they ladies, is, don't ask me whether I won't take none, or whether I will, but leave the bottle on the chimley-piece, and let me put my lips to it when I am so disposged.' "

The conclusion of this affecting narrative brought them to the house. In the passage they encountered Mr. Mould the undertaker: a little elderly gentleman, bald, and in a suit of black; with a note-book in his hand, a massive gold watch-chain dangling from his fob, and a face in which a queer attempt at melancholy was at odds with a smirk of satisfaction; so that he looked as a man might, who, in the very act of smacking his lips over choice old wine, tried to make believe it was physic.

"Well, Mrs. Gamp, and how are you, Mrs. Gamp?" said this gentle-man, in a voice as soft as his step.

"Pretty well, I thank you, sir," dropping a curtesy.

"You'll be very particular here, Mrs. Gamp. This is not a common case, Mrs. Gamp. Let everything be nice and comfortable, Mrs. Gamp, if you please," said the undertaker, shaking his head with a solemn air.

"It shall be, sir," she replied, curtesying again. "You knows me of old, sir, I hope."

[Chapter XXIX]

Arriving at the tavern, Mrs. Gamp (who was full-dressed for the journey, in her latest suit of mourning) left her friends to entertain themselves in the yard, while she ascended to the sick room, where her fellow-labourer Mrs. Prig, was dressing the invalid.

He was so wasted, that it seemed as if his bones would rattle when they moved him. His cheeks were sunken and his eyes unnaturally large. He lay back in the easy-chair like one more dead than living; and rolled his languid eyes toward the door when Mrs. Gamp appeared, as painfully as if their weight alone were burdensome to move.

"And how are we by this time?" Mrs. Gamp observed. "We looks charming."

"We looks a deal charminger than we are, then," returned Mrs. Prig, a little chafed in her temper. "We got out of bed backwards, I think, for we're as cross as two sticks. I never see sich a man. He wouldn't have been washed, if he'd had his own way."

"She put the soap in my mouth," said the unfortunate patient, feebly.

"Couldn't you keep it shut then?" retorted Mrs. Prig. "Who do you think's to wash one feater, and miss another, and wear one's eyes out with all manner of fine work of that description, for half-a-crown a day! If you wants to be tittivated, you must pay accordin."

"Oh dear me!" cried the patient, "Oh dear, dear!"

"There!" said Mrs. Prig, "that's the way he's been a-conducting himself, Sarah, ever since I got him out of bed, if you'll believe it."

"Instead of being grateful," Mrs. Gamp observed, "for all our little ways. Oh, fie for shame, sir, fie for shame!"

Here Mrs. Prig seized the patient by the chin, and began to rasp his unhappy head with a hair-brush.

"I suppose you don't like that, neither!" she observed, stopping to look at him.

It was just possible that he didn't for the brush was a specimen of the hardest kind of instrument, producible by modern art; and his very eye-lids were red with the friction. Mrs. Prig was gratified to observe the correctness of her supposition, and said triumphantly, "she know'd as much."

When his hair was smoothed down comfortably into his eyes, Mrs. Prig and Mrs. Gamp put on his neckerchief: adjusting his shirt-collar with great nicety, so that the starched points should also invade those organs, and afflict them with an artificial ophthalmia. His waistcoat and coat were next arranged: and as every button was wrenched into a wrong button-hole, and the order of his boots was reversed, he presented on the whole rather a melancholy appearance.

"I don't think it's right," said the poor weak invalid. "I feel as if I was in somebody else's clothes. I'm all on one side; and you've made one of my legs shorter than the other. There's a bottle in my pocket too. What do you make me sit upon a bottle for?"

"Deuce take the man!" cried Mrs. Gamp, drawing it forth. "If he ain't been and got my night-bottle here. I made a little cupboard of his coat when it hung behind the door, and quite forgot it, Betsey. You'll find a ingun or two, and a little tea and sugar in his t'other pocket, my dear, if you'll just be good enough to take 'em out."

Betsey produced the property in question, together with some other articles of general chandlery; and Mrs. Gamp transferred them to her own pocket, which was a species of nankeen pannier. Refreshment then arrived in the form of chops and strong ale for the ladies, and a basin of beef-tea for the patient: which refection was barely at an end when John Westlock appeared.

[From *Pickwick Papers*]

The medical students at Guy's Hospital were very much the same in the early nineteenth century as medical students are today. In *Pickwick Papers* Dickens describes their conversation.

"Well Sam," said Mr. Pickwick as that favoured servitor entered his bed-chamber with his warm water, on the morning of Christmas Day, "Still frosty?"

"Water in the wash-hand basin's a mask o' ice, Sir," responded Sam.

"Severe weather, Sam," observed Mr. Pickwick.

"Fine time for them as is well wropped up, as the Polar Bear said to himself, ven he was practising his skating," replied Mr. Weller.

"I shall be down in quarter of an hour, Sam," said Mr. Pickwick, untying his nightcap.

"Wery good, Sir," replied Sam. "There's a couple o' Sawbones down stairs."

"A couple of what!" exclaimed Mr. Pickwick, sitting up in bed.

"A couple of Sawbones," said Sam.

"What's a Sawbones?" inquired Mr. Pickwick, not quite certain whether it was a live animal, or something to eat.

"What! don't you know what a Sawbones is, Sir?" inquired Mr. Weller; "I thought everybody know'd as a Sawbones was a Surgeon."

"Oh, a Surgeon, eh?" said Mr. Pickwick with a smile.

"Just that, Sir," replied Sam. "These here ones as is below, though, ain't thorough-bred Sawbones; they're only in trainin."

"In other words they're Medical Students, I suppose?" said Mr. Pickwick.

Sam Weller nodded assent.

"I am glad of it," said Mr. Pickwick, casting his nightcap energetically on the counterpane. "They are fine fellows; very fine fellows, with judgements matured by observation and reflection; and tastes refined by reading and study. I am very glad of it."

"They're a smokin' cigars by the kitchen fire," said Sam.

"Ah!" observed Mr. Pickwick rubbing his hands, "Overflowing with kindly feelings and animal spirits. Just what I like to see!"

"And one on 'em," said Sam, not noticing his master's interruption, "one on 'em's got his legs on the table, and is drinkin' brandy neat, vile the t'other one—him in the barnacles—has got a barrel o' oysters atween his knees, vich he's a openin' like stream, and as fast as he eats 'em, he takes a aim vith the shells at young dropsy, who's a sittin' down fast asleep, in the chimbley corner."

"Eccentricities of genius, Sam," said Mr. Pickwick. "You may retire."

Sam did retire accordingly; and Mr. Pickwick, at the expiration of the quarter of an hour, went down to breakfast.

"Here he is at last," said old Wardle. "Pickwick, this is Miss Allen's brother, Mr. Benjamin Allen—Ben we call him, and so may you if you like. This gentleman is his very particular friend, Mr. ——"

"Mr. Bob Sawyer," interposed Mr. Benjamin Allen, whereupon Mr. Bob Sawyer and Mr. Benjamin Allen laughed in concert.

Mr. Pickwick bowed to Bob Sawyer and Bob Sawyer bowed to Mr. Pickwick; Bob and his very particular friend then applied themselves most assiduously to the eatables before them; and Mr. Pickwick had an opportunity of glancing at them both.

Mr. Benjamin was a coarse, stout, thick-set young man, with black hair cut rather short, and a white face cut rather long. He was embellished with spectacles, and wore a white neckerchief. Below his single-breasted black surtout, which was buttoned up to his chin, appeared the usual number of pepper-and-salt coloured legs, terminating in a pair of imperfectly polished boots. Although his coat was short in the sleeves, it disclosed no vestige of a linen wristband; and although there was quite enough of his face to admit of the encroachment of a shirt collar, it was not graced by the smallest approach to that appendage. He presented altogether rather a mildewy appearance, and emitted a fragrant odour of full-flavoured Cubas.

Mr. Bob Sawyer, who was habited in a coarse blue coat, which, without being either a great coat or a surtout, partook of the nature and qualities of both, had about him that sort of slovenly smartness, and swaggering gait, which is peculiar to young gentlemen who smoke in the streets by day, shout and scream in the same by night, call waiters by their christian names, and do various other acts and deeds of an facetious description. He wore a pair of plaid trousers, and a large rough double-breasted waistcoat; and out of doors, carried a thick stick with a big top. He eschewed gloves, and looked, upon the whole, something like a dissipated Robinson Crusoe.

Such were the two worthies to whom Mr. Pickwick was introduced, as he took his seat at the breakfast table on Christmas morning.

"Splendid morning, gentlemen," said Mr. Pickwick.

Mr. Bob Sawyer slightly nodded his assent to the proposition, and asked Mr. Benjamin Allen for the mustart.

"Have you come far this morning, gentlemen?" inquired Mr. Pickwick.

"Blue Lion at Muggleton," briefly responded Mr. Allen.

"You should have joined us last night," said Mr. Pickwick.

"So we should," replied Bob Sawyer, "but the brandy was too good to leave in a hurry; wasn't it, Ben?"

"Certainly," said Mr. Benjamin Allen; "and the cigars were not bad, or the pork chops either: were they, Bob?"

"Decidedly not," said Bob. And the particular friends resumed their attack upon the breakfast, more freely than before, as if the recollection of last night's supper had imparted a new relish to the meal.

"Peg away, Bob," said Mr. Allen to his companion, encouragingly.

"So I do," replied Bob Sawyer. And so, to do him justice, he did.

"Nothing like dissecting, to give one an appetite," said Mr. Bob Sawyer, looking around the table.

Mr. Pickwick slightly shuddered.

"By the bye, Bob," said Mr. Allen, "have you finished that leg, yet?"

"Nearly," replied Sawyer, helping himself to half a fowl as he spoke. "It's a very muscular one for a child's."

"Is it?" inquired Mr. Allen, carelessly.

"Very," said Bob Sawyer, with his mouth full.

"I've put my name down for an arm, at our place," said Mr. Allen. "We're clubbing for a subject, and the list is nearly full, only we can't get hold of any fellow that wants a head. I wish you'd take it."

"No," replied Bob Sawyer, "Can't afford expensive luxuries."

"Nonsense!" said Allen.

"Can't indeed," rejoined Bob Sawyer, "I wouldn't mind a brain, but I couldn't stand a whole head."

"Hush, hush, gentlemen, pray," said Mr. Pickwick, "I hear the ladies."

BOB SAWYER'S PARTY

[From Pickwick Papers, Chap. XXXII.]

There is a repose about Lant Street, in the Borough, which sheds melancholy upon the soul. There are always a good many houses to let in the street; it is a bye-street too, and its dulness is soothing. A house in Lant Street would not come within the denomination of a first-rate residence, in the strict acceptation of the term; but it is a most desirable spot nevertheless. If a man wished to abstract himself from the world—to remove himself from within the reach of temptation—to place himself beyond the possibility of any inducement to look out of the window—he should by all means go to Lant Street.

In this happy retreat are colonised a few clear-starchers, a sprinkling of journeymen bookbinders, one or two prison agents for the Insolvent Court, several small housekeepers who are employed in the Docks, a handful of mantua-makers, and a seasoning of jobbing tailors. The majority of the inhabitants either direct their energies to the letting of furnished apartments, or devote themselves to the healthful and invigorating pursuit of mangling. The chief features in the still of life of the street are green shutters, lodging-bills, brass door-plates, and bell-handles; the principal specimens of animated nature, the pot-boy, the muffin youth, and the baked-potato man. The population is migratory, usually disappearing on the verge of quarter-day, and generally by night. His Majesty's revenues are seldom collected in this happy valley; the rents are dubious; and the water communication is very frequently cut off.

Mr. Bob Sawyer embellished one side of the fire, in the first-floor front, early on the evening for which he had invited Mr. Pickwick; and Mr. Ben Allen the other. The preparations for the reception of visitors appeared to be completed. The umbrellas in the passage had been heaped into the little corner outside the back-parlour door; the bonnet and shawl of the landlady's servant had been removed from the banisters; there were not more than two pairs of pattens on the street-door mat, and a kitchen candle, with a very long snuff, burnt cheerfully on the ledge of the staircase window. Mr. Bob Sawyer had himself purchased the spirits at a wine vaults in High Street, and had returned home preceding the bearer thereof, to preclude the possibility of their delivery at the wrong house. The punch was ready-made in a red pan in the bed-room; a little table, covered with a green baize cloth, had been borrowed from the parlour, to play at cards on, and the glasses of the establishment, together with those which had been borrowed for the occasion from the public-house, were all drawn up in a tray, which was deposited on the landing outside the door. . . .

"Does Mr. Sawyer live here?" said Mr. Pickwick, when the door was opened.

"Yes," said the girl, "first floor. It's the door straight afore you, when you gets to the top of the stairs." Having given this instruction, the handmaid, who had been brought up among the aboriginal inhabitants of Southwark, disappeared, with the candle in her hand, down the kitchen stairs: perfectly satisfied that she had done everything that could be required of her under the circumstances.

Mr. Snodgrass, who entered last, secured the street door, after several ineffectual efforts, by putting up the chain; and the friends stumbled up stairs, where they were received by Mr. Bob Sawyer . . .

"How are you?" said the . . . student. "Glad to see you,—take care of the glasses." This caution was addressed to Mr. Pickwick, who had put his hat in the tray.

"Dear me," said Mr. Pickwick, "I beg your pardon."

"Don't mention it, don't mention it," said Bob Sawyer. "I'm rather confined for room here, but you must put up with all that, when you come to see a young bachelor. Walk in. You've seen this gentleman before, I think?" Mr. Pickwick shook hands with Mr. Benjamin Allen, and his friends followed his example. They had scarcely taken their seats when there was another double knock.

"I hope that's Jack Hopkins!" said Mr. Bob Sawyer. "Hush. Yes, it is. Come up, Jack; come up."

A heavy footstep was heard upon the stairs, and Jack Hopkins presented himself. He wore a black velvet waistcoat, with thunder-and-lightning buttons; and a blue striped shirt, with a white false collar.

"You're late, Jack?" said Mr. Benjamin Allen.

"Been detained at Bartholomew's," replied Hopkins.

"Anything new?"

"No, nothing particular. Rather a good accident brought into the casualty ward."

"What was that sir?" inquired Mr. Pickwick.

"Only a man fallen out of a four pair of stairs' window;—but it's a very fair case—very fair case indeed."

"Do you mean that the patient is in a fair way to recover?" inquired Mr. Pickwick.

"No," replied Hopkins, carelessly. "No, I should rather say he wouldn't. There must be a splendid operation though, to-morrow—magnificent sight if Slasher does it."

"You consider Mr. Slasher a good operator?" said Mr. Pickwick.

"Best alive," replied Hopkins. "Took a boy's leg out of the socket last week—boy ate five apples and a gingerbread cake—exactly two minutes after it was all over, boy said he wouldn't lie there to be made game of, and he'd tell his mother if they didn't begin."

"Dear me!" said Mr. Pickwick, astonished.

"Pooh! That's nothing, that ain't," said Jack Hopkins. "Is it, Bob?"

"Nothing at all," replied Mr. Bob Sawyer.

"By the bye, Bob," said Hopkins, with a scarcely perceptible glance at Mr. Pickwick's attentive face, "we had a curious accident last night. A child was brought in, who had swallowed a necklace."

"Swallowed what, sir?" interrupted Mr. Pickwick.

"A necklace," replied Jack Hopkins. "Not all at once, you know, that would be too much—you couldn't swallow that, if the child did—eh, Mr. Pickwick, ha! ha!" Mr. Hopkins appeared highly gratified with his own pleasantry; and continued, "No, the way was this. Child's parents were poor people who lived in a court. Child's eldest sister bought a necklace; common necklace, made of large black wooden beads. Child, being fond of toys, cribbed the necklace, hid it, played with it, cut the string, and swallowed a bead. Child thought it capital fun, went back next day, and swallowed another bead."

"Bless my heart," said Mr. Pickwick, "what a dreadful thing! I beg your pardon, sir. Go on."

"Next day, child swallowed two beads; the day after that, he treated himself to three, and so on, till a week's time he had got through the necklace—five and twenty beads in all. The sister, who was an industrious girl, and seldom treated herself to a bit of finery, cried her eyes out, at the loss of the necklace; looked high and low for it; but, I needn't say, didn't find it. A few days afterwards, the family were at dinner—baked shoulder of mutton, and potatoes under it—the child, who wasn't hungry, was playing about the room, when suddenly there was heard a devil of a noise, like a small hailstorm. 'Don't do that, my boy,' said the father. 'I ain't a doin' nothing,' said the child. 'Well, don't do it again,' said the father. There was a short silence, and then the noise began again, worse than ever. 'If you don't mind what I say, my boy,' said the father, 'you'll find yourself in bed, in something less than a pig's whisper.' He gave the child a shake to make him obedient, and such a rattling ensued as nobody ever heard before. 'Why, damme, it's in the child!' said the father, 'he's got the croup in the wrong place!' 'No, I haven't, father,' said the child, beginning to cry, 'it's the necklace; I swallowed it, father.' The father caught the child up, and ran with him to the hospital: the beads in the boy's stomach rattling all the way with the jolting; and the people looking up in the air, and down in the cellars to see where the unusual sound came from. He's in the hospital now," said Jack Hopkins, "and he makes such a devil of a noise when he walks about, that they're obliged to muffle him in a watchman's coat, for fear he should wake the patients!"

"That's the most extraordinary case I ever heard of," said Mr. Pickwick, with an emphatic blow on the table.

"Oh, that's nothing," said Jack Hopkins; "is it, Bob?"

"Certainly not," replied Mr. Bob Sawyer.

"Very singular things occur in our profession, I can assure you, sir," said Hopkins.

JOHN BROWN

John Brown was born in Lanarkshire, September 22, 1810. On both his mother's and father's side he was descended from a line of eminent surgeons. It was natural therefore that he should pursue the same studies. At Edinburgh he was apprenticed to the famous Dr. James Syme, professor of surgery and probably the greatest surgeon of his day in the British Isles. John Brown was clinical clerk at the Minto House Hospital in 1830 when the incidents described in *Rab and His Friends* actually happened. Syme was the surgeon. Syme's daughter married Joseph Lister and it is possible as Dr. Robert E. Schleuter suggests (*Peoria Medical News,* September 1937) that the recital or reading of "Rab" turned Lister's mind towards the terrible surgical tragedies of septic infection. Brown and the Listers were very close friends. Dr. John Brown after a busy professional career died in 1862. The manuscript of *Rab and His*

Friends was purchased at auction by William Osler and donated by him and a group of Edinburgh physicians to the Royal College of Physicians of Edinburgh of which Dr. Brown had been president.

In *Rab and His Friends*, Dr. John Brown gives us an idea of surgery before the days of anesthesia, asepsis, or pathology. As a piece of prose the medical profession may well be proud of it. Its simple limpidity does not endure comment, nor all the years blur the pain or dim the poignancy of the inevitable tragedy.

[From *Rab and His Friends*]

Six years have passed,—a long time for a boy and a dog: Bob Ainslie is off to the wars; I am a medical student, and Clerk at Minto House Hospital.

Rab I saw almost every week, on the Wednesday; and we had much pleasant intimacy. I found the way to his heart by frequent scratching of his huge head, and an occasional bone. When I did not notice him he would plant himself straight before me, and stand wagging that bud of a tail, and looking up, with his head a little to the one side. His master I occasionally saw; he used to call me "Maister John," but was laconic as any Spartan.

One fine October afternoon, I was leaving the hospital, when I saw the large gate open, and in walked Rab, with that great and easy saunter of his. He looked as if taking general possession of the place; like the Duke of Wellington entering a subdued city, satiated with victory and peace. After him came Jess, now white with age, with her cart; and in it a woman, carefully wrapped up,—the carrier leading the horse anxiously, and looking back. When he saw me, James (for his name was James Noble) made a curt and grotesque "boo," and said, "Maister John, this is the mistress; she's got a trouble in her breast—some kind o' an income we're thinkin'."

By this time I saw the woman's face; she was sitting on a sack filled with straw, her husband's plaid round her, and his big-coat, with its large white metal buttons, over her feet. I never saw a more unforgetable face—pale, serious, lonely,[1] delicate, sweet, without being what we call fine. She looked sixty, and had on a mutch, white as snow, with its black ribbon; her silvery smooth hair setting off her dark-grey eyes—eyes such as one sees only twice or thrice in a lifetime, full of suffering, but full also of the overcoming of it; her eye-

[1] It is not easy giving this look by one word; it was expressive of her being so much of her life alone.

brows black and delicate, and her mouth firm, patient, and contented, which few mouths ever are.

As I have said, I never saw a more beautiful countenance, or one more subdued to settled quiet. "Ailie," said James, "this is Maister John, the young doctor; Rab's freend, ye ken. We often speak aboot you, doctor." She smiled, and made a movement, but said nothing; and prepared to come down, putting her plaid aside and rising. Had Solomon, in all his glory, been handing down the Queen of Sheba at his palace gate, he could not have done it more daintily, more tenderly, more like a gentleman, than did James the Howgate carrier, when he lifted down Ailie, his wife. The contrast of his small, swarthy, weatherbeaten, keen, worldly face to hers—pale, subdued, and beautiful—was something wonderful. Rab looked on concerned and puzzled; but ready for anything that might turn up,—were it to strangle the nurse, the porter, or even me. Ailie and he seemed great friends.

"As I was sayin', she's got a kind o' trouble in her breest, doctor; wull ye tak' a look at it?" We walked into the consulting-room, all four; Rab grim and comic, willing to be happy and confidential if cause could be shown, willing also to be quite the reverse, on the same terms. Ailie sat down, undid her open gown and her lawn handkerchief round her neck, and, without a word, showed me her right breast. I looked at and examined it carefully,—she and James watching me, and Rab eyeing all three. What could I say? there it was, that had once been so soft, so shapely, so white, so gracious and bountiful, "so full of all blessed conditions,"—hard as a stone, a centre of horrid pain, making that pale face, with its grey, lucid, reasonable eyes, and its sweet resolved mouth, express the full measure of suffering overcome. Why was that gentle, modest, sweet woman, clean and lovable, condemned by God to bear such a burden?

I got her away to bed. "May Rab and me bide?" said James. "*You* may; and Rab, if he will behave himself." "I'se warrant he's do that, doctor"; and in slunk the faithful beast. I wish you could have seen him. There are no such dogs now: he belonged to a lost tribe. As I have said, he was brindled, and grey like Aberdeen granite; his hair short, hard, and close, like a lion's; his body thick set, like a little bull—a sort of compressed Hercules of a dog. He must have been ninety pounds' weight, at the least; he had a large blunt head; his muzzle black as night; his mouth blacker than any night, a tooth or two—being all he had—gleaming out of his jaws of darkness. His head was scarred with the records of old wounds, a sort of series of fields of battle all over it; one eye out; one ear cropped as close as was Archbishop Leighton's father's—but for different reasons,—the remaining eye had the power of two; and above it, and in constant communication with it, was a tattered rag of an ear, which was for ever unfurling itself, like an old flag; and then that bud of a tail, about one inch long, if it could in any sense be said to be long, being as broad as long—the mobility, the instantaneousness of that bud was very funny and surprising, and its expressive twinklings and winkings, the intercommunications between the eye, the ear, and it, were of the subtlest and swiftest. Rab had the dignity and simplicity of great size; and having fought his way all along the road to absolute supremacy, he was as mighty in his own line as Julius Caesar or the Duke of Wellington; and he had the gravity[1] of all great fighters.

Next day, my master, the surgeon, examined Ailie. There was no doubt it must kill her, and soon. It could be removed—it might never return—it would give her speedy relief—and she should have it done. She curtsied, looked at James, and said, "When?" "Tomorrow," said the kind surgeon, a man of few words. She and James and Rab and I retired. I noticed that he and she spoke little, but seemed to anticipate everything in each other. The following day, at noon, the students came in, hurrying up the great stair. At the first landing place, on a small well-known black board, was a bit of paper fastened by wafers, and many remains of old wafers beside it. On the paper were the words, "An operation to-day. J. B. *Clerk.*"

Up ran the youths, eager to secure good places: in they crowded, full of interest and talk. "What's the case?" "Which side is it?"

Don't think them heartless; they are neither better nor worse than you or I: they get over their professional horrors, and into their proper work; and in them pity—as an *emotion*, ending in itself or at best in tears and a long-drawn breath, lessens, while pity as a *motive*, is quickened, and gains power and purpose. It is well for poor human nature that it is so.

---

[1] A Highland game-keeper, when asked why a certain terrier, of singular pluck, was so much graver than the other dogs, said, "Oh, Sir, life's full o' sairiousness to him—he just never can get enuff o' fechtin'."

The operating theatre is crowded; much talk and fun, and all the cordiality and stir of youth. The surgeon with his staff of assistants is there. In comes Ailie: one look at her quiets and abates the eager students. That beautiful old woman is too much for them; they sit down, and are dumb, and gaze at her. These rough boys feel the power of her presence. She walks in quickly, but without haste; dressed in her mutch, her neckerchief, her white dimity shortgown, her black bombazeen petticoat, showing her white worsted stockings and her carpet-shoes. Behind her was James, with Rab. James sat down in the distance, and took that huge and noble head between his knees. Rab looked perplexed and dangerous; for ever cocking his ear and dropping it as fast.

Ailie stepped up on a seat, and laid herself on the table, as her friend the surgeon told her; arranged herself, gave a rapid look at James, shut her eyes, rested herself on me, and took my hand. The operation was at once begun; it was necessarily slow; and chloroform—one of God's best gifts to his suffering children—was then unknown. The surgeon did his work. The pale face showed its pain, but was still and silent. Rab's soul was working within him; he saw that something strange was going on,—blood flowing from his mistress, and she suffering; his ragged ear was up, and importunate; he growled and gave now and then a sharp impatient yelp; he would have liked to have done something to that man. But James had him firm, and gave him a glower from time to time and an intimation of a possible kick;—all the better for James, it kept his eye and his mind off Ailie.

It is over: she is dressed, steps gently and decently down from the table, looks for James, then, turning to the surgeon and the students, she curtsies,—and in a low, clear voice, begs their pardon if she has behaved ill. The students—all of us—wept like children; the surgeon wrapped her up carefully,—and, resting on James and me, Ailie went to her room, Rab following. We put her to bed. James took off his heavy shoes, crammed with tackets, heel-capt and toe-capt, and put them carefully under the table, saying, "Maister John, I'm for nane o' yer strynge nurse bodies for Ailie. I'll be her nurse, and on my stockin' soles I'll gang about as canny as pussy." And so he did; and handy and clever, and swift and tender as any woman, was that horny-handed, snell, peremptory little man. Everything she got he gave her: he seldom slept: and often I saw his small, shrewd eyes out of the darkness, fixed on her. As before, they spoke little.

Rab behaved well, never moving, showing us how meek and gentle he could be, and occasionally, in his sleep, letting us know that he was demolishing some adversary. He took a walk with me every day, generally to the Candlemaker Row; but he was sombre and mild; declined doing battle, though some fit cases offered, and indeed submitted to sundry indignities; and was always very ready to turn, and came faster back, and trotted up the stair with much lightness, and went straight to *that* door.

Jess, the mare—now white—had been sent, with her weather-worn cart, to Howgate, and had doubtless her own dim and placid meditations and confusions, on the absence of her master and Rab, and her unnatural freedom from the road and her cart.

For some days Ailie did well. The wound healed "by the first intention"; as James said, "Oor Ailie's skin's ower clean to beil." The students came in quiet and anxious, and surrounded her bed. She said she liked to see their young, honest faces. The surgeon dressed her, and spoke to her in his own short kind way, pitying her through his eyes. Rab and James outside the circle,—Rab being now reconciled, and even cordial, and having made up his mind that as yet nobody required worrying, but, as you may suppose, *semper paratus.*

So far well: but, four days after the operation, my patient had a sudden and long shivering, a "groofin'," as she called it. I saw her soon after; her eyes were too bright, her cheek coloured; she was restless, and ashamed of being so; the balance was lost; mischief had begun. On looking at the wound, a blush of red told the secret: her pulse was rapid, her breathing anxious and quick, she wasn't herself, as she said, and was vexed at her restlessness. We tried what we could. James did everything, was everywhere; never in the way, never out of it; Rab subsided under the table into a dark place, and was motionless, all but his eye, which followed every one. Ailie got worse; began to wander in her mind, gently; was more demonstrative in her ways to James, rapid in her questions, and sharp at times. He was vexed, and said, "She was never that way afore; no, never." For a time she knew her head was wrong, and was always asking our pardon—the dear, gentle old woman: then delirium set in strong, without pause. Her brain gave way, and that terrible spectacle.

"The intellectual power, through words and things,
    Went sounding on its dim and perilous way;"

she sang bits of old songs and Psalms, stopping suddenly, mingling
the Psalms of David, and the diviner words of his Son and Lord,
with homely odds and ends and scraps of ballads.

Nothing more touching, or in a sense more strangely beautiful,
did I ever witness. Her tremulous, rapid, affectionate, eager Scotch
voice,—the swift, aimless, bewildered mind, the baffled utterance,
the bright and perilous eye; some wild words, some household cares,
something for James, the names of the dead, Rab called rapidly and
in a "fremyt" voice, and he starting up, surprised, and slinking off
as if he were to blame somehow, or had been dreaming he heard.
Many eager questions and beseechings which James and I could
make nothing of, and on which she seemed to set her all and then
sink back ununderstood. It was very sad, but better than many things
that are not called sad. James hovered about, put out and miserable,
but active and exact as ever; read to her, when there was a lull, short
bits from the Psalms, prose and metre, chanting the latter in his
own rude and serious way, showing great knowledge of the fit
words, bearing up like a man, and doating over her as his "ain Ailie."
"Ailie, ma woman!" "Ma ain bonnie wee dawtie!"

The end was drawing on: the golden bowl was breaking; the
silver cord was fast being loosed—that *animula, blandula, vagula,
hospes, comesque,* was about to flee. The body and the soul—com-
panions for sixty years—were being sundered, and taking leave. She
was walking, alone, through the valley of that shadow, into which
one day we must all enter,—and yet she was not alone, for we know
whose rod and staff were comforting her.

One night she had fallen quiet, and as we hoped, asleep; her eyes
were shut. We put down the gas, and sat watching her. Suddenly
she sat up in bed, and taking a bed-gown which was lying on it
rolled up, she held it eagerly to her breast,—to the right side. We
could see her eyes bright with a surprising tenderness and joy,
bending over this bundle of clothes. She held it as a woman holds
her sucking child; opening out her night-gown impatiently, and
holding it close, and brooding over it, and murmuring foolish little
words, as over one whom his mother comforteth, and who is sucking,
and being satisfied. It was pitiful and strange to see her wasted dying
look, keen and yet vague—her immense love. "Preserve me!"

groaned James, giving way. And then she rocked back and forward, as if to make it sleep, hushing it, and wasting on it her infinite fondness. "Wae's me, doctor: I declare she's thinkin' it's that bairn." "What bairn?" "The only bairn we ever had; our wee Mysie, and she's in the Kingdom, forty years and mair." It was plainly true: the pain in the breast, telling its urgent story to a bewildered, ruined brain; it was misread and mistaken; it suggested to her the uneasiness of a breast full of milk, and then the child; and so again once more they were together, and she had her ain wee Mysie in her bosom.

This was the close. She sunk rapidly; the delirium left her; but as she whispered, she was clean silly; it was the lightening before the final darkness. After having for some time lain still—her eyes shut, she said "James!" He came close to her, and lifting up her calm, clear, beautiful eyes, she gave him a long look, turned to me kindly but shortly, looked for Rab but could not see him, then turned to her husband again, as if she would never leave off looking, shut her eyes, and composed herself. She lay for some time breathing quick, and passed away so gently, that when we thought she was gone, James, in his old-fashioned way, held the mirror to her face. After a long pause, one small spot of dimness was breathed out; it vanished away, and never returned, leaving the blank clear darkness of the mirror without a stain. "What is our life? it is even a vapour, which appeareth for a little time, and then vanisheth away."

Rab all this time had been full awake and motionless: he came forward beside us: Ailie's hand, which James had held, was hanging down; it was soaked with his tears; Rab licked it all over carefully, looked at her, and returned to his place under the table.

James and I sat, I don't know how long, but for some time,— saying nothing: he started up abruptly, and with some noise went to the table, and putting his right fore and middle fingers each into a shoe, pulled them out, and put them on, breaking one of the leather latchets, and muttering in anger, "I never did the like o' that afore!"

I believe he never did; nor after either. "Rab!" he said roughly, and pointing with his thumb to the bottom of the bed. Rab leapt up, and settled himself; his head and eye to the dead face. "Maister John, ye'll wait for me," said the carrier; and disappeared in the darkness, thundering down stairs in his heavy shoes. I ran to a front window: there he was, already round the house, and out at the gate, fleeing like a shadow.

I was afraid about him, and yet not afraid; so I sat down beside Rab, and being wearied, fell asleep. I woke from a sudden noise outside. It was November, and there had been a heavy fall of snow. Rab was in *statu quo*; he heard the noise too, and plainly knew it, but never moved. I looked out; and there, at the gate, in the dim morning—for the sun was not up, was Jess and the cart,—a cloud of steam rising from the old mare. I did not see James; he was already at the door, and came up the stairs, and met me. It was less than three hours since he left, and he must have posted out—who knows how?—to Howgate, full nine miles off; yoked Jess, and driven her astonished into town. He had an armful of blankets, and was streaming with perspiration. He nodded to me, spread out on the floor two pairs of old clean blankets, having at their corners, "A. G., 1794," in large letters in red worsted. These were the initials of Alison Graeme, and James may have looked in at her from without—unseen but not unthought of—when he was "wat, wat, and weary," and had walked many a mile over the hills, and seen her sitting, while "a' the lave were sleepin' "; and by the firelight putting her name on the blankets for her ain James's bed. He motioned Rab down, and taking his wife in his arms, laid her in the blankets, and happed her carefully and firmly up, leaving the face uncovered; and then lifting her, he nodded again sharply to me, and with a resolved but utterly miserable face, strode along the passage, and down stairs, followed by Rab. I also followed, with a light; but he didn't need it. I went out, holding stupidly the light in my hand in the frosty air; we were soon at the gate. I could have helped him, but I saw he was not to be meddled with, and he was strong, and did not need it. He laid her down as tenderly, as safely, as he had lifted her out ten days before—as tenderly as when he had her first in his arms when she was only "A. G.,"—sorted her, leaving that beautiful sealed face open to the heavens: and then taking Jess by the head, he moved away. He did not notice me, neither did Rab, who presided along behind the cart.

I stood till they passed through the long shadow of the College, and turned up Nicolson Street. I heard the solitary cart sound through the streets, and die away and come again; and I returned, thinking of that company going up Liberton brae, then along Roslin muir, the morning light touching the Pentlands and making them like on-looking ghosts; then down the hill through Auchindinny woods, past "haunted Woodhouselee"; and as daybreak came sweeping up the bleak Lammermuirs, and fell on his own door, the company would stop, and James would take the key, and lift Ailie up again, laying her on her own bed, and, having put Jess up, would return with Rab and shut the door.

James buried his wife, with his neighbours mourning, Rab inspecting the solemnity from a distance. It was snow, and that black ragged hole would look strange in the midst of the swelling spotless cushion of white. James looked after everything; then rather suddenly fell ill, and took to bed; was insensible when the doctor came, and soon died. A sort of low fever was prevailing in the village, and his want of sleep, his exhaustion, and his misery, made him apt to take it. The grave was not difficult to re-open. A fresh fall of snow had again made all things white and smooth; Rab once more looked on and slunk home to the stable.

And what of Rab? I asked for him next week at the new carrier's who got the goodwill of James's business, and was now master of Jess and her cart. "How's Rab?" He put me off, and said rather rudely, "What's *your* business wi' the dowg?" I was not to be so put off. "Where's Rab?" He, getting confused and red, and intermeddling with his hair, said, " 'Deed, sir, Rab's died." "Dead! what did he die of?" "Weel, sir," said he, getting redder, "he didna exactly die; he was killed. I had to brain him wi' a rack-pin; there was nae doin' wi' him. He lay in the treviss wi' the mear, and wadna come oot. I tempit him wi' kail and meat, but he wad tak' naething, and keepit me frae feedin' the beast, and he was aye gur gurrin', and grup gruppin' me by the legs. I was laith to mak' awa wi' the auld dowg, but his like wasna atween this and Thornill,—but 'deed, sir I could do naething else." I believed him. Fit end for Rab, quick and complete. His teeth and his friends gone, why should he keep the peace and be civil?

XXXI
ANESTHESIA

DURING THE EARLY DECADES OF THE NINETEENTH CENTURY A GREAT MANY people were seeking for some way to deaden pain during surgical operations. The idea undoubtedly was in the air. It is not surprising, therefore, that when the successful trials were made, almost simultaneously, conflicting claims for priority and credit were advanced. The controversy is the most acrimonious recorded in medical history.

Crawford Williamson Long, of Jefferson, Georgia, undoubtedly deserves credit for priority in using ether deliberately and successfully, for the purpose of inducing anesthesia in a surgical operation in March, 1842. He did not, however, publish a description of his feat until 1848. In the meanwhile a Hartford dentist, Horace Wells, induced insensibility by the use of nitrous oxide (1844) in dental operations, but he failed miserably when it was tried publicly for surgical anesthesia. It remained for another dentist, W. T. G. Morton, of Boston, a former partner of Wells, to prove the efficacy of ether in surgical operations. This was done publicly before the medical class of Harvard University at the Massachusetts General Hospital on October 16, 1846, Dr. John C. Warren operating and Dr. Henry J. Bigelow assisting. Bigelow announced the technique to the scientific world in the *Boston Medical and Surgical Journal* for November 18, 1846. His report is reprinted in full in the following pages. Within a month ether was used for anesthesia in London by Robert Liston (December, 1846), for an amputation of the thigh. Syme used it in Edinburgh and Pirogoff in the Crimea. Afterwards a Boston chemist, Dr. Charles T. Jackson, claimed that Morton had learned about the anesthetic properties of ether from him, which was undoubtedly true. Wells, in the meantime, had committed suicide and Jackson and Morton both applied independently to Congress for a patent. The question was hotly debated until Jackson learned of Long's claims, when the whole matter was dropped. Morton's work supported by Warren and Bigelow, undoubtedly introduced anesthesia to the scientific world. Sir James Young Simpson, of Edinburgh, introduced the use of chloroform on January 19, 1847, and for a time it was widely used in surgery and obstetrics. Cocaine as a local anesthetic was introduced for eye operations by Carl Koller in 1884.

REFERENCES

FULOP-MILLER, R. Triumph over Pain. Indianapolis-New York, Bobbs-Merrill, 1938.
LEAKE, C. D. The Historical Development of Surgical Anaesthesia. *Scient. Monthly*, 1925.
TAYLOR, F. L. Crawford W. Long and the Discovery of Ether Anaesthesia. New York, Paul B. Hoeber, 1928.
WELLS, C. J. Horace Wells. *Anesth. & Analg.* July-August, 1935.

CRAWFORD WILLIAMSON LONG*

FIRST SURGICAL OPERATION UNDER ETHER

In the month of December, 1841, or January, 1842, the subject of the inhalation of nitrous oxide gas was introduced in a company of young men assembled at night, in the village of Jefferson, Ga., and the party requested me to prepare them some. I informed them that I had not the requisite apparatus for preparing or using the gas, but that I had an article (sulphuric ether), which would produce equally exhilarating effects and was as safe. The company was anxious to witness its effects: the ether was produced, and all present, in turn, inhaled. They were so much pleased with its effects that they afterwards frequently used it and induced others to use it, and the practice became quite fashionable in the country and some of the contiguous counties. On numerous occasions I inhaled the ether for its exhilarating properties and would frequently at some short time subsequently discover bruises or painful spots on my person which I had no recollection of causing, and which I felt satisfied were received while under the influence of ether. I noticed my friends while etherized, receive falls and blows, which I believed sufficient to cause pain on a person not in a state of anaesthesia, and, on questioning them they uniformly assured me that they did not feel the least pain from these accidents.

Observing these facts I was led to believe that anaesthesia was produced by the inhalation of ether and that its use would be applicable in surgical operations.

The first person to whom I administered ether in a surgical operation was Mr. James M. Venable, who then resided within two miles

* *Tr. Georgia M. & S. Assn.*, 1853.

of Jefferson, and at the present time in Cobb county, Ga. Mr. Venable consulted me on several occasions as to the propriety of removing two small tumors on the back part of his neck, but would postpone from time to time having the operation performed from dread of pain. At length I mentioned to him the fact of my receiving bruises while under the influence of the vapor of ether, without suffering, and, as I knew him to be fond of and accustomed to inhale ether, I suggested to him the probability that the operation might be performed without pain, and suggested to him operating while he was under its influence. He consented to have one tumor removed and the operation was performed the same evening. The ether was given to Mr. Venable on a towel and fully under its influence, I extirpated the tumor. It was encysted and about one-half an inch in diameter. The patient continued to inhale ether during the time of the operation, and seemed incredulous until the tumor was shown to him. He gave no evidence of pain during the operation and assured me after it was over that he did not experience the least degree of pain from its performance. . . .

My third case was a negro boy who had a disease of the toe which rendered amputation necessary, and the operation was performed July 3, 1842, without the boy evincing the slightest degree of pain.

These were all the surgical operations performed by me in the year 1842 upon patients etherized, no other cases occurring in which I believed the inhalation of ether applicable. Since 1842, I have performed one or more operations, annually, on patients in a state of etherization. I procured some certificates in regard to these operations, but not with the same particularity as in regard to the first operations, my sole object being to establish my claim to priority of discovery of the power of ether to produce anaesthesia. However, these certificates can be examined.

Mr. Venable's statement under oath is as follows:

I, James Venable, of the county Cobb and State of Georgia, on oath depose and say, that in the year 1842 I resided at my mother's in Jackson County about two miles from the village of Jefferson, and attended the village academy that year. In the early part of the year the young men of Jefferson and the country adjoining were in the habit of inhaling ether for its exhilarating powers, and I inhaled it frequently for that purpose, and was very fond of its use.

While attending the academy I was frequently in the office of Dr.

C. W. Long, and having two tumors on the back of my neck, I several times spoke to him about the propriety of cutting them out, but postponed the operation from time to time. On one occasion we had some conversation about the probability that the tumors might be cut while I was under the influence of ether, without my experiencing pain, and he proposed operating on me while under its influence. I agreed to have one tumor cut out, and had the operation performed that evening after school was dismissed. This was in the early part of the spring of 1842.

I commenced inhaling the ether before the operation was commenced and continued it until the operation was over. I did not feel the slightest pain from the operation and could not believe the tumor was removed until it was shown to me.

A month or two after this time Dr. C. W. Long cut out the other tumor situated on the same side of my neck. In this operation I did not feel the least pain until the last cut was made, when I felt a little pain. In this operation I stopped inhaling the ether before the operation was finished.

I inhaled the ether, in both cases, from a towel, which was the common method of taking it.

JAMES VENABLE

Georgia
Cobb Co.,
July 23rd, 1849
}
Sworn to before me
Alfred Manes, J. P.

HENRY JACOB BIGELOW

INSENSIBILITY DURING SURGICAL OPERATIONS PRODUCED BY INHALATION*

It has long been an important problem in medical science to devise some method of mitigating the pain of surgical operations. An efficient agent for this purpose has at length been discovered. A patient has been rendered completely insensible during an amputation of the thigh, regaining consciousness after a short interval. Other severe operations have been performed without the knowledge of

---

* Read before the Boston Society of Medical Improvement, Nov. 9th, 1846, an abstract having been previously read before the American Academy of Arts and Sciences, Nov. 3d, 1846. Published in the *Boston M. & S. J.,* Vol. XXXV, No. 16, November 18, 1846.

the patients. So remarkable an occurrence will, it is believed, render the following details relating to the history and character of the process, not uninteresting.

On the 16th of Oct., 1846, an operation was performed at the hospital, upon a patient who had inhaled a preparation administered by Dr. Morton, a dentist of this city, with the alleged intention of producing insensibility to pain. Dr. Morton was understood to have extracted teeth under similar circumstances, without the knowledge of the patient. The present operation was performed by Dr. Warren, and though comparatively slight, involved an incision near the lower jaw of some inches in extent. During the operation the patient muttered, as in a semi-conscious state, and afterwards stated that the pain was considerable, though mitigated; in his own words, as though the skin had been scratched with a hoe. There was, probably, in this instance, some defect in the process of inhalation, for on the following day the vapor was administered to another patient with complete success. A fatty tumor of considerable size was removed, by Dr. Hayward, from the arm of a woman, near the deltoid muscle. The operation lasted four or five minutes, during which time the patient betrayed occasional marks of uneasiness; but upon subsequently regaining her consciousness, professed not only to have felt no pain, but to have been insensible to surrounding objects, to have known nothing of the operation, being only uneasy about a child left at home. No doubt, I think, existed in the minds of those who saw this operation, that the unconsciousness was real; nor could the imagination be accused of any share in the production of these remarkable phenomena.

I subsequently undertook a number of experiments, with the view of ascertaining the nature of this new agent, and shall briefly state them, and also give some notice of the previous knowledge which existed of the use of the substances I employed.

The first experiment was with sulphuric ether, the odor of which was readily recognized in the preparation employed by Dr. Morton. Ether inhaled in vapor is well known to produce symptoms similar to those produced by the nitrous oxide. In my own experience, the exhilaration has been quite as great, though perhaps less pleasurable, than that of this gas, or the Egyptian *haschish*.* It seemed probable that the ether might be so long inhaled as to produce excessive

* Extract of Indian hemp.

inebriation and insensibility; but in several experiments the exhilaration was so considerable that the subject became uncontrollable, and refused to inspire through the apparatus. Experiments were next made with the oil of wine (ethereal oil). This is well known to be an ingredient in the preparation known as Hoffman's anodyne, which also contains alcohol, and this was accordingly employed. Its effects upon the three or four subjects who tried it, were singularly opposite to those of the ether alone. The patient was tranquillized, and generally lost all inclination to speak or move. Sensation was partially paralyzed, though it was remarkable that consciousness was always clear, the patient desiring to be pricked or pinched, with a view to ascertain how far sensibility was lost. A much larger proportion of oil of wine, and also chloric ether, with and without alcohol, were tried, with no better effect. . . .

This variety of evidence tends to show that the knowledge of its effects, especially those of its inhalation, was of uncertain character. Anthony Todd Thomson well sums up what I conceived to have been the state of knowledge at the time upon this subject, in his London Dispensatory of 1818. "As an antispasmodic, it relieves the paroxysm of spasmodic asthma, whether it be taken into the stomach, or its vapor only be inhaled into the lungs. Much caution, however, is required in inhaling the vapor of ether, as the imprudent inspiration of it has produced lethargic and apoplectic symptoms." In his Materia Medica and Therapeutics, of 1832, however, omitting all mention of inhalation, he uses the following words: "Like other diffusible excitants, its effects are rapidly propagated over the system, and soon dissipated. From its volatile nature its exciting influence is probably augmented; as it produces distension of the stomach and bowels, and is thus applied to every portion of their sensitive surface. It is also probable that it is absorbed in its state of vapor, and is therefore directly applied to the nervous centres. It is the diffusible nature of the stimulus of ether which renders it so well adapted for causing sudden excitement, and producing immediate results. Its effects, however, so soon disappear, that the dose requires to be frequently repeated."

Nothing is here said of inhalation, and we may fairly infer that the process had so fallen into disrepute, or was deemed to be attended with such danger, as to render a notice of it superfluous in a work treating, in 1832, of therapeutics.

It remains briefly to describe the process of inhalation by the new method, and to state some of its effects. A small two-necked glass globe contains the prepared vapor, together with sponges to enlarge the evaporating surface. One aperture admits the air to the interior of the globe, whence, charged with vapor, it is drawn through the second into the lungs. The inspired air thus passes through the bottle, but the expiration is diverted by a valve in the mouth piece, and escaping into the apartment is thus prevented from vitiating the medicated vapor. A few of the operations in dentistry, in which the preparation has as yet been chiefly applied, have come under my observation. The remarks of the patients will convey an idea of their sensations.

A boy of 16, of medium stature and strength, was seated in the chair. The first few inhalations occasioned a quick cough, which afterwards subsided; at the end of eight minutes the head fell back, and the arms dropped, but owing to some resistance in opening the mouth, the tooth could not be reached before he awoke. He again inhaled for two minutes, and slept three minutes, during which time the tooth, an inferior molar, was extracted. At the moment of extraction the features assumed an expression of pain, and the hand was raised. Upon coming to himself, he said he had a "first rate dream—very quiet," he said, "and had dreamed of Napoleon—had not had the slightest consciousness of pain—the time had seemed long"; and he left the chair, feeling no uneasiness of any kind, and evidently in a high state of admiration. The pupils were dilated during the state of unconsciousness, and the pulse rose from 130 to 142.

A girl of 16 immediately occupied the chair. After coughing a little, she inhaled during three minutes, and fell asleep, when a molar tooth was extracted, after which she continued to slumber tranquilly during three minutes more. At the moment when force was applied she flinched and frowned, raising her hand to her mouth, but said she had been dreaming a pleasant dream, and knew nothing of the operation.

A stout boy of 12, at the first inspiration coughed considerably, and required a good deal of encouragement to induce him to go on. At the end of three minutes from the first fair inhalation, the muscles were relaxed and the pupils dilated. During the attempt to force open the mouth he recovered his consciousness, and again inhaled during two minutes, and in the ensuing one minute two teeth were extracted, the patient seeming somewhat conscious, but upon actually awaking he declared "it was the best fun he ever saw," avowed his intention to come there again, and insisted upon having another tooth extracted upon the spot. A splinter which had been left, afforded an opportunity of complying with his wish, but the pain proved to be considerable. Pulse at first 110, during sleep 96, afterwards 144; pupils dilated.

The next patient was a healthy-looking, middle-aged woman, who inhaled the vapor for four minutes; in the course of the next two minutes a back tooth was extracted; and the patient continued smiling in her sleep for three minutes more. Pulse 120, not affected at the moment of the operation, but smaller during sleep. Upon coming to herself, she exclaimed that "it was beautiful—she dreamed of being at home—it seemed as if she had been gone a month." These cases, which occurred successively in about an hour, at the room of Dr. Morton, are fair examples of the average results produced by the inhalation of the vapor, and will convey an idea of the feelings and expressions of many of the patients subjected to the process. Dr. Morton states that in upwards of two hundred patients, similar effects have been produced. The inhalation, after the first irritation has subsided, is easy, and produces a complete unconsciousness, at the expiration of a period varying from two to five or six, sometimes eight minutes; its duration varying from two to five minutes; during which the patient is completely insensible to the ordinary tests of pain. The pupils, in the cases I have observed, have been generally dilated; but with allowance for excitement and other disturbing influences, the pulse is not affected, at least in frequency; the patient remains in a calm and tranquil slumber, and wakes with a pleasurable feeling. The manifestation of consciousness or resistance I at first attributed to the reflex function, but I have since had cause to modify this view.

It is natural to inquire whether no accidents have attended the employment of a method so wide in its application, and so striking in its results. I have been unable to learn that any serious consequences have ensued. One or two robust patients have failed to be affected. I may mention as an early and unsuccessful case, its administration in an operation performed by Dr. Hayward, where an elderly woman was made to inhale the vapor for at least half an hour without effect. Though I was unable at the time to detect any imperfection in the process, I am inclined to believe that such existed. One woman became much excited, and required to be confined to the chair. As this occurred to the same patient twice, and in no other case as far as I have been able to learn, it was evidently owing to a peculiar susceptibility. Very young subjects are affected with nausea and vomiting, and for this reason Dr. M. has refused to administer it to children. Finally, in a few cases, the patient has continued to sleep tranquilly for eight or ten minutes, and once, after a protracted inhalation, for the period of an hour.

The following case, which occurred a few days since, will illustrate the probable character of future accidents. A young man was made to inhale the vapor, while an operation of limited extent, but somewhat protracted duration, was performed by Dr. Dix upon the tissues near the eye. After a good deal of coughing, the patient succeeded in inhaling the vapor, and fell asleep at the end of about ten minutes. During the succeeding two minutes the first incision was made, and the patient awoke, but unconscious of pain. Desiring to be again inebriated, the tube was placed in his mouth and retained there about twenty-five minutes, the patient being apparently half affected, but, as he subsequently stated, unconscious. Respiration was performed partly through the tube and partly with the mouth open. Thirty-five minutes had now elapsed, when I found the pulse suddenly diminishing in force, so much so, that I suggested the propriety of desisting. The pulse continued decreasing in force, and from 120 had fallen to 96. The respiration was very slow, the hands cold, and the patient insensible. Attention was now of course directed to the return of respiration and circulation. Cold affusions, as directed for poisoning with alcohol, were applied to the head, the ears were syringed, and ammonia presented to the nostrils and administered internally. For fifteen minutes the symptoms remained stationary, when it was proposed to use active exercise, as in case of narcotism from opium. Being lifted to his feet, the patient soon made an effort to move his limbs, and the pulse became more full, but again decreased in the sitting posture, and it was only after being compelled to walk during half an hour that the patient was able to lift his head. Complete consciousness returned only at the expiration of an hour. In this case the blood was flowing from the head, and rendered additional loss of blood unnecessary. Indeed, the probable hemorrhage was previously relied on as salutary in its tendency.

Two recent cases serve to confirm, and one I think to decide, the great utility of this process. On Saturday, the 7th Nov., at the Mass. General Hospital, the right leg of a young girl was amputated above the knee, by Dr. Hayward, for disease of this joint. Being made to inhale the preparation, after professing her inability to do so from the pungency of the vapor, she became insensible in about five minutes. The last circumstance she was able to recall, was the adjustment of the mouth piece of the apparatus, after which she was unconscious until she heard some remark at the time of securing the vessels—one of the last steps of the operation. Of the incision she knew nothing, and was unable to say, upon my asking her, whether or not the limb had been removed. She refused to answer several questions during the operation, and was evidently completely insensible to pain or other external influences. This operation was followed by another, consisting of the removal of a part of the lower jaw, by Dr. Warren. The patient was insensible to the pain of the first incision, though she recovered her consciousness in the course of a few minutes.

The character of the lethargic state, which follows this inhalation, is peculiar. The patient loses his individuality and awakes after a certain period, either entirely unconscious of what has taken place, or retaining only a faint recollection of it. Severe pain is sometimes remembered as being of a dull character; sometimes the operation is supposed by the patient to be performed upon somebody else. Certain patients, whose teeth have been extracted, remember the application of the extracting instruments; yet none have been conscious of any real pain.

As before remarked, the phenomena of the lethargic state are not such as to lead the observer to infer this insensibility. Almost all patients under the dentist's hands scowl or frown; some raise the hand. The patient whose leg was amputated, uttered a cry when the sciatic nerve was divided. Many patients open the mouth, or raise themselves in the chair, upon being directed to do so. Others manifest the activity of certain intellectual faculties. An Irishman objected to the pain, said that he had been promised an exemption from it. . . .

The duration of the insensibility is another important element in the process. When the apparatus is withdrawn at the moment of unconsciousness, it continues, upon the average, two or three minutes, and the patient then recovers completely or incompletely, without subsequent ill effects. In this sudden cessation of the symptoms, this vapor in the air tubes differs in its effects from the narcotics or stimulants in the stomach, and, as far as the evidence of a few experiments of Dr. Morton goes, from the ethereal solution of opium when breathed. Lassitude, headache and other symptoms lasted for several hours, when this agent was employed.

It is natural to inquire with whom this invention originated. Without entering into details, I learn that the patent bears the name of Dr. Charles T. Jackson, a distinguished chemist, and of Dr. Morton, a skilful dentist, of this city, as inventors—and has been issued to the latter gentleman as proprietor.

It has been considered desirable by the interested parties that the character of the agent employed by them, should not at this time be announced; but it may be stated that it has been made known to those gentlemen who have had occasion to avail themselves of it.

I will add, in conclusion, a few remarks upon the actual position of this invention as regards the public.

No one will deny that he who benefits the world should receive from it an equivalent. The only question is, of what nature shall the equivalent be? Shall it be voluntarily ceded by the world, or levied upon it? For various reasons, discoveries in high science have been unusually rewarded indirectly by fame, honor, position, and occasionally, in other countries, by funds appropriated for that purpose. Discoveries in medical science, whose domain approaches so nearly that of philanthropy, have been generally ranked with them; and many will assent with reluctance to the propriety of restricting by letters patent the use of an agent capable of mitigating human suffering. There are various reasons, however, which apologize for the arrangement which I understand has been made with regard to the application of the new agent.

1st. It is capable of abuse, and can readily be applied to nefarious ends.

2nd. Its action is not yet thoroughly understood, and its use should be restricted to responsible persons.

3d. One of its greatest fields is the mechanical art of dentistry, many of whose processes are, by convention, secret, or protected by patent rights. It is especially with reference to this art, that the patent has been secured. We understand, already, that the proprietor has ceded its use to the Mass. General Hospital, and that his intentions are extremely liberal with regard to the medical profession generally, and that so soon as necessary arrangements can be made for publicity of the process, great facilities will be offered to those who are disposed to avail themselves of what now promises to be one of the important discoveries of the age.

WILLIAM T. G. MORTON

REMARKS ON THE PROPER MODE OF ADMINISTERING ETHER BY INHALATION*

Although various publications have appeared since the new application of sulphuric ether was discovered which have made it evident that it can be used, both safely and effectually, for the relief of much of the suffering to which the human race is liable, I believe that a manual, containing an account of the mode of administering it, the effect which it produces, the symptoms of insensibility, the difficulties and dangers attending its use, and the best means of obviating and removing these, as far as possible, is still a desideratum. This is particularly the case with those who have not had an opportunity to witness its administration, but who may wish to make use of it in their own practice. To supply this want, and to avoid the necessity of replying to the letters frequently addressed to me for information upon these subjects, the following pages have been written. To those who have used the ether, many of the directions may appear tediously minute; but to those who have not, they will afford desirable information.

In the first place, it is of the utmost consequence that the ether which is used should be not only free from all impurities, but as highly concentrated as possible; as some of these impurities would prove injurious if taken into the system, and as, of course, the stronger the ether, the sooner the patient comes under its influence. Unrectified sulphuric ether contains, as impurities, alcohol, water, sulphurous acid, and oil of wine; and is unfit for use internally.

In order to make it fit for inhalation unrectified sulphuric ether

* Boston, Dutton and Wentworth, Printers, 1847.

must be redistilled and washed, and then dried with chloride of calcium. This will free it from the impurities above mentioned, and render it more concentrated than the original article.

The next point I have to treat of is the best mode of administering ether. The earliest experiments were mostly made by pouring ether upon cloths and inhaling it from them. The results obtained in this way were somewhat uncertain and not always satisfactory, and this mode of administering it was, before long, exchanged for that by means of an apparatus, which rendered the experiments more uniformly successful. Some alterations and improvements were afterward made in this apparatus, but, substantially, it remained the same as long as it continued in use; and, as many persons may read these pages who have never seen the apparatus, or one like it, a few words by way of description will not, perhaps, be unacceptable.

The apparatus first used consisted of a glass vessel about six inches square, with rounded corners; one opening, two inches in diameter, was left on the top, through which a sponge was inserted and the ether poured, and another, an inch and a half in diameter, on one side for the admission of external air. On the side opposite the last-named opening was a glass tube, two inches in diameter and an inch in length, terminating in a metal mouth-piece three inches long,[1] and of the same calibre as the glass tube. This mouth-piece was provided with two valves, one covering a circular opening, three-quarters of an inch in diameter, on the top, and the other extending across it. These valves were so arranged that, when the patient filled his lungs, the upper valve shut down, closing the aperture in the top of the mouth-piece, while the one across the mouth-piece opened and allowed the ethereal vapor, mixed with atmospheric air, to pass into the lungs; and, when he emptied his lungs, the pressure of the expired air closed the valve across the tube, while the same pressure opened the upper valve and allowed the vapor, which had been once breathed, to pass into the room instead of returning into the reservoir. Thus, at each inspiration, the patient had a fresh supply of air thoroughly charged with the vapor of ether, which vapor was continually given off by the sponge which was placed in the reservoir and thoroughly saturated with ether.

---

[1] In the early administration of ether I sometimes made use of a flexible tube, about four inches long, with a mouth-piece at its end; but I soon discontinued the use of this, as patients were not so soon brought under the influence of the ether when at this distance from the apparatus.

Before going on to speak of the mode of administering ether, and to describe the effects produced by it, it is proper to state that it is not necessary for patients who take it for different purposes to be at all put in precisely the same condition. That is, it is not necessary that all who take it should be brought equally under the influence of the ether, but that different degrees of etherization should be produced, according to the end proposed to be effected in different classes of cases. For instance, most surgical operations occupy a longer time than does the extraction of teeth, and the pain is, necessarily, of longer duration; of course, then, a patient must be brought more completely under the influence of ether to render him insensible to pain which is to last a number of minutes, than it is necessary to prevent him from feeling that which lasts only a few seconds. There are some conditions, too, usually attending operations in dentistry, which do not belong to most operations in surgery. Teeth are usually extracted at the house of the dentist, and the patient expects, as soon as the tooth is out, to be able to leave the room and go about his ordinary avocations. Now this may be done in a large majority of the cases where teeth are extracted while the patient is under the influence of ether, provided only enough has been given to render him insensible to the short, and comparatively slight, pain caused by the extraction of a tooth, and not as much as is required to prevent him from feeling a severe surgical operation. Many dentists have, no doubt, been prevented from using ether from a fear that its effects, after the tooth was extracted, would prove annoying both to themselves and their patients; which fear arose, I think, from an erroneous belief that a patient must be brought as much under the influence of ether to have a tooth extracted, as to undergo the amputation of a limb. That some cases have occurred where nausea and vomiting have followed the use of ether I have neither the wish, nor the intention, to deny; but they are by no means frequent, and have never, to my knowledge, been followed by any permanent ill effects. Ought we, then, to be deterred from using ether,—which we know to be an antidote to a positive evil, pain,—from the fear of consequences which but rarely follow its use, and which, when they do occur, seem to be of little moment?

Having made the necessary previous examination, decided upon the operation to be performed, and arranged the instruments near at hand, so as to lose no time when the ether takes effect, proceed as follows: Place your patient in the position which he is to occupy during the operation (which position should be made as comfortable as possible), and direct him to close his eyes and remain perfectly still. Direct, also, any one who may be present to say nothing to the patient, as talking materially retards the process. Where a tooth is to be extracted, or any operation performed upon the mouth, the patient should be directed to inhale the ether with his mouth open; for, if he commences this process with his mouth closed, he will sometimes keep it firmly shut after he is under the influence of the ether. Should the patient, at any time, close his teeth in this way, the operator must press the palm of his left hand firmly upon the patient's forehead, so as to fix the head steadily against the head-piece of the chair; then, placing his right hand upon the chin, overcome the resistance of the muscles by a sudden, but firm, downward pressure. Having made these preparations, pour upon the inside of such a bell-shaped sponge as I have above described two ounces of pure sulphuric ether; then bring the hollow part of the sponge to within an inch of the patient's mouth, and allow him to take four or five inspirations with it in this position. If no cough is produced, the sponge may then be applied so as to cover the nose and mouth of the patient, so that all the air he breathes must pass through it. If any cough is produced by the first breathing of the ether, the sponge should not be applied so as to cover the mouth and both nostrils, but should be inclined a little to one side, so that one nostril may remain open for the air to pass through until the patient becomes so accustomed to the ether that it does not produce coughing. Any forced or irregular breathing should be discouraged, and the patient told to breathe in a steady, regular, and natural manner.

Inhalation should be thus persevered in for three minutes, without anything being said to the patient, unless there is reason to believe, before that time, that he is under the influence of the ether; this can only be known by carefully observing the symptoms of etherization. These are:[1] redness of the countenance, dilatation of the pupils, which are also sometimes fixed or turned up, increased

[1] This is the usual appearance, although the countenance is sometimes pale; but this can easily be distinguished from the purple or livid appearance which too long a use of the ether produces.

action of the heart, so that the pulse is quicker, and usually fuller, relaxation of the muscles generally (sometimes accompanied by a flow of saliva from the corners of the mouth), and, finally, loss of consciousness and slowness of the pulse. It is not necessary, nor is it desirable, that, in every case where ether is inhaled, the patient should exhibit all these symptoms. When, therefore, a dentist is giving ether to a patient for the purpose of extracting a tooth, if, at the end of a minute from the time when the sponge is placed at the patient's lips, he perceives that the countenance is flushed, the pulse quickened, and, particularly, if there is any relaxation of the muscular system (which, I think, first manifests itself in a heavy dropping of the eyelids, similar to that which precedes natural sleep), he should be told to open his eyes. If the patient opens his eyes quickly, and if, when he does so, the pupils appear clear, bright and natural, he should be directed to close them again, and nothing more should be said to him until another minute has elapsed, when the same process should be repeated. If, on the contrary, when the patient is directed to open his eyes, he either neglects doing so entirely, or else partially raises the lids in a heavy, languid manner, disclosing dilated pupils, with a dull, lack-lustre expression, he should be asked whether he will have his tooth extracted; if he makes no reply, or drawls out a slow, hesitating assent, the instrument should be applied, and the tooth extracted immediately. This may surprise those dentists who are not in the habit of using ether; but I assure them that patients frequently retain sufficient consciousness to comprehend what is said to them, and even to express their willingness to have a tooth extracted, although, when it is done, they feel no pain. Sometimes, when in this state, they know when the instrument is applied, and know that a tooth is extracted, but regard it as belonging to some one else; or, if they are aware that it is their own, declare that it causes them no pain. Sometimes a patient will scream when a tooth is extracted, and very frequently will raise his hand to his head when the instrument is applied to it; but these must not be considered as indications that he is conscious of suffering, for I have very often had patients do so, and yet declare, when they fully recovered, that they had felt no pain.[1] As patients, however,

[1] Sometimes, where it has been necessary to extract several teeth, when I have removed one, I have had the patient say to me, "Be quick! Be quick!" and yet declare positively, when he recovered, that he had felt no pain.

sometimes seize the hand or instrument of the operator, and thus prevent the extraction of the tooth, it is well to have an assistant near, who can hold the hands of the patient if necessary; and those persons who appear to be uneasy and restless while inhaling ether, should be directed to grasp firmly the hands of an assistant, as this frequently tranquillizes and quiets them.

By carefully following the above directions, dentists may be assured that they will rarely fail to render their patients insensible to the extraction of teeth, and will usually be able to do this in from one to three minutes. If the ether does not take effect in three minutes, they will please to follow the directions given below for administering it in surgical operations; excepting that a dentist need not wait until the patient does not open his eyes when directed to do so, but may proceed to extract the tooth if the patient opens his eyes in the heavy, languid manner, or speaks in the slow, drawling tone before described.

I should mention here that it has been proposed, and practised, to allow patients to inhale ether with the eyes open, in order to observe the effect which it produces upon the pupils, and judge of the time for extracting the tooth by the appearance which they present. I have tried this method, but, on the whole, prefer the one which I have mentioned above as being more certain.

When a surgical operation is to be performed, the inhalation should be steadily continued for three minutes without speaking to the patient. If, at the end of this time, the pulse is quickened and the muscles relaxed, so that the head has a tendency to fall on one side, the patient should be told, in a loud, distinct tone, to open his eyes; and, if he does *not* do so, the operation should be immediately commenced. If he does open his eyes, even in a slow and languid manner, he should be directed to close them, and the inhalation should be continued two minutes longer, when the same question may be repeated; and it will usually be found that, by this time, the patient is unconscious. Should this not occur, however, the surgeon should place his hand over about one-half of the sponge, so as to prevent loss of ether by evaporation, and continue the inhalation until ten minutes have elapsed from the time when the patient first began to breathe it, calling upon him to open his eyes at intervals of about one minute each.[1] If, at the end of ten minutes, he still continues to open his eyes when directed to do so, the inhalation should be discontinued, and not resumed again for at *least* five minutes. At the end of that time two ounces more of ether should be poured upon the sponge, and the inhaling resumed as before; but, if, after inhaling a second time for ten minutes, it does not produce its effect, an *interval* of *ten* minutes must be allowed to the patient, and then, the ether having been again renewed, the inhaling may be resumed once more. If, at the end of the third trial of inhaling ether, the patient still remains unaffected by it, the operation had better be deferred until another day; but I can hardly suppose that this will ever happen where the ether is pure and highly concentrated, and has been administered in the manner above described.

WASHINGTON AYER

Account of an Eye Witness of the First Public Demonstration of Ether Anaesthesia at the Massachusetts General Hospital, October 16, 1846*

The day arrrived; the time appointed was noted on the dial, when the patient was led into the operating-room, and Dr. Warren and a board of the most eminent surgeons in the State were gathered around the sufferer. "All is ready—the stillness oppressive." It had been announced "that a test of some preparation was to be made for which the *astonishing* claim had been made that it would render the person operated upon free from pain." These are the words of Dr. Warren that broke the stillness.

Those present were incredulous, and, as Dr. Morton had not arrived at the time appointed and fifteen minutes had passed, Dr. Warren said, with significant meaning, "I presume he is otherwise engaged." This was followed with a "derisive laugh," and Dr. Warren grasped his knife and was about to proceed with the operation. At that moment Dr. Morton entered a side door, when Dr. Warren turned to him and in a strong voice said, "Well, sir, your

---

[1] Throughout the whole of the inhalation the operator should bear constantly in mind the symptoms indicating danger, which I have given below, and remove the sponge from the patient's mouth as soon as any of them appear.

* From The Semi-Centennial of Anaesthesia. Massachusetts General Hospital, 1897.

patient is ready." In a few minutes he was ready for the surgeon's knife, when Dr. Morton said, "*Your* patient is ready, sir."

Here the most sublime scene ever witnessed in the operating-room was presented, when the patient placed himself voluntarily upon the table, which was to become the altar of future fame. Not that he did so for the purpose of advancing the science of medicine, nor for the good of his fellow-men, for the act itself was purely a personal and selfish one. He was about to assist in solving a new and important problem of therapeutics, whose benefits were to be given to the whole civilized world, yet wholly unconscious of the sublimity of the occasion or the part he was taking.

That was a supreme moment for a most wonderful discovery, and, had the patient died under the operation, science would have waited long to discover the hypnotic effects of some other remedy of equal potency and safety, and it may be properly questioned whether chloroform would have come into use as it has at the present time.

The heroic bravery of the man who voluntarily placed himself upon the table, a subject for the surgeon's knife, should be recorded and his name enrolled upon parchment, which should be hung upon the walls of the surgical amphitheatre in which the operation was performed. His name was Gilbert Abbott.

The operation was for a congenital tumor on the left side of the neck, extending along the jaw to the maxillary gland and into the mouth, embracing a margin of the tongue. The operation was successful; and when the patient recovered he declared he had suffered no pain. Dr. Warren turned to those present and said, "Gentlemen, this is no humbug."

SIR JAMES YOUNG SIMPSON

[From *Account of a New Anaesthetic Agent as a Substitute for Sulphuric Ether**]

From the time at which I first saw Ether-Inhalation successfully practised in January last, I have had the conviction impressed upon my mind, that we would ultimately find out that other therapeutic agents were capable of being introduced with equal rapidity and

* Edinburgh, 1847.

success into the system, through the same extensive and powerful channel of pulmonary absorption.

With various professional friends, more conversant with chemistry than I am, I have, since that time, taken opportunities of talking over the idea which I entertained of the probable existence or discovery of new therapeutic agents, capable of being introduced into the system of respiration, and the possibility of producing for inhalation vaporizable or volatile preparations of some of our more active and old established medicines; and I have had, during the summer and autumn, etheral tinctures, etc., of several potent drugs, manufactured for me, for experiment, by Messrs. Duncan, Flockhart, & Co., the excellent chemists and druggists of this city.

Latterly, in order to avoid, if possible, some of the inconveniences and objections pertaining to sulphuric ether (particularly its disagreeable and very persistent smell, its occasional tendency to irritation of the bronchi during its first inspirations, and the large quantity of it occasionally required to be used, more especially in protracted cases of labour), I have tried upon myself and others the inhalation of different other volatile fluids, with the hope that some one of them might be found to possess the advantages of ether, without its disadvantages. For this purpose, I selected for experiment and have inhaled several chemical liquids of a more fragrant or agreeable odour, such as the chloride of hydrocarbon (or Dutch liquid), acetone, nitrate of oxide of ethyle (nitrate ether), benzin, the vapour of iodoform, etc. I have found, however, one infinitely more efficacious than any of the others, viz., Chloroform, or the Perchloride of Formyle, and I am enabled to speak most confidently of its superior anaesthetic properties, having now tried it upon upwards of thirty individuals. The liquid I have used has been manufactured for me by Mr. Hunter, in the laboratory of Messrs. Duncan, Flockhart, & Co.

Chloroform was first discovered and described at nearly the same time by Soubeiran (1831) and Liebig (1832); its composition was first accurately ascertained by the distinguished French chemist, Dumas, in 1835.—See the Annales de Chimie et de Physique, vols. xlviii, xlix, and lviii. It has been used by some practitioners internally; Guillot prescribed it as an anti-spasmodic in asthma, exhibiting it in small doses, and diluted 100 times.—(See Bouchardat's Annuaire de Therapeutique for 1844, p. 35). But no person, so far as

I am aware, has used it by inhalation, or discovered its remarkable anaesthetic properties till the date of my own experiments.

It is a dense, limpid, colourless liquid, readily evaporating, and possessing an agreeable, fragrant, fruit-like odour, and a saccharine pleasant taste.

As an inhaled anaesthetic agent, it possesses over sulphuric Ether the following advantages:—

1. A greatly less quantity of Chloroform than of Ether is requisite to produce the anaesthetic effect; usually from a hundred to a hundred and twenty drops of Chloroform only being sufficient; and with some patients much less. I have seen a strong person rendered completely insensible by six or seven inspirations of thirty drops of the liquid.

2. Its action is much more rapid and complete, and generally more persistent. I have almost always seen from ten to twenty inspirations suffice. Hence the time of the surgeon is saved; and that preliminary stage of excitement, which pertains to all narcotizing agents, being curtailed, or indeed practically abolished, the patient has not the same degree of tendency to exhilaration and talking.

3. Most of those who know from previous experience the sensations produced by ether inhalation, and who have subsequently breathed the Chloroform, have strongly declared the inhalation and influence of Chloroform to be far more agreeable and pleasant than those of Ether.

4. I believe, that considering the small quantity requisite, as compared with Ether, the use of Chloroform will be less expensive than that of Ether; more especially, as there is every prospect that the means of forming it may be simplified and cheapened.

5. Its perfume is not unpleasant, but the reverse; and the odour of it does not remain, for any length of time, obstinately attached to the clothes of the attendant,—or exhaling in a disagreeable form from the lungs of the patient, as so generally happens with Sulphuric Ether.

6. Being required in much less quantity, it is much more portable and transmissable than Sulphuric Ether.

7. No special kind of inhaler or instrument is necessary for its exhibition. A little of the liquid diffused upon the interior of a hollow-shaped sponge, or a pocket-handkerchief, or a piece of linen

or paper, and held over the mouth and nostrils, so as to be fully inhaled, generally suffices in about a minute or two to produce the desired effect.

I have not yet had an opportunity of using Chloroform in any capital surgical operation, but have exhibited it with perfect success, in tooth-drawing, opening abscesses, for annulling the pain of dysmenorrhoea and of neuralgia, and in two or three cases where I was using deep, and otherwise very painful galvano-puncture for the treatment of ovarian dropsy etc. I have employed it also in obstetric practice with entire success. The lady to whom it was first exhibited during parturition, had been previously delivered in the country by perforation of the head of the infant, after a labour of three days' duration. In this, her second confinement, pains supervened a fortnight before the full time. Three hours and a-half after they commenced, and, ere the first stage of the labour was completed, I placed her under the influence of the Chloroform, by moistening, with half a tea-spoonful of the liquid, a pocket handkerchief, rolled up into a funnel shape, and with the broad or open end of the funnel placed over her mouth and nostrils. In consequence of the evaporation of the fluid, it was once more renewed in about ten or twelve minutes. The child was expelled in about twenty-five minutes after the inhalation was begun. The mother subsequently remained longer soporose than commonly happens after Ether. The squalling of the child did not, as usual, rouse her; and some minutes elapsed after the placenta was expelled, and after the child was removed by the nurse into another room, before the patient awoke. She then turned round and observed to me that she had "enjoyed a very comfortable sleep, and indeed required it, as she was so tired, but would now be more able for the work before her." I evaded entering into conversation with her, believing, as I have already stated, that the most complete possible quietude forms one of the principal secrets for the successful employment of either Ether or Chloroform. In a little time she again remarked that she was afraid her "sleep had stopped the pains." Shortly afterwards, her infant was brought in by the nurse from the adjoining room, and it was a matter of no small difficulty to convince the astonished mother that the labour was entirely over, and that the child presented to her was really her "own living baby." . . .

I have never had the pleasure of watching over a series of better or more rapid recoveries; nor once witnessed any disagreeable result follow to either mother or child; whilst I have now seen an immense amount of maternal pain and agony saved by its employment. And I most conscientiously believe that the proud mission of the physician is distinctly twofold—namely, to alleviate human suffering, as well as preserve human life.

XXXII

BACTERIOLOGY

THE IDEA OF CONTAGION IS AN OLD ONE (SEE FRACASTORIUS). LEEUWENHOEK found bacteria in his mouth scrapings, but the complete proof of our modern conception of bacteriology was the work of Louis Pasteur (1833-1895). The *Etudes sur le vin* (1866), *Etudes sur les maladies des vers a soie* (1870), *Etudes sur la bière* (1876), papers on Anthrax, Chicken Cholera (1877), Preventive Inoculation Particularly of Hydrophobia (1885) successively laid down the principles of man's most remarkable conquest of Nature.

Almost simultaneously Robert Koch (1843-1910), in Germany, was inventing and perfecting the technique of bacteriology—cultures, staining of bacilli, hanging drop cultures, incubation, and so forth. His discovery of the tubercle bacillus and his demonstration of the unity of all forms of tuberculosis are described in his paper on the etiology of tuberculosis, which was published in 1882. This article is one of the great medical classics.

To Paul Ehrlich (1854-1915) belongs the credit of developing the side-chain theory of immunity, the method of staining the tubercle bacillus, and specific arsenic therapy for syphilis. Hundreds of other men contributed details to the new science of bacteriology, but these three made it a completely integrated structure.

REFERENCES

BULLOCH, W. The History of Bacteriology. London, Oxford University Press, 1938.

FORD, W. W. Bacteriology. Clio Medica series. New York, Paul B. Hoeber, 1939.

VALLERY-RADOT, R. The Life of Pasteur. London, Constable, 1902.

LOUIS PASTEUR

It is difficult to quote from Pasteur. His style is diffuse, and he does not state in any one single place the doctrine that we have come to regard as the foundation of bacteriology. I present for their historic interest Pasteur's own history of the theory of spontaneous generation and part of his paper on the prevention of rabies, including the story of Joseph Meister.

ON THE HISTORY AND THEORIES OF SPONTANEOUS GENERATION

[From *Memoirs on the Organic Corpuscles Which Exist in the Atmosphere**]

CHAPTER ONE—HISTORICAL

Up to the end of the Middle Ages everyone believed in the existence of spontaneous generation. Aristotle says that all dry bodies which become moist and all moist bodies which become dry engender animals.

Van Helmont describes the means of accomplishing the birth of mice.

Many authors in the 17th Century pointed out the method of producing frogs in the marsh mud, or eels in the water of our rivers.

Similar errors could not long withstand the spirit of research which seized Europe in the 16th and 17th Centuries.

Redi, a celebrated member of the Academy del Cimento, showed that the maggots in putrefying flesh were the larvae eggs of flies. His proofs were as simple as they were conclusive, for he showed that it sufficed to wrap the putrefying flesh in fine gauze to prevent the birth of the larvae.

Redi was also the first to recognize males, females and eggs of animals which live in other animals.

Later Réaumur said that in his researches he discovered flies who laid their eggs in fruits, and he knew when he found a worm in an apple that it was pollution which had engendered it, but that, on the contrary, it was the worm who had polluted the fruit.

In the second part of the 17th Century and the first part of the 18th Century microscopic observation increased. The doctrine of spontaneous generation revived then.

Those who could not explain the origin of the various creatures which the microscope revealed in infusions of vegetable and animal matter and saw nothing in them which resembled sexual generation concluded that living matter retained some vital property after death under the influence of which with favorable conditions these disjointed parts reunited with various structures and organisms.

* Memoire sur les corpuscles organises qui existent dans l'atmosphere, examen de la doctrine des generations spontanées, Paris, 1862. Translation by Dorothy H. Clendening.

Others, on the contrary, adding imagination to the wonders which the microscope revealed to them, believed that they saw copulation in the *Infusoria*—male, female and eggs—and became avowed adversaries to the theory of spontaneous generation.

The question ended there until there appeared in London in 1745 a book by Needham—a research worker and devout Catholic priest, which fact would be a sufficient guarantee of his sincerity and convictions.

In his works the doctrine of spontaneous generation was applied on entirely new facts; I refer to experiments in hermetically sealed containers previously exposed to heat. In fact, it was Needham who first conceived the idea of such experiments.

Less than two years after the publication of Needham's researches the Royal Society of London admitted him to membership. Later he became one of eight associates of the Academy of Sciences.

But it was chiefly through the support it was given by Buffon's system of generation that Needham's work received so much acclaim.

The first three volumes of Buffon appeared in 1749. It is in the second volume of this edition, four years after Needham's book, that Buffon announced his system of organic molecules and defended the hypothesis of spontaneous generation. It is probable that Needham's findings greatly influenced Buffon, for at the time the illustrious Naturalist was draughting the first volumes of his work Needham visited Paris where he was frequently Buffon's guest and, so to speak, his collaborator.

Needham's and Buffon's ideas had their partisans and their detractors. They were opposed to another famous system—that of Bonnet, on the pre-existence of germs. Today we know that the truth existed in neither system.

Needham's conclusions were soon put to experimental verification. In Italy there was one of the most able physiologists honored by science, the most ingenious and hardest to convince—the Abbé Spallanzani.

Needham had based his doctrine of spontaneous generation on well conceived direct experiments. Experiment alone could either prove or disprove his opinions. Spallanzani fully understood this. "In several Italian cities," he said, "there are groups who are opposed to Needham's opinion, but I do not believe that anyone has considered examining it by experiment."

In 1765 Spallanzani published, at Modena, a treatise in which he refuted Needham's and Buffon's systems. This treatise was translated into French, probably at Needham's request, as the edition which appeared in 1769 is accompanied by notes written by Needham, in which he replies to all Spallanzani's objections.

Impressed by the justice of Needham's criticisms Spallanzani set to work again and soon published that wonderful set of works "*Opuscules Physiques.*"

It would be useless to give a complete history of the quarrel between these two learned Naturalists. But it is important to point out the experimental difficulties to which they applied their efforts, and to consider whether this long debate dispelled all the doubts. This is generally considered to be the case. Spallanzani is acknowledged as the victorious adversary of Needham. If this decision was established is it not surprising that to this day there are still many partisans to the doctrine of spontaneous generation?

As I said, Needham was the author of experiments relative to what one observes in closed containers which have previously been exposed to heat.

"Needham," says Spallanzani, "assures us that experiments thus prepared have always turned out well in his hands—that is to say that the infusions showed *Infusoria*, and it was that that put the seal of proof on his system."

"If," adds Spallanzani, "after sterilizing by heat and placing the substances in the containers in which air remains, and taking precautions to prevent all contact with circulating air, in spite of all this one still finds living animals in the containers when they are opened, would this not be strong proof against the system of ovaries? *I do not know what even his supporters can reply to this.*"

I underline the last words in order to show that Spallanzani placed the test of truth or error in the result of experiments thus conducted. But we will see from the following passage, taken from Needham's notes, that this was also Needham's opinion. Here, in fact, is a passage from Needham's remarks on Chapter X of Spallanzani's first dissertation:

"It only remains for me," says Needham, "to speak of Spallanzani's last experiment which he himself considers the only one of all his dissertations to have some weight against my principles.

"He hermetically sealed nineteen containers filled with vegetable substances and boiled them, thus closed, for one hour. But, by the manner in which he treated and tortured these nineteeen vegetable infusions, it is obvious that not only did he weaken or perhaps totally annihilate the vegetative strength of the infused substances, but also entirely spoiled, by effluvia and intense heat, the small amount of air which remained in the empty part of the containers. Consequently it is not surprising that the infusions thus treated gave no sign of life. It should be so.

"In a few words here is my final proposition and the result of my work: That he use in repeating his experiment substances sufficiently cooked to destroy all the so-called germs which are believed to be attached to the substances, or on their internal walls, or floating in the air in the container; that he seal the container hermetically, leaving in a certain portion of air without disturbing it; that he then plunge the container into boiling water for a few minutes—only the length of time necessary to harden a hen's egg and to kill the germs; in a word, that he take all the precautions he wishes, provided he attempts only to destroy the so-called foreign germs which come from without, and I affirm that he will always find a sufficient number of these living microscopical beings to prove my principles. If, having conformed to these conditions, he does not find, on opening his containers after having left them undisturbed the necessary length of time to generate these bodies, something vital or some sign of life I abandon my system and renounce my ideas. I believe it is all that a fair adversary could exact of me."

There is a clear statement of the argument between Needham and Spallanzani. In Chapter III, Volume I, of his *Opuscles* Spallanzani broached the decisive difficulty. And what is his conclusion? To prevent all growth of *Infusoria* it is necessary to keep the infusions at the boiling point for three-quarters of an hour. But does not this enforced period of three-quarters of an hour at a temperature of 100° C. justify Needham's fears of possible alteration in the air in the containers? In his experiments Spallanzani would at least have had to include an analysis of this air. But science was not yet sufficiently advanced; eudiometry had not yet been established. The composition of air was scarcely known.

The results of Spallanzani's experiments on the most delicate point of the question maintained all the value of Needham's objections. Furthermore these were found legitimate, at least in appearance, by the subsequent progress of science.

Appert applied to Domestic Science Spallanzani's experiments carried out according to Needham's methods. For example: One of the Italian scientist's experiments consisted in putting peas and water in a glass container and hermetically closing it, after which it was kept in boiling water for three-quarters of an hour. This is exactly Appert's procedure. But Gay-Lussac, wishing to investigate the process, submitted it to various experiments the result of which he put in one of his most frequently quoted *Memoirs*.

The following extracts from Gay-Lussac's work leave no doubt as to the illustrious physician's opinions, opinions which passed to science whole and uncontested.

"One can convince one's self," says Gay-Lussac, "by analysing the air in bottles in which substances (beef, mutton, fish, mushrooms, clusters of raisins) have been carefully preserved, that they do not contain oxygen, *and that consequently the absence of this gas is a necessary condition for the conservation of animal and vegetable substances.*"

Needham's fears about an alteration in the air in the containers in Spallanzani's experiments are found to be justified by this discovery of the absence of oxygen in Appert's conserves.

One of Schwann's experiments brought a very notable advance to the question. In February, 1837, Schwann published the following facts: An infusion of muscular flesh was placed in a glass balloon; the balloon was then sealed. Then it was entirely exposed to the temperature of boiling water and, after it had cooled, it was left alone. The liquid did not putrefy. Thus far nothing very new. It is one of Spallanzani's experiments, or rather one of Appert's conserves. "It was desirable," adds Schwann, "to modify the test in such a manner that renewal of the air was possible; in this procedure the fresh air was always previously heated like the air originally in the balloon." Then Schwann repeated the preceding experiment by fitting into the neck of the balloon a cork pierced with two holes which were connected by glass tubes which were bent and curved so that their curves were submerged in baths of a fusible mixture kept at a temperature similar to that of the boiling point of mercury. With the help of an aspirator the air was renewed; it entered the balloon cold, but after having been heated in passing through the part of the tubes which were surrounded by the fusible mixture. The experiment was begun by having the liquid boil. The result was the same as in Spallanzani and Appert's experiments. There was no change in the organic liquid.

Air heated and then cooled leaves intact the juice of meat which has been brought to the boiling point. That was a great step forward because it advanced Spallanzani's arguments against Needham. This confirmed all Needham's fears on the possible alteration in the air in Spallanzani's experiments; it finally destroyed Gay-Lussac's statement on the role of oxygen in Appert's process of conserving and in alcoholic fermentation.

Meanwhile there were doubts about this last point. In this same work of Schwann's, in addition to the experiment on meat broth, which related to the cause of putrefaction, there was another relative to alcoholic fermentation which must be remembered. The author filled four bottles with a solution of cane sugar mixed with beer yeast; then, after tightly corking the bottles, he placed them in boiling water and then reversed them on the mercury vat. After they had cooled he introduced air—ordinary air into two of them, calcinated in two others. At the end of a month there was fermentation in the bottles which had received ordinary air; it had not appeared in the other two at the end of two months. "But in repeating these experiments I found," he says, "that they do not always turn out so well, and that sometimes the fermentation does not appear in any of the bottles. For example—after they have been kept in boiling water too long, and also sometimes the liquid ferments in the bottles which have received the calcinated air."

In summing up, Schwann's experiment relative to the putrefaction of bouillon is very clear. But in those which concern alcoholic fermentation, the only well known fermentation in 1837 at the time of Schwann's work, the learned physiologist's experiments were contradictory, and meanwhile we learned from Cagniard de Latour's observations and Schwann's own that wine fermentation is determined by a fermentation organism.

This is the conclusion which Schwann deduced from the experiments I have just reported: "In alcoholic fermentation," he says, "as in putrefaction, it is not the oxygen, at least not oxygen alone, which causes it, but a principle stored in ordinary air which heat can destroy."

Schwann's experiments have been repeated and modified by several researchers. Ure and Helmholtz have confirmed his results by experiments analogous to his. Schultze, instead of calcinating the air before putting it in contact with Appert conserves, passed it through reactive chemicals: potash and sulphuric acid concentrates. Schroeder and Dusch thought of filtering the air through cotton instead of modifying it by heat as Schwann did, or by strong reactive chemicals as Schultze did.

Schroeder and Dusch's first *"Memoire"* appeared in 1854; the second in 1859. These are excellent works which have, among others, the historical merit of showing the state of the question which we are discussing in 1859.

Before the first discussions on spontaneous generation it had long been known that a fine gauze, already used by Redi with much success in his researches on the origin of larvae in putrefying meat, sufficed to prevent, or at least considerably modify alterations in the infusions. This fact was among those stressed by adversaries to the doctrine of spontaneous generation.

Guided no doubt by these facts, and especially, as they distinctly say, by the ingenious experiments of Loevel who observed that ordinary air was incapable of promoting the crystallisation of sodium sulphate after it had been filtered through cotton, Schroeder and Dusch proceeded in the following manner.

The organic matter was placed in a glass balloon. The cork of the balloon was crossed by two tubes curved at right angles. One of the tubes connected with a water aspirator, the other with a large tube one inch in diameter and twenty inches long filled with cotton. When the connections had been secured, the cock of the aspirator closed and the organic matter placed in the balloon, it was heated to the boiling point, the boiling continued long enough for all the connecting tubes to be thoroughly heated by the steam; then the cock of the aspirator was opened and left day and night.

Here are the results of the first test conducted in this manner:

Schroeder and Dusch experimented:

    1st—On meat with the addition of water.

2nd—On beer yeast.

3rd—On milk.

4th—On meat without the addition of water.

In the first two cases the air filtered over cotton left the liquids intact, even after several weeks. But the milk curdled and rotted as quickly as in ordinary air, and the meat without water promptly started putrefying.

Schroeder returned to the subject in 1859 alone, in a *"Memoire."* which covered, among other things, the cause of crystallization. This new work led its author no nearer to conclusions devoid of uncertainty. He introduced some new organic liquids which do not putrefy when placed in contact with filtered air, such as urine, starch paste and various isolated milk factors; but he adds the yolk of egg to the list of bodies which, like milk and meat without water, putrefy in air filtered through cotton.

"I will not venture," says Schroeder, "to attempt the theoretic explanation of these facts. One could admit that fresh air contains an active substance which causes the phenomena of alcoholic fermentation and putrefaction—a substance which heat destroys or that cotton stops."

I have intentionally reviewed in detail these very learned works because they give the exact expression of the difficulties which, in 1859, besieged all impartial minds free from preconceived ideas and desirous of forming a well founded opinion on this serious question of spontaneous generation. It can be affirmed that all those who believed at this time that the question had been settled were ignorant of the facts.

Spallanzani had not overcome Needham's objections. Schwann, Schultze and Schroeder had only demonstrated the existence of an unknown principle in the air which was the cause of life in the infusions. Those who affirmed that this principle was nothing more than germs had no more proof to support their opinion than those who thought it might be a gas, a fluid, noxious fumes, etc., and who, consequently, were inclined to believe in spontaneous generation. In this respect Schwann and Schroeder's conclusions cannot leave the least doubt in the reader's mind. The very terms of these conclusions arouse doubt and support the doctrine of spontaneous generation. And also Schwann, Schultze and Schroeder's experiments were only successful with certain liquids. Moreover they constantly failed with all liquids while the experiments were performed on the mercury vat, no one knowing the reason for the failure or being able to figure out a reason for it.

When, some time after the experiments about which I have just been speaking, an able Naturalist from Rouen, Pouchet—a corresponding member of the Académie des Sciences—announced to the Académie certain results upon which he believed that he could definitely establish the causes of the confusion no one could explain the true reasons for the errors in these experiments, so the Académie, realizing how much remained to be done, proposed the following question as a prize subject:

Try, by careful experiment, to throw a new light on the question of spontaneous generation.

The question seemed so obscure that Biot was distressed to see me engaged in these researches and implored my obedience to his advice, to accept a time limit, at the end of which I would abandon the subject if I had not conquered the difficulties which beset me. Dumas said to me at the same time, "I would never advise anyone to spend too much time on that subject."

What need was there for me to stick to it?

Twenty years ago the chemists discovered a really extraordinary group of phenomena, classified under the generic name of *fermentations*. All demanded the concurrence of two substances: one *fermentable*—like sugar, the other *nitrogenizable*—always an albuminoid substance. Here is the theory which was universally accepted: the albuminoid substances showed a change when they were exposed to contact with air, a particular oxidation of an unknown nature which gives them the character of a *ferment*. That is to say, the property of acting on fermentable substances by contact.

There was a ferment, the oldest, the most remarkable of all, that was known to be an organic being—beer yeast. But because in all the fermentations discovered more recently than the discovery of the fact of the composition of beer yeast (1836), the physiologists were unable to recognize the existence of organic beings, even by careful research, they had gradually abandoned (some with much regret) Cagniard de Latour's hypothesis of a probable relation of the composition of this ferment to its property of being a ferment, and applied to beer yeast a general theory by saying: "It is not because it is organic that beer yeast is active, it is because it has been in contact with air. It is the dead part of the yeast, that which was alive and which is in the course of alteration, which acts on the sugar."

My studies led me to entirely different conclusions. I found that all properly called fermentations—viscous, lactic, butyric, fermentation of tartaric acid, malic acid, urine—were always correlative to the presence and multiplication of organic beings. And so far as the composition of beer yeast being a hindrance to the theory of fermentation, it was, on the contrary, from that that it became the common rule and was the type of all properly called ferments. In my opinion albuminoid substances were never ferments but the food of the ferments. The true ferments were organic beings.

PREVENTION OF RABIES*

[From *Compt. rend. Acad. d. sc.*, Paris 1885, 1886]

We announced a positive advance in the study of rabies in the papers appearing under my own name and under the names of my fellow-workers; this was a method of prevention of the disease. The evidence was acceptable to the scientific mind, but had not been given practical demonstration. Accidents were liable to occur in its application.

Of twenty dogs treated, I could not render more than fifteen or sixteen refractory to rabies. Further, it was desirable, at the end of the treatment, to inoculate with a very virulent virus—a control virus—in order to confirm and reinforce the refractory condition. More than this, prudence demanded that the dogs should be kept under observation during a period longer than the period of incubation of the disease produced by the direct inoculation of this last virus. Therefore, in order to be quite sure that the refractory state had been produced, it was sometimes necessary to wait three or four months. The application of the method would have been very much limited by these troublesome conditions.

Another objection was that the method did not lend itself easily to the emergency treatment rendered necessary by the accidental and unforeseen way in which bites are inflicted by rabid animals.

It was necessary, therefore, to discover, if possible, a more rapid

* Translation adopted by Dorothy Hixon Clendening, from translation by D. Berger, MA., The Founders of Modern Medicine, Walden Publications. New York, 1939. By permission of Wehman Bros., Publishers.

method. Otherwise who would have the temerity, before this progress had been achieved, to make any experiment on man?

After making almost innumerable experiments, I have discovered
a prophylactic method which is practical and prompt, and which
has already in dogs afforded me results sufficiently numerous, certain, and successful, to warrant my having confidence in its general
applicability to all animals, and even to man himself.

This method depends essentially on the following facts:

The inoculation of the infective spinal cord of a dog suffering
from ordinary rabies under the dura mater of a rabbit, always produces rabies after a period of incubation having a mean duration
of about fifteen days.

If, by the above method of inoculation, the virus of the first rabbit
is passed into a second, and that of the second into a third, and so
on, in series, a more and more striking tendency is soon manifested
towards a diminution of the duration of the incubation period of
rabies in the rabbits successively inoculated.

After passing twenty or twenty-five times from rabbit to rabbit,
inoculation periods of eight days are met with, and continue for
another interval, during which the virus is passed twenty or twenty-
five times from rabbit to rabbit. Then an incubation period of
seven days is reached, which is encountered with striking regularity
throughout a new series extending as far as the ninetieth animal.
This at least is the number which I have reached at the present
time, and the most that can be said is that a slight tendency is manifested towards an incubation period of a little less than seven days.

Experiments of this class, begun in November, 1882, have now
lasted for three years without any break in the continuity of the
series, and without our ever being obliged to have recourse to any
other virus than that of the rabbits successively dead of rabies. Consequently, nothing is easier than to have constantly at our disposal,
over considerable intervals of time, a virus of rabies, quite pure,
and always quite or very nearly identical. This is the central fact
in the practical application of the method.

The virus of rabies at a constant degree of virulence is contained
in the spinal cords of these rabbits throughout their whole extent.

If portions, a few centimeters long, are removed from these spinal
cords with every possible precaution to preserve their purity, and
are then suspended in dry air, the virulence slowly disappears, until

at last it entirely vanishes. The time within which this extinction of virulence is brought about varies a little with the thickness of the morsels of spinal cord, but chiefly with the external temperature. The lower the temperature the longer is the virulence preserved. These results form the central scientific point in the method.

These facts being established, a dog may be rendered refractory to rabies in a relatively short time in the following way:

Every day morsels of fresh infective spinal cord from a rabbit which has died of rabies developed after an incubation period of seven days, are suspended in a series of flasks, the air in which is kept dry by placing fragments of potash at the bottom of the flask. Every day also a dog is inoculated under the skin with a Pravaz' syringe full of sterilized broth, in which a small fragment of one of the spinal cords has been broken up, commencing with a spinal cord far enough removed in order of time from the day of the operation to render it certain that the cord was not at all virulent. (This date had been ascertained by previous experiments.) On the following days the same operation is performed with more recent cords, separated from each other by an interval of two days, until at last a very virulent cord, which has only been in the flask for two days, is used.

The dog has now been rendered refractory to rabies. It may be inoculated with the virus of rabies under the skin, or even after trephining, on the surface of the brain, without any subsequent development of rabies.

Never having once failed when using this method, I had in my possession fifty dogs, of all ages and of every race, refractory to rabies, when three individuals from Alsace unexpectedly presented themselves at my laboratory, on Monday the 6th of last July.

Théodore Vone, grocer, of Meissengott, near Schlestadt, bitten in the arm, July 4th, by his own dog, which had gone mad.

Joseph Meister, aged 9 years, also bitten on July 4th, at eight o'clock in the morning, by the same dog. This child had been knocked over by the dog and presented numerous bites, on the hands, legs, and thighs, some of them so deep as to render walking difficult. The principal bites had been cauterized at eight o'clock in the evening of July 4th, only twelve hours after the accident, with phenic acid, by Dr. Weber, of Villé.

The third person, who had not been bitten, was the mother of little Joseph Meister.

At the examination of the dog, after its death by the hand of its master, the stomach was found full of hay, straw, and scraps of wood. The dog was certainly rabid. Joseph Meister had been pulled out from under him covered with foam and blood.

M. Vone had some severe contusions on the arm, but he assured me that his shirt had not been pierced by the dog's fangs. As he had nothing to fear, I told him that he could return to Alsace the same day, which he did. But I kept young Meister and his mother with me.

The weekly meeting of the Académie des Sciences took place on July 6th. At it I met our colleague Dr. Vulpian, to whom I related what had just happened. M. Vulpian, and Dr. Grancher, Professor in the Faculté de Médicine, had the goodness to come and see little Joseph Meister at once, and to take note of the condition and the number of his wounds. There were no less than fourteen.

The opinion of our learned colleague, and of Dr. Grancher, was that, owing to the severity and the number of the bites, Joseph Meister was almost certain to take rabies. I then communicated to M. Vulpian and to M. Grancher the new results which I had obtained from the study of rabies since the address which I had given at Copenhagen a year earlier.

The death of this child appearing to be inevitable, I decided, not without lively and sore anxiety, as may well be believed, to try upon Joseph Meister the method which I had found constantly successful with dogs. . . .

Consequently, on July 6th, at 8 o'clock in the evening, sixty hours after the bites on July 4th, and in the presence of Drs. Vulpian and Grancher, young Meister was inoculated under a fold of skin raised in the right hypochondrium, with half a Pravaz' syringeful of the spinal cord of a rabbit, which had died of rabies on June 21st. It had been preserved since then, that is to say, fifteen days, in a flask of dry air.

In the following days fresh inoculations were made. I thus made thirteen inoculations, and prolonged the treatment to ten days. I shall say later on that a smaller number of inoculations would have been sufficient. But it will be understood how, in the first attempt, I would act with a very special circumspection. . . .

On the last days, therefore, I had inoculated Joseph Meister with the most virulent virus of rabies, that, namely, of the dog, reinforced by passing a great number of times from rabbit to rabbit, a virus which produces rabies after seven days incubation in these animals, after eight or ten days in dogs. . . .

Joseph Meister, therefore, has escaped, not only the rabies which would have been caused by the bites he received, but also the rabies with which I have inoculated him in order to test the immunity produced by the treatment, a rabies more virulent than ordinary canine rabies.

The final inoculation with very virulent virus has this further advantage, that it puts a period to the apprehensions which arise as to the consequences of the bites. If rabies could occur it would declare itself more quickly after a more virulent virus than after the virus of the bites. Since the middle of August I have looked forward with confidence to the future good health of Joseph Meister. At the present time, three months and three weeks have elapsed since the accident, his state of health leaves nothing to be desired. . . .

ROBERT KOCH

THE AETIOLOGY OF TUBERCULOSIS*

Villemin's discovery that tuberculosis is transmissible to animals has, as is well known, found varied confirmation, but also apparently well-grounded opposition, so that it remained undecided until a few years ago whether tuberculosis is or is not an infectious disease. Since then, however, inoculations into the interior ocular chamber, first performed by Cohnheim and Salomonsen, and later by Baumgarten, and furthermore the inhalation experiments done by Tappeiner and others have established the transmissibility of tuberculosis beyond any doubt, and in future tuberculosis must be classed as an infectious disease.

If the number of victims which a disease claims is the measure of its significance, then all diseases, particularly the most dreaded in-

* Die Aetiologie der Tuberculose. Paper read before the Physiological Society in Berlin, March 24, 1882, and published in the *Berliner klinische Wochenschrift,* 1882, XIX, 221. Translation by Dr. and Mrs. Max Pinner. *Am. Rev. Tuber.,* March, 1932. By permission of the National Tuberculosis Association.

fectious diseases, such as bubonic plague, Asiatic cholera, etc., must rank far behind tuberculosis. Statistics teach that one-seventh of all human beings die of tuberculosis, and that, if one considers only the productive middle-age groups, tuberculosis carries away one-third and often more of these. Public hygiene has therefore reason enough to devote its attention to so destructive a disease, without taking into any account that still other conditions, such as the relations of tuberculosis to Perlsucht, engage the interest of Public Health.

Since it is part of the task of the *Gesundheitsamt* to investigate infectious diseases from the point of view of public health, that is, primarily as regards their aetiology, it seemed an urgent duty to make thorough studies on tuberculosis particularly.

There have been repeated attempts to fathom the nature of tuberculosis, but thus far without success. The so frequently successful staining methods for the demonstration of pathogenic micro-organisms have failed in regard to this disease, and, to date, the experiments designed to isolate and cultivate a tubercle virus cannot be considered successful, so that Cohnheim, in the recently published and newest edition of his lectures on general pathology had to designate "the direct demonstration of the tuberculous virus as a still unsolved problem."

In my studies on tuberculosis I first used the known methods without elucidating the nature of the disease. But by reason of several incidental observations I was prompted to abandon these methods and to follow other paths which finally led to positive results.

The aim of the study had to be directed toward the demonstration of some kind of parasitic forms, which are foreign to the body and which might possibly be interpreted as the cause of the disease. This demonstration became successful, indeed, by means of a certain staining process, which disclosed characteristic and heretofore unknown bacteria in all tuberculous organs. It would take us too far afield to tell of the road by which I arrived at this new process, and I shall therefore immediately give its description.

The objects for study are prepared in the usual fashion for the examination for pathogenic bacteria. They are either spread on the cover-slip, dried and heated, or they are cut in sections after being hardened in alcohol. The cover-slips or sections are put in a staining solution of the following formula: 200 cc. of distilled water are mixed with 1 cc. of a concentrated alcoholic methylene-blue solution, and with repeated shaking 0.2 cc. of a 10 per cent potassium-hydrate solution is added. This mixture should not produce a precipitate even after standing for several days. The objects to be stained remain in it from 20 to 24 hours. By heating the staining solution to 40° C. in a water-bath this time can be shortened to from one-half to one hour. The cover-slips are then covered with a concentrated aqueous solution of vesuvin, which must be filtered each time before use and rinsed after one to two minutes with distilled water. When the cover-slips are removed from the methylene-blue the smear looks dark blue and is much overstained, but upon the treatment with vesuvin the blue color disappears and the specimen appears faintly brown. Under the microscope all constituents of animal tissue, particularly the nuclei and their disintegration products, appear brown, with the tubercle bacilli, however, beautifully blue. With the exception of leprosy bacilli, all other bacteria which I have thus far examined in this respect assume a brown color with this staining method. The color contrast between the brown-stained tissue and the blue tubercle bacilli is so striking, that the latter, which are frequently present only in very small numbers, are nevertheless seen and identified with the greatest certainty.

Sections are treated quite similarly. They are transferred from the methylene-blue solution into the filtered vesuvin solution, in which they remain from fifteen to twenty minutes, and are then rinsed in distilled water until the blue color has disappeared and a more or less distinct brown stain remains. They are then dehydrated with alcohol and cleared in oil of cloves, and are either at once microscopically examined directly in this fluid or are finally embedded in canada-balsam. In these preparations the tissue constituents appear brown and the tubercle bacilli a vivid blue.

Incidentally, the bacteria can be stained not only with methylene-blue, but they take also other aniline dyes, with the exception of brown dyes, under the simultaneous action of an alkali; but their staining is not so beautiful by far as with methylene-blue. In the staining process mentioned the potassium-hydrate solution can be substituted by sodium hydroxide or ammonia water, from which one may conclude that the potassium does not play an essential role, but it is the strongly alkaline reaction of the solution which counts.

This is further confirmed by the fact that a stronger addition of potassium will stain bacteria in such places where a weaker potassium solution fails. But the tissues in sections shrink and change so much under the influence of stronger potassium solutions that the latter are only exceptionally of advantage.

In several respects the bacteria made visible by this process exhibit a characteristic behavior. They are rod-shaped, and they belong to the group of bacilli. They are very thin and one-fourth to one-half as long as the diameter of a red blood-corpuscle, although they may sometimes reach a greater length,—up to the full diameter of an erythrocyte. In shape and size they bear a striking similarity to leprosy bacilli. They are differentiated from the latter by being a bit more slender and by having tapered ends. Further, leprosy bacilli are stained by Weigert's nuclear stain, while the tubercle bacilli are not. Wherever the tuberculous process is in recent evolution and is rapidly progressing, the bacilli are present in large quantities; they usually form, then, densely bunched and frequently small braided groups, often intracellular; and they present at times the same picture as leprosy bacilli accumulated in cells. In addition, numerous free bacilli are found. It is particularly at the margin of larger caseous foci that there occur practically only shoals of bacilli which are not enclosed in cells. As soon as the height of tubercle-development is passed the bacilli become rarer, and occur only in small groups or quite singly, in the margin of the tuberculous focus and side by side with weakly stained and sometimes hardly recognizable bacilli which are presumably dying or dead. Finally, they may disappear completely; but they are but seldom entirely absent and, if so, only in such places in which the tuberculous process has come to a standstill.

If giant cells occur in the tuberculous tissue the bacilli are by predilection within these formations. In very slowly progressing tuberculous processes, the interior of giant cells is usually the only place in which bacilli are to be found. In this case the majority of giant cells enclose one or a few bacilli; and it produces a surprising impression to find repeatedly in large areas of the section groups of giant cells, most of which contain one or two tiny blue rods in the centre, and within the wide space enclosed by brown-stained nuclei. Frequently, the bacilli are seen only in small groups of giant cells, sometimes only in single cells, while simultaneously many other giant cells are free from them. Then, by their size and position, the bacilliferous cells are recognized as the younger ones, while those free from bacilli, which have died or changed into a resting state, soon to be mentioned. Analogous to the formation of giant cells around foreign bodies, such as vegetable fibres and strongylus eggs, observed by Weiss, Friedlander and Laulamie, it may be assumed that the bacilli, acting as foreign bodies, are enclosed by giant cells; therefore it seems justifiable to assume that, even when the giant cell is found empty while all other features indicate processes as tuberculous, the giant cell was formerly the host of one or several bacilli, and that the latter have been responsible for their formation.

The bacilli are also observable unstained in unprepared specimens. For this it is necessary to take a little material from such places as contain considerable numbers of bacilli, for example, from a gray tubercle from the lung of a guinea pig, dead of inoculation tuberculosis. This material must be examined after the addition of a little distilled water, or preferably, blood-serum, and best in a hollow slide, in order to avoid streaming in the fluid. The bacilli appear then as very fine rods which show only molecular, but not the slightest trace of intrinsic, movement.

Under certain conditions to be mentioned later the bacilli form spores even in the animal body. Individual bacilli contain several, usually 2 to 4 spores, oval in shape, and distributed at even intervals along the entire length of the bacillus.

In regard to the occurrence of bacilli in the various tuberculous manifestations in human beings and animals it has been possible to examine the following material thus far:

1. From Human Beings: Eleven cases of miliary tuberculosis: The bacilli never failed of demonstration in miliary tubercles of the lungs; frequently, however, they could not be found any more in those tubercles whose centre no longer stained with nuclear dyes. However, they were then still present, and all the more numerous, in small groups in the margin of the tubercle and in younger nodules as yet without central caseation. They were demonstrable, also, in miliary tubercles of the spleen, liver and kidneys, and quite numerous in the gray nodules of the pia mater in basilar meningitis. The caseated bronchial lymph nodes which were examined in several cases contained, in part, dense shoals of bacilli and among them many spore-bearing ones. Tubercles, embedded partly in the lym-

phoid tissue with a central giant cell surrounded by epithelioid cells, showed some tubercle bacilli within the giant cells.

Twelve cases of caseous bronchitis and pneumonia (in 6 cases cavity-formation): The presence of bacilli was usually limited to the margin of the caseously infiltrated tissue, where several times, however, they were very numerous. Also, within the infiltrated portions of the lung one occasionally encounters nests of bacilli. In most cavities the bacilli are abundant, and the well-known small caseous particles in the cavity contents consist almost completely of bacillary masses. Among those bacilli found in soft caseous foci and in cavities there were, now and then, numerous spore-bearing bacilli.

In larger cavities the bacilli occur mixed with other bacteria. However, they were easily distinguishable because, with the staining-method mentioned, only tubercle bacilli stain blue while other bacteria assume a brown color.

One case of solitary tubercle in the brain, larger than a hazel-nut: The caseous part of the tubercle was enclosed by a cellular tissue in which were embedded many giant cells. Most of these did not contain any parasites, but here and there were encountered groups of giant cells, each of which contained one or two bacilli.

Two cases of intestinal tuberculosis: In tuberculous nodules which were grouped around the intestinal ulcers the bacilli were demonstrable with particular ease. Here again they were predominantly numerous in the youngest and smallest nodules. In the mesenteric lymph nodes of these two cases bacilli were present in great numbers.

Three cases of freshly excised scrofulous lymph nodes: In only two of them could bacilli be demonstrated in giant cells.

Four cases of fungoid arthritis: In two cases separate small groups of giant cells contained bacilli.

2. From Animals: Ten cases of Perlsucht with calcified nodules in the lungs, in several cases also in the peritoneum and once on the pericardium: In all cases bacilli were found predominantly within giant cells in the tissue surrounding the calcareous masses. The distribution of the bacilli is most frequently so even that among numerous giant cells there is hardly one which does not contain one or several bacilli, and sometimes as many as twenty. In one of these cases the bacilli could also be demonstrated in the bronchial lymph nodes and in the second case in the mesenteric lymph nodes.

Three cases in which the lungs of cattle contained, not the well-known calcified nodules with knobby surface, as usually seen in Perlsucht, but smooth-walled spherical nodes, filled with a thick, cheesy material: Usually this form is not regarded as tuberculosis but is interpreted as bronchiectasis. In the neighborhood of these nodules were found giant cells containing tubercle bacilli.

A caseated cervical lymph node from a hog likewise contained the bacilli.

Large amounts of tubercle bacilli were found in the organs of a chicken dead of tuberculosis, both in tubercles in the bone-marrow and in the peculiar large lymph nodes of the intestines, the liver, and the lungs.

Of three monkeys which had died spontaneously of tuberculosis, the lungs, spleen, liver and omentum, all studded with innumerable nodules, and the caseated lymph nodes were examined; and bacilli were found everywhere in the nodules or in immediate proximity to them.

Of spontaneously diseased animals, nine guinea pigs and seven rabbits were examined; all showed bacilli in the tubercles.

In addition to these cases of spontaneous tuberculosis, I could avail myself of a not inconsiderable number of animals which had been infected by inoculation with the most varied tuberculous materials: as, for instance, with gray and caseated tubercles from human lungs, with sputum from consumptives, with tuberculous masses from spontaneously diseased monkeys, rabbits and guinea pigs, with masses of calcified or caseated lesions from Perlsucht in cattle, and finally with material from lesions obtained by animal-passage. The number of animals so infected amounted to 172 guinea pigs, 32 rabbits and 5 cats. In the majority of these cases the demonstration of bacilli had to be limited to the examination of tubercles in the lungs which were always present in large numbers. Here bacilli never failed to be found: frequently they were extraordinarily numerous, and sometimes spore-bearing, but in some preparations only a few yet unmistakable individual forms were observed.

Considering the regularity of the presence of tubercle bacilli it is striking that so far they have not been seen by anyone. But this is explained by the fact that the bacilli are extraordinarily small formations, and are usually so scanty in number, particularly when their occurrence is limited to the interior of giant cells, that for this reason alone they are not detectable by the most attentive observer without the use of quite specific staining-reactions. When present in large numbers, they are mixed with a finely granular detritus, and obscured by it in such a manner that their visualization is made difficult in the highest degree. . . .

On the basis of my numerous observations I consider it established that, in all tuberculous affections of man and animals, there occur constantly those bacilli which I have designated tubercle bacilli and which are distinguishable from all other microorganisms by characteristic properties. However, from the mere coincidental relation of tuberculous affections and bacilli it may not be concluded that these two phenomena have a causal relation, notwithstanding the not inconsiderable degree of likelihood for this assumption that is derivable from the fact that the bacilli occur by preference where tuberculous processes are incipient or progressing, and that they disappear where the disease comes to a standstill.

To prove that tuberculosis is a parasitic disease, that it is caused by the invasion of bacilli and that it is conditioned primarily by the growth and multiplication of the bacilli, it was necessary to isolate the bacilli from the body; to grow them in pure culture until they were freed from any disease-product of the animal organism which might adhere to them; and, by administering the isolated bacilli to animals, to reproduce the same morbid condition which, as known, is obtained by inoculation with spontaneously developed tuberculous material.

Disregarding the many preliminary experiments which served for the solution of this task, here again the finished method will be described. Its principle rests on the use of a solid transparent medium, which retains its solid consistence at incubator temperature. The advantages of this method of pure culture which I have introduced into bacteriology I have explained in detail in an earlier publication. That the really complicated task of growing tubercle bacilli in pure culture was achieved by this method is to me a new proof of its efficiency.

Serum from sheep-or-cattle-blood, separated as pure as possible, is put into test-tubes closed with a cotton stopper and heated every day to 58° C. for six subsequent days. It is not always possible to sterilize the serum completely by this process, but in most cases it suffices. Then the serum is heated to 65° C. or if it lasts too long, the serum becomes opaque. In order to obtain a large surface for the preparation of the cultures the serum is solidified in test-tubes slanted as much as possible. For those cultures intended for direct microscopical examination the serum is solidified in flat watch-crystals or in small hollow glass blocks.

Upon this solidified blood-serum, which forms a transparent medium that remains solid at incubator temperature, the tuberculous materials are applied in the following manner:

The simplest case in which the experiment is successful is presented, almost without exception, when an animal which has just died of tuberculosis, or a tuberculous animal which has just been killed for this purpose, is at one's disposal. First, the skin is deflected over the thorax and abdomen with instruments flamed before use. With similarly prepared scissors and forceps, the ribs are cut in the middle, and the anterior chest-wall is removed without opening the abdominal cavity, so that the lungs are to a large extent laid free. Then the instruments are again exchanged for freshly disinfected ones, and single tubercles or particles of them, of the size of a millet-seed, are quickly excised with scissors from the lung tissue, and immediately transferred to the surface of the solidified blood-serum with a platinum wire, which has been melted into a glass rod which must be flamed immediately before use. Of course, the cotton stopper may be removed for only a minimal time. In this manner a number of test-tubes, about six to ten, are implanted with tuberculous material, because, with even the most cautious manipulation, not all test-tubes remain free from accidental contamination.

Lymph nodes in a state of incipient caseation are as well suited for this experiment as are pulmonary tubercles; less so, however, the pus from liquefied lymph nodes which usually contain very few bacilli or none at all.

The direct isolation of bacilli from tuberculous human organs or from lungs with Perlsucht is more difficult. Objects of this kind, whose excision from the body I could not attend to with all the precautions just mentioned, I have washed carefully and repeatedly with a solution of bichloride of mercury, and have removed their superficial layers with flamed instruments and taken the substance for inoculation from a depth which putrefactive bacteria had presumably not yet invaded.

The test-tubes, provided with tuberculous substance in the described manner, are kept in the incubator at a constant temperature of 37° or 38° C. In the first week no noticeable alteration occurs. If some change does occur and if, already during the first days, bacterial growth develops, starting from the inoculation material or even remotely from it, and spreading, and appearing usually as whitish-gray or yellowish drops, and often liquefying the solid serum, then one is dealing with contaminations and the experiment has failed.

Cultures that result from a growth of tubercle bacilli do not appear to the naked eye until the second week after the seeding, and ordinarily not until after the tenth day. They come into view as very small points and dry-looking scales. Depending upon whether the tuberculous material was more or less crushed in seeding and whether it was brought into contact with a large surface of the medium by rubbing motions, the colonies surround the explanted bit of tissue in smaller or larger areas. If only very few bacilli were present in the inoculum it is hardly possible to free the bacilli from the tissue and bring them into immediate contact with the medium. In this case the colonies develop in the fragments of explanted tissue; and one sees, if the tissue is transparent enough (for example, in bits from scrofulous lymph nodes), dark points in transmitted light and white points in direct light.

With the aid of a 30 to 40 times magnification one can perceive the bacterial colonies as soon as toward the end of the first week. They appear as very neat spindle and usually S-shaped or similarly curved formations, which consist of the well-known most tenuous bacilli when spread on a cover-slip, and stained and examined with high magnifications. Up to a certain degree their growth proceeds for a period of 3 to 4 weeks, as they enlarge to flat scale-like bits, usually not reaching the size of a poppy seed, and lie loosely on the medium, which they never invade or liquefy. Furthermore, the bacillary colony forms such a compact mass that its small scale can easily be removed with a platinum wire from the solidified blood-serum as a whole, and can be crushed only upon the application of a certain pressure. The exceedingly slow growth which can be obtained only at incubator temperature, and the peculiar scale-like dry and firm texture of these bacillary colonies are not met with in any other known bacterial species, so that it is impossible to confuse cultures of tubercle bacilli with those of other bacteria; and, even with but little experience, nothing is simpler than to recognize accidental contamination.

The tubercle bacilli can also be cultivated on other nutritive media, provided the latter possess properties similar to those of the solidified blood-serum. For example, they grow on agar-agar, which remains solid at incubator temperature and which contains an addition of meat-infusion and peptone. But on this medium there take form only amorphous small crumbs and never, as on blood-serum, the characteristic vegetations.

Up to this point it was established by my studies that the occurrence of characteristic bacilli is regularly coincidental with tuberculosis and that these bacilli can be obtained and isolated in pure cultures from tuberculous organs. It remained to answer the important question whether the isolated bacilli when again introduced into the animal body are capable of reproducing the morbid process of tuberculosis.

In order to exclude every error in the solution of this question, which contains the principal point in the whole study of the tubercle virus, many different series of experiments were done, which, on account of the significance of the point at issue, will be enumerated.

First, were done experiments involving the simple inoculation of bacilli in the previously described manner.

First Experiment: Of 6 recently bought guinea pigs which were kept in the same cage, four were inoculated on the abdomen with bacillary culture material derived from human lungs with miliary tubercles and grown in five transfers for fifty-four days. Two animals remained uninoculated. In the inoculated animals the inguinal lymph nodes swelled after fourteen days, the site of inoculation changed into an ulcer, and the animals became emaciated. After thirty-two days one of the inoculated animals died, and after thirty-five days the rest were killed. The inoculated guinea pigs, the one that had died spontaneously as well as the three killed ones, showed far-advanced tuberculosis of the spleen, liver and lungs; the inguinal nodes were much swollen and caseated; the bronchial lymph nodes were but little swollen. The two noninoculated animals displayed no trace of tuberculosis in lungs, liver or spleen.

Second Experiment: Of 8 guinea pigs, six were inoculated with bacillary culture material, derived from the tuberculous lung of an ape, and cultivated in eight transfers for ninety-five days. Two animals remained uninoculated as controls. The course was exactly the same as in the first experiment. At autopsy the 6 inoculated animals were found with far-advanced tuberculosis, while the two non-inoculated ones were healthy when they were killed, after thirty-two days.

Third Experiment: Of 6 guinea pigs, five were inoculated with culture material, derived from a Perlsucht lung, and seventy-two days old and transferred six times. After thirty-four days all animals were killed. The five inoculated ones were tuberculous, the non-inoculated one was healthy.

Fourth Experiment: A number of animals (mice, rats, hedgehogs, hamsters, pigeons, frogs) whose susceptibility to tuberculosis is not known, were inoculated with cultures derived from the tuberculous lung of a monkey which had been cultivated for 113 days outside the animal body. Four field mice, killed 53 days after the inoculation, had numerous tubercles in the spleen, liver and lung; a hamster, killed 53 days after inoculation, showed the same result.

In these four experiments the inoculation of bacillary cultures on the abdomen of the experimental animals had, then, produced exactly the same kind of inoculation tuberculosis as if fresh tuberculous materials had been inoculated.

In the next experiment the inoculum was introduced into the anterior chamber of rabbit's eyes, in order to find out whether, in the so modified inoculation method, the same effect would be obtained by the artificially cultivated tubercle virus as with the natural virus.

Fifth Experiment: Three rabbits were inoculated with a small crumb of a culture (derived from a caseous pneumonia in a human lung and cultivated for 89 days) in the anterior ocular chamber. An intense iritis developed after a few days, and the cornea soon became clouded and discolored to a yellowish-gray. The animals rapidly became emaciated. They were killed after 25 days and their lungs were found studded with countless tubercles.

Sixth Experiment: Of three rabbits, one received an injection of pure blood-serum into the anterior chamber of the eye, and the two others an injection of the same blood-serum, in which, however, a small crumb of a culture (originating from a lung with Perlsucht and cultivated 91 days) had been suspended. In the latter two rabbits, the same phenomena occurred as in the preceding experiment

—rapidly progressing iritis and clouding of the cornea. After 28 days the animals were killed. The first rabbit, injected with pure blood-serum, was completely healthy, while the lungs of the other two were studded with innumerable tubercles.

Seventh Experiment: Of 4 rabbits, the first received pure blood-serum in the anterior eye chamber; in the case of the second the needle, which contained blood-serum with a bacillary culture (from monkey tuberculosis, cultivated 132 days), was introduced into the anterior chamber of the eye, but the plunger was not moved, so that only a minimal amount of the fluid could get into the aqueous humor. The third and fourth rabbits were injected in the anterior chamber with several drops of the blood-serum with bacillary culture. Iritis and panophthalmitis developed in the latter two animals, and very rapid emaciation followed.

In the case of the second rabbit, on the other hand, the eye remained at first unchanged, but in the course of the second week single whitish-yellow nodules appeared on the iris near the site of the puncture, and there developed, growing out from this centre, a typical tuberculosis of the iris. New nodules kept forming on the iris, which became wrinkled, while the cornea clouded slowly and the further changes were obscured to further observation. After 30 days these four animals were killed. The first was entirely healthy; in the second, besides the formerly noted changes in the eye, the lymph nodes near the mandible and beside the root of the ear were swollen and studded with yellowish-white foci. The lungs and the other organs were still free from tuberculosis. The two latter rabbits, again, had countless tubercles in the lungs.

All these facts, taken together, justify the statement that the bacilli present in tuberculous substances are not only coincidental with the tuberculous process, but are the cause of the process, and that we have in the bacilli the real tuberculous virus.

This establishes the possibility of defining the boundaries of the diseases to be understood as tuberculosis, which could not be done with certainty until now. A definite criterion for tuberculosis was lacking. One author would reckon miliary tuberculosis, phthisis, scrofulosis, Perlsucht, etc., as tuberculosis; another would hold, perhaps with quite as much right, that all these morbid processes were different. In future it will not be difficult to decide what is tuberculous and what is not tuberculous. The decision will be established, not by the typical structure of the tubercle, nor its avascularity, nor the presence of giant cells, but by the demonstration of tubercle bacilli, whether in the tissues by staining-reactions or by culture on coagulated blood-serum. Taking this criterion as decisive, miliary tuberculosis, caseous pneumonia, caseous bronchitis, intestinal and lymph-node tuberculosis, Perlsucht in cattle, spontaneous and infectious tuberculosis in animals, must, according to my investigations, be declared identical. My investigations of scrofulosis and fungoid joint affections are not numerous enough to make a decision possible. In any event, a large number of scrofulous lymph nodes and joint affections belong to true tuberculosis. Perhaps they belong entirely to tuberculosis. The demonstration of tubercle bacilli in the caseated lymph node of a hog, or in the tubercles of a hen, permits the inference that tuberculosis has a wider dissemination among domestic animals than is commonly supposed. It is very desirable to learn exactly the distribution of tuberculosis in this respect.

Since the parasitic nature of tuberculosis is proved, it is still necessary for the completion of its aetiology to answer the questions of where the parasites come from and how they enter the body.

In regard to the first question it must be decided whether the infectious materials can propagate only under such conditions as prevail in the animal body or whether they may undergo a development independent of the animal organism, somewhere in free nature, such as, for example, is the case with anthrax bacilli.

In several experiments it was found that the tubercle bacilli grow only at temperatures between 30° and 41° C. Below 30° and above 42° not the slightest growth occurred within three weeks, while anthrax bacilli, for example, grew vigourously at 20° and between 42° and 43° C. The question mentioned can already be decided on the basis of this fact. In temperate climates there is no opportunity offered outside the animal body for an even temperature of above 30° C. of at least two weeks' duration. It may be concluded that, in their development, tubercle bacilli are dependent exclusively upon the animal organism; that they are true and not occasional parasites; and that they can be derived only from the animal organism.

Also the second question, as to how the parasites enter the body, can be answered. The great majority of all cases of tuberculosis begin in the respiratory tract, and the infectious material leaves its mark first in the lungs or in the bronchial lymph nodes. It is therefore very likely that tubercle bacilli are usually inspired with the air, attached to dust particles. There can hardly be any doubt about the manner by which they get into the air, considering in what excessive numbers tubercle bacilli present in cavity-contents are expectorated by consumptives and scattered everywhere.

In order to gain an opinion about the occurrence of tubercle bacilli in phthisical sputum I have examined repeatedly the sputum of a large series of consumptives and have found that in some of them no bacilli are present, and that, however, in approximately one-half of the cases, extraordinarily numerous bacilli are present, some of them sporogenic. Incidentally, it may be remarked that, in a number of specimens of sputum of persons not diseased with phthisis, tubercle bacilli were never found. Animals inoculated with fresh bacilliferous sputum become tuberculous as certainly as following inoculations with miliary tubercles.

In regard to milk from cows with Perlsucht it is noteworthy that the extension of the tuberculous process to the mammary gland has been observed not rarely by veterinarians, and it is therefore quite possible that in such cases the tuberculous virus may be mixed directly with milk.

Still further viewpoints might be mentioned in regard to measures which could serve to limit the disease on the basis of our present knowledge of the aetiology of tuberculosis but the discussion here would lead too far. When the conviction that tuberculosis is an exquisite infectious disease has become firmly established among physicians, the question of an adequate campaign against tuberculosis will certainly come under discussion and it will develop by itself.

PAUL EHRLICH

ON IMMUNITY WITH ESPECIAL REFERENCE TO THE RELATIONS EXISTING BETWEEN THE DISTRIBUTION AND THE ACTION OF ANTIGENS*

[Harben Lectures. Lecture 1]

There can be no doubt that the three great fields of knowledge, Pharmacology, Toxicology and Therapeutics, in their theoretical

* London, 1908.

and practical aspects form the most important branches of medicine. It is matter, therefore, for no surprise that in the study of the various substances with which these sciences are concerned, the mode of action and the reasons for such call for much consideration, and theory and speculation necessarily form a great part of our study.

Besides, pharmacology has but just emerged from the stage of pure observation and description. One was content formerly with describing the physiological effects and the secondary action of substances which act pharmaco-dynamically, as well as the morphological changes which they bring about in the organs and tissues of the body. Observations made on an empirical basis such as this, formed a mass of most needful knowledge and even to-day we have no hesitation in admitting that the study of the symptomatology of drugs is still a work of absolute necessity and must yield very fruitful results. Indeed, by such means we learn not only how to make use of known drugs in a purposeful manner, but also how to avoid their undesired secondary actions. But merely to increase the contents of our pharmacopoeia is not to add to our resources in this desirable direction: for such an increase may depend on accidents, which, in their turn, may be the outcome of empiricism. It is to the great influence which chemistry exerts on medical science that we owe the change in this state of affairs; for it is especially necessary to have clear ideas of the relations between chemical constitution and pharmacological action.

About the middle of the last century the influence of these inquiries made itself especially felt, and this influence is chiefly evident in the mass of drugs with which the united efforts of synthetic chemistry and pharmacology have enriched us. But observers were content with an advance in this direction, based on rational grounds. They recognised a limited number of atom-groupings, which were of importance either for their therapeutic or their toxic action; but the drugs used were directed not against the causes of disease but against the symptoms to which these gave rise; it was not the causes but their effects which were combatted. Therapeutics were chiefly symptomatic, and so it is in many cases to-day. Since the search after the seat and cause of disease has, from the time of Morgagni, and especially under the leadership of Virchow's genius, influenced our entire field of thought, the effect of these considerations has become more evident in our treatment. The features of an aetiological treatment, directed against the causes and the seat of disease, were not satisfactorily brought out by merely insisting on the relationship which existed between the constitution of drugs and their action; the fact was overlooked that between chemical constitution and pharmacological action another and important bond of union exists, which influences the relations between the pharmaco-dynamic agent and the substance on which it is intended to act. This bond of union is the mode of distribution, and represents the sum of the peculiarities of the cells and tissues and of the drug. In this we have to do with a principle so obvious that it should at once be accepted as an axiom, but even when accepted as an axiom it is scarcely ever applied to the study of practical questions. The reason for this, I think, lies partly in a certain disinclination to attempt to master the difficulties of the problem, but mainly in the fact that, in view of the triumphs of synthetic chemistry, *per se,* the biological factor of the pharmaco-dynamic action is somewhat lost sight of. I may mention that from the beginning, as the result of my studies on dyes, I have endeavoured to point out the necessity of the study of localisation; in a word, to give pharmacology an aetiologically therapeutic tendency; regarding the details of this I shall speak further in my second lecture. Excepting the case of dyes, which, by reason of their easily appreciable properties allow their distribution in the organism to be followed, the study of the laws which govern distribution is exceedingly difficult. Besides, owing to the great number of chemicals at our disposal, a large amount of empirical work must be carried out before we are able to find those substances which in any given case would give the desired mode of distribution. All the greater, then, must be our admiration for the powers of nature, in view of the fact that the living organism, when it takes upon itself the production of curative agents, does this in such a manner as to form ideal aetiological remedies. The protective substances of the blood, with which this lecture is concerned, completely fulfil the requirements of the case, and the study of *antigens* and *antibodies* may form the basis of relationships which must exist between constitution, distribution, and action, in order that our treatment may be successful.

Since the conditions of substances treated of in the study of immunity are specially clear and matters of common knowledge, I shall begin the discussion of my views regarding distribution and localisation with the consideration of this class of phenomena of pharmaco-dynamic and toxicological action.

The discovery of antitoxins by von Behring, fundamental in itself, has opened to pharmacology and therapeutics this new field in which the principle of distribution is exemplified in an ideal manner: for antitoxins and antibacterial substances are, so to speak, charmed bullets which strike only those objects for whose destruction they have been produced by the organism. I call these substances mono-tropic; the monotropism of these antibodies is characterised by the fact that they are bacteriotropic or generally speaking aetiotropic, *i.e.*, they are directed against bacteria or against those products of their metabolism (toxins) which cause disease. The definition of monotropism is, therefore, here overlapped by that of specificity, which, in the language of the study of immunity, is the characteristic of monotropic action. As the cause of this specificity we must note, in my opinion, only the effect of chemical relations which exist between the agents of infection, or their products, and the antibodies. From the very beginning my standpoint has been, that all those substances which have the power to bring about the creation of antibodies, I mean the antigens, must be distinguished, as a matter of principle, from the other pharmaco-dynamical and poisonous sub-stances. That distinction I consider to be of the greatest importance, and this is further borne out by the fact that in spite of the most strenuous endeavours we have been unable to find any antigen of a known chemical constitution.

I have always held the view that the antitoxins do not in any way destroy the toxin, but that they merely limit its sphere of action by combining with it. If further proof of the correctness of this view be required, we have it in the researches carried out by Morgenroth, who showed that in suitable cases the toxin may be entirely regained from a perfectly neutral union of toxin and antitoxin, just as glu-cosides may by suitable treatment be resolved into their two com-ponents. The antitoxin, then, exercises its curative influence merely by anchoring the distributive group of the toxins.

When we speak of monotropism or specificity in the case of an-tigens and antibodies, we of course mean only the chemical relations between haptophoric groups and receptors. For example, toxins may be specific and yet act upon the cells of all species of animals, if the toxin-anchoring receptors be widespread amongst them. The other extreme we find in those cases in which the organism itself reacts to elements introduced into it and forms substances, antibodies, which act on the corresponding antigens; we are here dealing with antibodies which are specifically monotropic, just as the antitoxins are, and which affect only those substances to which they owe their origin. The only difference is that the antitoxins act by localisation alone— when they have anchored the toxins their work is done. The other antibodies, it is true, act at first in a similar way on the substances sensitive to them; but they have a further action on the anchored substances, an action which is either direct, as in those cases (agglutinins and precipitins) in which, like the toxins, they have special ergophoric groups, or indirect, in that the union is merely a preliminary to their further action on their prey. Thus one class of these antibodies has the power of rendering the cells assimilable by phagocytes (opsonins, bacteriotropic substances); another class (amboceptors) has the power of rendering the cells liable to the action of toxin-like constituents of the blood serum (complements). In the latter case by the simultaneous action of two substances a destructive effect is produced. These substances are also called cytotoxins. For the study of these the way has been prepared by the work of Pfeiffer, Metchnikoff, and Bordet, and by the discovery of Metchnikoff and Bordet that haemolysins are produced by immunisation against blood corpuscles.

These haemolysins are of special importance in considering the question of the relations between constitution, distribution, and action, because, in their case, the haptophoric and toxophoric groups are distinct, distribution and toxic effect being dependent upon two different substances, the more stable amboceptor controlling the distribution, and the labile complement the toxic effect. The complement has no direct relations with the cell, on which it acts only through the medium of the amboceptor. It is a normal constituent of the blood-serum, and its quantity undergoes no change as a result of the process of immunisation. On the other hand the amboceptor is a new formation brought about by immunisation, and, on the ground of the principle already admitted, it must appear probable that the amboceptor, like the antitoxins, possesses a marked monotropism for the corresponding antigen. It was therefore not the result of accident, but of logical sequence, that my first researches, carried out with Morgenroth, on the mechanism of haemolysis led us to the fundamental conclusion that the amboceptor alone stands in direct relation to the cell, and is quantitatively anchored by it. This anchoring of the amboceptor takes place with a maximum of chemical energy—it occurs even at $0°$ C.

The union of the amboceptor and the cell has no harmful results on the latter. On the other hand, the amboceptor-laden cell is exposed to the action of the complement, which by itself is harmless. As regards complements, what I have said about toxins holds true; they may be regarded as toxin-like substances which possess a haptophoric group, and a toxophoric group which I call the "zymotoxic" group. That these groups are independent of each other, is well seen in the case of modified complement—"complementoid," as numerous test-tube experiments have proved. Of special interest are the conditions of their distribution. The real state of affairs, which has been thoroughly investigated, especially in the case of haemolysins, is this: the intact erythrocytes do not unite with the complement, which, however, is anchored by the complex of erythrocyte and amboceptor. A closer acquaintance with the conditions which govern distribution in this matter, cannot be arrived at by the hypothesis that the erythrocyte is sensitised by the amboceptor in such manner that an action of the complement is rendered possible. If one accepts the theory of Bordet, which really consists of the denial of the existence of direct relations between amboceptor and complement, one enters the realm of pure speculation; for one must then presume new affinities between the erythrocyte and complement to arise under the influence of the amboceptor. For this assumption we have no grounds. Bordet's method of proof must, therefore, limit itself to indirect conclusions, and consists merely of objections to the view held by Morgenroth and myself, that the amboceptor and the complement stand in direct relationship to one another. As a matter of fact, from all sides proofs of the existence of this direct relationship have been advanced, and I think that the great majority of my colleagues to-day accept my view, which is known as the amboceptor theory. It is true, as we have from the beginning insisted, that the distributive relation of the complement in the presence of the amboceptor is not that of maximum chemical affinity; indeed, we have, on the contrary, as a rule an exceedingly loose relationship which perhaps corresponds to a reversible reaction. To show that this relation is purposeful, we need only the following proof—amboceptors are already present in large quantity and of various kinds in the blood-serum of normal animals. What, then, would happen if the entire mass of normal amboceptors reacted with pronounced avidity with the complement? Obviously the entire mass of complement would be anchored by the complementophile groups of the amboceptors, and there would be no free complement present in the living body. The grave results of such a state of affairs are evident; as soon as the necessity for the action of complements with a special kind of amboceptor arose, there would be no complement available, all having previously been used up for the action of indifferent amboceptors. It is thus owing to the fact that the complement is free or only loosely united to the amboceptors in the circulation, that at a given moment it is ready for use. The maximum stimulus to action is rendered possible by the anchoring of the amboceptor to the erythrocyte, the avidity of the former toward the complement being thus increased. This increase of avidity which consists in the chemical affinity of the complementophile group for the complement being carried to its maximum, represents the gist of our knowledge of the action of the amboceptor.

The amboceptor, therefore, exercises the important function of bringing about a specific modification of those conditions existing in the organism which determine the distribution of complements, and which otherwise are not very evident. It causes the complements to become monotropic by its union with the given substance. The complements are thus localised by amboceptors which have previously become united to the substance—cell or otherwise. At the same time this action represents a purposive saving; if the complement were already a constituent part of the amboceptor, then—as the complement is easily destroyed—there would often be no action. The purposive saving is evident from the fact that only in case of need the amboceptor becomes able to combine with the complement, and from the fact that complementoids which, by reason of having lost the zymotoxic group, become incapable of action, have at the same time lost some of their avidity for the amboceptor. By this change in the distributive quality of the complement that is associated with the formation of complementoid, the sphere of action of the amboceptor is considerably extended. Of other possible influences which may govern distribution and action we have an indication given by the work of Ferrata. Under Morgenroth's direction he carried out a research, from which it appears that a complement is not a single body, one and indivisible; for in a salt-free medium it is split up into two components, which are capable of action only when they work in concert in a salt solution. As to the intimate relations of these two components, the researches which were carried out at the suggestion of Sachs by Dr. Brand appear to indicate that in the blood serum they are always united.

The increase of avidity, which forms the basis of the mode of action of the amboceptor, not only governs the phenomena which occur in the organism, but, since it lends itself to test-tube experiments, also opens up a wide field for serum diagnosis. If we remember that the amboceptors, by their union with the sensitive substances—cells or dissolved bodies—exert such a marked localising influence on the complements, it is evident that in a mixture of an amboceptor and its corresponding antigen, if the presence of the one constituent of the mixture be known, that of the other can be proved by the occurrence of the phenomenon of complement-fixation.

This method, which was elaborated by Neisser and Sachs on the basis of the work of Bordet, Gengou and Moreschi, for the medico-legal test of the source of blood by "complement deviation," depends solely on the principle of this increase of avidity. And, thanks to the genius of Wassermann, we possess a similar method of sero-diagnosis for some infectious diseases, the cause of which is at present unknown or invisible—a method which has yielded valuable results and gives great promise for the future.

As to the biological factor in the formation of antibodies, it has been shown, by the recent investigations of Pfeiffer and Wassermann, that a part of the process takes place in the obscure province of the physiology of stimulation. Light, however, has been thrown on the subject of the specificity of reactions by the view I have expressed that the receptors of the cell-protoplasm are the seat of the process, and that their regeneration and elimination are the consequences of this. Supposing these processes to occur in the normal organism, and to be merely intensified in the case of active immunisation, it is possible to understand that antibodies of the most varied kinds may exist in the serum of the normal organism, and also to comprehend the processes of immunity by viewing them from the standpoint of the physiology of nutrition and metabolism. Thus I am glad to be able, by my conception of the problem, to come into entire agreement with Metchnikoff. The immense number of exceedingly various substances which possess haptine characters in the blood serum, substances of whose existence at one time one did not even dream, is thus to be explained as an expression of a many sided and differentiated action of the most varied organs. One may already by simple means differentiate the multitude of serum substances into antitoxins, amboceptors, agglutinins, precipitins, opsonins, complements, ferments, antiferments, etc., but a deeper study of the subject shows that each of these divisions, in its turn, consists of a multitude of functionally different components, and thus a pluralistic view of the observed phenomena is the only justifiable one. Although in some quarters the endeavour is still made to reduce everything to the most simple form, I believe that such a rudimentary way of looking at things is not justified by the appreciably complex character of natural phenomena and vital processes; for we see, in the investigation of immunity, that when earlier opinions, based on experiment, must be replaced by newer ones, the process takes place always by the substitution of a complex conception in place of a simple one; I would remind you of Buchner's idea regarding alexin, which had to give way to the proved fact that all cytotoxins have a complex nature; and also I would remind you of the anti-complements, which we had supposed to be simple bodies, while it is now known that anticomplementary action is, as a rule, the result of the concerted action of two substances. In the case of the complements, too, in the light of Ferrata's researches, we must assume that the conception of these as simple substances is wrong. The more readily, then, may we assume the existence of a multitude of substances as the cause of the various actions which are exerted by one and the same blood serum, a multitude whose existence has in numerous cases been proved. Objections which have been raised by several observers, especially by Bordet, against our method of proof, and which consist in stating that in every experiment the substance (whose unity is assumed) has been injuriously affected, and in attributing any difference to the varying degree of sensitiveness of the various test-substances alone, cannot be justified; for, if one takes the trouble to work—as I have always urged that one should do—quantitatively, such sources of error are immediately excluded from our conclusions. And even in spite of these, especially in the case of the proof of the multitude of complements, in many cases one has been able, by employing means of the most varied kind, to obtain either the loss of a certain function, or an absolutely disproportionate change of degree in isolated functions. In the plurality of haptines present in the serum we have a wide field open for more profound observation of the mechanism of receptor-metabolism of the laws governing variations, and the influences which bring these about, a field from which new light will be shed on human pathology and clinical medicine. Successful work in this field must proceed on a broad basis. One would have first to make exact observations regarding a large number of functions of human blood serum, and then one would have to investigate systematically in cases of all sorts of diseases, anomalies of nutrition, etc., the causes of departure from the normal, as to whether these result from the failure of certain functions, or from the existence of new functions acting under pathological conditions. Thus, without doubt, would be detected differences in the sum of the functions of the cell, and this section of the physiology and pathology of the blood might be named the blood canon.

I firmly believe that by extensive research we shall find that there exist great differences, the result of biological laws, and these will permit us to come to correct conclusions as to the origin of certain substances in the cell, and to apply these conclusions to diagnosis and therapeutics. Of course, the united action of many observers in many institutes, and the closest relations between clinical and laboratory work, are needful in order that progress may be made in the direction indicated. The recent work of Wassermann, who by means of complement-fixation, was able to prove the existence in the blood serum of anti-bodies for certain nutritive substances (glycogen, albumoses, peptones, etc.), and to obtain an increased concentration of these by increased doses of the nutritive substances, appears to me to be very promising. In the case of pathogenetic or pathognomonic questions it does not appear to be of use to seek for those haptines which may be present in the blood as the result of immunisation; for in their case one would either find differences that are but slightly marked, or, if one found sufficiently-marked differences, one would have to be very careful in drawing conclusions from them. It would be better, therefore, to avoid using as test objects those substances which are already present in the normal body, or gain entrance to it by infection. In order to be able to draw absolutely correct conclusions as to the relations existing between certain substances in the serum and the normal or pathological activity of organs, one ought to choose cells or other elements, regarding which one may assume that they never come into relation with the human body in a natural way.

We have a small beginning in this direction, I think, in the research which I carried out with Wechsberg; we compared the behaviour of human blood-serum towards the trypanosomata in the case of healthy and diseased individuals, and we found that in cases of liver disease the amount of trypanosomicidal substances in the serum is markedly decreased, as will be seen from the subsequent table.

After Laveran had by his researches shown that the serum of man and of a few species of monkeys had a trypanosomicidal influence which was not possessed by the sera of other animals, it appeared worth while to pursue these researches further, and to study this action quantitatively, and when influenced by various conditions. For this purpose we employed a strain of the trypanosoma of *mal de caderas*, two or three control animals being also employed in each case, which succumbed at the latest on the fifth, and generally on the fourth day. The animals were injected with equal quantities in a similar manner, and as soon as the parasites appeared in the blood (on the second day) the curative serum was injected. By this means we obtained a titration of a number of human sera, regarding which I shall only give the following particulars as to the results obtained with small doses.

Looking back upon what I have said, you will see that it is the principle of distribution which governs the processes resulting in active immunisation. The antibodies—the protective substances in the serum—all possess the power of reacting, with maximum chemical energy, with their corresponding antigen, as, *e.g.*, in the anchoring of bacterial cells. This anchoring is a necessary preliminary for further reaction, which may be of the most varied kind. We are acquainted with a number of such haptines or antibodies, of which the following are examples:

(1). The agglutinins, which cause clumping of the cells.

(2) The amboceptors, which play the part of carriers in the action of the complement.

(3) The opsonins, which render bacteria liable to be seized by the

phagocytes; also those haptines which are directed against the contents, or metabolic products, of bacteria.

(4) The antitoxins.

(5) The anti-endotoxins, which are directed against the endotoxins, to our knowledge of which Macfadyen, whose early death the scientific world deplores, has contributed so much.

(6) The precipitins.

(7) The antiferments, which are directed against certain ferments of the bacterial cells, *e.g.*, pyocyanase.

There is no doubt, however, that many more haptines exist. In order to obtain an idea of the extreme diversity of the phenomena which cause immunity, one must look for other substances whose functions are those which characterise the haptines—the substances, for example, which prevent cell division, or those which combat the biological adaptation of bacteria. The possibilities which I have here indicated appear to be limitless, and the need for study in this direction is evident.

I would here express my dissent from a prejudice which often makes itself felt, to the effect that in this matter there exists a profound contradiction between humoral and cellular immunity. As a matter of fact, to assume that the action of antibodies is merely a process of humoral pathology, is to put an artificial construction on the facts observed, for the side-chain theory is founded upon the view that the antibodies are purely and simply the product of cellular secretion, and that with their appearance in the blood there are associated changes in the cells which correspond to the phenomenon of serum immunity. That the action of antibodies takes place in the juices of the organism is an incontestable fact, which, however, in view of the cellular processes which give rise to it, cannot with justice be claimed to be evidence in proof of the correctness of humoral pathology, otherwise we must consider the action of ferments to be one of humoral physiology.

In the Protean forms of the phenomena of immunity, of course, the action of haptines by no means excludes phagocytosis; destruction of the bacteria outside the cells and their assimilation by the phagocytes are processes which may take place alongside each other, and, by their simultaneous action, increase the protective power. A special proof of the importance of the study of haptines appears to me to be the fact that—as the opsonin theory, which we owe to Sir

Almroth Wright, has made more evident—specific haptine reactions form the basis also of phagocytosis, which Metchnikoff has studied in so masterly a manner. The opsonins and cytotropic substances render the bacteria liable to attack by the phagocytes, and here we have a field in which humoral and cellular processes meet. One cannot, however, say that the possible causes of immunity are confined to haptine action and phagocytosis. Perhaps the atreptic view, by which differences of degree in avidity on the one side or on the other are presumed, is correct in many cases in which other influences are at work. This view of the case I shall treat more fully when I come to speak of carcinoma.

The immunity of an animal is, therefore, explained as being due to the great energy of the cells of its body, which are able to appropriate nutritious substances for themselves, and in so doing to deprive parasites of them. The opposite condition must be due to a certain disposing influence, and immunity of the parasites must be a condition of the cause of infectivity. The bacterial cells may in the same way be immune against haptine substances, and may withstand the action of the serum.

Thus there exist unstable relations between immunity and infection, and between parasite and host, relations which may depend on the most varying influences, and which lead up to the phenomena of reversible action, which calls for further study.

One cannot, therefore, go to work in a one-sided way when analysing and judging the various forms of phenomena, but must carefully consider together all the factors in question. The study of every possibility will bear fruit and make for an understanding of the processes of infection and immunity. I believe, however, that I have shown the influence exerted by the haptines upon the cause of infection is of great importance, not only when these are viewed as destroyers of the cause of infection outside the cells, but also when viewed in connection with the results of their anchoring power, chief among which at present stands phagocytosis.

It is our task to advance by a more accurate and more extensive study of *all* the haptines and their actions, and in the first place we must gain a knowledge of the influences exerted upon the causes of infection by the distribution and action of dissolved substances whose action is cytotropic, so that we may obtain a nearer insight into the manifold secondary phenomena which arise from them.

XXXIII

MODERN THERAPEUTICS

IN THE FOLLOWING PAGES WILL BE FOUND EARLY ACCOUNTS OF THE USE OF three natural remedies—a drug, water, and air, and the first account of a therapeutic instrument—the hypodermic syringe.

William Withering (1741-1799) of Birmingham, published his *Account of the Foxglove* in 1785. It is not only a masterpiece of clinical observation, but it also gives the best insight in literature into the methods used to determine the value of a drug, and to place its use on a sound basis.

James Currie (1756-1805) settled into practice in Liverpool, after experience as a ship's surgeon. He used the cold water douche as a treatment in continued fevers (probably typhoid). Weir Mitchell said his book, *Medical Reports on the Effects of Water in Fever,* had "absolute genius." After him, Priessnitz, an Austrian peasant, and Brand of Stettin (immersion baths for typhoid) and Winternitz (1835-1917) put hydrotherapy on an empirical if not a scientific basis.

George Bodington (1799-1882) of Sutton Coldfield, England, pointed out the value of fresh air in pulmonary complaints—a revolutionary idea in his time. His ideas were neglected in his own day.

Some controversy surrounds the question of priority in hypodermic medication. Francis Rynd (1845), Charles-Gabriel Pravaz (1851), and Alexander Wood (1855) are the dates given in Garrison's *History of Medicine.* Rynd was physician to the Meath Hospital, Dublin, a colleague of that brilliant group that included Stokes, Graves, and Corrigan.

REFERENCES

WITHERING, W. *Medical Classics,* Vol. 2, No. 4, December, 1937.

RODDIS, L. H. William Withering: The Introduction of Digitalis into Medical Practice. New York, Paul B. Hoeber, 1936.

SCHATZ, W. J. How Old is the Syringe in Parenteral Procedure? *Tr. Am. Therap. Soc.,* 1932.

FRANCIS RYND

ON THE SUBCUTANEOUS INTRODUCTION OF FLUIDS

[From *Description of an Instrument for the Subcutaneous Introduction
of Fluids in Affections of the Nerves.** Art. II]

The canula (a) screws on the instrument at (b); and when the
button (c), which is connected with the needle (f), and acted on by
a spring, is pushed up (as in Fig. 2), the small catch (d) retains it in
place. The point of the needle then projects a little beyond the
canula (Fig. 2). The fluid to be applied is now to be introduced
into the canula through the hole (e), either from a common writ-
ing-pen or the spoon-shaped extremity of a silver director; a small
puncture through the skin is to be made with a lancet, or the point
of the instrument itself is to be pressed through the skin, and on
to the depth required; light pressure now made on the handle raises
the catch (d), the needle is released, and springs backwards, leaving
the canula empty, and allowing the fluid to descend. If the instrument be slowly withdrawn, the parts it passes through, as well as the
point to which it has been directed, receive the contained fluid; and
still more may be introduced, if deemed expedient.

The subcutaneous introduction of fluids, for the relief of neuralgia, was first practised in this country by me, in the Meath Hospital, in the month of May, 1844. The cases were published in the
"Dublin Medical Press" of March 12, 1845. Since then, I have
treated very many cases, and used many kinds of fluids and solutions,
with variable success. The fluid I have found most beneficial is a
solution of morphia in creosote, ten grains of the former to one
drachm of the latter; six drops of this solution contain one grain
of morphia, and a grain or two, or more, may be introduced in cases
of sciatica at one operation, with the very best effects, particularly
if they are of long standing; or even in cases of tic in the head and
face, with equally beneficial results. The small instrument is for
operations on superficial nerves, the larger one for deep-seated
nerves; for though it is not necessary to introduce the fluid to the
nerve itself to ease pain, still the nearer to the seat of pain it is

---

* *Dublin Quarterly Journal,* 32:13, 1861.

conveyed, the more surely relief is given. They were manufactured, and completed entirely to my satisfaction, by the celebrated surgical instrument-maker, Mr. Weiss, of London, and are faithfully represented in the accompanying lithograph, by Foster & Co., of this city.

WILLIAM WITHERING

An Account of the Foxglove

[From *An Account of the Foxglove and Some of Its Medical Uses; With Practical Remarks on Dropsy and Other Diseases**]

As the more obvious and sensible properties of plants such as colour, taste and smell have but little connection with the diseases they are adapted to cure: so their peculiar qualities have no certain dependence upon their external configuration. Their chemical examination by fire, after an immediate waste of time and labour having been found useless is now abandoned by general consent. Possibly other modes of analysis will be found out which may turn to better account, but we have hitherto made only a very small progress in the chemistry of animal and vegetable substances. Their virtues must therefore be learnt either from observations of their effects upon insects and quadrupeds: from analogy, deduced from already known powers of some of their congenera or from empirical usages and experience of the populace.

The first method has not been much attended to: the second can only be perfected in proportion as we approach towards the discovery of a truly natural system: but the last, as far as it extends, lies within the reach of everyone who is open to information regardless of the source from which it springs.

It was a circumstance of this kind which first fixed my attention on the Foxglove.

In the year 1775 my opinion was asked concerning a family receipt for the cure of the Dropsy. I was told that it had long been kept a secret by an old woman in Shropshire who had sometimes made cures after the more regular practitioners had failed. I was informed also that the effects produced were violent vomiting and purging: for the diuretic effects seem to have been overlooked.

* London, 1785.

This medicine was composed of twenty or more different herbs; but it was not very difficult for one conversant with these subjects to perceive that the active herb could be no other than the Foxglove.

My worthy predecessor in this place, the very humane and ingenious Dr. Small had made it a practice to give his advice to the poor during one hour in the day. This practice which I continued until we had a hospital opened for the reception of the sick poor, gave me an opportunity of putting my ideas into execution in a variety of cases: for the number of poor who thus applied for advice amounted to between two and three thousand annually. I soon found the Foxglove to be a very powerful diuretic: but then and for a considerable time afterwards I gave it in doses very much too large, and urged its continuance too long; for misled by reasoning from the effects of the squill, which generally acts best on the kidneys when it excites nausea, I wanted to produce the same effect by the Foxglove. In this mode of prescribing, when I had so many patients to attend to in the space of one or at most of two hours, it will not be expected that I could be very particular, much less could I take notes of all the cases which occurred. Two or three of them only in which the medicine succeeded I find mentioned among my papers. It was from this kind of experience that I ventured to assert, in the Botanical Arrangement published in the course of the following spring that the Digitalis purpurea "merited more attention than modern practice bestowed upon it."

I had not however introduced it into the more regular mode of prescription: but a circumstance happened which accelerated that event. My truly valuable and respectable friend, Dr. Ash, informed me that Dr. Carvley, then principal of Brazen Nose College, Oxford, had been cured of a Hydrops Pectoris, by an empirical exhibition of the root of the Foxglove, after some of the first physicians of the age had declared that they could do no more for him. I was now determined to pursue my former ideas more vigorously than before, but was too well aware of the uncertainty which must attend on the exhibition of the *root* of a *biennial* plant and therefore continued to use the *leaves*. These I found to vary much as to dose, at different seasons of the year: but I expected, if gathered always in one condition of the plant, viz. when it was in its flowering state, and carefully dried, that the dose might be ascertained as exactly as that of any other medicine; nor have I been disappointed in this expectation. The more I saw of the great powers of this plant, the more it seemed necessary to bring the doses of it to the greatest possible accuracy. I suspected that this accuracy was not reconcilable with the use of a *decoction*, as it depended not only upon the care of those who had the preparation of it, but it was easy to conceive from the analogy of another plant of the same natural order, the tobacco, that its active properties might be impaired by long boiling. The decoction was therefore discarded, and the *infusion* substituted in its place. After this I began to use the leaves in *powder* but I still very often prescribed the infusion.

Further evidence convinced me that the diuretic effects of this medicine do not at all depend upon its exciting a nausea or vomiting: but on the contrary, that through the increased secretion of urine will frequently succeed to or exist along with these circumstances yet they are far from being friendly or necessary that I have often known the discharge of urine checked, when the doses have been imprudently urged so as to occasion sickness.

If the medicine purges, it is almost certain to fail in its desired effect: but this having been the case I have seen it afterwards succeed when joined with small doses of opium, so as to restrain its action on the bowels.

In the summer of 1776 I ordered a quantity of the leaves to be dried, and as it then became profitable to ascertain its doses, it was gradually adopted by the medical practitioners in the circle of my acquaintance.

In the month of November, 1777, in consequence of an application from that very celebrated surgeon, Mr. Russell of Worcester, I sent him the following account, which I choose to introduce here, as shewing the ideas I then entertained of the medicine, and how much I was mistaken as to its real dose:—

"I generally order it in decoction. Three drams of the dried leaves, collected at the time of the blossoms expanding, boiled in twelve to eight ounces of water. Two spoonfuls of medicine given every two hours, will sooner or later excite nausea. I have sometimes used the green leaves, gathered in the winter, but then I order three times the weight; and in one instance I used three ounces to a pint decoction before the desired effect took place. I consider the Foxglove thus given, as the most certain diuretic I know, nor do its diuretic effects depend merely upon the nausea it produces, for in cases where squill and ipecac have been so given as to keep up a nausea several days together, and the flow of urine not taken place, I have found the Foxglove to succeed: and I have in more than one instance given the Foxglove in smaller and more distant doses so that the flow of urine has taken place without any sensible affection of the stomach; but in general I give it in the manner first mentioned, and order one dose to be given after the sickness commences. I have then omitted all medicine, except those of the cordial kind are wanted, during the space of three, four or five days. By this time the nausea abates, and the appetite becomes better than it was before. Sometimes the brain is considerably affected by the medicine, and indistinct vision ensues: but I have never found any permanent bad effects from it.

"I use it in the Ascites, Anasaca, and Hydrops Pectoris, and so far as the removal of the water will contribute to cure the patient so far may be expected from this medicine: but I wish it not to be tried in ascites of female patients, believing that many of these cases are dropsies of the ovaria: and no sensible man will ever expect to see these encysted fluids removed by any medicine.

"I have often been obliged to evacuate the water repeatedly in the same patient, by repeating the decoction; but then this has been at such distance of time as to allow of the interference of other medicines and a proper regimen, so that the patient obtained in the end a perfect cure. In these cases the decoction becomes at length so very disagreeable that a much smaller quantity will produce the effect, and I often find it necessary to alter its taste by the addition of Ag. Cinnam. fp. or Ag. Juniper composite.

"I allow and indeed enjoin my patients to drink very plentifully of small liquors through the whole course of the cure, and sometimes, where the evacuations have been very sudden, I have found a bandage as necessary as in the use of trocar." . . .

CASE IV

July 25th—Mrs. H. of A. near N. between forty and fifty years of age, a few weeks ago, after some previous indisposition, was attacked by a severe cold, shivering fit succeeded by fever: great pain in her left side, shortness of breath, perpetual cough, and after some days copious expectoration. On the 4th of June, Dr. Darwin was called to see her. I have not heard what was then done for her, but between the fifteenth of June and the twenty-fifth of July, the doctor, at his different visits, at his various medicines of the deobstruent, tonic, antispasmodic, diuretic and evacuant kinds.

On the twenty-fifth of July I was desired to meet Dr. Darwin at the lady's house. I found her nearly in a state of suffocation, her pulse extremely weak and irregular, her breath very short and laborious, her countenance sunk, her arms of a leaden colour, clammy and cold. She could not lie down in bed and had neither strength nor appetite, but was extremely thirsty. Her stomach, legs and thighs were greatly swollen: her urine very small in quantity, not more than a spoonful at a time, and that very seldom. It had been proposed to scarify her legs, but the proposition was not acceded to.

She had experienced no relief from any means that had been used, except from the ipecacuanha vomits; the dose of which had been gradually increased from 15 to 40 grains, but such was the insensible state of her stomach for the last few days, that even those very large doses failed to make her sick and consequently purged her. In this situation of things I knew of nothing likely to avail us, except the digitalis: but this I hesitated to propose, from an apprehension that little could be expected from anything; that an unfavorable termination would tend to discredit a medicine which promised to be of great benefit to mankind, and I might be censured for a prescription which could not be countenanced by the experience of any other regular practitioner. But these considerations soon gave way to the desire of saving the life of this valuable woman, and accordingly I proposed that the Digitalis be tried: adding that I sometimes had found it to succeed when other, even the most judicious methods, had failed. Dr. Darwin very politely acceded immediately to my proposition, and as he had never seen it given, left the preparation and the dose to my direction. We therefore prescribed as follows:

Rx. Fol. Digital. purp. recent. oz. iv. coque ex Aq. fontan.
     purae lb iss ad lb i. et cola.
Rx. Decoct. Digital. oz. iss.
     Aq. Nuc. Moschat. oz. ii. M. fiat, haust. 2dis horis
     sumend.

The patient took five of these draughts, which made her very sick, and acted very powerfully upon the kidneys, for within the first twenty-four hours she made upwards of eight quarts of water. The sense of fulness and oppression across her stomach was greatly diminished, her breath was eased, her pulse became more full and regular, and the swellings of her legs subsided.

26th—Our patient being thus snatched from impending dissolution, Dr. Darwin proposed to give her a decoction of pareira brava, and guiacum shavings, with pills of myrrh and white vitriol: and, if costive, a pill with calomel and aloes. To these propositions I gave a ready assent.

30th—This day Dr. Darwin saw her, and directed a continuation of the medicines last prescribed.

August 1st—I found the patient perfectly free from every appearance of dropsy, her breath quite easy, her appetite much improved, but still very weak. Having some suspicion of a diseased liver I directed pills of soap, rhubarb, tartar of vitriol, and calomel to be taken twice a day with a neutral saline draught.

9th—We visited our patient together and repeated the draughts directed on the 26th of June, with the addition of tincture of bark, and also ordered pills of aloe, guiacum, and fol martis, to be taken if costive.

September 10th—From this time the management of the case fell entirely under my direction, and perceiving symptoms of effusion going forwards, I desired that a solution of merc. subl. corr. might be given twice a day.

19th—The increase of the dropsical symptoms now made it necessary to repeat the Digitalis. The dried leaves were used in infusion, and the water was presently evacuated as before.

It is now almost nine years since the Digitalis was first prescribed for this lady, and notwithstanding every preventive method I could devise, the dropsy still continues to recur at times: but is never allowed to increase so as to cause much distress, for she occasionally takes the infusion and relieves herself whenever she chooses. Since the first exhibition of that medicine, very small doses have been always sufficient to promote the flow of urine. . . .

The Foxglove when given in very large and quickly repeated doses occasions sickness, vomiting, purging, giddiness, confused vision, objects appearing green and yellow; increased secretion of urine, with frequent motions to part with it; and sometimes inability to retain it; slow pulse, even as slow as 35 in a minute, cold sweats, convulsions, syncope, death.

When given in a less violent manner, it produces most of these effects in a lower degree; and it is curious to observe that the sickness, with a certain dose of the medicine, does not take place for many hours after its exhibition has been discontinued; that the flow of urine will often proceed, sometimes accompany, frequently follow the sickness at the distance of some days, and not infrequently be checked by it. The sickness thus excited, is extremely different from that occasioned by any other medicine: it is peculiarly distressing to the patient; it ceases, it recurs again as violent as before; and thus it will continue to recur for three or four days, at distant and more distant intervals.

But these sufferings are not at all necessary; they are the effects of our inexperience, and would in similar circumstances, more or less attend the exhibition of almost every active and powerful medicine we use.

Perhaps the reader will better understand how it ought to be given, from the following detail of my own improvement, than from precepts peremptorily delivered, and their source veiled in obscurity.

At first I thought it necessary *to bring on and continue the sickness in order to insure the diuretic effects.*

I soon learned that the nausea being once excited it was unnecessary to repeat the medicine, as it was certain to recur frequently at intervals more or less distant.

Therefore my patients were ordered *to persist until the nausea came on and then to stop.* But it soon appeared that the diuretic effects would often take place first, and sometimes be checked when the sickness or a purging supervened.

The direction was therefore enlarged thus—*Continue the medicine until the urine flows, or sickness or purging takes place.*

I found myself safe under this regulation for two or three years; but at length cases occurred in which the pulse would be retarded to an alarming degree, without any other preceding effect.

The directions therefore required an additional attention to the state of the pulse, and it was moreover of consequence not to re-

peat the doses too quickly but to allow sufficient time for the effects
of each to take place, as it was found possible to pour an injurious
quantity of the medicine, before any of the signals for forbearance
appeared.

*Let the medicine therefore be given in the doses and at the inter-
vals mentioned above:—let it be continued until it either acts on
the kidneys, the stomach, the pulse, or the bowels; let it be stopped
upon the first appearance of any of these effects,* and I will maintain
that the patient will not suffer from its exhibition, nor the practitioner disappointed in any reasonable expectation.

JAMES CURRIE

THE EFFECTS OF COLD AND WARM WATER IN TREATMENT

[From *Medical Reports on the Effects of Water, Cold and Warm, as a
remedy in Fever and Other Diseases, whether applied to the Surface of the
Body or used Internally. Including an Inquiry into the Circumstances
that render Cold Drink, or the Cold Bath dangerous in Health. To which
are added Observations on the Nature of Fever and on the Effects of
Opium, Alcohol and Inanition**]

[Chapter 1]

In the London Medical Journal for the year 1768 Dr. William
Wright,[1] formerly of the Island of Jamaica gave an account of the
successful treatment of some cases of fever by ablution of the patient
with cold water.

"On the 1st of August, 1777," says Dr. Wright, "I embarked in
a ship bound to Liverpool and sailed the same evening from Mon-
tego Bay. The master told me he had hired several sailors on the
same day we took our departure, one of whom had been sick at
quarters on shore, and was now but in a convalescent state. On the
23rd of August we were in the latitude of Bermudas, and had had
a very heavy gale of wind for three days when the above mentioned
man relapsed and had a fever with symptoms of the greatest malignity. I attended this person often, but could not prevail with him
to be removed from a dark and confined situation, to a more airy

---

* Second edition. Liverpool, 1798 (or earlier).
[1] Now I believe physician to the Army in the West Indies.

and convenient part of the ship; and as he refused medicines, and even food, he died on the eighth day of his illness.

"By my attention to the sick man I caught the contagion, and began to be indisposed on the 5th of September and the following is a narrative of my case, extracted from notes daily marked down: I had been many years in Jamaica but except being somewhat relaxed by the climate, and fatigue of business, I ailed nothing when I embarked. This circumstance, however, might perhaps dispose me more readily to receive the infection.

"September 5th, 6th, 7th—Small rigors now and then—a preternatural heat of the skin—a dull pain in the forehead—the pulse small and quick—a loss of appetite but no sickness at stomach—the tongue white and shiny—little or no thirst—the belly regular—the urine pale and rather scanty—in the night restless, with starting and delirium.

"September 8th,—Every symptom aggravated, with pains in the loins and lower limbs and stiffness in the thighs and hams.

"I took a gentle vomit in the second day of this illness, and next morning a decoction of tarnarinds; at bedtime an opiate, joined with autimonial wine, but this did not procure sleep, or open the pores of the skin. No inflammatory symptoms being present, a drachm of Peruvian Bark was taken every hour for six hours successively, and now and then a glass of Port Wine, but with no apparent benefit. When upon deck, my pains were greatly instigated and the colder the air the better. This circumstance and the failure of every means I had tried, encouraged me to put in practice on myself what I had often wished to try on others, in fevers similar to my own.

"September 9th,—Having given the necessary directions, about three in the afternoon, I stripped off all my clothes, and threw a sea-cloak loosely about me till I got upon the deck, when the cloak also was laid aside; three buckets full of salt water were then thrown at once on me; the shock was great but I felt immediate relief. The headach [sic] and other pains instantly abated and a fine glow and diaphoresis succeeded. Towards evening, the febrile symptoms threatened a return, and I had again recourse to the same method as before, with the same good effect. I now took food with an appetite, and for the first time had a sound nights rest.

"September 10th—No fever, but a little uneasiness in the hams and thighs—used the cold bath twice.

"September 11th—Every symptom vanished, but to prevent a relapse, I used the cold bath twice."

Having before experienced that Dr. Wright was a safe guide, I immediately on reading this narrative determined on following his practice in the present instance; and before an opportunity occurred of carrying my intention into effect, I was farther encouraged by learning that my respectable colleague, Dr. Brandreth, had employed cold water externally in some recent cases of fever with happy results.

HISTORY OF A FEVER WHICH BROKE OUT IN THE LIVERPOOL INFIRMARY

[Chapter 11]

On the 9th of December, 1787, a contagious fever made its appearance in the Liverpool Infirmary. For some time previously the weather had been extremely cold, and the discipline of the house, owing to causes which it is unnecessary to mention, had been much relaxed. The intensity of the cold prevented the necessary degree of ventilation, and the regulations for the preservation of cleanliness had been in some measure, neglected. These circumstances operated particularly on one of the wards of the eastern wing. The contagion spread rapidly and before its progress could be arrested, sixteen persons were affected of which two died. Of these sixteen, eight were under my care. On this occasion I used for the first time the affusion of cold water, in the manner described by Dr. Wright. It was first tried in two cases only the one in the second, the other in the fourth day of fever. The effects corresponded exactly with those mentioned by him to have occurred in his own case and thus encouraged the remedy was employed in five other cases. It was repeated daily, and of these seven patients, the whole recovered.

From this time I have constantly wished to employ the affusion of cold water in every case of the low contagious fever, in which the strength was already not too much exhausted; and I have preserved a register of a hundred and fifty three cases in which the cure was chiefly trusted to this remedy. Of these ninety four occurred in the hospital in the four years subsequent to the period already men-

tioned, twenty seven in private practice and thirty two in the 30th
regiment of foot when quartered in Liverpool in the year 1792.

THE MANNER IN WHICH THE AFFUSION OF COLD WATER OUGHT TO BE USED IN FEVER

[Chapter IV]

Having given this general account of my experience of this remedy in fever, it will now be necessary to enter more particularly on
the rules which ought to govern its application, and on the different
effects to be expected from it, according to the different stages of
the disease in which it is employed. It will be proper to premise that
when the term fever is used in the present work without any adjunc-
tive, it is the low contagious fever that is meant. This is the Typhus
of Dr. Cullen; the contagious fever of Dr. Lind; the Febris inirri-
tativa of Dr. Darwin. In popular language it is generally called the
nervous fever, and where particular symptoms occur the putrid fever.

Whoever has watched the progress of fever, must have observed
the justness of the observation made by Cullen, Vogel, DeHaen, and
others that even those genera which are denominated continued are
not strictly such, but have pretty regular and distinct exacerbations
and remissions in each diurnal period. In this space of time Dr.
Cullen contends that an attentive observer may commonly distinguish two separate paroxysms. My observations do not enable me to
confirm his position in its full extent—but one exacerbation and
one remission in the twenty-four hours seem generally observable.
The exacerbation usually occurs in the afternoon, or evening, and
remission towards morning. These exacerbations are marked by increased flushing, thirst and restlessness. If the heat of the patient
be, at such times taken by the thermometer it will be found to have
risen one or two degrees in the central parts of the body, above the
average heat of the fever, and still more to the extremities. The safest
and most advantageous time for using the aspersion or affusion of
cold water is when the exacerbation is at its height, or immediately
after its declination is begun; and this has led me almost always to
direct it to be employed from six to nine o'clock in the evening; but
it may be safely used at any time of the day, *when there is no sense
of chilliness present when the heat of the surface is steadily above*

what is natural and when there is no general or profuse perspiration.
These particulars are of utmost importance.

Under these restrictions the cold affusion may be used at any period
of fever; but its effects will be more salutary in proportion as it is
used more early. When employed in the advanced stages of fever
where the heat is reduced, and the debility great some cordial should
be given immediately after it and the best is warm wine. The general effects of the cold affusion will be more clearly illustrated by the
following cases.

CASES IN WHICH THE AFFUSION OF COLD WATER WAS USED IN THE
DIFFERENT STAGES OF FEVER

[Chapter v, Case 1]

January 1, 1790.

A nurse in the fever ward of the Infirmary having several patients
under her care, caught the infection. She was seized with violent
rigors, chilliness, and wandering pains, succeeded by great heat,
thirst and head-ache. Sixteen hours after the first attack, her heat in
the axilla was 103° of Fah, her pulse 112 in the minute and strong,
her thirst great, her tongue furred, and her skin dry.

Five gallons of salt water, of the temperature of 44° were poured
over her naked body, at five o'clock in the afternoon, and after being
hastily dried with towels, she was replaced in bed, when the agitation and sobbing had subsided her pulse was found to beat at the
rate of 96 strokes in the minute and in half an hour afterward it had
fallen to 80. The heat was reduced to 98° by the ablution, and half
an hour afterwards it remained stationary. The sense of heat and
head-ache were gone, and the thirst nearly gone. Six hours afterwards she was found perfectly free of fever, but a good deal of debility remained.

Small doses of colombo were ordered for her, with a light nourishing diet, and for several days the cold affusion was repeated at the
same hour of the day as the first; the fever never returned.

In taking the heat of the patient I have generally used a small
mercurial thermometer of great sensibility with a movable scale,
made for me by Mr. Ramsdea, after a form invented by the late Mr.
Hunter, and used by him on his experiments on the heat of animals
and I have introduced the bulb under the tongue with the lips close,

or under the axilla indifferently; having found by repeated experiments that the heat in these two places corresponds exactly.

GEORGE BODINGTON

[From *An Essay on the Treatment and Cure of Pulmonary Consumption**]

A uniform and complete success having resulted in the treatment of several cases of tuberculous Consumption upon the principles and plan explained in the following pages, the author deems it his duty to publish them, with his opinions and principles of treatment. It would not accord with the brevity and conciseness of the plan of this treatise to enter at length into the nature and causes of Consumption, the diagnostic symptoms, physical signs, morbid anatomy, etc.; these are subjects which have been elaborately handled by several eminent authors, whilst little has yet been done by way of improvement in the treatment of the disease. Consumptive patients are still lost as heretofore; they are considered hopeless and desperate cases by most practitioners, and the treatment commonly is conducted upon such an inefficient plan as scarcely to retard the fatal catastrophe. One mode of treatment prevailing consists in shutting the patients up in a close room, to exclude as far as possible the access of the atmospheric air, and then forcing them to breathe over and over again the same foul air contaminated with the diseased effluvia of their own persons. But what could rationally be expected to be the result from such practice than that of the conversion of a slow or moderate consumption into an intense or galloping one? This is, indeed, a treatment founded on the most erroneous principles, and is much more deserving of reprobation than is even the apathetic indifference and desperate hopelessness generally entertained with regard to this disease. . . .

There is nothing gained by resorting to the coast; in truth, the interior of the island is the best; the air is just as pure and much milder, and more suitable for the lungs of consumptive people, if they will but breathe it. There is but one other proposition in the way of treatment to which I have to allude—I mean to the inhalation of gases of various kinds, by which means it is proposed to convert the cough of consumption into a catarrhal cough, which catarrh is to continue so long as the patient lives, or, discontinuing, the

* London, 1840.

consumption would supervene. We have not heard what success has attended this method of treatment, but it may be fairly inferred that such an artificial mode of proceeding, so contrary to the dictates of common sense and sound principles could not sustain itself for long, and must have perished nearly at its birth. The only gas fit for the lungs is the pure atmosphere freely administered, without fear; its privation is the most constant and frequent cause of the progress of the disease. To live in and breathe freely the open air, without being deterred by the wind or weather, is one important and essential remedy in arresting its progress—one about which there appears to have generally prevailed a groundless alarm lest the consumptive patient should take cold. Thus one of the essential measures necessary for the cure of this fatal disease is neglected, from the fear of suffering or incurring another disease of trifling import. No two diseases can be more distinct from each other than consumption and catarrh. It is the latter only which might be caught by exposure to atmospheric causes; with the former they have nothing to do. Farmers, shepherds, ploughmen, etc., are rarely liable to consumption, living constantly in the open air; whilst the inhabitants of the towns, and persons living much in close rooms, or whose occupations confine them many hours within doors, are its victims. The habits of these latter ought, in the treatment of the disease, to be made to resemble as much as possible those of the former class, as respects air and exercise, in order to effect a cure. How little does the plan of shutting up the patients in close rooms accord with this simple and obvious principle! As to the result of such a practice, it is known to all; one-fifth of the deaths annually in England are from consumption, whilst cures are scarcely ever heard of, and never expected. Despair seems to have taken full possession of the medical profession as regards this destructive disease, and none but the feeblest efforts are exerted to oppose its progress. The successful treatment of several cases successively of severe, decided, and genuine tubercular consumption on principles, I believe, differing from the usual routine of practice, and from the doctrines and theories of the present day, which form the basis of medical practice, induces me to lay those cases before the public, and to explain my views and principles of treatment on which that success was founded.

. . . I come now to the most important remedial agent in the cure of consumption, that of the free use of a pure atmosphere; not the impure air of a close room, or even that of the house generally, but the air out of doors, early in the morning, either by riding or walking; the latter when the patients are able, but generally they are unable to continue sufficiently long in the open on foot, therefore riding or carriage exercise should be employed for several hours daily, with intervals of walking as much as the strength will allow of, gradually increasing the length of the walk until it can be maintained easily several hours every day. The abode of the patient should be in an airy house in the country; if on an eminence the better. The neighborhood chosen should be dry and high; the soil, generally of a light loam, a sandy or gravelly bottom; the atmosphere is in such situations comparatively free from fogs and dampness. The patient ought never to be deterred by the state of the weather from exercise in the open air; if wet and rainy, a covered vehicle should be employed with open windows. The cold is never too severe for the consumptive patient in this climate; the cooler the air which passes into the lungs, the greater will be the benefit the patient will derive. Sharp frosty days in the winter season are most favorable. The application of cold pure air to the interior surface of the lungs is the most powerful sedative that can be applied, and does more to promote the healing and closing of cavities and ulcers of the lungs than any other means that can be employed; for it is by the use of the means which have the power of restoring to a healthy condition the nervous system, interwoven with and forming a portion of the substance of the lungs, that healthy actions can be induced in the remaining tissues. This, then is to be aimed,—a healthy nervous system, which will embrace in its consequences, due sensibility, motive power, nutritive and reparative power,—conditions necessary to resist and overcome the morbid influence arising from the presence of tuberculous matter. Many persons are alarmed and deterred from taking much exercise in the open air, from the circumstance of their coughing much on their first emerging from the warm room of a house; but this shows that the air of the room was too warm, not that the common atmosphere was too cold. To live in a temperature nearly equal to the latter at all times should be the aim of the patient, who should avoid warm close rooms as much as possible, and always keep away from the fire, taking care to keep the surface of the body warm by sufficient clothing. Thus the equal temperature so much considered, and said to be necessary, should be that of the external air, instead of that so commonly employed, the warmth of a close room.

In order effectually to overcome consumptive disease, all these several circumstances will be required to be adopted and followed up with the greatest attention, regularity, assiduity, and patience. Of those cases which I have treated upon these principles, having had some of the patients under my own roof, by which I secured all the advantages of situations, etc., before spoken of, and some in my immediate neighborhood, so that I could closely watch them, I have met with signal success, and scarcely an instance in which this mode of treatment has been fully carried out in all its particulars wherein the consumptive symptoms have not gradually yielded, and the patients restored to complete health. I shall now proceed to give an outline of the history of the treatment of several cases.

One occurred in the person of an awl-blade grinder, living in the country, in the year 1833. He was of a consumptive family; a sister of his had died at about the age of twenty years, and others of his nearest relatives had died from the same disease. There could be no stronger exciting cause for the development of the disease than that which arose from his daily occupation; he was about thirty years of age, of fair complexion, florid, shoulders high, chest narrow, and his general figure rather spare and slender. His finger nails were incurvated; he was troubled with a pain in his side; and a cough more or less with intermission. It was upon the accession of a sudden attack of consumption that I was called in to attend. A feeling of suffocation affected him, which was distressing, arising from the pressure of an abscess in the bronchial passages, attended with irritative fever; the breathing was relieved by the bursting of the abscess, and the free expectoration of pus and mucus. A cavity was formed in the upper portion of the substance of the lungs; the pulse beat 140 in a minute; he had profuse night perspirations; and his respiration was exceedingly quickened. He was much exhausted, and fully impressed with a belief that his life was about to terminate. He had no inclination for food of any kind; his muscles were relaxed and powerless, and his whole frame collapsed. Under these circumstances, had the antiphlogistic treatment, or even any part of it, been adopted, I believe he would have sunk past recovery; and yet. would not this be called acute inflammation of the substance of the lungs? and are not the remedies for this said to be bleeding, blisters, calomel, antimony, digitalis, purgatives, etc.? But any of these, I firmly believe, would have hazarded his existence; the application of the antiphlogistic routine would have destroyed him. The treatment adopted was this. Seeing that nutrition was at a standstill, that the muscular power was collapsed, and the sanguiferous system running away, at the rate of 140 beats per minute: to counteract these dangerous symptoms, he took, first, a wine glass of port wine, and repeated it in a few hours; at bedtime he took a sedative draught, and slept well; he continued to cough, and expectorated freely pus and mucus; he took at intervals small doses of hydrochlor. morph., about a tenth of a grain; this, and the full dose he had taken on the previous night, allayed, in a great degree, the nervous excitement in the lungs, and the irritative fever subsided; but the cough, debility, and expectoration continued; there was a cavity in the lungs to be healed. I told him that could not be done without a strenuous effort on his part; and explained to him my views as to the beneficial effects to be obtained by early rising and remaining out of doors a considerable time in the open air; that this would soothe, expand, and invigorate the lungs, so that the sores would heal, and that by no other means could he be cured; that if he remained within doors, shut up in the house, more abscesses would be likely to form, and the irritative fever again attack him. He saw the force of this advice, and determined to follow it, being a man of much firmness of character. All this occurred on the second day after the acute attack. On the next day following he related to me, nearly in these words, the particulars of his morning's walk: "I got up at four o'clock, and crawled out of the house as well as I could, and felt, and, I believe, looked, the most miserable, weak, and pitiable wretch in the world. I crept along, panting for breath, towards the common; I thought I must have died on the road; at last I reached Welchman's Hill, and when I began to walk round it I felt my lungs open, my breathing free, and my strength increase fast. I was now sure it was doing me good; I went quite round the hill, and then home and was so hungry that I ordered a beefsteak for breakfast, and ate heartily of it." The distance he walked would be about three miles. The spot called Welchman's Hill is said to be equal in elevation to any table-land in the island. The soil lying on a sandy or gravelly bottom, the air is very pure and mild. He continued for some time daily to pursue the same course, and became convalescent in a week, losing his cough entirely. I wished him to change his employment, but his circumstances forbade that. He resumed, after a short interval of rest, his trade of an awl-blade grinder, and continues it to this time. He has had symptoms of a return of his disorder on several occasions since, and informs me that, when that is the case, he betakes himself early in the morning to the common, and that always prevents any serious attack. The cure in this case was obtained by means applied to stimulate and invigorate the nutritive, sanguiferous, and muscular powers; wine and such nourishing diet as the stomach could bear, and by means applied to soothe and allay nervous excitement, locally and generally; first, by a full dose at night of the muriate of morphine, followed by small alterative doses given every five or six hours; secondly, by the application of the early morning air to the internal surface of the lungs, continued for several hours, accompanied with muscular exertion. The change in the character of the expectorated matter is very striking. As soon as the full effects of the morning air are experienced, it becomes light, white, more transparent, and devoid of puriform matter; it has more of the nature of mucus, and is no longer heavy, yellow, and solid. So powerfully does this remedy affect the lungs as a sedative, allaying and subduing nervous disturbance, at the same time inducing a vigorous tone of the digestive apparatus, and of the nutrient functions generally, that it will, if boldly and thoroughly applied, directly and entirely change the character of the cough, and completely remove the wasting irritative fever.

The next opportunity I had of witnessing the advantages of the mode of treatment described occurred in the case of a young lady, about sixteen years of age, whose parents, brothers, and sisters were all at this time healthy generally; consumption was not known in the family previous to her case, but at the present time her brother suffers from the disease. For several years she had suffered occasionally from pain in the side, cough and debility. In 1835 she returned home from a boarding-school, where she had been placed under medical treatment for these complaints; she was still ill, and her friends thought it advisable she should go to the sea-coast. She went near to Liverpool; the sea-air had a bad effect, the pain and cough increased; she was placed under medical care, and went through a long course of treatment. She continued to get worse in every respect, and her friends saw the necessity of her removal home; and she came to her native air in Warwickshire in October 1835, after an absence of several months. Her friends were impressed with a notion that the iodine which she had been taking, if persevered with, would be ultimately successful. This very interesting patient came under my care. Her parents, relatives, and numerous friends were watching her with the deepest solicitude; for she was, by all who knew her, most highly and justly esteemed. I found it necessary, at least for a short time, to acquiesce in the treatment by iodine, although there was but little hope of any advantage from it. I met several medical men in consultation, and a treatment was pursued in the usual manner; the patient being confined to her room, and consumption gradually wearing her away. I had explained my views to her friends respecting air and exercise out of doors, but could not succeed in gaining their consent to the plan. The two months of November and December were thus lost to the patient, or rather, during that period every symptom of the disease had become aggravated; she was now extremely emaciated, suffered from profuse night perspiration, violent cough, and difficulty of breathing; the expectoration was abundant, consisting of mucus, mixed with opaque solid portions frequently tinged with blood, most of which sank in water, some floated. There was a dull sound on percussion of the upper portion of the lungs, mucous rattle, with a gurgling noise, and a hoarseness, and weakness of voice; the physical signs, in combination with the general symptoms, were clearly indicative of the existence of cavities in the upper portion of the lungs. In the month of January, 1836, the case was left entirely to my management; and, having urged my views strongly to her friends, I gained their consent to their being adopted. A donkey was procured, on which the patient began to take exercise out of doors, notwithstanding the inclemency of the season, in the depth of winter. The first trial was unpromising; the cough appearing to be much increased in coming into the open air from the warm bedroom. This arose from the undue closeness and heat of the bedroom, and not the external air. There cannot be a more fatal error than that which arises from the supposition of there being something deleterious in the external atmosphere, because persons cough when first brought into it out of unwholesome heated apartments. The latter should be especially avoided, and apartments kept cool and airy, corresponding in temperature nearly to the external atmosphere, whilst the former should be courted and indulged in to the utmost. The surface of the body may and should always be kept warm by sufficient clothing, the lungs cool by the constant access of cold pure air to them; thus undue heat is driven from the interior to the surface. In the present instance it was soon found that by continuing a long time out of doors the cough abated materially; every day some improvement was observed to take place, very gradual, but constant. A sedative draught was given every night, which, together with the exercise of the day, procured sleep and warded off the cough till morning. In the daytime an emulsion mixture was taken at intervals, and very small doses of morphine, to subdue by degrees the irritation arising from the presence of tubercles in the lungs. The diet was nourishing, consisting of boiled egg, fresh meat, milk and bread, and two glasses of sherry in water daily. This treatment was continued very strictly through the winter and spring months of the year 1836; by June the patient had entirely lost her cough, with all the other symptoms of the disease, regained her health and strength, and passed through the succeeding winter in very good health, accustoming herself to go out of doors, walking or riding almost daily. At this time, July 1839, she is in perfect health.

. . . In conclusion, I have to add that the natural, rational, and, so far as to my knowledge of it has been tried, the successful treatment of pulmonary consumption appertains exclusively neither to the theory of phlogiston, or inflammation, nor to that of the Brunonian system; but it is a mixture of both. As I believe, both theories have truth in them, but are not exclusively true, and independent one of the other. Further, physiological investigations into the nature of nervous power, and the influence it exercises over the sanguiferous and other tissues, by its presence or absence, or undue exhaustion or irritation, will probably develop the true nature of those changes of structure which occur under the influence of disease, which are designated by the term "phlogosis" or inflammation, language which not improbably is destined at some future period to be expunged from medical science and literature; or, at least, to be understood as conveying very different ideas of the nature of disease than are commonly implied in those terms at present, as well as to effect a great change in the mode and application of remedial agents generally. The experimental labours of Majendie in France, in relation to the operation of the nervous power in animal life, and the investigation of Kiernan and others in England, as to the condition of the capillary vessels in diseased parts, have both a direct tendency to weaken the faith hitherto so universally and implicitly placed in the old theory.

XXXIV

HUMANITARIAN MEDICINE

MAN'S CALLOUS INDIFFERENCE TO HIS UNFORTUNATE FELLOW CREATURES began to be overcome in the middle of the eighteenth century, with the introduction of the liberal ideas of Rousseau, Voltaire, Kant, Hume, Thomas Jefferson, and Thomas Paine. It would be unfair to attempt to excerpt any of the writings of that great humanist, John Howard (1726-1790), who was responsible for the reform of the conditions in prisons and lazarettos (contagious disease hospitals, pest houses). Jails were the endemic reservoirs of fevers which spread to armies and ships.

The condition of the insane evoked the enlightened sympathy of an Englishman, William Tuke (1732-1822), and a Frenchman, Philippe Pinel (1745-1826). Tuke, a Quaker, founded the Retreat, at York, in 1794, for the accommodation of those of his faith afflicted with mental illness. This move was the result of a communication brought him by relatives of Hannah Mills, inmate of the Lunatick Asylum of York: they were denied permission to see her; no reason was given, but a few days later they learned that she was dead. Investigation of conditions in asylums revealed conditions unbelievably inhumane.

Pinel's experience was much the same. Appointed physician to the Bicêtre, he found insane patients chained to the walls of filthy cells, treated as if they were wicked animals. In 1798 he obtained permission from the National Assembly to take off their fetters and allow them some reasonable liberty. Couthon, the warder, said to him—"Citizen, are not you yourself crazy that you would unchain these wild beasts?" But to the astonishment of all the poor creatures reacted to kindness. Pinel, besides his work, *L'alienation.mentale,* wrote an extensive practice of medicine, *Nosographie.* He was known among his colleagues for his acumen in diagnosis.

Included in humanitarian medicine must be classed the first account of the hazards of industry, by Bernardino Ramazzini (1633-1714), *De morbis artificum diatriba.* Ramazzini was a good observer: he noted and understood an epidemic of spastic paraplegia (lathyrism) from the use of meal contaminated with a form of bean, and described stone mason's consumption (silicosis), potter's sciatica, gilder's ophthalmia, lead poisoning, etc. He is the spiritual ancestor of Alice Hamilton.

Theodor Fliedner (1800-1864), a Protestant pastor, first had the idea that women should be especially trained to nurse the sick. He and his wife Friederike established a school for nurses at Kaiserwerth on the

Rhine in 1833. This was visited by Florence Nightingale and it became the model for the nursing schools she established in England.

Florence Nightingale (1823-1910), an Englishwoman, was induced by Sidney Herber to go to the Crimea (1854) and investigate the conditions of the military hospitals. She was shocked at what she saw, instituted reforms, and when she returned to England she persuaded her countrymen of the need for organized instruction of nurses for the sick. Her *Notes on Nursing* was published in 1859. She defined nursing as "helping the patient to live." Elizabeth Fry (1780-1902), Elizabeth Blackwell (1821-1910), and Marie Zakrzewska (1829-1902) were also pioneers in the field of nursing.

SAMUEL TUKE*

ON THE TREATMENT OF THE INSANE

[From *Description of the Retreat,*† Chapter 1]

HISTORICAL ACCOUNT

Origin—Difficulties—First Meeting of Friends on the subject in 1792—Resolutions—Subscriptions—Meetings of Subscribers— —Amount of Subscriptions, and general Opinion of Friends respecting the Institution—Resolutions of a Meeting in 1793—Determination to build—Land purchased and Building commenced in 1794—Amount of Subscriptions—Necessary to Money—Rules agreed upon in 1795—Additional Subscriptions at this time— House very nearly completed—Committee appointed to engage Servants, and to admit Patients—West Wing ordered to be built First Month, 1796.

The history of the rise and progress of establishments, which have been peculiarly serviceable to society, like the biography of eminent men, is both interesting and useful. The inquisitive and speculative mind, loves to trace the causes of every striking object; and the practical philanthropist may derive considerable advantage, from an account of benevolent experiments that have been made by others.

The origin of the Institution which forms the subject of the following pages, has much the appearance of accident. In the year 1791, a female, of the Society of Friends, was placed at an establishment for insane persons, in the vicinity of the City of York; and her family, residing at a considerable distance, requested some of their

* Grandson of William Tuke.
† 1813.

acquaintance in the City to visit her. The visits of these Friends were
refused, on the ground of the patient not being in a suitable state to
be seen by strangers: and, in a few weeks after her admission, death
put a period to her sufferings.

The circumstance was affecting, and naturally excited reflections
on the situation of insane persons, and on the probable improvements which might be adopted in establishments of this nature. In
particular, it was conceived that peculiar advantage would be derived to the Society of Friends, by having an Institution of this kind
under their own care, in which a milder and more appropriate system of treatment, than that usually practised, might be adopted; and
where, during lucid intervals or the state of convalescence, the patient might enjoy the society of those who were of similar habits and
opinions. It was thought, very justly, that the indiscriminate mixture, which must occur in large public establishments, of persons of
opposite religious sentiments and practices; of the profligate and the
virtuous; the profane and the serious; was calculated to check the
progress of returning reason, and to fix, still deeper, the melancholy
and misanthropic train of ideas, which, in some descriptions of insanity, impresses the mind. It was believed also, that the general
treatment of insane persons was, too frequently, calculated to depress
and degrade, rather than to awaken the slumbering reason, or correct its wild hallucinations.

In one of the conversations to which the circumstance before-
mentioned gave rise, the propriety of attempting to form an Establishment for persons of our own Society, was suggested to William
Tuke, whose feelings were already much interested in the subject,
and whose persevering mind, rendered him peculiarly eligible to
promote such an undertaking. After mature reflection, and several
consultations with his most intimate Friends[1] on the subject, he was
decidedly of opinion, that an Establishment for the insane of our
own Society, of every class in regard to property, was both eligible
and highly desirable. It was necessary to excite a general interest in
the Society on the subject. He therefore, after the close of the Quarterly Meeting at York, in the 3d Month, 1792, requested Friends to
allow him to introduce to them a subject, connected with the welfare
of the Society. He then stated the views which he, and those whom
he had consulted, had taken of this subject; the circumstance which

[1] Amongst the most early and strenuous friends of this Establishment, I wish to
particularize the name of the excellent Lindley Murray; to whose steady endeavours,
for promoting its welfare, the Institution is much indebted.

had given rise to their interest respecting it, and the conviction which had resulted in their minds, in favour of an institution under the government of Friends, for the care and accommodation of their own Members, labouring under that most afflictive dispensation—the loss of reason.

Few objections were then made, and several persons appeared to be impressed with the importance of the subject, and the propriety of the proposed measure. The Friends with whom the proposal originated, were requested to prepare the outline of a plan, for the consideration of those who might attend the next Quarterly Meeting. Several objections, however, on a variety of grounds, soon afterward appeared. Many Friends were acquainted with but few, if any, objects for such an Establishment; and they seemed to forget that there might probably be many cases with which they were not acquainted. Some were not sensible that any improvement could be made in the treatment of the insane; supposing that the privations, and severe treatment, to which they were generally exposed, were necessary in their unhappy situation; and others, seemed rather averse to the concentration of the instances of this disease amongst us.

It was not, however, at all surprising that considerable diversity of opinion, should prevail upon a subject which was entirely new, and foreign to the general inquiries of those to whom it was proposed; and we must not forget that there was a respectable number, who duly appreciated the advantages likely to accrue to the Society from the proposed Establishment, and who cordially engaged in the promotion of the design. To these persons, and to the steady exertions of its chief promoter, whose mind was not to be deterred by ordinary difficulties, the Society of Friends, may justly be said to owe the advantages it derives from this admirable Institution.

Proposals for raising money and forming the Establishment, were prepared and laid before Friends, at the conclusion of the next Quarterly Meeting; which were generally approved. A subscription was immediately entered into; and the contributions were one hundred pounds, for a life-annuity of five per cent, per annum; annual subscriptions £11:0:6 for three years certain, and donations amounting to £192:3s. The following minutes were also made at this Meeting, viz.

"At a Meeting of Friends held at York the 28th of 6th Month, 1792, for the purpose of taking into consideration the propriety of providing a retired Habitation, with necessary advice, attendance, &c.,

*for the Members of our Society, and others in profession with us,
who may be in a state of Lunacy, or so deranged in mind* (not
Idiots) *as to require such a provision,*

RESOLVED,

"*That persons of this description, (who are truly objects of great
sympathy and compassion;) are often, from the peculiar treatment
which they require, necessarily committed, wholly, to the government of people of other Societies; by which means the state of
their own minds, and the feelings of their near connexions, are
rendered more dissatisfied and uncomfortable than would probably be the case, if they were under the notice and care of those,
with whom they are connected in Religious Society. It appears,
therefore, very desirable that an Institution should be formed,
wholly under the government of Friends, for the relief and accommodation of such Persons of all ranks, with respect to property. This would doubtless, in some degree, alleviate the anxiety
of the relatives, render the minds of the Patients more easy in their
lucid intervals, and consequently tend to facilitate and promote
their recovery ——*

IT IS THEREFORE PROPOSED,

1st. "THAT, in case proper encouragement be given, Ground be
purchased, and a Building be erected, sufficient to accommodate
Thirty Patients, in an airy situation, and at as short a distance from
York as may be, so as to have the privilege of retirement; and that
there be a few acres for keeping cows, and for garden ground for the
family; which will afford scope for the Patients to take exercise, when
that may be prudent and suitable.

2d. "THAT the Institution be established and supported by annui-
ties, donations, and annual subscriptions; and that the same (which
should be altogether voluntary) be promoted amongst Friends,
within the compass of this, and any other Quarterly Meeting.

3d. "THAT each Subscriber, by way of annuity contributing a sum
not less than Twenty Pounds, shall receive an interest of five per
cent. per ann. during life; and as the undertaking may not be able
to pay this interest, and otherwise maintain itself for the first three
years, those entered as Subscribers for annual payments, be engaged
for three years certain, in case the Subscriber should so long live.

4th. "THAT a contribution of One Hundred Pounds, from any

Quarterly Meeting in its collective capacity, paid to the Treasurer of this Institution before the year 1794; or a donation, at any time, of Twenty-five Pounds from any Friend; or a subscription of Fifty Pounds for an annuity, shall entitle such Quarterly Meeting, Donor, or Annuitant, respectively, to the privilege of nominating one *poor* patient at a time on the lowest terms of admission.

5th. "THAT the name of every annuitant, donor, and subscriber, be recorded in a book to be kept for the purpose; and that every Annuitant, Donor of not less than Two Guineas, and Subscriber of sums in any manner equal to Two Guineas, in the first three years, (being and continuing a member of our Society,) shall be a Member of the Meetings which are to be held for the government and superintendence of the Institution.

6th. "THAT there be paid for board, washing, medical advice, medicines, and all other things necessary except clothing, according to the circumstances of the Patients or their friends, from four shillings to fifteen shillings per week, or higher in particular cases; and six shillings per week for the board of the Servant of a Patient, in case the friends of any patient should incline to send one; which servant must be approved by the Committee.

7th. "THAT eight shillings per week and upwards, according to circumstances, be the terms for patients who come from the compass of any other Quarterly Meeting than Yorkshire, unless privileged agreeably to the 4th proposal.—These terms for patients to be subject to future alteration, if found necessary.

"WILLIAM TUKE is desired to get one thousand copies of these proposals printed, and circulated amongst Friends; with an account of the Subscriptions which have been or may be made previously to the printing thereof."

PHILIPPE PINEL

PROPER TREATMENT OF THE INSANE

[From *Treatise on Insanity**]

GENERAL PLAN OF THE WORK

Nothing has more contributed to the rapid improvement of modern natural history, than the spirit of minute and accurate observa-

* Traité médico-philosophique sur l'alienation mentale, 1801. Translation by D. D. Davis, M.D., Physician to the Sheffield General Infirmary, 1806.

tion which has distinguished its votaries. The habit of analytical investigation, thus adopted, has induced an accuracy of expression and a propriety of classification, which have themselves, in no small degree, contributed to the advancement of natural knowledge. Convinced of the essential importance of the same means in the illustration of a subject so new and so difficult as that of the present work, it will be seen that I have availed myself of their application, in all or most of the instances of this most calamitous disease, which occured in my practice at the Asylum de Bicêtre. On my entrance upon the duties of that hospital, every thing presented to me the appearance of chaos and confusion. Some of my unfortunate patients laboured under the horrors of a most gloomy and desponding melancholy. Others were furious, and subject to the influence of a perpetual delirium. Some appeared to possess a correct judgment upon most subjects, but were occasionally agitated by violent sallies of maniacal fury; while those of another class were sunk into a state of stupid ideotism and imbecility. Symptoms so different, and all comprehended under the general title of insanity, required, on my part, much study and discrimination; and to secure order in the establishment and success to the practice, I determined upon adopting such a variety of measures, both as to discipline and treatment, as my patients required, and my limited opportunity permitted. From systems of nosology, I had little assistance to expect; since the arbitrary distributions of Sauvages and Cullen were better calculated to impress the conviction of their insufficiency to simplify my labour. I, therefore, resolved to adopt that method of investigation which has invariably succeeded in all the departments of natural history, viz. to notice successively every fact, without any other object than that of collecting materials for future use; and to endeavour, as far as possible, to divest myself of the influence, both of my own prepossessions and the authority of others. With this view, I first of all took a general statement of the symptoms of my patients. To ascertain their characteristic peculiarities, the above survey was followed by cautious and repeated examinations into the condition of individuals. All our new cases were entered at great length upon the journals of the house. Due attention was paid to the changes of the seasons and the weather, and their respective influences upon the patients were minutely noticed. Having a peculiar attachment for the more general method of descriptive history, I did not confine myself to any exclusive mode of arranging my observations, nor to any one system of nosography. The facts which I have thus collected are now submitted to the consideration of the public, in the form of a regular treatise.

Few subjects in medicine are so intimately connected with the history and philosophy of the human mind as insanity. There are still fewer, where there are so many errors to rectify, and so many prejudices to remove. Derangement of the understanding is generally considered as an effect of an organic lesion of the brain, consequently as incurable; a supposition that is, in a great number of instances, contrary to anatomical fact. Public asylums for maniacs have been regarded as places of confinement for such of its members as are become dangerous to the peace of society. The managers of those institutions, who are frequently men of little knowledge and less humanity, (a) have been permitted to exercise towards their innocent prisoners a most arbitrary system of cruelty and violence; while experience affords ample and daily proofs of the happier effects of a mild, conciliating treatment, rendered effective by steady and dispassionate firmness. Availing themselves of this consideration, many empirics have erected establishments for the reception of lunatics, and have practiced this very delicate branch of the healing heart with singular reputation. A great number of cures have undoubtedly been effected by those base born children of the profession; but, as might be expected, they have not in any degree contributed to the advancement of science by any valuable writings. It is on the other hand to be lamented, that regular physicians have indulged in a blind routine of inefficient treatment, and have allowed themselves to be confined within the fairy circle of antiphlogisticism, and by that means to be diverted from the more important management of the mind. Thus, too generally, has the philosophy of this disease, by which I mean the history of its symptoms, of its progress, of its varieties, and of its treatment in and out of hospitals, been most strangely neglected.

Intermittent or periodical insanity is the most common form of the disease. The symptoms which mark its accessions, correspond with those of continued mania. Its paroxysms are of a determined duration, and it is not difficult to observe their progress, their highest development, and their termination. The present essay will, therefore, not improperly commence with an historical exposition of periodical insanity. The leading principles of our moral treatment will then be developed. Attention to these principles alone will, frequently, not only lay the foundation of, but complete a cure: while neglect of them may exasperate each succeeding paroxysm, till, at length, the disease becomes established, continued in its form, and incurable. The successful application of moral regimen exclusively, gives great weight to the supposition, that, in a majority of instances, there is no organic lesion of the brain nor of the cranium. In order however to ascertain the species, and to establish a nosology of insanity, so far as it depends upon physical derangement, I have omitted no opportunities of examination after death. I, therefore, flatter myself, that my treatment of this part of the subject will not discredit my cautious and frequently repeated observations. By these and other means, which will be developed in the sequel, I have been enabled to introduce a degree of method into the services of the hospital, and to class my patients in a great measure according to the varieties and inveteracy of their complaints. An account of our system of interior police, will finish this part of the enquiry. The last section will comprehend the principles of our medical treatment.

In the present enlightened age, it is to be hoped, that something more effectual may be done towards the improvement of the healing art, than to indulge with the splenetic Montaigne, in contemptuous and ridiculous sarcasms upon the vanity of its pretensions. I flatter myself, that the perusal of the following work will not excite the sentiment of that celebrated censor of human extravagance and folly, when he said, "that of whatever of good and salutary fortune or nature, or any other foreign cause may have bestowed upon the human frame, it is the privilege of medicine to arrogate to itself the merit."

THE ESTIMABLE EFFECTS OF COERCION ILLUSTRATED IN THE CASE OF A SOLDIER

A soldier, who for sometime had been insane, and a patient at the Hôtel Dieu, was suddenly seized with a vehement desire to join his regiment. All fair means to appease him being exhausted, coercive measures became indispensable to convey him to his chamber, and to secure him for the night. This treatment exasperated his phrenzy, and before morning he broke to pieces every thing that he could lay his hands upon. He was then bound and closely confined. For some days he was allowed to vent his fury in solitude: but he continued to be agitated by the most violent passions, and to use the language of imprecation and abuse against every body that he saw, but especially against the governor, whose authority he affected to despise. In about a week, however, he began to feel that he was not his own master; and, as the governor was going his round one morning, he assumed a more submissive air and tone, advanced with looks of mildness and contrition, and kissing his hand, said, "You have promised, upon my engaging to be peaceable and quiet, to permit me to go into the interior court. Now, Sir, have the goodness to keep your word." The governor, with a countenance full of sweetness and affability, expressed the very great pleasure which he felt, congratulated him on his returning health, and instantly ordered him to be set at liberty. Further constraint would have been superfluous, and probably injurious. In seven months from the date of his admission into the hospital, he was restored to his family and to his country, and has since experienced no relapse.

VARIETIES OF BODILY EXERCISES INCLUDING LABORIOUS OCCUPATIONS RECOMMENDED FOR CONVALESCENTS

Convalescent maniacs, when, amidst the languors of an inactive life, a stimulus is offered to their natural propensity to motion and exercise, are active, diligent and methodical. Laborious or amusing occupations arrest their delirious wanderings, prevent the determination of blood to the head by rendering the circulation more uniform, and induce tranquil and refreshing sleep. I was one day deafened by the tumultuous cries and riotous behaviour of a maniac. Employment of a rural nature, such as I knew would meet his taste, was procured for him. From that time I never observed any confusion nor extravagance in his ideas. It was pleasing to observe the silence and tranquility which prevailed in the Asylum de Bicêtre, when nearly all the patients were supplied by the tradesmen of Paris with employments which fixed their attention, and allured them to exertion by the prospect of a trifling gain. To perpetuate those advantages, and to ameliorate the condition of the patients, I made, at that time, every exertion in my power to obtain from the government an adjacent piece of ground, the cultivation of which, might employ the convalescent maniacs, and conduce to the re-establishment of their health. The disturbances which agitated the country in the second and third years of the republic, prevented the accomplishment of my wishes, and I was obliged to content myself with the subsidiary means which had been previously adopted by the governor; that of choosing the servants from among the convalescents. The same method is still continued at the mad-house at Amsterdam.[1] The accomplishment of this scheme would be most effectually obtained by combining with every lunatic asylum, the advantages of an extensive enclosure, to be converted into a sort of farm, which might be cultivated at the expence of the patients, and the profits of which might be devoted to their support. A principal hospital of Spain, presents in this respect an excellent example for our imitation. The maniacs, capable of working, are distributed every morning into separate parties. An overlooker is appointed for each class, who apportions to them all, individually, their respective employments, directs their exertions, and watches over their conduct. The whole day is thus occupied in salutary and refreshing exercises, which are interrupted only by short intervals of rest and relaxation. The fatigues of the day prepare the labourers for sleep and repose during the night. Hence it happens, that those whose condition does not place them above the necessity of submission to toil and labour, are almost always cured; whilst the grandee, who would think himself degraded by any exercises of this description, is generally incurable.

BERNARDINO RAMAZZINI

DISEASES OF POTTERS

[From *On the Diseases of Trades*,* Chapter v]

In almost all cities there are other workers who habitually incur serious maladies from the deadly fumes of metals. Among these are

---

[1] "It is remarkable," says Thouin, "that in a house containing so many residents there should be so few hired servants. I never saw more than four or five permanent domestics there. All the others are taken from among the convalescents, who, impressed by respect for the governor, are eager in the offer of their services to those who stand in need of them. Having themselves experienced similar attentions from their predecessors, they are the more zealous in the fulfilment of this duty. Servants of this description are never wanting, as there are almost as many able convalescents as there are patients who require their assistance. This economical practice is adopted in all the hospitals of Holland. Hence it happens that maniacs are there better treated and at much less expence of officers and servants than in the hospitals of this country."

* De Morbis Artificum, 1700. Translation by Dr. Wilmer Cave Wright. Diseases of Workers, the Latin text of 1713, revised and translated with notes. Chicago, The University of Chicago Press, 1940. Copyright 1940 by the New York Academy of Medicine. By permission of Dr. Wright, The University of Chicago Press, and the New York Academy of Medicine.

the potters. What city or town is there in which men do not follow the potter's craft, the oldest of all the arts? Now when they need roasted or calcined lead for glazing their pots, they grind the lead in marble vessels, and in order to do this they hang a wooden pole from the roof, fasten a square stone to its end, and then turn it round and round. During this process or again when they use tongs to daub the pots with molten lead before putting them into the furnace, their mouths, nostrils, and the whole body take in the lead poison that has been melted and dissolved in water; hence they are soon attacked by grievous maladies. First their hands became palsied, then they become paralytic, splenetic, lethargic, cachectic, and toothless, so that one rarely sees a potter whose face is not cadaverous and the color of lead. In the Copenhagen Transactions there is a case of a potter whose corpse was dissected; his right lung was found to adhere to his ribs; it was on the way to become fibrous and phthisical; the trade that he had followed was held to have caused this diseased condition of the lungs. For he had been trained in the potter's craft and, falling sick, had himself decided that it was unhealthy for him, but not soon enough. Pierre de la Poterie records a case of a potter whose right side became paralyzed, and his vertebrae were so distorted that his neck became rigid; he says that he cured him with a decoction of sassafras wood and bayberries, and he records the case of another potter who died suddenly.

DISEASES OF PAINTERS

[From On the Diseases of Trades, Chapter v]

Painters too are attacked by various ailments such as palsy of the limbs, cachexy, blackened teeth, unhealthy complexions, melancholia, and loss of the sense of smell. It very seldom happens that painters look florid or healthy, though they usually paint the portraits of other people to look handsomer and more florid than they really are. I have observed that nearly all the painters whom I know, both in this and other cities, are sickly; and if one reads the Lives of painters it will be seen that they are by no means long-lived, especially those who were the most distinguished. We read that Raphael of Urbino, the famous painter, was snatched from life in the very flower of his youth; Baldassarre Castiglione lamented his untimely death in an elegant poem. Their sedentary life and melancholic temperament may be partly to blame, for they are almost entirely cut off from intercourse with other men and constantly absorbed in the creations of their imagination. But for their liability to disease there is a more immediate cause, I mean the materials of the colors that they handle and smell constantly, such as red lead, cinnabar, white lead, varnish, nut-oil and linseed oil which they use for mixing colors; and the numerous pigments made of various mineral substances. The odors of varnish and the above-mentioned oils make their workrooms smell like a latrine; this is very bad for the head and perhaps accounts for the loss of the sense of smell. Moreover, painters when at work wear dirty clothes smeared with paint, so that their mouths and noses inevitably breathe tainted air; this penetrates to the seat of the animal spirits, enters by the breathing passages the abode of the blood, disturbs the economy of the natural functions, and excites the disorders mentioned above. We all know that cinnabar is a product of mercury, that *cerussa* is made from lead, verdigris from copper, and ultramarine from silver; for colors derived from metals are far more durable than those from vegetables and hence are in greater demand by painters. In fact, the mineral world supplies the materials of almost every color in use, and this accounts for the really serious ailments that ensue. Painters, then, are inevitably attacked by the same disorders as others who work with metals, though in a milder form.

Fernel illustrates this and records a rather curious case of a painter of Anjou who was seized first with palsy of the fingers and hands, later with spasms in these parts, and the arm too was similarly affected; the disorder next attacked his feet; finally he began to be tormented by pain in the stomach and both hypochondria, so violent that it could not be relieved by clysters, fomentations, baths or any other remedy. When the pain came on, the only thing that gave him any relief was for three or four men to press with their whole weight on his abdomen; this compression of the abdomen lessened the torture. At last, after about three years of this cruel suffering he died consumptive. Fernel says that the most eminent physicians disagreed violently as to the true cause of this terrible disorder, both before and after the autopsy, since nothing abnormal in the viscera came to light. When I read this case history I admired the frank confession of Fernel: "We were all beside the mark and completely off the track." As Celsus says, it is only great men who speak out like this. However, Fernel goes on to say that, since this painter was in the habit of squeezing the color from his brush with his fingers and worse still was imprudent and rash enough to suck it, it is probable that the cinnabar was carried from the fingers to the brain by direct communication and so to the whole nervous system; while that which he took in by the mouth "infected the stomach and intestines with its mysterious and malignant qualities and was the occult cause of those violent pains."

FLORENCE NIGHTINGALE

[From *Notes on Nursing**]

Shall we begin by taking it as a general principle—that all disease, at some period or other of its course, is more or less a reparative process, not necessarily accompanied with suffering: an effort of nature to remedy a process of poisoning or of decay, which has taken place weeks, months, sometimes years beforehand, unnoticed, the termination of the disease being then, while the antecedent process was going on, determined?

If we accept this as a general principle we shall be immediately met with anecdotes and instances to prove the contrary. Just so if we were to take, as a principle—all the climates of the earth are meant to be made habitable for man, by the efforts of man—the objection would be immediately raised—Will the top of Mont Blanc ever be made habitable? Our answer would be, it will be many thousands of years before we have reached the bottom of Mont Blanc in making the earth healthy. Wait till we have reached the bottom before we discuss the top.

In watching disease, both in private houses and in public hospitals, the thing which strikes the experienced observer most forcibly is this, that the symptoms or the sufferings generally considered to be inevitable and incident to the disease are very often not symptoms of the disease at all, but of something quite different—of the want of fresh air, or of light, or of warmth, or of quiet, or of cleanliness, or of punctuality and care in the administration of diet, of each or of all of these. And this quite as much in private as in hospital nursing.

The reparative process which Nature has instituted and which we

* Not dated, but published December, 1859.

call disease has been hindered by some want of knowledge or attention, in one or in all of these things, and pain, suffering, or interruption of the whole process sets in.

If a patient is cold, if a patient is feverish, if a patient is faint, if he is sick after taking food, if he has a bed-sore, it is generally the fault not of the disease, but of the nursing.

I use the word nursing for want of a better. It has been limited to signify little more than the administration of medicines and the application of poultices. It ought to signify the proper use of fresh air, light, warmth, cleanliness, quiet, and the proper selection and administration of diet—all at the least expense of vital power to the patient.

It has been said and written scores of times, that every woman makes a good nurse. I believe, on the contrary, that the very elements of nursing are all but unknown.

By this I do not mean that the nurse is always to blame. Bad sanitary, bad architectural, and bad administrative arrangements often make it impossible to nurse. But the art of nursing ought to include such arrangements as alone make what I understand by nursing, possible.

The art of nursing, as now practised, seems to be expressly constituted to unmake what God had made disease to be, viz., a reparative process.

To recur to the first objection. If we are asked, Is such or such a disease a reparative process? Can such an illness be unaccompanied with suffering? Will any care prevent such a patient from suffering this or that?—I humbly say, I do not know. But when you have done away with all that pain and suffering, which in patients are the symptoms not of their disease, but of the absence of one or all of the above-mentioned essentials to the success of Nature's reparative processes, we shall then know what are the symptoms of and the sufferings inseparable from the disease.

Another and the commonest exclamation which will be instantly made is—Would you do nothing, then, in cholera, fever, etc.?—so deep-rooted and universal is the conviction that to give medicine is to be doing something, or rather everything; to give air, warmth, cleanliness, etc., is to do nothing. The reply is, that in these and many other similar diseases the exact value of particular remedies and modes of treatment is by no means ascertained. while there is universal experience as to the extreme importance of careful nursing in determining the issue of the disease.

The very elements of what constitutes good nursing are as little understood for the well as for the sick. The same laws of health or of nursing, for they are in reality the same, obtain among the well as among the sick. The breaking of them produces only a less violent consequence among the former than among the latter—and this sometimes, not always.

It is constantly objected—"But how can I obtain this medical knowledge? I am not a doctor. I must leave this to doctors."

Oh, mothers of families! You who say this, do you know that one in every seven infants in this civilized land of England perishes before it is one year old? That, in London, two in every five die before they are five years old? And, in the other great cities of England, nearly one out of two? "The life duration of tender babies" (as some Saturn, turned analytical chemist, says) "is the most delicate test" of sanitary conditions. Is all this premature suffering and death necessary? Or did Nature intend mothers to be always accompanied by doctors? Or is it better to learn the piano-forte than to learn the laws which subserve the preservation of offspring?

Macaulay somewhere says, that it is extraordinary that, whereas the laws of the motions of the heavenly bodies, far removed as they are from us, are perfectly well understood, the laws of the human mind, which are under our observation all day and every day, are no better understood than they were two thousand years ago.

But how much more extraordinary is it that, whereas what we might call the coxcombries of education—e.g., the elements of astronomy—are now taught to every school-girl, neither mothers of families of any class, nor school-mistresses of any class, nor nurses of children, nor nurses of hospitals, are taught anything about those laws which God has assigned to the relations of our bodies with the world in which He has put them. In other words, the laws which make these bodies, into which He has put our minds, healthy or unhealthy organs of those minds, are all but unlearnt. Not but that these laws—the laws of life—are in a certain measure understood, but not even mothers think it worth their while to study them—to study how to give their children healthy existences. They call it medical or physiological knowledge, fit only for doctors.

Another objection.

We are constantly told,—"But the circumstances which govern our children's healths are beyond our control. What can we do with winds? There is the east wind. Most people can tell before they get up in the morning whether the wind is in the east."

To this one can answer with more certainty than to the former objections. Who is it who knows when the wind is in the east? Not the Highland drover, certainly, exposed to the east wind, but the young lady who is worn out with the want of exposure to fresh air, to sunlight, etc. Put the latter under as good sanitary circumstances as the former, and she too will not know when the wind is in the east.

1. VENTILATION AND WARMING

The very first canon of nursing, the first and the last thing upon which a nurse's attention must be fixed, the first essential to the patient, without which all the rest you can do for him is as nothing, with which I had almost said you may leave all the rest alone, is this: TO KEEP THE AIR HE BREATHES AS PURE AS THE EXTERNAL AIR, WITHOUT CHILLING HIM. Yet what is so little attended to? Even where it is thought of at all, the most extraordinary misconceptions reign about it. Even in admitting air into the patient's room or ward, few people ever think where that air comes from. It may come from a corridor into which other wards are ventilated, from a hall, always unaired, always full of the fumes of gas, dinner, of various kinds of mustiness; from an underground kitchen, sink, washhouse, water-closet, or even, as I myself have had sorrowful experience, from open sewers loaded with filth; and with this the patient's room or ward is aired, as it is called—poisoned, it should rather be said. Always air from the air without, and that, too, through those windows, through which the air comes freshest. From a closed court, especially if the wind do not blow that way, air may come as stagnant as any from a hall or corridor.

Again, a thing I have often seen both in private houses and institutions. A room remains uninhabited; the fire place is carefully fastened up with a board; the windows are never opened; probably the shutters are kept always shut; perhaps some kind of stores are kept in the room; no breath of fresh air can by possibility enter into that room, nor any ray of sun. The air is as stagnant, musty, and corrupt as it can by possibility be made. It is quite ripe to breed small-pox, scarlet fever, diphtheria, or anything else you please.

Yet the nursery, ward, or sick room adjoining will positively be aired (?) by having the door opened into that room. Or children will be put into that room, without previous preparation, to sleep.

A short time ago a man walked into a back-kitchen in Queen square, and cut the throat of a poor consumptive creature, sitting by the fire. The murderer did not deny the act, but simply said, "It's all right." Of course he was mad.

But in our case, the extraordinary thing is that the victim says, "It's all right," and that we are not mad. Yet, although we "nose" the murderers, in the musty, unaired, unsunned room, the scarlet fever which is behind the door, or the fever and hospital gangrene which are stalking among the crowded beds of a hospital ward, we say, "It's all right."

With a proper supply of windows, and a proper supply of fuel in open fire places, fresh air is comparatively easy to secure when your patient or patients are in bed. Never be afraid of open windows then. People don't catch cold in bed. This is a popular fallacy. With proper bed-clothes and hot bottles, if necessary, you can always keep a patient warm in bed, and well ventilate him at the same time.

But a careless nurse, be her rank and education what it may, will stop up every cranny and keep a hot-house heat when her patient is in bed,—and, if he is able to get up, leave him comparatively unprotected. The time when people take cold (and there are many ways of taking cold, besides a cold in the nose) is when they first get up after the two-fold exhaustion of dressing and of having had the skin relaxed by many hours, perhaps days, in bed, and thereby rendered more incapable of re-action. Then the same temperature which refreshes the patient in bed may destroy the patient just arisen. And common sense will point out that, while purity of air is essential, a temperature must be secured which shall not chill the patient. Otherwise the best that can be expected will be a feverish re-action.

To have the air within as pure as the air without, it is not necessary, as often appears to be thought, to make it as cold.

In the afternoon again, without care, the patient whose vital powers have then risen often finds the room as close and oppressive as he found it cold in the morning. Yet the nurse will be terrified, if a window is opened.

I know an intelligent humane house surgeon who makes a practice of keeping the ward windows open. The physicians and surgeons invariably close them while going their rounds; and the house surgeon very properly as invariably opens them whenever the doctors have turned their backs.

In a little book on nursing, published a short time ago, we are told, that "with the proper care it is very seldom that the windows cannot be opened for a few minutes twice in the day to admit fresh air from without." I should think not; nor twice in the hour either. It only shows how little the subject has been considered.

Of all methods of keeping patients warm the very worst certainly is to depend for heat on the breath and bodies of the sick. I have known a medical officer keep his ward windows hermetically closed, thus exposing the sick to all the dangers of an infected atmosphere, because he was afraid that, by admitting fresh air, the temperature of the ward would be too much lowered. This is a destructive fallacy.

To attempt to keep a ward warm at the expense of making the sick repeatedly breathe their own hot, humid, putrescing atmosphere is a certain way to delay recovery or to destroy life.

Do you ever go into the bed-rooms of any persons of any class, whether they contain one, two, or twenty people, whether they hold sick or well, at night, or before the windows are opened in the morning, and ever find the air anything but unwholesomely close and foul? And why should it be so? And of how much importance it is that it should not be so? During sleep, the human body, even when in health, is far more injured by the influence of foul air than when awake. Why can't you keep the air all night, then, as pure as the air without in the rooms you sleep in? But for this, you must have sufficient outlet for the impure air you make yourselves to go out; sufficient inlet for the pure air from without to come in. You must have open chimneys, open windows, or ventilators; no close curtains round your beds; no shutters or curtains to your windows, none of the contrivances by which you undermine your own health or destroy the chances of recovery of your sick.

A careful nurse will keep a constant watch over her sick, especially weak, protracted, and collapsed cases, to guard against the effects of the loss of vital heat by the patient himself. In certain diseased states much less heat is produced than in health; and there is a constant tendency to the decline and ultimate extinction of the vital powers by the call made upon them to sustain the heat of the body. Cases where this occurs should be watched with the greatest care from hour to hour, I had almost said from minute to minute. The feet and legs should be examined by the hand from time to time, and whenever a tendency to chilling is discovered, hot bottles, hot bricks, or warm flannels, with some warm drink, should be made use of until the temperature is restored. The fire should be, if necessary, replenished. Patients are frequently lost in the latter stages of disease from want of attention to such simple precautions. The nurse may be trusting to the patient's diet, or to his medicine, or to the occasional dose of stimulant which she is directed to give him, while the patient is all the while sinking from want of a little external warmth. Such cases happen at all times, even during the height of summer. This fatal chill is most apt to occur towards early morning at the period of the lowest temperature of the twenty-four hours, and at the time when the effect of the preceding day's diets is exhausted.

Generally speaking, you may expect that weak patients will suffer cold much more in the morning than in the evening. The vital powers are much lower. If they are feverish at night, with burning hands and feet, they are almost sure to be chilly and shivering in the morning. But nurses are very fond of heating the foot-warmer at night, and of neglecting it in the morning, when they are busy. I should reverse the matter.

All these things require common sense and care. Yet perhaps in no one single thing is so little common sense shown, in all ranks, as in nursing.

The extraordinary confusion between cold and ventilation, in the minds of even well educated people, illustrates this. To make a room cold is by no means necessarily to ventilate it. Nor is it at all necessary, in order to ventilate a room, to chill it. Yet, if a nurse finds a room close, she will let out the fire, thereby making it closer, or she will open the door into a cold room, without a fire, or an open window in it, by way of improving the ventilation. The safest atmosphere of all for a patient is a good fire and an open window, excepting in extremes of temperature. (Yet no nurse can ever be made to understand this.) To ventilate a small room without draughts of course requires more care than to ventilate a large one.

Another extraordinary fallacy is the dread of night air. What air can we breathe at night but night air? The choice is between pure night air from without and foul night air from within. Most people prefer the latter. An unaccountable choice. What will they say if it is proved to be true that fully one-half of all the diseases we suffer from is occasioned by people sleeping with their windows shut? An open window most nights in the year can never hurt any one. This is not to say that light is not necessary for recovery. In great cities, night air is often the best and purest air to be had in the twenty-four hours. I could better understand in towns shutting the windows during the day than during the night, for the sake of the sick. The absence of smoke, the quiet, all tend to making night the best time for airing the patients. One of our highest medical authorities on Consumption and Climate has told me that the air in London is never so good as after ten o'clock at night.

Always air your room, then, from the outside air, if possible. Windows are made to open; doors are made to shut—a truth which seems extremely difficult of apprehension. I have seen a careful nurse airing her patient's room through the door, near to which were two gaslights (each of which consumes as much air as eleven men), a kitchen, a corridor, the composition of the atmosphere in which consisted of gas, paint, foul air, never changed, full of effluvia, including a current of sewer air from an ill-placed sink, ascending in a continual stream by a well-staircase, and discharging themselves constantly into the patient's room. The window of the said room, if opened, was all that was desirable to air it. Every room must be aired from without—every passage from without. But the fewer passages there are in a hospital the better.

If we are to preserve the air within as pure as the air without, it is needless to say that the chimney must not smoke. Almost all smoky chimneys can be cured—from the bottom, not from the top. Often it is only necessary to have an inlet for air to supply the fire, which is feeding itself, for want of this, from its own chimney. On the other hand, almost all chimneys can be made to smoke by a careless nurse, who lets the fire get low and then overwhelms it with coal; not, as we verily believe, in order to spare herself trouble (for very rare is unkindness to the sick), but from not thinking what she is about.

In laying down the principle that the first object of the nurse must be to keep the air breathed by her patient as pure as the air without, it must not be forgotten that everything in the room which can give off effluvia, besides the patient, evaporates itself into his air. And it follows that there ought to be nothing in the room, excepting him, which can give off effluvia or moisture. Out of all damp towels, etc., which become dry in the room, the damp, of course, goes into the patient's air. Yet this "of course" seems as little thought of, as if it were an obsolete fiction. How very seldom you see a nurse who acknowledges by her practice that nothing at all ought to be aired in the patient's room, that nothing at all ought to be cooked at the patient's fire! Indeed the arrangements often make this rule impossible to observe.

If the nurse be a very careful one, she will, when the patient leaves his bed, but not his room, open the sheets wide, and throw the bed clothes back, in order to air his bed. And she will spread the wet towels or flannels carefully out upon a horse, in order to dry them. Now either these bed-clothes and towels are not dried and aired, or they dry and air themselves into the patient's air. And whether the damp and effluvia do him most harm in his air or in his bed, I leave to you to determine, for I cannot.

Even in health people cannot repeatedly breathe air in which they live with impunity, on account of its becoming charged with unwholesome matter from the lungs and skin. In disease where everything given off from the body is highly noxious and dangerous, not only must there be plenty of ventilation to carry off the effluvia, but everything which the patient passes must be instantly removed away, as being more noxious than even the emanations from the sick.

Of the fatal effects of the effluvia from the excreta it would seem unnecessary to speak, were they not so constantly neglected. Concealing the utensils behind the valance to the bed seems all the precaution which is thought necessary for safety in private nursing. Did you but think for one moment of the atmosphere under that bed, the saturation of the under side of the mattress with the warm evaporations, you would be startled and frightened too!

The use of any chamber utensil *without a lid* should be utterly abolished, whether among sick or well. You can easily convince yourself of the necessity of this absolute rule, by taking one with a lid, and examining the under side of that lid. It will be found always covered, whenever the utensil is not empty, by condensed offensive moisture. Where does that go, when there is no lid?

Earthenware, or if there is any wood, highly polished and varnished wood, are the only materials fit for patients' utensils. The very lid of the old abominable close-stool is enough to breed a pestilence. It becomes saturated with offensive matter, which scouring is only wanted to bring out. I prefer an earthenware lid as being always cleaner. But there are various good new-fashioned arrangements.

A slop pail should never be brought into a sick room. It should be a rule invariable, rather more important in the private house than elsewhere, that the utensil should be carried directly to the water-closet, emptied there, rinsed there, and brought back. There should always be water and a cock in every water-closet for rinsing. But even if there is not, you must carry water there to rinse with. I have actually seen, in the private sick room, the utensils emptied into the foot-pan, and put back unrinsed under the bed. I can hardly say which is most abominable, whether to do this or to rinse the utensil *in* the sick room. In the best hospitals it is now a rule that no slop pail shall ever be brought into the wards, but that the utensils shall be carried direct to be emptied and rinsed at the proper place. I would it were so in the private house.

Let no one ever depend upon fumigations, "disinfectants," and the like, for purifying the air. The offensive thing, not its smell, must be removed. A celebrated medical lecturer began one day "Fumigations, gentlemen, are of essential importance. They make such an abominable smell that they compel you to open the window." I wish all the disinfecting fluids invented made such an "abominable smell" that they forced you to admit fresh air. That would be a useful invention.

XXXV

PREVENTIVE MEDICINE

PREVENTION OF DISEASE IMPLIES AN UNDERSTANDING OF THE CAUSE OF disease. To this end empiric knowledge may prove as useful as specific. Thus Lind found that fresh fruits prevented scurvy, although he had no conception of vitamin C, and Snow showed that contaminated water caused cholera although he had no knowledge of the vibrio.

Scurvy was a dangerous malady in the sixteenth and seventeenth centuries: its ravages had very serious economic consequences. The crews of vessels on long voyages were universally affected with it, and there were many deaths. Since this was the great period of exploration and the beginning of world wide sea trading the effects were matters of public moment. Jacques Cartier, as reported in Hakluyt's Voyages (1600), had cured an epidemic among his crew on his voyage to the St. Lawrence by the use of a native remedy, the juice from the leaves and bark of the Ameda tree (sassafras or spruce).

Cartier's account came to the attention of James Lind (1716-1794), of Edinburgh, who had become familiar with scurvy during his service as surgeon with the British Navy. In his *Treatise on the Scurvy* (1753), Lind recommended the use of lemon juice for the prevention of scurvy, but it was not until years later that his recommendation was officially adopted by the British Navy.

John Snow (1813-1858), a Yorkshireman, was the pioneer in the development of the idea that diseases could be water borne. The story of the Broad Street epidemic of cholera is one of the romantic tales of epidemiology. Snow noted that in an epidemic that occurred, in 1854, in St. James parish the victims were those who depended on the Broad Street pump for drinking water. He showed that sewage from the houses where cholera cases dwelt seeped into the well, and he suggested that the vestrymen of the parish take the pump off the Broad Street well in order to stop the epidemic.

Sir Edward Chadwick (1800-1890) worked for poor law reform and the health of the laboring classes. William Farr (1807-1883) laid the foundation of all public health administration by his work on *Vital Statistics* (1885). The interested reader should consult the authors directly as their writings cannot be adequately abstracted.

To Theobald Smith (1859-1934), of Harvard University, belongs the credit for originating the idea of the insect transmission of infectious disease.

Walter Reed (1851-1902) of the United States Army Medical Corps, with his associates Carroll, Agramonte, and Lazear, demonstrated the practical value of this knowledge of insect-borne diseases when he proved that yellow fever was transmitted by the Stegomyia mosquito, and by simple means protected the army in Cuba from contracting the disease (1902). Reed's success in preventing yellow fever made possible the completion of the Panama Canal under the sanitary direction of William C. Gorgas (1854-1920).

REFERENCES

NEWSHOLME, Sir A. Evolution of Preventive Medicine. Baltimore, Williams & Wilkins, 1927.
BAUMGARTNER, L. John Howard (1726-1790) Hospital and Prison Reformer: A Bibliography. Baltimore, Johns Hopkins Press, 1939.
RICHARDSON, W. Snow on Cholera, Being a Reprint of Two Papers by John Snow Together With a Biographical Memoir. New York, Commonwealth Fund, 1936.
SMITH, T. Medical Classics, Vol. 1, No. 5, January, 1937.

JAMES LIND

[From A Treatise of the Scurvy *]

I come in the next place, to an additional and extremely powerful cause, observed at sea to occasion this disease, and which concurring with the former, in progress of time, seldom fails to breed it. And this is, the want of fresh vegetables and greens; either, as may be supposed, to counteract the bad effects of their before mentioned situation; or rather, and more truly, to correct the quality of such hard and dry food as they are obliged to make use of. Experience indeed sufficiently shews, that as greens or fresh vegetables, with ripe fruits, are the best remedies for it, so they prove the most effectual preservatives against it. And the difficulty of obtaining them at sea, together with a long continuance in the moist sea-air, are the true causes of its so general and fatal malignity upon that element.

The diet which people are necessarily obliged to live upon while at sea, was before assigned as the occasional cause of the disease; as in a particular manner it determines the effects of the before mentioned predisposing causes to the production of it. And there will be no difficulty to conceive the propriety of this distinction, or understand how the most innocent and wholesome food, at times,

* Edinburgh, 1753.

and in peculiar situations, will with great certainty form a disease. Thus, if a man lives on a very slender diet, and drinks water, in the fens of *Lincolnshire*, he will almost infallibly fall into an ague.

All rules and precepts of diet, as well as the distinction of ailments into wholesome and unwholesome, are to be understood only as relative to the constitution or state of the body. We find a child and a grown person, a valetudinarian and a man of health, require aliment of different kinds; as does even the same person in the heat of summer and in the depth of winter, during a dry or rainy season. Betwixt the tropics, the natives live chiefly on fruits, seeds and vegetables; whereas northern nations find a flesh and solid diet more suitable to their climate. In like manner it appears, I think, very plainly, that such hard dry food as ship's provisions, or the sea-diet, is extremely wholesome: and that no better nourishment could be well contrived for labouring people, or any person in perfect health, using proper exercise in a dry pure air; and that, in such circumstances, seamen will live upon it for several years, without any inconvenience. But where the constitution is predisposed to the scorbutic taint, by the causes before assigned (the effect of which, as shall be shown in a proper place, are a weakening of the animal powers of digestion), the influence of such diet in bringing on this disease, sooner or later, according to the state and constitution of the body, becomes extremely visible. . . .

The first indication of the approach of this disease, is generally a change of colour in the face, from the natural and usual look, to a pale and bloated complexion; with a listlessness to action, or an aversion to any sort of exercise. When we examine narrowly the lips, or the caruncles of the eye, where the blood-vessels lie most exposed, they appear of a greenish cast. Meanwhile the person eats and drinks heartily, and seems in perfect health; except that his countenance and lazy inactive disposition, portend a future scurvy.

This change of colour in the face, although it does not always precede the other symptoms, yet constantly attends them when advanced. Scorbutic people for the most appear at first of a pale or yellowish hue, which becomes afterwards more darkish or livid.

Their former aversion to motion degenerates soon into an universal lassitude, with a stiffness and feebleness of their knees upon using exercise; with which they are apt to be much fatigued, and upon that occasion subject to a breathlessness or panting. And this lassitude, with a breathlessness upon motion, are observed to be among the most common concomitants of the distemper.

Their gums soon after become itchy, swell, and are apt to bleed upon the gentlest friction. Their breath is then offensive; and upon looking into their mouth, the gums appear of an unusual livid redness, are soft and spongy, and become afterwards extremely putrid and fungous; the pathognomonic sign of the disease. They are subject not only to a bleeding from the gums, but prone to fall into hemorrhages from other parts of the body.

Their skin at this time feels dry, as it does through the whole course of the malady. In many, especially if feverish, it is extremely rough; in some it has an anserine appearance; but most frequently it is smooth and shining. And, when examined, it is found covered with several reddish, bluish, or rather black and livid spots, equal with the surface of the skin, resembling an extravasation under it, as it were from a bruise. These spots are of different sizes, from the bigness of a lentil to that of a handbreadth, and larger. But the last are more uncommon in the beginning of the distemper; they being usually then but small, and of an irregular roundish figure. They are to be seen chiefly on the legs and thighs; often on the arms and breasts, and trunks of the body; but more rarely on the head and face.

Many have a swelling of their legs; which is first observed on their ankles towards the evening, and hardly to be seen next morning; but after continuing a short time in this manner, it gradually advances up the leg, and the whole member becomes oedematous; with this difference only in some, that it does not so easily yield to the finger, and preserves the impression of it longer afterwards than a true oedema.

Let the squeezed juice of these fruits [lemons and oranges] be well cleared from the pulp, and depurated by standing for some time; then poured off from the gross sediment: or, to have it still purer, it may be filtered. Let it then be put into any clean open earthen vessel, well glazed; which should be wider at the top than bottom, so that there may be the largest surface above to favour the evaporation. For this purpose a china basin or punch-bowl is proper; or a common earthen basin used for washing, if well glazed, will be sufficient, as it is generally made in the form required. Into this pour the purified juice; and put it into a pan of water, upon a clear fire. Let the water come almost to a boil, and continue nearly in a state of boiling (with the basin containing the juice in the middle of it) for several hours, until the juice is found to be the consistence of oil when warm, or of a syrup when cold. It is then to be corked up in a bottle for use. Two dozen of good oranges, weighing five pounds four ounces, will yield one pound nine ounces and a half of depurated juice; and when evaporated, there will remain about five ounces of the extract; which in bulk will be equal to less than three ounces of water. So that thus the acid, and virtues of twelve dozen of lemons or oranges, may be put into a quart bottle, and preserved for several years.

I have some of the extract of lemons now by me, which was made four years ago. And when this is mixed with water, or made into punch, few are able to distinguish it from the fresh squeezed juice mixed up in like manner; except when both are present, and their different tastes compared at the same time; when the fresh fruits discover a greater degree of smartness and fragrancy.

JOHN SNOW

CHOLERA

[From *On the Mode of Communication of Cholera**]

The most terrible outbreak of cholera which ever occurred in this kingdom, is probably that which took place in Broad Street, Golden Square, and the ajoining streets, a few weeks ago. Within two hundred and fifty yards of the spot where Cambridge Street joins Broad Street, there were upwards of five hundred fatal attacks of cholera in ten days. The mortality in this limited area probably equals any that was ever caused in this country, even by the plague; and it was much more sudden, as the greater number of cases terminated in a few hours. The mortality would undoubtedly have been much greater had it not been for the flight of the population. Persons in furnished lodgings left first, then other lodgers went away, leaving their furniture to be sent for when they could meet with a place to put it in. Many houses were closed altogether, owing to the death of the proprietors; and, in a great number of instances, the tradesmen who remained had sent away their families: so that in less than six days from the commencement of the outbreak, the

*First published in 1849; second edition (excerpted here) appeared in 1855.

most afflicted streets were deserted by more than three-quarters of their inhabitants.

There were a few cases of cholera in the neighbourhood of Broad Street, Golden Square, in the latter part of August; and the so-called outbreak, which commenced in the night between the 31st August and the 1st September, was, as in all similar instances only a violent increase of the malady. As soon as I became acquainted with the situation and extent of this irruption of cholera, I suspected some contamination of the water of the much-frequented street-pump in Broad Street, near the end of evening of the 3rd September, I found so little impurity in it of an organic nature, that I hesitated to come to a conclusion. Further inquiry, however, showed me that there was no other circumstance or agent common to the circumscribed locality in which this sudden increase of cholera occurred, and not extending beyond it, except the water of the above mentioned pump. I found, moreover, that the water varied, during the next two days, in the amount of organic impurity, visible to the naked eye, on close inspection, in the form of small white, flocculent particles; and I concluded that, at the commencement of the outbreak, it might possibly have been still more impure. I requested permission, therefore, to take a list, at the General Register Office, of the deaths from cholera, registered during the week ending 2nd September, in the sub-districts of Golden Square, Berwick Street, and St. Ann's, Soho, which was kindly granted. Eighty-nine deaths from cholera were registered, during the week, in the three sub-districts. Of these, only six occurred in the four first days of the week; four occurred on Thursday, the 31st August; and the remaining seventy-nine on Friday and Saturday. I considered, therefore, that the outbreak commenced on the Thursday; and I made inquiry, in detail, respecting the eighty-three deaths registered as having taken place during the last three days of the week.

On proceeding to the spot, I found that nearly all the deaths had taken place within a short distance of the pump. There were only ten deaths in houses situated decidedly nearer to another street pump. In five of these cases the families of the deceased persons informed me that they always sent to the pump in Broad Street, as they preferred the water to that of the pump which was nearer. In three other cases, the deceased were children who went to school near the pump in Broad Street. Two of them were known to drink the water; and the parents of the third think it probable that it did so. The other two deaths, beyond the district which this pump supplied, represent only the amount of mortality from cholera that was occurring before the irruption took place.

With regard to the deaths occurring in the locality belonging to the pump, there were sixty-one instances in which I was informed that the deceased persons used to drink the pump-water from Broad Street, either constantly or occasionally. In six instances I could get no information, owing to the death or departure of every one connected with the deceased individuals; and in six cases I was informed that the deceased persons did not drink the pump-water before their illness.

The result of the inquiry then was, that there had been no particular outbreak or increase of cholera, in this part of London, except among the persons who were in the habit of drinking the water of the above-mentioned pump-well.

I had an interview with the board of Guardians of St. James's parish, on the evening of Thursday, 7th September, and represented the above circumstances to them. In consequence of what I said, the handle of the pump was removed on the following day.

Besides the eighty-three deaths mentioned above as occurring on the three last days of the week ending September 2nd, and being registered during that week in the sub-districts in which the attacks occurred, a number of persons died in Middlesex and other hospitals, and a great number of deaths which took place in the locality during the last two days of the week, were not registered till the week following. The deaths altogether, on the 1st and 2nd of September, which have been ascertained to belong to this outbreak of cholera, were one hundred and ninety-seven; and many persons who were attacked about the same time as these, died afterwards. I should have been glad to inquire respecting the use of the water from Broad Street pump in all these instances, but was engaged at the time in an inquiry in the south districts of London, which will be alluded to afterwards; and when I began to make fresh inquiries in the neighbourhood of Golden Square, after two or three weeks had elapsed, I found that there had been such a distribution of the remaining population that it would be impossible to arrive at a complete account of the circumstances. There is no reason to suppose, however, that a more extended inquiry would have yielded a different result from that which was obtained respecting the eighty-three deaths which happened to be registered within the district of the outbreak before the end of the week in which it occurred.

The additional facts that I have been able to ascertain are in accordance with those above related; and as regards the small number of those attacked, who were believed not to have drank the water from Broad Street pump, it must be obvious that there are various ways in which the deceased persons may have taken it without the knowledge of their friends. The water was used for mixing with spirits in all the public houses around. It was used likewise at dining-rooms and coffee-shops. The keeper of a coffee-shop in the neighbourhood, which was frequented by mechanics, and where the pump-water was supplied at dinner time, informed me (on 6th September) that she was already aware of nine of her customers who were dead. The pump-water was also sold in various little shops, with a teaspoonful of effervescing powder in it, under the name of sherbet; and it may have been distributed in various other ways with which I am unacquainted. The pump was frequented much more than is usual, even for a London pump in a populous neighbourhood.

There are certain circumstances bearing on the subject of this outbreak of cholera which require to be mentioned. The Workhouse in Poland Street is more than three-fourths surrounded by houses in which deaths from cholera occurred, yet out of five hundred and thirty-five inmates only five died of cholera, the other deaths which took place being those of persons admitted after they were attacked. The workhouse has a pump-well on the premises, in addition to the supply from the Grand Junction Water Works, and the inmates never sent to Broad Street for water. If the mortality in the workhouse had been equal to that in the streets immediately surrounding it on three sides, upwards of one hundred persons would have died.

There is a Brewery in Broad Street, near to the pump, and on perceiving that no brewer's men were registered as having died of cholera, I called on Mr. Huggins, the proprietor. He informed me that there were above seventy workmen employed in the brewery, and that none of them had suffered from cholera,—at least in a severe form,—only two having been indisposed, and that not seriously, at the time the disease prevailed. The men are allowed a certain quantity of malt liquor, and Mr. Huggins believes they do not drink at all; and he is quite certain that the workmen never obtained water from the pump in the street. There is a deep well in the brewery, in addition to the New River water.

At the percussion-cap manufactory, 37 Broad Street, where, I understand, about two hundred workpeople were employed, two tubs were kept on the premises always supplied with water from the pump in the street, for those to drink who wished; and eighteen of these workpeople died of cholera at their own homes, sixteen men and two men.

Mr. Marshall, surgeon, of Greek Street, was kind enough to inquire respecting seven workmen who had been employed in the manufactory of dentists' materials, at Nos. 8 and 9 Broad Street, and who died at their homes. He learned that they were all in the habit of drinking water from the pump, generally drinking about half-a-pint once or twice a day; while two persons who reside contantly on the premises, but do not drink the pump-water, only had diarrhoea. Mr. Marshall also informed me of the case of an officer in the army, who lived at St. John's Wood, but came to dine in Wardour Street, where he drank the water from Broad Street pump at his dinner. He was attacked with cholera, and died in a few hours.

I am indebted to Mr. Marshall for the following cases, which are interesting as showing the period of incubation, which in these three cases was from thirty-six to forty-eight hours. Mrs. —, of 13 Bentinck Street, Berwick Street, aged 28, in the eighth month of pregnancy, went herself (although they were not usually water drinkers), on Sunday, 3rd September, to Broad Street pump for water. The family removed to Gravesend on the following day; and she was attacked with cholera on Tuesday morning at seven o'clock, and died of consecutive fever on 15th September, having been delivered. Two of her children drank also of the water, and were attacked on the same day as the mother, but recovered.

Dr. Fraser, of Oakley Square, kindly informed me of the following circumstance. A gentleman in delicate health was sent for from Brighton to see his brother at 6 Poland Street, who was attacked with cholera and died in twelve hours, on 1st September. The gentleman arrived after his brother's death, and did not see the body. He only stayed about twenty minutes in the house, where he took a hasty and scanty luncheon of rumpsteak, taking with it a small tumbler of brandy and water, the water being from Broad Street pump. He went to Pentonville, and was attacked with cholera on the evening of the following day, 2nd September, and died the next evening.

THEOBALD SMITH AND F. L. KILBORNE

ON INSECT TRANSMISSION OF DISEASE

[From *Investigations into the Nature, Causation and Prevention of Southern Cattle Fever**]

PRACTICAL OBSERVATIONS AND CONCLUSIONS

It will undoubtedly be conceded by all impartial readers of the foregoing pages that the economic value of the results derived from these investigations is very promising. As yet they are undeveloped, however, and their true importance can not be estimated. Experiments must be built upon them in various directions. These we have thus far been unable to undertake, owing to the large amount of labor involved in determining the relation of ticks to the disease. In the following pages, in addition to deductions immediately available in the control of this disease, a few suggestions are made in regard to the objects to be attained by further investigations and the manner in which they should be conducted. Those readers technically interested in carrying on such investigations will undoubtedly have read between the lines of the foregoing chapters all that can be suggested here.

DIAGNOSIS

One of the immediate results of the work is the simplicity and ease with which an outbreak of Texas fever can be positively determined. Most veterinarians and pathologists are able to recognize Texas fever when an acute case presents itself for postmortem examination. The greatly enlarged spleen, the peculiar coloration of the liver, the thick bile, and especially the haemoglobinuria, are so obvious that no one trained to a knowledge of the appearance of the healthy organs and excretions in cattle can make a mistake. But all cases are not in the acute stage at the time of death, and one or several of these important pathological changes may be missing or

* Eighth and Ninth Annual Reports of the Bureau of Animal Industry, 1893.

barely recognizable when present. In fact, there may be no animals which can be sacrificed, and all may be on the road to recovery. In such cases even the clinical signs, such as the high temperature, may be missing.

Among the diagnostic characters to be added to the list are the examination of the blood and the presence or absence of the cattle tick (*Boöphilus bovis*). We may now consider it demonstrated that Texas . . . fever outbreaks in the North are not possible without the cattle tick. Isolated cases may occur through other agencies, perhaps, but no general infection of fields or pastures is possible without the cattle tick. Hence, in any doubtful disease where Texas fever is suspected, ticks should be looked for, and in doing so all those facts concerning the size of the ticks on animals in the acute stage and during recovery and their location on the body must be borne in mind. On animals which have passed through the disease the ticks are nearly or quite full-grown, and therefore easily detected. But even when great care is exercised the ticks may be overlooked, or in a late fall infection they may have speedily disappeared. In such cases the examination of the blood will give the necessary information. This requires some skill, and a good miscroscope with objectives and oculars giving a magnification of not less than 500 diameters is necessary. The method of examination as well as the pitfalls to be avoided in interpreting appearances under the microscope have been discussed at length, and need not be again referred to here. While the presence of the microparasite within the red blood corpuscles, and the changed size and appearance of many of the corpuscles themselves, are usually of sufficient diagnostic value, it is always desirable that the number of red corpuscles be estimated at the same time.

In the miscroscopic examination of the blood attention should be paid, first of all, to the presence of the various stages of the microparasite. In the mild type, the minute coccus-like body will be found within the corpuscle, near its periphery. As it is rarely seen in fresh preparations, stained preparations should invariably be examined. In the acute type of midsummer, associated with high fever, the larger, paired, pyriform bodies are always present but usually in very small numbers. They may be detected as readily in fresh blood carefully mounted as in died and stained preparations. Next in importance to the microparasite of the disease are the changes induced in the blood corpuscles by the anaemia. In fresh blood the variation in size of the individual corpuscles and the very large size of many (from one and one-half to one and three-quarters times the diameter of the normal red corpuscles) is at once apparent. In properly stained preparations the peculiar granulations and the diffusely stained appearance of a greater or smaller number of the larger corpuscles, is quite characteristic. These changes may, of course, be the result of very severe, repeated hemorrhages, and these must be excluded first before the former can be considered as due to Texas fever. The changes in the blood corpuscles may be directly associated with the parasite in the mild type, but they usually follow the parasite in the acute type. Hence they may be the only indication of disease recognizable under the microscope in some cases.

A reduction in the number of corpuscles is a very reliable sign of Texas fever. If we except the occurrence of severe hemorrhages and the feeding of chemical poisons, their number is but slightly, if at all, influenced by disease of various kinds. In several cases of advanced tuberculosis no reduction was noticed. In fact, there seems to be but little specific action of bacterial poisons on the red corpuscles, while the Texas-fever microbe limits its destructive action entirely to them. Anaemia in cattle seems to be rare, as we found it but once among the many cases under observation. Hence the counting apparatus is of great service in detecting Texas fever in all its phases, and should be used whenever possible.

A summary of the diagnostic characters to be looked for when this . . . disease is suspected would include among others the following salient ones:

(1) Cattle ticks.
(2) Gross pathological changes: Haemoglobinuria; enlarged spleen; enlarged, yellowish liver; thick, flaky bile; ecchymoses on the external and internal surfaces of the heart.
(3) The microparasite within the red corpuscles.
(4) Modified or changed corpuscles (enlargement, the presence of stainable granules, etc.).
(5) The reduction in the number of red corpuscles.

PREVENTION

Texas fever in the territory outside of the enzoötic region is the result of the distribution of ripe egg-laying ticks by cattle from the enzoötic region. Hence such cattle should not be allowed on uninfected territory during the warmer half of the year. It is also evident that all cars carrying Southern cattle contain a larger or smaller number of ticks which have dropped off during the journey, and which are ready to lay their eggs. The sweepings of such cars, wherever deposited, may give rise to a crop of young ticks, and these, when they have access to cattle, will produce the disease. Wherever Southern tick-bearing cattle are kept within twenty-five to thirty days after their departure from their native fields they are liable to infect such places, since it requires the period mentioned for the smaller ticks to ripen and drop off. But under special conditions even this period is too short and the Southern cattle may remain dangerous a longer time. This would occur when such cattle remain in any one inclosure long enough (four to five weeks) for the progeny of the first ticks which drop off to appear on the same cattle.

The above points are covered in the regulations of the Department of Agriculture concerning cattle transportation. These regulations insist on the complete isolation of cattle coming from the permanently infected territory between March 1 and December 1 of each year, and on the proper disinfection of the litter and manure from such cattle during transportation. Furthermore, such cattle can only be transported into uninfected territory for immediate slaughter during the prescribed period. These regulations, if properly carried out, would prevent the appearance of Texas fever at any time in those areas north of the enzoötic territory. The only question which now presents itself with reference to them is the efficiency of the prescribed disinfection. It has been shown that the infection resides only in the cattle ticks and their eggs; hence the destruction of these is absolutely eessential to make the disinfection of any value. In the present report this question has not been touched upon; therefore, pending the trial of various disinfectants, which is now going on, any discussion or any suggestions are of little value.

The harmlessness of Southern cattle after being deprived of the cattle tick brings up the very important question whether such cattle can not by some means be freed from ticks so that their transportation may go on without any restriction during the entire year. There are several ways in which experiments might be undertaken. Cattle might be subjected to disinfecting washes of various kinds, or else they might be run through disinfecting baths which expose the whole body to the action of the liquid used. Such processes would require careful attention. . . . The survival of a very few ticks might lead to serious consequences, since a ripe tick averages about 2,000 eggs.

Cattle may be deprived of ticks on a large scale without the use of any disinfection if the following plan be adopted: Two large fields in a territory naturally free from cattle ticks are inclosed. The tick-bearing cattle are put into the first inclosure and kept there about fifteen days. They are then transferred to the second inclosure for the same length of time. Thirty days after the beginning of their confinement they may be considered free from infection. The reason for this procedure is simple enough. The cattle drop the ticks as they ripen in the inclosures. By being transferred to a second (or even a third) inclosure they are removed from the possible danger of a reinfection by the progeny of the ticks which dropped off first. It is evident that such inclosures can only be used once a season, since the young ticks subsequently hatched remain alive on the ground for an indefinite length of time. Such inclosures must not be located where there is a possibility that the ticks might survive the winter.

For cattle which are introduced into the enzoötic territory two modes of prevention may be adopted. Either they are kept entirely free from ticks by confinement in stables or upon pastures known to be free from ticks, or else they are exposed to the infection in such a way as to become insusceptible to it after a time. The first method is open to the objection that ticks may at some time accidentally gain access to such cattle and produce a fatal disease. The second method seems the more rational, provided it can be successfully carried out. We know that Southern cattle are insusceptible to the disease, and the way in which this insusceptibility has been acquired has been already discussed. . . . Young animals seem to be largely proof against a fatal infection, although they are by no means insusceptible. The repeated mild attacks to which they are subjected finally make the system indifferent to the virus. The introduction of young animals into the permanently infected territory though not without danger, is far safer than the introduction of animals older than one year. The danger of a fatal infection increases with the age of the animal, and is very great in cows over 5 or 6 years old, as is distinctly shown by the experiments recorded in this report.

The subject of preventive inoculation has already been discussed and experiments cited on another page. It has been shown that while in general two mild attacks may not prevent a third attack, this will not be fatal. One very acute attack will usually prevent a second severe attack. Hence it is possible to prevent cattle, even when fairly along in years, from succumbing to a fatal attack by several preliminary carefully guarded exposures to a mild infection. This infection may be produced by scattering ripe ticks in an inclosure, or by placing young ticks on cattle in the fall of the year. . . . Protective inoculation of this kind should be carried on at some locality outside of the enzoötic territory carefully chosen for the purpose. A few years of careful experimentation would probably lead to an efficient method which, when definitely formulated in all its details, could be applied in different parts of the country. Such experimentation should, of course, pay special attention to the relative susceptibility of the various higher grades of cattle, a matter which we have been unable to touch upon thus far.

What can the individual farmer or stock-owner do in the event that Texas fever has been introduced into his pastures? From what has been said thus far pastures which have been infected by Southern cattle . . . or ticks from the litter and manure of infected cattle cars should be avoided during the entire summer season. While we know that young ticks may remain alive in jars for two or three months without food, it would be premature to conclude that such is the case on pastures, as the conditions are quite different. Yet everything seems to point to a long sojourn of young ticks on infected fields, and pending the carrying out of experiments to test this question we would recommend that native cattle be not allowed to graze on infected fields until after the first frosts, for even a mild attack in fall before the ticks have been destroyed by frosts is debilitating to cattle. The period of time during which infected localities remain dangerous varies, of course, with the latitude, and would be shorter the colder the climate.

The infection of stables, stalls, and other structures with the ticks should be counteracted by thorough disinfection. The adult ticks and the eggs must be destroyed. As stated above, we know as yet very little concerning the agents which will destroy the vitality of the eggs of ticks, but the use of water near the boiling point may be sufficient, if liberally applied, to destroy the life of the embryos. In the case of litter and manure heaps the thorough saturation with some strong mineral acid in dilution may accomplish the purpose. Ordinary lime, slaked or unslaked, densely sprinkled over infected places, so as to form a continuous layer, may be recommended. The slow incrustation of the egg massed with carbonate of lime may be expected, provided the manure is under cover. Otherwise it will be washed away and may leave the eggs unharmed. In regions outside of the enzoötic territory the absence of ticks may be accounted for by the severity of the winter; hence in unprotected localities disinfection is unnecessary after the winter has set in. But it may occur that in sheltered places the eggs will winter over and the ticks reappear the following spring. Whether such ticks are likely to produce any serious trouble in the absence of Southern cattle we are unable to state definitely. All that we know is that disease may break out when Southern cattle of the preceding year are in the pasture, as was demonstrated accidentally in our investigations during 1891. Hence all infected material should be freely exposed to the frost, even though treated with disinfectants beforehand.

WALTER REED

YELLOW FEVER

[From *The Propagation of Yellow Fever; Observations Based on Recent Researches**]

. . . In the hope that what I have to say to-night may contribute to the solution of a somewhat obscure problem, and may assist us hereafter in the struggle with a grave epidemic disease, I have concluded to present for your consideration "The Propagation of Yellow Fever, Based on Recent Researches."

Before proceeding to the discussion of this subject, it is fitting that I should pay brief tribute to the memory of a former member of this Faculty, the late Dr. Jesse W. Lazear, United States Army. I can hardly trust myself to speak of my late colleague, since the mention of his name brings back such scenes of anxiety and depression as one

* Address delivered at the One Hundred and Third Annual Meeting of the Medical and Chirurgical Faculty of the State of Maryland, held at Baltimore, April 24-27, 1901. *Med. Rec.*, Vol. 60, No. 6, August 10, 1901.

recalls only with pain. Along with these sad memories, however, come other recollections of a manly and fearless devotion to duty such as I have never seen equaled. In the discharge of the latter, Dr. Lazear seemed absolutely tireless and quite oblivious of self. Filled with an earnest enthusiasm for the advancement of his profession and for the cause of science, he let no opportunity pass unimproved. Although the evening might find him discouraged over the difficult problem at hand, with the morning's return he again took up the task full of eagerness and hope. During a service of less than one year in Cuba, he won the good-will and respect of his brother officers, and the affection of his immediate associates. Almost at the beginning of what promised to be a life full of usefulness and good works he was suddenly stricken, and, dying, added one more name to that imperishable roll of honor to which none others belong than martyrs to the cause of humanity. . . .

I do not propose to set before you this evening so much the views of others in relation to the etiology of yellow fever or to the conditions under which it originates and spreads, as to give you the results of my own experience with regard to the manner in which this disease propagates itself. In the ordinary course of army administration, I found myself brought in contact with yellow fever during the summer of 1900, under such circumstances as permitted me to give my entire time to the study of its etiology and propagation. . . .

At the time of our arrival in Cuba—June, 1900—the situation as regards the etiology of yellow fever may be briefly stated as follows: The claims of all investigators for the discovery of the specific agent of this disease have been disproved by the exhaustive work of Sternberg (Report on the Etiology and Prevention of Yellow Fever, 1890), published in 1890, except that made by Dr. Sanarelli, in June, 1897, for his *bacillus icteroides*. I need not take up time here with mention of those who had investigated Sanarelli's claim, except to state that the confirmation of his discovery came chiefly from workers in the United States, of whom I may speak of Achinard and Woodson, of New Orleans, and especially of Wasdin and Geddings of the Marine Hospital Service. . . .

The theory of Finlay for the propagation of yellow fever by the mosquito, concerning which I shall presently have something to say, had either gained no credence, or been rejected by reason of the absence of any results that had been presented by its author in support of it. In the meantime, and before we had completed our search for Sanarelli's bacillus, certain facts had been cropping out, as it were, which served to arrest the attention. Just here, while mentioning the first fact, let me say that one does not like to confess his ignorance of such a well-known disease as yellow fever, especially before an audience some of whose members may have already treated cases during the last epidemic in this city, in the year 1876; and yet candor compels me to record my very great surprise, when brought face to face with yellow fever, to learn that attendance on patients by non-immune nurses, in every stage of the malady, involved no danger. In other words, that yellow fever, both in the wards at Columbia Barracks, as well as in the wards of Las Animas Hospital, Havana, was not contracted by the attendants under circumstances such as those in which typhoid fever and Asiatic cholera are too often conveyed. Further inquiry along this line seemed to indicate that the disease had not been contracted in hospitals, even during the earlier epidemics, when disinfection of articles of clothing and bedding was unknown. . . .

At first the results obtained at this station were not encouraging. From November 20, 1900, the date of the establishment of the station, until December 4—a period of two weeks—we had tried to infect four individuals with entirely negative results. Two of these had been bitten twice, at intervals of three days, by contaminated mosquitoes which had been kept from ten to fourteen days after they had fed on yellow fever cases; while the other two non-immunes had been thrice bitten, at the same intervals, by mosquitoes that had bitten cases of yellow fever ten to eighteen days before. As the weather during this time was cool and the insects had been kept at room temperature (and this is practically out-door temperature in Cuba), we conjectured that the negative results might, perhaps, be thus explained. We remembered that Daniels (On Transmission of Proteosoma to Birds by the Mosquito—Royal Society, Reports of Malarial Committee, London, 1900) in repeating, during the winter season Ross' observations with proteosoma infection of birds, had obtained a much smaller percentage of successes than had Ross, who worked during summer weather. We would have been glad to draw some encouragement from our negative experiments, also, with infected bedding, but as at this date (December 4) our three subjects had been sleeping with fomites only four nights (which is within the period of incubation of the disease), this comfort was debarred.

On the fifteenth day of our encampment, therefore—December 5, at 2 P.M. o'clock—we concentrated our insects, so to speak, on one of these non-immunes—Kissinger by name—selecting five of our most promising mosquitoes for the purpose. These had been contaminated as follows: two fifteen days; one nineteen days, and two twenty-one days previously. This inoculation was more successful, for at the expiration of three days and nine and one-half hours the subject, who had been under strict quarantine during fifteen days, was suddenly seized with a chill about midnight, December 8, which was the beginning of a well-marked attack of yellow fever.

I cannot let this opportunity pass without expressing my admiration of the conduct of this young Ohio soldier, who volunteered for this experiment, as he expressed it, "solely in the interest of humanity and the cause of science," and with the only proviso that he should receive no pecuniary reward. In my opinion this exhibition of moral courage has never been surpassed in the annals of the Army of the United States.

The following morning—December 9—(Sunday at 10.30 A.M. o'clock) we selected from those insects that had bitten Case I, one mosquito that seemed to us to possess the best record of contamination, as it had bitten a fatal case of yellow fever, on the second day of the disease, nineteen days before. This insect was applied to a Spanish immigrant, who had been strictly quarantined at our station for nineteen days. At the expiration of three days and eleven hours (December 9, 9.30 P.M.) this individual was also seized with an attack of yellow fever.

In the meanwhile, on December 8, 1900, at 4 o'clock P.M., we had applied to a young Spaniard three of the mosquitoes that had, three days previously, bitten Case I, together with an additional mosquito contaminated seventeen days before. At the end of four days and twenty hours (December 13, noon), this Spaniard suddenly lost his vivacity and took to his bed. The following morning, at 9 A.M., his febrile paroxysm began. His case, which was the mildest of our series, was also marked by a long period of incubation, viz., five days and seventeen hours. He had been in quarantine nine days.

December 11, at 4.30 P.M. o'clock, the identical four insects which had bitten Case III were fed on a Spanish immigrant who had been in quarantine for the past twenty-one days. At the expiration of three days and nineteen and one-half hours (December 15, noon) he was likewise seized with yellow fever.

Thus within the period of one week—December 9 to December 15—we had succeeded in producing an attack of yellow fever in each of the four individuals whom we had caused contaminated insects to bite, and in all save one of the five non-immunes whom we had originally selected for experimentation.

It can readily be imagined that the concurrence of four cases of yellow fever in our small command of twelve non-immunes, within the space of one week, while giving rise to feelings of exultation in the hearts of the experimenters, in view of the vast importance attaching to these results might inspire quite other sentiments in the bosoms of those who had previously consented to submit themselves to the mosquito's bite. In fact several of our good-natured Spanish friends who had jokingly compared our mosquitoes to "the little flies that buzzed harmlessly about their tables," suddenly appeared to lose all interest in the progress of science, and forgetting, for the moment, even their own personal aggrandizement, incontinently severed their connection with Camp Lazear. Personally, while lamenting, to some extent, their departure, I could not but feel that in placing themselves beyond our control they were exercising the soundest judgment. In striking contrast to the want of confidence shown by these Andalusians who had agreed to be bitten by mosquitoes was the conduct now displayed by the three young Americans, who had consented to jeopardize their lives by exposures to fomites, and who, as a matter of fact, had already spent fifteen nights in a small, illy-ventilated building, breathing in an atmosphere dreadfully contaminated by the soiled garments of yellow fever patients. With the occurrence of these cases of mosquito infection, the countenance of these men, which had before borne the serious aspect of those who were bravely facing an unseen foe, suddenly took on the glad expression of "school boys let out for a holiday," and from this time their contempt for "fomites" could not find sufficient expression. Thus illustrating once more, gentlemen, the old adage that familiarity, even with fomites, may breed contempt!

As the continued good health of those who were occupying the "Infected Clothing Building" pointed strongly to the harmlessness of fomites, the next experiment at this station was undertaken for the purpose of demonstrating that the essential factor in the infection of a building with yellow fever is the presence therein of mosquitoes that have bitten cases of yellow fever.

Accordingly at 11.55 A.M., December 21, 1900, fifteen mosquitoes were freed in the larger room of the "Infected Mosquito Building," which, as I have said, was divided into two compartments by a wire-screen partition. The interval that had elapsed since the contamination of these insects was as follows: one, twenty-four days; three, twelve days; four, eight days; and seven, five days. The only articles of furniture in this building consisted of three beds, one being placed in the mosquito room and two beyond the wire screen, these latter intended to be occupied by two "control" non-immunes. The articles of bedding as well as the bedsteads had been carefully disinfected with steam. At noon on the same day, five minutes after the mosquitoes had been placed therein, a plucky Ohio boy, Moran by name, clad only in his night shirt, and fresh from a bath, entered the room containing the mosquitoes, where he lay down for a period of thirty minutes. On the opposite of the screen were the two "controls" and one other non-immune. Within two minutes from Moran's entrance, he was being bitten about the face and hands by the insects that had promptly settled down upon him. Seven in all bit him at this visit. At 4.30 P.M., the same day, he again entered and remained twenty minutes, during which time five others bit him. The following day at 4.30 P.M., he again entered and remained fifteen minutes, during which time three insects bit him, making the number fifteen that had fed at these three visits. The building was then closed, except that the two non-immune "controls" continued to occupy the beds on the non-infected side of the screen. On Christmas morning at 11 A.M., this brave lad was stricken with yellow fever, and had a sharp attack, which he bore without a murmur. The period of incubation in this case was three days and twenty-three hours, counting from his first visit, or two days and seventeen and a half hours, if reckoned from his last visit. The two "controls" who had slept each night in this house, only protected by the wire screen, but breathing the common atmosphere of the building, had remained in good health. They continued to so remain, although required to sleep here for thirteen additional nights.

XXXVI

MODERN CLINICAL DESCRIPTIONS

AN ORIGINAL DESCRIPTION OF A NEW DISEASE REQUIRES JUST AS UNIQUE A SORT of genius as the successful accomplishment of a planned research. Osler's *Practice of Medicine* was a scientific work because it dealt, in the scientific spirit, with natural phenomena—a disease was discussed under the headings of etiology, pathology, symptoms, signs, prognosis, prevention, and treatment.

The great clinical descriptions given in the following pages need no comment. They are models for any medical student who wishes to advance his science.

PERCIVAL POTT

Percival Pott (1714-1788) of London was surgeon to St. Bartholomew's Hospital. He fell in the street and sustained the fracture of the ankle which he so perfectly described in 1769. In 1779 he published *Remarks on that Kind of Palsy of the Lower Limbs which is frequently found to accompany a curvature of the spine.*

REFERENCE

Percival Pott. *Medical Classics,* Vol. I, No. 4, December, 1936.

POTT'S FRACTURE

[From *General Remarks on Fractures and Dislocations**]

Whoever will take a view of the leg of a skeleton, will see that although the fibula be a very small and slender bone, and very inconsiderable in strength, when compared with the tibia, yet the support of the lower joint of that limb (the ankle) depends so much on this slender bone, that without it the body would not be upheld, nor locomotion performed, without hazard of dislocation every moment. The lower extremity of this bone, which descends considerably below that end of the tibia, is by strong and inelastic ligaments

* London, 1769.

firmly connected with the last-named bone, and with the astragalus, or that bone of the tarsus which is principally concerned in forming the joint of the ankle. This lower extremity of the fibula has, in its posterior part, a superficial sulcus for the lodgment and passage of the tendons of the peronei muscles, which are here tied down by strong ligamentous capsulae, and have their action so determined from this point or angle, that the smallest degree of variation from it, in consequence of external force, must necessarily have considerable effect on the motions they are designed to execute, and consequently distort the foot. Let it also be considered, that upon the due and natural state of the joint of the ankle, that is, upon the exact and proper disposition of the tibia and fibula, both with regard to each other and to the astragulus, depend the just disposition and proper action of several other muscles of the foot and toes; such as the gastrocnemii, the tibialis anticus, and posticus, the flexor pollicis longus, and the flexor digitorum pedis longus, as must appear demonstrably to any man who will first dissect and then attentively consider these parts.

If the tibia and fibula be both broken, they are both generally displaced in such manner, that the inferior extremity, or that connected with the foot, is drawn under that part of the fractured bone which is connected with the knee; making by this means a deformed, unequal tumefaction in the fractured part, and rendering the broken limb shorter than it ought to be, or than its fellow, and this is generally the case, let the fracture be in what part of the leg it may.

If the tibia only be broken, and no act of violence, indiscretion, or inadvertence be committed, either on the part of the patient or of those who conduct him, the limb most commonly preserves its figure and length; the same thing generally happens if the fibula only be broken, in all part of it, which is superior to letter A in the annexed figure, or in any part of it between its upper extremity, and within two or three inches of its lower one.

I have already said, and it will obviously appear to every one who examines it, that the support of the body and the due and proper use and execution of the office of the joint of the ankle depend almost entirely on the perpendicular bearing of the tibia upon the astragalus, and on its firm connection with the fibula. If either of these be perverted or prevented, so that the former bone is forced from its just and perpendicular position on the astragalus, or if it be separated by violence from its connection with the latter, the joint of the ankle will suffer a partial dislocation internally; which partial dislocation cannot happen without not only a considerable extension, or perhaps laceration of the bursal ligament of the joint, which is lax and weak, but a laceration of those strong tendinous ligaments, which connect the lower end of the tibia with the astragalus and os calcis, and which constitute in great measure the ligamentous strength of the joint of the ankle.

This is the case, when by leaping or jumping the fibula breaks in the weak part already mentioned, that is within two or three inches of its lower extremity. When this happens, the inferior fractured end of the fibula falls inward toward the tibia, that extremity of the bone which forms the outer ankle is turned somewhat outward and upward, and the tibia having lost its proper support, and not being of itself capable of steadily preserving its true perpendicular bearing, is forced off from the astragulus inwards, by which means the weak bursal, or common ligament of the joint is violently stretched, if not torn, and the strong ones, which fasten the tibia to the astragalus and os calcis, are always lacerated, thus producing at the same time a perfect fracture and a partial dislocation, to which is sometimes added a wound in the integuments, made by the bone at the inner ankle. By this means, and indeed as a necessary consequence, all the tendons which pass behind or under, or are attached to the extremities of the tibia and fibula, or os calcis, have their natural direction and disposition so altered, that instead of performing their appointed actions, they all contribute to the distortion of the foot, and that by turning it outward and upward.

When this accident is accompanied, as it sometimes is, with a wound of the integuments of the inner ankle, and that made by the protrusion of the bone, it not infrequently ends in a fatal gangrene, unless prevented by timely amputation, though I have several times seen it do very well without. But in its most simple state, unaccompanied with any wound, it is extremely troublesome to put to rights, still more so to keep it in order, and unless managed with address and skill, is very frequently productive both of lameness and deformity ever after.

JOHN HUNTER

John Hunter (1728-1793) was born in Scotland. He went to London in 1748, where he studied with his brother William, and with Cheselden and Pott, later becoming the leading London surgeon. He was an indefatigable investigator of all aspects of nature, and his collections may be seen in the Hunterian Museum of the Royal College of Surgeons. Among his important observations are those on the venereal diseases—he infected himself with the virus from a chancre and went through the first two stages of syphilis. His description of the chancre is remembered in the eponym "Hunterian chancre." He confused syphilis with gonorrhea. His work, *On Venereal Diseases,* appeared in 1786.

REFERENCES

John Hunter. *Medical Classics,* Vol. IV, No. 5, January, 1940.
PAGET, S. John Hunter, Man of Science and Surgeon (1728-1793). London, T. Fisher Unwin, 1897.

[From *A Treatise on the Venereal Disease**]

When the venereal matter has affected the constitution in any of the ways before mentioned, it has the whole body to work upon, and shows itself in a variety of shapes, many of which putting on the appearance of a different disease, we are often obliged to have recourse to the preceding history of the case before we can form any judgement of it. Probably the varieties in the appearances may be referred to the three following circumstances: the different kinds of constitutions, the different kinds of solids affected, and the different dispositions which the solids affected, and the different dispositions which the solids are in at the time; for I can easily conceive that a peculiarity of constitution may make a very material difference in the appearance of the same specific complaint, and I am certain that the solids, according to their different natures, produce a very different appearance when attacked with this disease; and I can also easily conceive that a different disposition from the common in the solids at the time may make a considerable difference in the appearances.

The difference of constitution, and of the same parts at different times, may have considerable effects in the disease with respect to its

* The Works of John Hunter, edited by J. F. Palmer. London, Longman, 1837. Vol. 11.

appearing sooner or later. This I am certain of, that the different parts of the body produce a very considerable difference in the times of appearance of this disease. That it appears much sooner in some parts than in others is best seen where different parts are affected in the same person; for I have already endeavoured to show that it is most probable that all the parts affected are contaminated nearly at the same time. This difference in the times is either owing to some parts being naturally put into action more easily by the poison than others, or they are naturally more active in themselves, and therefore probably will admit more quickly the action of every disease that is capable of affecting them.

When on the general history of the lues venerea, I divided the parts into two orders, according to the time of their appearance. I also observed that the first were commonly the external parts, as the skin, nose, tonsils; and that the second were more internal, as the bones, periosteum, fasciae, and tendons.

The time necessary for its appearance, or for producing its local effects in the several parts of the body most readily affected, after it has got into the constitution, is uncertain, but in general it is about six weeks, in many cases, however, it is much later, and in others much sooner. In some cases it appears to produce its local effects within a fortnight after the possibility of the absorption of the matter. In one case a gentleman had a chancre, and a swelling in the groin came on, and with the before-mentioned time he had venereal eruptions all over the body. He could not impute this to any former complaint, yet there is a possibility of its having arisen from the first mode of catching the disease, by simple contact, at the time he got the local or chancre, which might extend the time to a week or more, although this is not probable. In another case, three weeks after the healing of a chancre, eruptions broke out all over the body, and this happened only a fortnight after leaving off the course of mercury that cured the chancre. The effects on other parts of the body, that are less susceptible of this irritation, or are slower in their action, are of course much later in appearing; and in those cases where both orders of parts are contaminated it is in general not till after the first has made its appearance for a considerable time, and even perhaps after it has been cured; for while the parts first in order of action were contaminated and under cure, the second in order are only in a state of contamination, and go on with the disease afterwards, although it may never again appear in the first.

From this circumstance of the parts second in order coming later into action, we can plainly see the reason why it shall appear in them, although the first in order may have been cured; for if the external parts, or first in order, have been cured, and the internal, or second, such as the tendons, bones, periosteum, &c., have not been cured, then it becomes confined solely to these parts. The order of parts may sometimes be inverted; for I have seen cases where the periosteum, or bone, was affected prior to any other part. Whether in the same case it might in the end have affected the skin or throat I will not pretend to say, as it was not allowed to go on; but it is possible that the second order of parts may be affected without the first having ever been contaminated.

Its effects on the deeper-seated parts are not like those produced in the external, and the difference is so remarkable as to give the appearance of another disease; and a person accustomed to see it in the first parts only would be entirely at a loss about the second.

The parts which come first into action go on with it, probably on the same principle, much quicker than the others; and this arises from the nature of the parts, as has already been observed.

Each succeeding part that becomes affected is slower and slowei in its progress, and more fixed in its symptoms when produced; this arises also from the natural disposition of such parts, all their actions being slow, which indolent action may be assisted by the absence of the great disposing cause, that is cold. I should, however, suspect that warmth does not contribute much to their indolence of action; for if it did it would assist in the cure, which it appears not to do, these parts being as slow in their operations of restoration as they are in their actions of disease. We may also observe that similar parts come sooner into action, and appear to go on more rapidly with it, as they are nearer the source of the circulation. It appears earlier on the face, head, shoulders, and breast than on the legs, and the eruptions come sooner to suppuration in the before-mentioned parts.

The circumstance of its being very late in appearing in some parts, when it had been only cured in its first appearances, as mentioned. has made many suppose that the poison lurked somewhere in the solids; and others that it kept circulating in the blood for years.

It is not, however. easy to determine this point; but there can be no good reason for the first hypothesis, as the lurking disposition never takes place prior to its first appearance; for instance, we never find that a man had a chancre a twelve-month ago, and that it broke out after in venereal scurfs upon the skin, or ulcers in the throat. The slowness of its progress is only when the parts less susceptible of its irritation have been affected by it.

1. OF THE SYMPTOMS OF THE FIRST STAGE OF THE LUES VENEREA

The first symptoms of the disease, after absorption, appear either on the skin, throat, or mouth. These differ from one another according to the nature of the parts affected. I shall therefore divide them into two kinds, although there appears to be difference in the nature of the disease itself.

The appearance on the skin I shall call the first, although it is not always the first appearance; for that in the throat is often as early a symptom as any. The appearances upon the skin generally show themselves in every part of the body, no part being more susceptible than another, first in discolorations, making the skin appear mottled, many of them disappearing, while others continue and increase with the disease.

In others it will come on in distinct blotches; often not observed till scurfs are forming; at other times they appear in small distinct inflammations, containing matter and resembling pimples, but not so pyramidal, nor so red at the base.

Venereal blotches, at their first coming out, are often attended with inflammation, which gives them a degree of transparency, which I think is generally greater in the summer than in the winter, especially if the patient be kept warm. In a little time this inflammation disappears, and the cuticle peels off in the form of a scurf. This sometimes misleads the patient and the surgeon, who look upon this dying away of the inflammation as a decay of the disease, till a succession of scurfs undeceives them.

These discolorations, of the cuticle arise from the venereal irritation, and are seldom to be reckoned a true inflammation, for they seldom have any of its characteristics, such as tumefaction and pain; but this is true only on those parts most exposed, for in parts well covered, and in parts constantly in contact with other parts, there is more of the true inflammatory appearance, especially about the anus.

The appearance of the parts themselves next begins to alter, forming a copper-coloured dry inelastic cuticle, called a scurf; this is thrown off, and new ones are formed. These appearances spread to the breadth of a sixpence or shilling, but seldom broader, at least for a considerable time, every succeeding scurf becoming thicker and thicker, till at last it becomes a common scab, and the disposition for the formation of matter takes place in the cutis under the scab, so that at last it turns out a true ulcer, in which state it commonly spreads, although but slowly.

These appearances arise first from the gradual loss of the true sound cuticle, the diseased cutis having lost the disposition to form one; and, as a kind of substitute for this want of cuticle, an exudation takes place, forming a scale, and afterwards becoming thicker, and the matter acquiring more consistence, it at last forms a scab; but before it has arrived at this state the cutis has given way, and ulcerated, after which the discharge becomes more of a true pus. When it attacks the palms of the hands and the soles of the feet, where the cuticle is thick, a separation of the cuticle takes place, and it peels off, a new one is immediately formed, which also separates, so that a series of new cuticles takes place, from its not so readily forming scurfs as on the common skin. If the disease is confined to those parts it becomes more difficult to determine whether or not it be venereal, for most diseases of the cutis of these parts produce a separation of the cuticle attended with the same appearances in all, and having nothing characteristic of the venereal disease.

Such appearances are peculiar to that part of the common skin of the body which is usually exposed; but when the skin is opposed by another skin which keeps it in some degree more moist, as between the nates, about the anus, or between the scrotum and the thigh, or in the angle between the two thighs, or upon the prolabium of the mouth, and in the armpits, the eruptions never acquire the above-described appearances, and instead of scurfs and scabs we have the skin elevated, or, as it were, tumefied by the extravasated lymph into a white, soft, moist, flat surface, which discharges a white matter. This may perhaps arise from there being more warmth, more perspiration, and less evaporation, as well as from the skin being thinner in such places. What strengthens this idea still more is, that in many venereal patients I have seen an approach towards such appearances on the common skin of the body; but this has been on such parts as were covered with the clothes, for on those parts of the skin that were not covered there was only the flat scurf; these, however, were redder that the above-described appearances, but hardly so high. How far this is peculiar to the venereal disease I know not. It may take place in most scurfy eruptions of the skin. From a supposition of this not being venereal, I have destroyed them at the side of the anus with a caustic, and the patient has got well; however, from my idea of the disease, that every effect from the constitution is truly local, and therefore may be cured locally, a cure effected by this treatment does not determine the question.

This disease, on its first appearance, often attacks that part of the fingers upon which the nail is formed, making that surface red which is seen shining through the nail, and, if allowed to continue, a separation of the nail takes place, similar to the cuticle in the before-described symptoms; but here there cannot be that regular succession of nails as there is of cuticle.

It also attacks the superficies of the body which is covered with hair, producing a separation of the hair. A prevention of the growth of young hair is also the consequence while the disease lasts.

The second part in which it appears is most commonly the throat, sometimes the mouth and tongue. In the throat, tonsils, and inside of the mouth, the disease generally shows itself at once, in the form of an ulcer, without much previous tumefaction, so that the tonsils are not much enlarged; for when the venereal inflammation attacks these parts, it appears to be always upon the surface, and it very soon terminates in an ulcer.

These ulcers in the throat are to be carefully distinguished from all others of the same parts. It is to be remarked, that this disease, when it attacks the throat, always, I believe, produces an ulcer, although this is not commonly understood; for I have seen cases where no ulceration had taken place called, by mistake, venereal. It is therefore only this ulcer that is to be distinguished from other ulcers of these parts. This species of ulcer is generally tolerably well marked, yet it is perhaps in all cases not to be distinguished from others that attack this part, for some have the appearance of being venereal, and what are really venereal resemble those that are not. We have several diseases of this part which do not produce ulceration on the surface, one of which is common inflammation of the tonsils, which often suppurates in the centre, forming an abscess, which bursts by a small opening, but never looks like an ulcer begun upon the surface, as in the true venereal; this case is always attended with too much inflammation, pain, and tumefaction of the parts to be venereal; and if it suppurates and bursts it subsides directly, and it is generally attended with other inflammatory symptoms in the constitution.

There is another disease of these parts, which is an indolent tumefaction of the tonsils, and is peculiar to many people whose constitutions have something of the scrofula in them, producing a thickness in the speech. Sometimes the coagulable lymph is thrown out on the surface, and called by some ulcers, by others sloughs, and such are often called putrid sore throats. Those commonly swell to too large a size for the venereal; and this appearance is easily distinguished from an ulcer or loss of substance: however, where it is not plain at first sight it will be right to endeavour to remove some of it; and if the surface of the tonsil is not ulcerated, then we may be sure it is not venereal. I have seen a chink filled with this, appearing very much like an ulcer, but upon removing the coagulable lymph the tonsil has appeared perfectly sound. I have seen cases of a swelled tonsil where a slough formed in its centre, and that slough has opened a passage out for itself, and when it has been, as it were sticking in this passage, it has appeared like a foul ulcer.

The most puzzling stage of the complaint is when the slough is come out, for then it has most of the characters of the venereal ulcer; but when I have seen the disease in its first stages I have always treated it as of the crysipelatous kind, or as something of the nature of a carbuncle.

When I have seen them in their second state only I have been apt to suppose them venereal: however, no man will be so rash as to pronounce what a disease is from the eye only, but will make inquiries into all the circumstances before he forms a judgment. If there have been no preceding local symptoms within the proper date he will suspend his judgment, and wait a little to see how far Nature is able to relieve herself. If there has been any preceding fever, it will be still less probable that it is venereal. However, I will not say of what nature such cases are, but only that they are not venereal as they are often believed to be. I have seen a sore throat of this kind mistaken for venereal, and mercury given till it affected the mouth, which when it did, it brought on a mortification on all the parts concerned in the first disease. It would therefore appear that this species of the sore throat is aggravated by mercury.

There is another complaint of those parts which is often taken for venereal, which is an ulcerous excoriation, where the ulceration or excoriations run along the surface of the parts, becoming very broad, and sometimes foul, having a regular termination, but never going deep into the substance of the parts, as the venereal ulcer does. There is no part of the inside of the mouth exempted from this ulcerous excoriation, but I think it is most frequent about the root of the uvula, and spreads forwards along the palatum molle. That such are not venereal is evident, from their not giving way in general to mercury; and I have seen them continue for weeks without altering, and a true venereal ulcer appear upon the centre of the excoriated part.

The difference between the two is so strong that there can be no mistake; patients have gone through a course of mercury which has perfectly cured the venereal ulcers, but has had no effect upon the others, which have afterwards been cured by bark.

The true venereal ulcer in the throat is perhaps the least liable to be mistaken of any of the forms of the disease. It is a fair loss of substance, part being dug out, as it were, from the body of the tonsil, with a determined edge, and is commonly very foul, having thick white matter adhering to it like a slough, which cannot be washed away.

Ulcers in such situations are always kept moist, the matter not being allowed to dry and form scabs, as in those upon the skin; the matter is carried off the ulcers by deglutition, or the motion of the parts, so that no succession of scurfs or scabs can take place, as on the skin.

Their progress is also much more rapid than on the common skin, ulceration taking place very fast.

Like most other spreading ulcers, they are generally very foul, and for the most part have thickened or bordered edges, which is very common to venereal or cancerous sores, and indeed to most sores which have no disposition to heal, whatever the specific disease may be.

When it attacks the tongue it sometimes produces a thickening and hardness in the part; but this is not always the case, for it very often ulcerates, as in the other parts of the mouth.

They are generally more painful than those of the skin, although not so much so as common sore throats arising from inflamed tonsils. They oblige the person to speak thick, or as if his tongue was too large for his mouth, with a small degree of snuffling. These are the most common symptoms of this state of the disease, but it is perhaps impossible to know all the symptoms this poison produces when in the constitution. I knew a gentleman who had a teasing cough which he imputed to it; for it came on with the symptomatic fever, and continued with it, and by using mercury both disappeared.

There are inflammations of the eyes which are supposed to be venereal; for after the usual remedies against inflammation have been tried in vain, mercury has been given, on the supposition of the case being venereal, and sometimes with success, which had tended to establish this opinion. But if such cases are venereal, the disease is very different from what it is when attacking other parts of the constitution, for the inflammation is more painful than in venereal inflammation proceeding from the constitution; and I have never seen such cases attended with ulceration, as in the mouth, throat, and tongue, which makes me doubt much of their being venereal.

2. EXPERIMENTS MADE TO ASCERTAIN THE PROGRESS AND EFFECTS OF THE VENEREAL POISON

To ascertain several facts relative to the venereal disease, the following experiments were made. They were begun in May 1767.

Two punctures were made on the penis with a lancet dipped in venereal matter from a gonorrhea; one puncture was on the glans, the other on the prepuce.

This was on a Friday; on the Sunday following there was a teasing itching in those parts, which lasted till the Tuesday following. In the mean time, these parts being often examined, there seemed to be a greater redness and moisture than usual, which was imputed to the parts being rubbed. Upon the Tuesday morning the parts of the prepuce where the puncture had been made were redder, thickened, and had formed a speck; by the Tuesday following the speck had increased, and discharged some matter, and there seemed to be a little pouting of the lips of the urethra, also a sensation in it in making water, so that a discharge was expected from it. The speck was now touched with lunar caustic, and afterwards dressed with calomel ointment. On Saturday morning the slough came off, and it was again touched, and another slough came off on the Monday following. The preceding night the glans had itched a good deal, and on Tuesday a white speck was observed where the puncture had been made; this speck, when examined was found to be a pimple full of yellowish matter. This was now touched with the caustic, and dressed as the former. On the Wednesday the sore on the prepuce was yellow, and therefore was again touched with caustic. On the Friday both sloughs came off, and the sore on the prepuce looked red, and its basis was not so hard; but on the Saturday it did not look quite so well, and was touched again, and when that went off it was allowed to heal, as also the other, which left a dent in the glans. This dent on the glans was filled up in some months, but for a considerable time it had a bluish cast.

Four months afterwards the chancre on the prepuce broke out again, and very stimulating applications were tried; but these seemed not to agree with it, and nothing being applied, it healed up. This it did several times afterwards, but always healed up without any application to it. That on the glans never did break out, and herein also it differed from the other.

While the sores remained on the prepuce and glans a swelling took place in one of the glands of the right groin. I had for some time conceived an idea that the most effectual way to put back a bubo was to rub in mercury on that leg and thigh; that thus a current of mercury would pass through the inflamed gland. There was a good opportunity of making the experiment. I had often succeeded in this way, but now wanted to put it more critically to the test. The sores upon the penis were healed before the reduction of the bubo was attempted. A few days after beginning the mercury in this method the gland subsided considerably. It was then left off, for the intention was not to cure it completely at present. The gland some time after began to swell again, and as much mercury was rubbed in as appeared to be sufficient for the entire reduction of the gland; but it was meant to do no more than to cure the gland locally, without giving enough to prevent the constitution from being contaminated.

About two months after the last attack of the bubo, a little sharp pricking pain was felt in one of the tonsils in swallowing anything, and on inspection a small ulcer was found, which was allowed to go on till the nature of it was ascertained, and then recourse was had to mercury. The mercury was thrown in by the same leg and thigh as before, to secure the gland more effectually, although that was not now probably necessary.

As soon as the ulcer was skinned over the mercury was left off, it not being intended to destroy the poison, but to observe what parts it would next affect. About three months after, copper-coloured blotches broke out on the skin, and the former ulcer returned in the tonsil. Mercury was now applied the second time for those effects of the poison upon the constitution, but still only with a view to palliate.

It was left off a second time, and the attention was given to mark where it would break out next; but it returned again in the same parts. It not appearing that any further knowledge was to be procured by only palliating the disease a fourth time in the tonsil, and a third time in the skin, mercury was now taken in a sufficient quantity, and for a proper time, to complete the cure.

The time the experiments took up, from the first insertion to the complete cure, was about three years.

The above case is only uncommon in the mode of contracting the disease, and the particular views with which some parts of the treatment were directed; but as it was meant to prove many things which though not uncommon, are yet not attended to, attention was paid to all the circumstances. It proves many things, and opens a field for further conjectures.

It proves, first, that matter from a gonorrhoea will produce chancres.

It makes it probable that the glans does not admit the venereal irritation so quickly as the prepuce. The chancre on the prepuce infoamed and suppurated in somewhat more than three days, and that on the glans in about ten. This is probably the reason why the glans did not throw off its sloughs so soon.

It renders it highly probable that to apply mercury to the legs and thighs is the best method of resolving a bubo; and therefore also the best method of applying mercury to assist in the cure, even when the bubo suppurates.

It also shows that buboes may be resolved in this way, and yet the constitution not be safe; and therefore that more mercury should be thrown in, especially in cases of easy resolution, than what simply resolves the bubo.

It shows that parts may be contaminated, and may have the poison kept dormant in them while under a course of mercury for other symptoms, but break out afterwards.

It also shows that the poison having originally only contaminated certain parts, when not completely cured, can break out again only in those parts.

ABRAHAM COLLES

Abraham Colles (1773-1843), of Dublin, was the morning star of the great Dublin school of medicine, which includes Graves, Stokes, Corrigan, Rynd, Adams, and Cheyne. In 1814 he published his fine description of the fracture of the wrist known as "Colles' fracture." His observations on syphilis are remembered through "Colles' law."

REFERENCE

Abraham Colles. *Medical Classics,* Vol. IV, No. 10, June, 1940. [Biography, portrait and bibliography.]

COLLES'S LAW

[From *Practical Observations on the Venereal Disease, and on the Use of Mercury**]

One fact well deserving our attention is this: that a child born of a mother who is without any obvious venereal symptoms, and which, without being exposed to any infection subsequent to its birth, shows this disease when a few weeks old, this child will infect the most healthy nurse, whether she suckle it, or merely handle and dress it; and yet this child is never known to infect its own mother, even though she suckle it while it has venereal ulcers of the lips and tongue.

COLLES'S FRACTURE

[From *On the Fracture of the Carpal Extremity of the Radius†*]

The injury to which I wish to direct the attention of surgeons, has not, as far as I know, been described by an author; indeed, the

* 1837.
† *Edinburgh M. & S. J.,* 1814.

form of the carpal extremity of the radius would rather incline us to question its being liable to fracture. The absence of crepitus and of other common symptoms of fracture, together with the swelling which instantly arises in this, as in other injuries of the wrist, render the difficulty of ascertaining the real nature of the case very considerable.

This fracture takes place at about an inch and a half above the carpal extremity of the radius, and exhibits the following appearances.

The posterior surface of the limb presents a considerable deformity, for a depression is seen in the forearm, about an inch and a half above the end of this bone, while a considerable swelling occupies the wrist and the metacarpus. Indeed the carpus and base of metacarpus appear to be thrown backward so much, as on first view to excite a suspicion that the carpus has been dislocated forward.

On viewing the anterior surface of the limb, we observe a considerable fulness, as if caused by the flexor tendons being thrown forwards. The fulness extends upwards to about one-third of the length of the fore-arm, and terminates below at the upper edge of the annular ligament of the wrist. The extremity of the ulna is seen projecting towards the palm and inner edge of the limb; the degree, however, in which this projection takes place, is different in different instances.

If the surgeon proceed to investigate the nature of this injury, he will find that the end of the ulna admits of being readily moved backwards and forwards.

On the posterior surface, he will discover, by the touch, that the swelling on the wrist and metacarpus is not caused entirely by an effusion among the softer parts; he will perceive that the ends of the metacarpal, and second row of carpal bones, form no small part of it. This, strengthening the suspicion which the first view of the case had excited, leads him to examine, in a more particular manner, the anterior part of the joint; but the want of that solid resistance, which a dislocation of the carpus forward must occasion, forces him to abandon this notion, and leaves him in a state of perplexing uncertainty as to the real nature of injury. He will therefore endeavour to gain some information, by examining the bones of the fore-arm. The facility with which (as was before noticed) the ulna can be moved backward and forward, does not furnish him with any useful hint. When he moves his fingers along the anterior surface of the radius, he finds it more full and prominent than is natural; a similar examination of the posterior surface of this bone, induces him to think that a depression is felt about an inch and a half above its carpal extremity. He now expects to find satisfactory proofs of a fracture of the radius at this spot. For this purpose, he attempts to move the broken pieces of the bone in opposite directions: but, although the patient is by this examination subjected to considerable pain, yet, neither crepitus nor a yielding of the bone at the seat of fracture, nor any other positive evidence of the existence of such an injury is thereby obtained. The patient complains of severe pain as often as an attempt is made to give to the limb the motions of pronation and supination.

If the surgeon lock his hand in that of the patient's, and make extension, even with a moderate force, he restores the limb to its natural form; but the distortion of the limb instantly returns on the extension being removed. Should the facility with which a moderate extension restores the limb to its form, induce the practitioner to treat this as a case of sprain, he will find, after a lapse of time sufficient for the removal of similar swellings, the deformity undiminished. Or, should he mistake the case for a dislocation of the wrist, and attempt to retain the parts in situ by tight bandages and splints, the pain caused by the pressure on the back of the wrist will force him to unbind them in a few hours; and, if they be applied more loosely, he will find, at the expiration of a few weeks, that the deformity still exists in its fullest extent, and that it is now no longer to be removed by making extension of the limb. By such mistakes the patient is doomed to endure for many months considerable lameness and stiffness of the limb, accompanied by severe pains on attempting to bend the hand and fingers. One consolation only remains, that the limb will at some remote period again enjoy perfect freedom in all its motions and be completely exempt from pain; the deformity, however, will remain undiminished through life.

THOMAS ADDISON

Thomas Addison (1793-1860) of Guy's Hospital, London, was a brilliant clinician, whose peculiar personal traits prevented him from obtaining a large practice. His paper "On the Constitutional and Local Effects of Disease of the Supra-renal Capsules," published in 1855, describes both pernicious anemia and Addison's disease.

REFERENCE

Thomas Addison. *Medical Classics*, Vol. II, No. 3, November, 1937.

ON THE CONSTITUTIONAL AND LOCAL EFFECTS OF DISEASE OF THE
SUPRA-RENAL CAPSULES

It will hardly be disputed that at the present moment the functions of the supra-renal capsules, and the influence they exercise in the general economy, are almost or altogether unknown. The large supply of blood, which they receive from three separate sources; their numerous nerves, derived immediately from the semilunar ganglia and solar plexus; their early development in the foetus; their unimpaired integrity to the latest period of life; and their peculiar gland-like structure—all point to the performance of some important office: nevertheless, beyond an ill-defined impression, founded on a consideration of their ultimate organization, that, in common with the spleen, thymus, and thyroid body, they in some way or other minister to the elaboration of the blood, I am not aware that any modern authority has ventured to assign to them any special function or influence whatever.

To the physiologist and to the scientific anatomist, therefore, they continue to be objects of deep interest; and doubtless both the physiologist and anatomist will be inclined to welcome and regard with indulgence the smallest contribution calculated to open out any new source of inquiry respecting them. But if the obscurity which at present so entirely conceals from us the uses of these organs justify the feeblest attempt to add to our scanty stock of knowledge, it is not less true, on the other hand, that any one presuming to make such an attempt ought to take care that he do not, by hasty pretensions, or by partial and prejudiced observation, or by an overstatement of facts, incur the just rebuke of those possessing a sounder and more dispassionate judgement than himself.

Under the influence of these considerations I have for a considerable period withheld, and now venture to publish, the few facts bearing upon the subject that have fallen within my own knowledge, believing, as I do, that these concurring facts, in relation to each other, are not merely casual coincidences, but are such as admit of a fair and logical inference—an inference that, where these concurring facts are observed, we may pronounce with considerable confidence the existence of diseased supra-renal capsules.

As a preface to my subject, it may not be altogether without interest or unprofitable to give a brief narrative of the circumstances and observations by which I have been led to my present convictions.

For a long period I had from time to time met with a very remarkable form of general anaemia, occurring without any discoverable cause whatever—cases in which there had been no previous loss of blood, no exhausting diarrhoea, no chlorosis, no purpura, no renal, splenic, miasmatic, glandular, strumous, or malignant disease.

Accordingly, in speaking of this form in clinical lecture, I perhaps with little propriety applied to it the term "idiopathic," to distinguish it from cases in which there existed more or less evidence of some of the usual causes or concomitants of the anaemic state.

The disease presented in every instance the same general character, pursued a similar course, and, with scarcely a single exception, was followed, after a variable period, by the same fatal result.

It occurs in both sexes generally, but not exclusively, beyond the middle period of life, and, so far as I at present know, chiefly in persons of a somewhat large and bulky frame, and with a strongly-marked tendency to the formation of fat.

It makes its approach in so slow and insidious a manner that the patient can hardly fix a date to his earliest feeling of that languour which is shortly to become so extreme. The countenance gets pale, the whites of the eyes become pearly, the general frame flabby rather than wasted; the pulse, perhaps, large, but remarkably soft and compressible, and occasionally with a slight jerk, especially under the slightest excitement; there is an increasing indisposition to exertion, with an uncomfortable feeling of faintness or breathlessness on attempting it; the heart is readily made to palpitate; the whole surface of the body presents a blanched, smooth, and waxy appearance; the lips, gums and tongue seem bloodless; the flabbiness of the solids increase; the appetite fails; extreme langour and faintness supervene, breathlessness and palpitations being produced by the most trifling exertion or emotion; some slight oedema is probably perceived about the ankles; the debility becomes extreme. The patient can no longer rise from his bed, the mind occasionally wanders, he falls into a prostrate and half-torpid state, and at length expires. Nevertheless, to the very last, and after a sickness of, perhaps, several months' duration, the bulkiness of the general frame and the obesity often present a most striking contrast to the failure and exhaustion observable in every other respect.

With perhaps a single exception the disease, in my own experience, resisted all remedial efforts, and sooner or later terminated fatally.

On examining the bodies of such patients after death I have failed to discover any organic lesion that could properly or reasonably be assigned as an adequate cause of such serious consequences; nevertheless, from the disease having uniformly occurred in fat people, I was naturally led to entertain a suspicion that some form of fatty degeneration might have a share, at least, in its production; and I may observe that, in the case last examined, the heart had undergone such a change, and that a portion of the semilunar ganglion and solar plexus, on being subjected to microscopic examination, was pronounced by Mr. Quekett to have passed into a corresponding condition.

Whether any or all of these morbid changes are essentially concerned—as I believe they are—in giving rise to this very remarkable disease, future observation will probably decide.

The cases having occurred prior to the publication of Dr. Bennett's interesting essay on "Leucocythaemia," it was not determined by microscopic examination whether there did or did not exist an excess of white corpuscles in the blood of such patients.

It was whilst seeking in vain to throw some additional light upon this form of anaemia that I stumbled upon the curious facts which it is my more immediate object to make known to the profession; and however unimportant or unsatisfactory they may at first sight appear, I cannot but indulge the hope that, by attracting the attention and enlisting the co-operation of the profession at large, they may lead to the subject being properly examined. and sifted, and the inquiry so extended as to suggest, at least, some interesting physiological speculation, if not still more important practical indications.

The leading and characteristic features of the morbid state to which I would direct attention are, anaemia, general langour and debility, remarkable feebleness of the heart's action, irritability of the stomach, and peculiar change of colour in the skin, occurring in connection with a diseased condition of the "supra-renal capsules."

As has been observed in other forms of anaemic disease, this singular disorder usually commences in such a manner that the individual has considerable difficulty in assigning the number of weeks, or even months, that have elapsed since he first experienced indications of failing health and strength; the rapidity, however, with which the morbid change takes place varies in different instances.

In some cases that rapidity is very great, a few weeks proving sufficient to break up the powers of the constitution, or even to destroy life, the result, I believe, being determined by the extent, and by the more or less speedy development of the organic lesion.

The patient, in most of the cases I have seen, has been observed gradually to fall off in general health; he becomes languid and weak, indisposed to either bodily or mental exertion; the appetite is impaired or entirely lost; the whites of the eyes become pearly; the pulse small and feeble, or perhaps somewhat large, but excessively soft and compressible; the body wastes, without, however, presenting the dry and shrivelled skin and extreme emaciation usually attendant on protracted malignant disease; slight pain or uneasiness is from time to time referred to the region of the stomach, and there is occasionally actual vomiting, which in one instance was both urgent and distressing; and it is by no means uncommon for the patient to manifest indications of disturbed cerebral circulation.

Notwithstanding these unequivocal signs of feeble circulation, anaemia and general prostration, neither the most diligent inquiry nor the most careful physical examination tend to throw the slightest gleam of light upon the precise nature of the patient's malady; nor do we succeed in fixing upon any special lesion as the cause of this gradual and extra-ordinary constitutional change.

We may, indeed, suspect some malignant or strumous disease—we may be led to inquire into the condition of the so-called blood-making organs—but we discover no proof of organic change anywhere—no enlargement of the spleen, thyroid, thymus, or lymphatic glands—no evidence of renal disease, of purpura, of previous exhausting diarrhoea, or ague, or any long-continued exposure to miasmatic influences; but with a more or less manifestation of the symptoms already enumerated we discover a most remarkable and, so far as I know, characteristic discoloration taking place in the skin—sufficiently marked, indeed, as generally to have attracted the attention of the patient himself or of the patient's friends.

This discoloration pervades the whole surface of the body, but is commonly most strongly manifested on the face, neck, superior extremities, penis, and scrotum, and in the flexures of the axillae and around the navel.

It may be said to present a dingy or smoky appearance, or various tints or shades of deep amber or chestnut-brown; and in one instance the skin was so universally and so deeply darkened that but for the features the patient might have been mistaken for a mulatto.

In some cases the discoloration occurs in patches, or perhaps rather certain parts are so much darker than others as to impart to the surface a mottled or somewhat chequered appearance; and in one instance there were, in the midst of this dark mottling, certain insular portions of the integument presenting a blanched or morbidly white appearance, either in consequence of these portions having remained altogether unaffected by the disease, and thereby contrasting strongly with the surrounding skin, or, as I believe, from an actual defect of colouring matter in these parts. Indeed, as will appear in the subsequent cases, this irregular distribution of pigment-cells is by no means limited to the integument, but is occasionally also made manifest on some of the internal structures.

We have seen it in the form of small black spots, beneath the peritoneum of the mesentery and omentum—a form which in one instance presented itself on the skin of the abdomen.

This singular discoloration usually increases with the advances of the disease; the anaemia, langour, failure of appetite, and feebleness of the heart, become aggravated; a darkish streak usually appears on the commissure of the lips: the bodywastes, but without the emaciation and dry, harsh condition of the surface, so commonly observed in ordinary malignant diseases; the pulse becomes smaller and weaker; and without any special complaint of pain or uneasiness the patient at length gradually sinks and expires.

PIERRE-CHARLES-ALEXANDRE LOUIS

Pierre-Charles-Alexandre Louis (1787-1872) spent most of his professional life as a teacher at La Pitié Hospital, in Paris. He was one of the great inspirational teachers of all time. Oliver Wendell Holmes, who, with other American students in the early nineteenth century, sat under his instruction, speaks of him almost with reverence. His ground-breaking contribution consists of his introduction of the statistical method. By figures he proved the uselessness of blood-letting in pneumonia, which largely did away with the practice. Other important studies were on tuberculosis and typhoid fever.

REFERENCES

OSLER, W. C. P. A. Louis. Bull. Johns Hopkins Hosp., 8:161-166, 1897.
STEINER, W. R. Dr. Pierre-Charles-Alexandre Louis, A Distinguished Parisian Teacher of American Medical Students. Ann. M. Hist., Third Series, 2:451-460, 1940.

DISCUSSION OF FACTORS IN THE CAUSES OF TUBERCULOSIS

[From Researches on Phthisis, Anatomical, Pathological and Therapeutical,* Chapter v]

ETIOLOGY

We have now reached the most important point in the history of phthisis, though, unfortunately, at the same time that which has hitherto been most imperfectly studied. Not assuredly that there is any penury of assertions as to the causes which predispose to, or determine the actual development of, the disease;—but that facts accurately established, facts of such stamp as to qualify them for taking part in advancing the limits of knowledge, are wanting, in regard of almost every question that may be started. And in the small number of conclusions which I shall be enabled to derive from those I have myself collected, I shall rather find materials for combating error than establishing truth.

I shall commence—following the natural order of things—by the consideration of such facts as bear upon the class of causes called predisposing; I shall then undertake that of the exciting causes.

* Recherches anatomico-physiologiques sur la phthisie, Paris, 1825. Translation by Walter Hayle Walshe, M.D., 1876.

I. Age

Age is incontestably one of the circumstances most powerfully influencing the development of tubercles. The examples of phthisis in the foetus, on record, are few in numbers; and, according to Billard and M. Baron, tubercles are rare during the first months of extra-uterine life. M. Guyot, according to the statement of M. Papavoine, opened four hundred new-born infants, without meeting with a single example of the kind. According to the latter observer, it is at the period of the first dentition, more especially when accompanied with some morbid condition, that tubercles appear in the various organs; and nevertheless, he adds, they are far from being as frequent during the first two years of life, as those immediately following. According to M. Lombard, of Geneva, tubercles are most frequent from the age of four to five years; and of nine hundred and twenty children, aged from two to fifteen years—three hundred and eighty-eight of them boys, and five hundred and thirty-two girls—opened by M. Papavoine, three hundred and twenty-eight, or nearly the three fifths, presented tubercles. And in these five hundred and thirty-two cases tubercles were, if not the sole cause, the immediate cause of death in three hundred and twenty-seven, that is to say, in more than one third of the whole number. In the two hundred and eleven others the tubercles constituted a secondary alteration only.

The following table, drawn up by M. Papavoine, shows the numerous variations in the frequency of tubercles according to age. That is to say, that from the fourth to the thirteenth year the number of children affected with tubercles appears to be invariably greater than that of those free from them; and that the frequency of tubercles is more especially great between the ages of four and seven years.

It has also been seen, in a previous part of this work, that the proportional frequency of tubercles varies greatly at different ages. But, considering in a general manner the total mass of tuberculous individuals aged either less or more than fifteen years, the proportional frequency is found to differ less than would on first thought have been anticipated, inasmuch as it follows from my own investigations at the Hospital of La Charité that about two fifths of the patients who die in that hospital are tuberculous.

OF 709 CHILDREN, FEMALE AND MALE

Age	Number Tuberculous	Number not Tuberculous	Ratio of the First to the Second	Ratio of Tubercles at Each Age to Their Mortality
2 or less.............	73	110	7 to 11	1/5 and 1/2
3...................	64	64	1 to 1	1/6
4...................	46	24	2 to 1	1/9
5...................	35	13	2 2/3 to 1	1/12
6...................	32	14	2 1/4 to 1	1/13
7...................	29	10	3 to 1	1/14
8...................	24	14	1 5/7 to 1	1/17
9...................	16	8	2 to 1	1/25
10...................	18	13	1 1/2 to 1	1/23
11...................	12	8	1 1/2 to 1	1/34
12...................	24	8	3 to 1	1/16
13...................	10	5	2 to 1	1/41
14...................	18	10	1 to 1	1/41
Age not noted........	14	0		1/29
Total..............	408	301		

THOMAS HODGKIN

Thomas Hodgkin (1798-1866), after graduating in medicine at Edinburgh in 1823, and studying intensively in France and Italy, settled in practice in London and was appointed curator of the pathologic museum and demonstrator of pathology at Guy's Hospital, London. This was one of the first chairs of this particular subject to be created. Hodgkin held it for ten years and made many important studies of the pathologic collections. His paper on the diseases of the "absorbent glands and spleen" published in 1832 first described the disease which Wilks in 1865 named Hodgkin's disease. Hodgkin was noted for his philanthropic labors. He was a close friend of Sir Moses Montefiore, the Jewish philanthropist, and while traveling with Sir Moses in the Orient contracted dysentery and died at Joffa where he is buried.

REFERENCE

THOMAS HODGKIN. *Medical Classics,* Vol. I, No. 7, March, 1937.

[From *On Some Morbid Appearances of the Absorbent Glands and Spleen**]

The morbid alterations of structure which I am about to describe are probably familiar to many practical morbid anatomists, since

* *Tr. Roy. Med. Chir. Soc. Glasgow,* Vol. XVII, 1832.

they can scarcely have failed to have fallen under their observation in the course of cadaveric inspection. They have not, as far as I am aware, been made the subject of special attention, on which account I am induced to bring forward a few cases in which they have occurred to myself, trusting that I shall at least escape severe or general censure, even though a sentence or two should be produced from some existing work, couched in such concise but expressive language as to render needless the longer details with which I shall trespass on the time of my hearers.

CASE I

November 2, 1826. Joseph Linnot, a child of about nine years of age, in Lazarus's ward, under care of J. Morgan. His brother, his constant companion, with whom he had habitually slept, died of phthisis a few months previously; he was much reduced by an illness of about nine months, during which time he had been subject to pain in back, extending round to abdomen. On his admission his belly was much distended with ascites. He had also effusion into the prepuce and scrotum. On the latter was a large ulcer induced by a puncture made to evacuate the fluid.

Head—There was a considerable quantity of serous effusion under the arachnoid and within the ventricle. There were a few opaque spots in the arachnoid, but this membrane was in other respects healthy. The pia mater appeared remarkably thin and free from vessels. The substance of the brain was generally soft and flabby, but no local morbid change was observable.

Chest—The pleura on right side had contracted many strong and old adhesions in addition to which there were extensive marks of recent pleuritis. On the left the pleura was nearly or quite free from adhesions, but there was some fluid effused into the cavity. There was some little trace of tubercular cicatrix at the summit of the right lung, but the substance of both lungs was generally light and crepitant, with a very few exceedingly small tubercules scattered through them.

The mucous membrane exhibited an excess of vascularity; the bronchial glands were greatly enlarged and much indurated. The heart appeared quite healthy.

Abdomen—There was extensive recent inflammation of the peritoneum, in the cavity of which was a copious seropurulent effusion, and the viscera were universally overlayed with a very soft and light green coagulum, too feeble to effect their union, though evidently having a tendency to do so. The mucous membrane of the stomach and intestines was generally pale and of its ordinary appearance, but in some few spots it was softened and readily separated itself from the subjacent coat. The contents of the intestines were copious and of an unhealthy character, overcharged with bile. The mesenteric glands were generally enlarged, but one or two very considerably so, equalling in size a pigeon egg, of semi-cartilaginous hardness and streaked with black matter. The substance of the liver was generally natural, but contained a few tubercules somewhat larger than peas, while semi-cartilaginous, and of uneven surface. The pancreas was firmer than usual, more particularly at its head which was somewhat enlarged. The spleen was large and contained numerous tubercules. The absorbent glands about both the last two mentioned organs were much enlarged. Both kidneys were mottled with a light colour but were free from induration. A continuous chain of much enlarged and indurated absorbent glands of light colour accompanied the aorta throughout its course, closely adherent to the bodies of the vertebrae, and extended along the sides of the iliac vessels as far as they could be traced in the pelvis. None of these vessels had been sufficiently compressed to occasion the coagulation of the contained fluid. The coats of the thoracic ducts, which were everywhere perfectly transparent, were healthy.

CASE III

By H. Peacock, Esq.

November 28, 1829. William Burrows, aged about thirty years. He was admitted into the Naaman's ward on the twenty-sixth of September, 1829, under M. J. Morgan, for ulcers of a scrofulous character in the axilla and neck, accompanied with general cachexia; he had previously been a patient in the Samaritan's ward with secondary symptoms of syphilis and was supposed to have taken large quantities of mercury. About four months before his death which occurred on November 27th, abdominal dropsy made its appearance. The body was extremely emaciated, some ragged excavated ulcers were situated about the right axilla and thorax; the ulceration extended beneath the neighbouring skin and between the pectoral muscles. The muscles of the body were pale.

The head was not examined.

The left cavity of the chest contained about a pint of serum. The lung was rather oedematous, but otherwise healthy, with the exception of some puckering, and apparently chalky deposit at its apex. The lung at the right side adhered closely to the walls of the cavity, the adhesions being firm and cellular. The lung resembled that of the right side, and was also slightly disorganized at its apex. The pericardium contained about an ounce of clear and straw-coloured fluid. The heart was small and flabby. The abdomen contained about two pints of clear serum. The stomach and alimentary canal were much distended with flatus. The liver was of a shrunken irregular shape, and was connected through the diaphragm with a few adhesions. Its structure was indurated, pale, and thickly pervaded with a substance having a white hard tuberculous character, which in some parts had the form of defined surrounded masses of the size of large pinheads, but for the most part was diffuse. Some sections exhibited parts apparently stained with a dark ecchymosis as if from extravasated blood.

From some portions of liver seen after the inspection by Dr. Hodgkin, it appeared to him that the liver was in that state in which the acini became dense rounded, and of a light colour, resembling small tubercules and are readily detached; a condition of liver which is almost peculiar to those who have laboured under a cachectic condition from mercury. The gall-bladder was small and filled with a dark-coloured green bile. The pancreas was not diseased. The spleen had contracted several firm adhesions to the neighboring peritoneum; it was enlarged to about twice its usual size and was unusually firm. Sections exhibited its structure as dense, rather dry and of a dark red colour but homogeneous. Dr. Hodgkin examined this spleen a short time after its removal from the body, and found its substance generally resembling insipient miliary tubercules of the lung, but considerably smaller than these generally are.

The kidneys were pale, flabby, and slightly mottled.

A few small miliary tubercules were found in the peritoneum, about the inguinal region, resembling those which have been noticed above in the liver. Some of the mesentery glands were much enlarged, and filled with a firm white deposit. The inguinal, lumbar and aortic glands were similarly affected. The bronchial glands were in a similar state, and also extensively ossified (or loaded with earthy matter). The axillary glands were in a state of suppuration, and exposed by ulceration at the part. The thoracic duct presented nothing unusual.

CASE IV

(The glands of the axilla and neck as might have been expected, were prodigiously enlarged, the deepest seated being in general the largest. The cellular structure around those were loose and free from any morbid deposit. These glands were smooth and of a whitish colour externally, with a few small bloody spots. When cut into, their internal structure was likewise seen to be of a light nearly white colour, with a few interspersed vessels. They were of a soft consistence, which might be compared to a testicle. They possessed a slight translucence and were nearly or quite uniform throughout, exhibiting no trace of partial softening or suppuration. Although in appearance and consistence these enlarged glands bore considerable resemblances to some fungoid tumours, they presented nothing of the encysted formation. The alteration in this case seemed to consist in an interstitial deposit from a morbid hypertrophy of adventitious growth. The glands in the groin presented precisely the same character of those just described; the same may be said of those in the thorax and abdomen, the situation and extent of which will be presently stated.)

The pleurae were nearly if not all together free from adhesions and effusion. There were a few ecchymosed spots on the posterior part of the right lung; both lungs were spongy and crepitant but rather emphysematous, and of a light colour from the small quantity of blood which they contained. The bronchial tubes contained some thick mucous.

The pericardium was healthy. The heart was greatly enlarged, and the right cavities particularly dilated; but the left were also large and distended, with thickened parietes. The muscular structure however did not appear to be diseased. The blood in the heart was barely coagulated resembling that recently drawn into a basin. The glands along the subclavian arteries and about the roots of the bronchi were much enlarged.

In the abdomen nothing particular was noticed about the peritoneum. The glands at the small curvature of the stomach, several in Glisson's capsule, and a large mass of them along the entire course of the abdominal aorta and iliac arteries were greatly enlarged. There was a marked difference in the mesenteric glands, which though larger than is natural, were none of them of the prodigious size of those above mentioned; they were however of a light colour, and their increase of size evidently depended on an interstitial deposit similar to that of the other glands. One of the enlarged glands in the lumbar region had a good deal of superficial ecchymosis. The absorbent vessels connected with it were enlarged and distended with a bloody serum. A similar fluid less deeply tinged was found in the thoracic duct.

The liver was very large, pale, slightly granular. The spleen was very greatly enlarged, being at least nine inches long, five broad, and proportionately thick; its colour was lighter and redder than is natural, and more firm and close. On cutting into it, an almost indefinite number of small white nearly opaque spots were seen pervading its substance; they were of irregular figure, but a few appeared nearly circular. They appeared to depend on a deposit in the cellular structure of the organs. There were no tubercles in the spleen, but the spots just mentioned were perhaps a commencement of this kind of formation.

The pancreas was large and pale, but otherwise healthy. The mucous membrane of the stomach and bowels offered nothing remarkable.

CASE V

Inspection of a middle aged man who had latterly been a patient of Dr. Band. He had long been in bad health, and had been for some time a patient under Dr. Bright. His last most urgent symptoms were referable to the chest. When in the hospital the former time, he was observed to have the glands of the neck, and more particularly those near the upper part of the thyroid cartilage considerably enlarged.

The body was emaciated. The glands before mentioned were still much enlarged, those in the axillae were not observed to be particularly so, those in the groin were somewhat so. The abdomen was distended.

The head was not examined. The greater part of one lung was distended, solid and void of air, its texture was soft and readily lacerable. Its colour seemed to be the result of the acute white hepatization very deeply soiled with reddish-brown. The other lung was far from healthy, but it was rather engorged and softened than hepatized and still contained air. One if not both pleurae exhibited traces of recent inflammation with little or no effusion.

Nothing remarkable was noticed in the heart or pericardium.

In the abdomen there was a large quantity of serum with little appearance of coagulable lymph. In the stomach the mucous membrane was not quite healthy presenting some indications of chronic inflammation; it as well as the intestines contained unhealthy secretions. The liver was of remarkably large size, weight upward of seven pounds. Its form and smoothness of surface were little if at all altered. The colon was somewhat mottled with a mixture of darkish green and yellow. The acini were enlarged, and it was suspected that they had undergone the fatty degeneration; but on exposure to heat, they appeared to contain little if any fatty substance. The spleen was very large, its weight unknown, but it appeared four or five times the average size; its texture was rather more solid and compact than is natural; it contained tubercles, but the cellular structure interspersed through the parenchyma was more conspicuous than is usual, in some parts appearing in the form of specks in which it was soft and easily broken down. The absorbent glands accompanying the aorta were greatly enlarged, some equalling the size of a pullet's egg; some, but more especially those in the abdomen, were reddened by injected or ecchymosed blood. The receptaculum chlyi and some of the large lymphatic branches contained blood mixed with dark and almost black coagula. The thoracic duct which was large was filled in the same manner.

CASE VI

July 19, 1830. Thos. Black, aged about fifty years, 'admitted into Barnabas Ward, on June 30, 1830, under care of Dr. Bright. He was affected with large tuberose swellings of considerable firmness on both sides of the neck, in both axillae and in both groins. His abdomen was greatly distended, he suffered from difficulty in breathing and was pale and emaciated.

It appeared that two years before he had labored under fever. That, being exposed to cold shortly after he noticed the glands swell on one side of his neck; not long after on the other side, and in succession, those in the situations above mentioned.

The body presented considerable lividity, especially the extremities of the left side. The left side of neck and left axilla presented the largest tumours.

Head was not examined.

The tumours evidently depended on greatly enlarged absorbent glands along the course of the carotid and axillary arteries. On raising the sternum they were found to extend along the subclavians and internal mammaries; they were also found, though in less number and size, along the aorta in the posterior mediastinum; but it did not appear that the bronchial glands were at all similarly affected. There was some appearance of recent pleuritis and serous effusion in the chest.

In the peritoneal cavity there was a large quantity of yellow serum mixed with some flakes of lymph. A large and continued mass of nodular glandular tumours surrounded the aorta and iliac arteries, but the mesenteric glands were very slightly affected. The omentum was corrugated. The liver was rather small, with an irregular and uneven surface, its colour was lighter than natural and the acini were converted into rounded fleshy masses without any very great change in the interacinary cellular membrane. It also contained two or three white tubercles, which resembled fungoid tubercles of the liver and were situated at surface of the organ. The structure dependent in cysts was not demonstrable in them, but from the form it might be suspected. The spleen was of moderate size and seemed quite free from any adventitious deposit, which is a fact worthy of remark, as in very many cases of glandular diseases, bearing a resemblance to the present case, this organ has been affected and generally tubercular. The pancreas was imbedded in the tumours, but appeared healthy.

It may be observed that, notwithstanding some differences in structure, to be noticed hereafter, all these cases agree in the remarkable enlargement of the absorbent glands accompanying large arteries; namely the glandular concatenatae in the neck, the axillary and inguinal glands and those accompanying the aorta in the thorax and abdomen. That as far as could be ascertained from observation, or from what could be collected from the history of the cases, this enlargement of the glands appeared to be a primitive affection of those bodies, rather than the result of an irritation propagated to them from some ulcerated surface or other inflamed texture through the medium of their inferent vessels; and that although in some instances the glands as enlarged may contain a little concrete inorganizable matter such as is known to result from what is called scrofulous inflammation, it is obvious that this circumstance is not an essential character, but rather an accidental concomitant to the idiopathic interstitial enlargement of the absorbent glandular structure throughout the body.

That unless the word inflammation be allowed to have a more indefinite and loose meaning than is generally assigned to it, this affection of the glands can scarcely be attributed to that cause, since they are unattended with pain, heat, and other ordinary symptoms, of inflammation, and are not necessarily by any alteration in the cellular or other surrounding structure, and do not show any disposition to go on to the production of pus or any other acknowledged product of inflammation, except where, as in the cases above alluded to, inflammation may have supervened, as an accidental affection of the hypertrophied structure. Nor can the enlargement in question with any better reason, be attributed to the formation of any of those adventitious structures, the production of which I have already had occasion to describe, and have referred to the type of compound adventitious cysts. Notwithstanding the different characters which this enlargement may permit, it appears in nearly all cases to consist of a pretty uniform texture throughout and this rather to be the consequence of a general increase of every part of the gland than of a new structure developed within it, pushing the original structure aside, as when ordinary tuberculous matter is deposited in those bodies. At the same time it must be admitted that the new material by which the enlargement is affected presents various degrees of organizability, which in some instances is extremely slight, and appears incompetent to maintain the vitality of the affected gland. In such cases the new structure will generally become opaque, soften, or break down and, acting as a foreign irritating body, excite irritation and lead to the formation of abcess. The case of William Burrows (No. III) and also that of a native of Owhyhee, who died in Guy's Hospital with extensive abcess in the axilla, are I believe to be considered of this kind.

Another circumstance which has arrested my attention in conjunction with this affection of the absorbent glands is the state of the spleen which, with one exception, in all the cases that I have had the opportunity of examining has been found more or less diseased, and in some thickly pervaded with defined bodies of various sizes, in structure resembling that of the diseased glands. We might from this circumstance be induced to suspect that these bodies in the spleen, like the enlarged glands themselves, are the result of the morbid enlargement of a pre-existing structure, an idea which may derive some support from the fact that, although in human spleens no glandular is distinguishable, in those of some inferior animals a multitude of minute bodies exist which appear to be of that nature.

Besides the preceding cases, of which I have been enabled to obtain the inspections, I have met with other examples in the living subject which as far as the glands were concerned were evidently of the same character with those I have been describing. One of the most remarkable occurred in the person of a Jew, apparently between forty and fifty years of age; the glands in the neck were prodigiously enlarged, forming smooth ovoid masses, unaccompanied by inflammatory symptoms or thicking of the surrounding cellular structure. The glands in the axilla and groin were in the same state; in fact in this case the enlargement was more considerable than in any other I have witnessed. His general health was much impaired; I do not recollect that there were any dropsical symptoms at the time I saw him. I accidently lost sight of him, but afterwards learned that he died about two months from the time of my seeing him.

Most of the cases, it may be observed, were those of patients in the hospital, where they had not sought admission until the disease had reached an advanced and hopeless stage. The Jew was the only individual whom I had an opportunity of treating myself, and him only for a short period, when his case had already become hopeless. The cascarilla and soda which were given with a view to improve his general health, and the iodine employed as the agent most likely to affect the glands, appeared to be productive of no advantage, on which account it is probable the patient withdrew himself from my observation. Were patients thus affected to come under my care in an earlier and less hopeless period of their malady, I think I should be inclined to endeavor as far as possible to increase the general vigour of the system; to enjoin, as far as consistent with this object, the utmost protection from the inclemencies and vicissitudes of the weather, to employ iodine externally, and to push the internal use of caustic potash as far as circumstances might render allowable. I mention this last part of the treatment in consequence of the strong commendation which Brandrich has bestowed on the use of this caustic alkali in absorbent glandular affections. The views which I have been induced to take respecting the functions of the absorbent vessels would make me the more disposed to adopt it.

KARL A. VON BASEDOW

Karl A. von Basedow (1799-1854) was a general practitioner in the little town of Merseburg, where he observed the patients who form the clinical material for his paper on exophthalmic goiter which appeared in the *Wochenschrift für die gesamte Heilkunde*, Berlin, 1840. He was not the first to describe the condition; the accounts of Caleb Hillier Parry (1755-1822) of Bath, and Robert James Graves (1796-1853) of Dublin both preceded his—Parry's in 1825, and Graves' in 1835. But Basedow's description is the clearest and the best: it was also independently made, as he apparently had never read his predecessor's accounts.

[From *Exophthalmos Due to Hypertrophy of the Cellular Tissue in the Orbit**]

Exophthalmos is to be differentiated from prolapsus bulbi caused by injury to the muscular retention apparatus (always associated with swelling and disease of the neighboring soft and hard coats of the eye ball), from osteomalacia, periostitis, exostosis, polypous disease of the frontal, antrum and ethmoidal sinuses, tumors of the brain, cystic tumors of the orbit, scirrhous tumor of the lacrymal glands, trauma, ecchymosis and inflammatory swellings of the cellular tissue in the orbit.

I have, however, had occasion, many times to observe exophthalmos which was caused by other means than the above mentioned, namely by a peculiar hypertrophy of the cellular tissue, following a disease of the heart and the large vessel trunks, of the glands and cellular tissue.

St. Yves, 1722, reported a case of exophthalmos in which none of the above diseases was the cause, and Louis, Paris, 1821, Ganz, 1837, also reported similar cases.

---

* Exophthalmos durch Hypertrophie des Zellgeswebes in der Augenhöhle. *Wchnsch. f. d. ges. Heilk.* Berlin, March 28 and April 4, 1840. Translation under direction of the editor.

Case Report: Woman 33. I had known her for 14 years. She had had many scrophulous glands in the past, but was otherwise well. She was married and a mother. Two years after the birth of her child she had Intermittens quartana with hepatitis. This was followed by icterus, which left her with a noticeably enlarged liver. This was thought to be the cause of a disturbance in the portal circulation. I bled her, gave her mercury inunctions, quinine for the fever and belladonna for the enlarged liver. The patient made a fairly rapid recovery. A year later she became ill again, with a severe case of acute rheumatism which traveled through all the joints. She had profuse weakening sweats, but finally with sublimate 1/16 grain with 2 drops of Vinum Sem. colchicum every three hours I broke it up. She was left with oedema of the legs, general emaciation, amenorrhoea, palpitation of the heart, fast pulse, feeble pulse and a sense of depression about the chest. She had shortness of breath. There was a noticeable protrusion of the eyeballs. The visual apparatus was undisturbed. She slept with open eyes, had a frightened look. Her manner was light and pessimistic and she was known in our whole town as a crazy woman.

At the same time there was a strumous enlargement of the thyroid gland, presumably the same cellular enlargement that was taking place back of the eye. The enlargement of the thyroid invited the use of iodine and digitalis. There was from all appearances improvement. In the next five years she went through two pregnancies. She had a very sickly pale complexion and very wide open, protruding eyes.

Case 2. A woman, of decided phlegmatic temperament, was subject as a child to articular rheumatism. In the years of puberty she was subject to acute tonsillitis. She lost her mother of carcinoma of the uterus. She came to puberty at 14 and married at 19. She had a child and after weaning it, and her menses were re-established she made a trip to Leipzig. There she took sick. She felt weak in her limbs, had a feeling of continuous pressure in the stomach and sense of pressure in the chest, and suddenly, in one day had two attacks of severe vomiting. It was so severe that she became unconscious and appeared to be near death. After six weeks she returned to Merseburg to her home. The next year she bore another child, nursed him and remained well. Two years later had another child which she nursed for nine months. After weaning this child she had an illness of six weeks; the cause was acute rheumatic fever. She returned to health apparently and bore another child, the fourth, nursed it, but after weaning it her menses returned with only a scanty flow, and then was entirely suppressed through a gross error in her diet. She now felt very weak, had an obstinate diarrhoea, and night sweats. She lost weight, the eyeballs began to protrude from the sockets. The patient complained of shortness of breath, sense of compression of the chest, but she could take deep inspirations. She had a very quick, feeble pulse, a clinking heart sound, could not hold her hands still and spoke exceedingly fast. She would sit down while she was hot with chest and arms bared exposed to the cold air. She showed an unusual light-heartedness and carefree attitude toward her condition. Without embarrassment she made her appearance in the shops. She indulged herself to satisfy her big appetite. She slept well, but always with wide open eyes. Her symptoms became intensified after the attempt to re-establish her menses by medical treatment.

Her arms, neck, chest and breasts were very emaciated. The abdomen full and thick. There was no tympanites over the abdomen on percussion, no fluid inside. Her thighs posteriorly were enormously thickened but not oedematous. The cellular tissue was filled with a plastic jelly, which sometimes looked like chlorosis, but upon deep pressure did not pit and upon puncture did not give off any serum.

At the neck there appeared a strumous enlargement of the thyroid. The heart impulse was now spread out and enlargement noticeable. A sawing sound was heard over the carotids. Her pulse became faster and weaker and the quickness of speech and unnatural gaiety of the patient even more pronounced. She had night sweats of very foul odor, urine scanty and red, appetite always great. Whatever the eyes signified, they protruded so far that one could see the sclera above and below the cornea. The eye lids were spread wide apart and with much force was unable to bring them together, and she slept with wide open eyes.

The position of the eyeballs was not much changed. They were turned only slightly outward. The position of the cornea and its transparency, the texture of the sclera, the position and reaction of the iris were all quite normal. The pupils were clean. One could not push the tense, bulging eyeballs back. The patient complained of pain practically not at all. She minded only the tension in the eyes. She had to blink often and had a small stream of tears in order to keep the conjunctiva from shrinking, and because of inadequate cooling had eye infections. Her sight remained unaltered. She was near-sighted as she had been from childhood.

For a long time the patient was considered crazy in our town and was next brought to a lunatic asylum. At no time did she appear sick. Never did she present any abnormal manifestations of the will. There the striking lightheartedness over her really sad condition was blamed on her phlegmatic temperament. The hastiness of her speech, the unsightly condition of her neck and hands was looked upon lightly and as a coincidence, and full attention was put on her heart symptoms.

After many applications of leeches to the breast every 8 days and the use of "Adelheidsbrunnens," there was a real improvement. Before that, only Lapis infernalis ¼ grain three times a day had been given for suspected hypertrophied heart and for the hypertrophy of the glandular system with good results. Her menstrual periods appeared again. The wide space above and below the cornea disappeared, the patient was less irritable, spoke more quietly and the exophthalmos was less.

In the fall of 1837 suppression of the menses occurred again, with the appearance of all the old symptoms. Because of this a four weeks course of the Heilbrunnens from Heilbronn was given and great improvement followed.

In the winter the patient was the victim of an epidemic of nervous gastric fever, came through it under my care, but relapsed again into her earlier condition. For a third time she returned to the wonder-working Adelheidsbrunnen, and in eight weeks was better, and even noticed some change after the third or fourth bottle. The patient now feels fairly well. She menstruates regularly. The swelling of the legs and the enlarged thyroid disappeared. The chest, neck, breasts and arms are well nourished and full. The belly is yet too thick. Her digestion is normal, but not her circulation. While her pulse is still rapid and feeble and the impulse of the heart too disseminated, yet she does not complain after walking or climbing stairs of shortness of breath or discomfort. The exophthalmos is not much changed and during the last attack had to be treated with applications to the eye because of the inability to use normal means of cooling the eyeballs. But through it all the sight remained unchanged. She has clear vision.

SIR WILLIAM WITHEY GULL

Sir William Withey Gull (1816-1890) was physician to Guy's Hospital, London, and the foremost consultant in London of his day. His description of myxedema, *A Cretinoid State in Women*, was published in 1873.

REFERENCE

HALE-WHITE, *Sir* W. Great Doctors of the Nineteenth Century. London, Edward Arnold, 1935.

[From *On a Cretinoid State Supervening in Adult Life in Women**]

The remarks I have to make upon the above morbid state are drawn from the observation of five cases. Of two of these I am able to give many details, but the three others were only seen by me on one or two occasions.

CASE I

Miss B., after the cessation of the catamenial period, became insensibly more and more languid, with general increase of bulk. This change went on from year to year, her face altered from oval to round, much like the full moon at rising. With a complexion soft and fair, the skin presenting a peculiarly smooth and fine texture was almost porcelainous in aspect, the cheeks tinted of a delicate rose-purple, the cellular tissue under the eyes being loose and folded, and that under the jaws and in the neck becoming heavy, thickened, and folded. The lips large and of a rose-purple, alae nasi thick, cornea and pupil of the eye normal, but the distance between the eyes appearing disproportionately wide, and the rest of the nose depressed, giving the whole face a flattened broad character. The hair flaxen and soft, the whole expression of the face remarkably placid. The tongue broad and thick, voice gutteral, and the pronunciation as if the tongue were too large for the mouth (cretinoid). The hands peculiarly broad and thick, spade-like, as if the whole texture were infiltrated. The integuments of the chest and abdomen loaded with

* Reprinted from the *Tr. Clin. Soc. London*, 7:180, 1873, by the New Sydenham Society, 1894.

subcutaneous fat. The upper and lower extremities also large and fat, with slight traces of oedema over the tibiae, but this not distinct, and pitting doubtfully on pressure. Urine normal. Heart's action and sounds normal. Pulse, 72; breathing, 18.

Such is a general outline of the state to which I wish to call attention.

On the first aspect of such a case, without any previous experience of its peculiarity, one would expect to find some disease of the heart leading to venous obstruction, or a morbid state of the urine favouring oedema. But a further inquiry would show that neither condition was present; nor, when minutely studied, is the change in the body which I have described to be accounted for from either of these points of view.

Had one not proof that such a patient had been previously fine-featured, well-formed, and active, it would be natural to suppose that it was an original defect such as is common in mild cretinism. In the patient whose condition I have given above, there had been a distinct change in the mental state. The mind, which had previously been active and inquisitive, assumed a gentle, placid indifference, corresponding to the muscular langour, but the intellect was unimpaired. Although there was no doubt large deposit of subcutaneous fat on the extremities, chest and abdomen, the mere condition of corpulency, obesity, or fatness, would not in any way comprehend the entire pathology.

It is common to see patients with a very superabundant accumulation of fat in the subcutaneous adipose tissues, and on that ground more inactive, without the change in the texture of the skin, in the lips and nose, increased thickness of tongue and hands, &c., which I have enumerated. The change in the skin is remarkable. The texture being peculiarly smooth and fine, and the complexion fair, at a first hasty glance there might be supposed to be a general slight oedema of it, but this is not confirmed by a future examination, whilst the beautiful delicate rose-purple tint on the cheek is entirely different to what one sees in the bloated face of renal anasarca. This suspicion of renal disease failing, any one who should see a case for the first time might suppose that the heart was the faulty organ, and that this general change in the features and increase of bulk were owing to venous congestion. But neither would be confirmed by an exact inquiry into the cardiac condition.

I am not able to give any explanation of the cause which leads to the state I have described. It is unassociated with any visceral disease, and having begun appears to continue uninfluenced by remedies.

To those about such a patient the whole morbid condition is likely to be attributed to indolent habits, and the apparent incapacity for exertion to be deemed dependent upon mere inertness of the will. No doubt extreme circumstances have a distinct influence upon these as upon other patients, but I believe the disinclination to mental or muscular activity is largely pathological.

There is certainly a degree of habitual and mental indifference, though this may under occasional circumstances be obviated, since the intellect seems to be unimpaired. It will be noticed that I have designated this state *cretinoid*. My remarks are rather tentative than dogmatical, my hope being that once the attention of the profession is called to these cases, our clinical knowledge of them will in proportion improve. That the state is a substantive and definite one, no one will doubt who has had fair opportunity of observing it. And that it is allied to the cretin state would appear from the form of the features, the changes in the lips and tongue, the character of the hands, the alterations in the condition of locomotion, and the peculiarities, though slight, of the mental state; for, although the mind may be clear and the intellect unimpaired, the temper is changed.

The occasional occurrence of cretinism in children of healthy parents, and living in healthy districts in this country, is now well known. But our experience as to its development at different periods of childhood is of the most limited kind. The whole information on the point is contained, I believe, in Dr. Fagge's Paper, and is illustrated by the second case given.

In the cretinoid condition in adults which I have seen, the thyroid was not enlarged; but from the general fulness of the cutaneous tissues, and from the folds of skin about the neck, I am not able to state what the exact condition of it was. The supra-clavicular masses of fat first described by Mr. Curling, and specially drawn attention to by Dr. Fagge as occurring in cases of sporadic cretinism in children, did not attract my attention in adults. The mass of supra-clavicular fat are not infrequent in the adult, without any associated morbid changes whatever.

FRIEDRICH THEODOR VON FRERICHS

Friedrich Theodor von Frerichs (1819-1885) was professor of medicine at Berlin. He was the first to observe leucin and tyrosine in the urine of patients with acute yellow atrophy of the liver, and he added much to our knowledge of the pathology of cirrhosis of the liver and of nephritis. His treatise on diseases of the liver, *Klinik der Leberkrankheiten*, appeared in 1858.

VARIETIES OF GRANULAR INDURATION OF THE LIVER, AND ILLUSTRATIVE CASES

[From *A Clinical Treatise on Diseases of the Liver**]

Although the main symptoms of cirrhosis of the liver are always the same, its clinical history presents manifold varieties, according to the mode of origin of the morbid process, and the existence of complications, which are for the most part dependent upon the primary causes of the disease. The simplest form is that observed in drunkards; here the hepatic affection either remains uncomplicated or is associated with Bright's Disease, or sometimes with pneumonia, delirium tremens and so forth. The disorders of the digestive organs are as a rule very prominent, because in addition to the obstruction of the circulation, the mucous membrane of the stomach is kept in a state of constant irritation by the imbibition of spiritous liquors.

The cirrhosis which occurs in syphilitic patients is often accompanied by amyloid degeneration of the spleen and kidneys, and sometimes of the liver and the mucous membrane of the intestines. The cachexia attains a high grade at an early period. In addition to this, the remains of syphilitic inflammation are found in the liver; the gland is divided into lobes by bands of areolar tissue penetrating more or less deeply into its substance, whilst the cirrhotic induration is restricted to isolated masses.

The cirrhosis, which is developed in the course of intermittent fever, is usually accompanied by enlarged pigment-spleen.

In cases where chronic inflammation, originating in the capsule

* Klinik der Leberkrankheiten, 1858. Translated from the German by Charles Murchison, M.D., F.R.C.P. In Wood's Standard Library of Medical Authors.

or in the diaphragm, attacks the glandular substance, I have observed the portal vein or the hepatic veins implicated to a great extent, the glandular parenchyma at different places uniformly indurated, and the outer surface lobulated. The following observations will illucidate more fully many details:—

OBSERVATION XXIV

Disordered Gastric Digestion.—Vomiting.—Diarrhoea.—Ascites.— Oedema of the Feet.—Puncture of the Abdomen.—Splenic Tumor. —Liver small, with nodulated surface.—Death.

Autopsy.—Cirrhotic and lobulated liver.—Thickening of Glisson's Capsule.—Firm adhesion of the lower surface of the Liver to the adjoining parts, and also of the indurated Pancreas to the Vertebral Column and Retro-peritoneal Glands.—Recent Peritonitis.

Susanne Springer, a female day-laborer, aged 54, was admitted on July 30th, 1852. Up to three years before, the patient had enjoyed good health, and menstruated regularly; but, ever since, she had been in a sickly state. Her symptoms were pains in the upper part of the abdomen, particularly after eating, failing appetite, and constipation; while the abdomen became gradually enlarged to a considerable extent. In May, June and July of 1852, haemorrhages took place from the sexual organs, which in July became so copious, that the patient applied for medical relief. With these haemorrhages the swelling of the abdomen was reduced; but the patient was attacked with diarrhoea, and vomiting of a greenish-bitter substance, which persisted for a long period and greatly exhausted her. Eight days before admission, the vomiting and diarrhoea had both subsided. A fortnight before admission, the lower extremities became oedematous to above the knees; the ascites increased greatly; the respiration was impeded; the cutaneous veins upon the abdomen and chest became distended; and the urine was dismissed in quantity, but contained no albumen. The upper part of the body was much emaciated. There was a dullness over the lower third of the left side of the thorax, while above, a rough expiratory murmur was audible. There was no dullness in the epigastrium; owing to the anasarca, the boundaries of the spleen and of the right lobe of the liver could not be defined. Under the use of diuretics and bitter remedies, the quantity of urine increased and the appetite improved; but the ascites increased. On the 4th day of August, the patient was ordered Infusion of Rhubarb with Spirit of Nitric Ether. This was followed by a firm claylike stool, while the oedema of the feet diminished. On the 5th, paracentesis was performed; 12 quarts of clear, opalescent, highly albuminous fluid were drawn off. On the 8th, the abdomen was painful when touched, and by the 11th the ascites had increased to its former amount. Infusion of Rhubarb with Bitartrate of Potash was prescribed. On the 23rd, there had been an increase in the amount of urine for some days. On September 1st, paracentesis was repeated; after the fluid was drawn off, the liver could be felt along the lower margin of the right ribs, with its margin sharp and covered with nodules. After this the patient had from two to four thin, pale stools daily. Decoction of Cascarilla Bark with Tincture of Nux Vomica was prescribed, without any benefit; the diarrhoea increased, and the patient lost strength. The urine was of normal quantity and color, and free from albumen. Death from exhaustion occurred on the 24th of September.

AUTOPSY ON SEPTEMBER 26TH

Serous effusions in both pleural sacs, but most abundant in the left; the lungs emphysematous at their anterior margins, at other places oedematous; a pulpy calcareous deposit the size of a cherry stone, surrounded by gray indurated tissue, in either apex.

The pericardium and heart normal; the large flap of the mitral valve thickened but not shortened; the blood in the right side of the heart fluid, in the left, coagulated in clots.

The mucous membrane of the stomach near the pylorus was of a deep slaty-gray hue, but not thickened. There were patches of vascular injection at many places near the lower extremity of the small intestine; the caecum and the large intestine, throughout its entire extent, was of a slaty-color, the mucous membrane being slightly oedematous and the solitary glands enlarged. The contents of the bowels were grayish-yellow and pultaceous. The mesenteric glands contained pigment, and were hard and flattened. The lymphatic glands surrounding the large vessels of the pelvis and along the vena cava were enlarged, and on section exhibited the lustre of lardaceous deposit. The areolar tissue lying along the vertebral column was increased, and of a dense character, particularly in the region of the pancreas, which was almost immovably adherent to the vertebral column, and which appeared firmer and more finely granular than in the normal state. The increase of the areolar tissue extended to the porta hepatis; at this place, portions of the great omentum, the under surface of the liver, the duodenum, the pyloric end of the stomach and the right curvature of the colon were all drawn closely towards one another, and firmly adherent. The coats of the gall-bladder were thickened; its cavity would barely hold a pigeon's egg, and its contents consisted of a grayish-white mucus; its mucous surface was of a slaty-gray hue; the ductus hepaticus was much enlarged and of a bright yellow color. The liver was divided by deep fissures into large lobes, and exhibited throughout granulations the size of a pea, which, on section, appeared dry and grayish-yellow. The organ was reduced in volume, but not more than about one-third. On the cut surface, the divided extremities of the branches of the portal vein and of the bile ducts were imbedded in thick white layers of dense areolar tissue.

The kidneys were of normal size; their capsule was easily separable. Their outer surface was granular, and also exhibited deep and superficial cicatrix-like depressions. The cortical substance was much shrivelled; the parenchyma was firm and tenacious.

The spleen was five inches long and three inches broad; it was dark-brown and contained but little blood. The uterus and ovaries were atrophied. The peritoneal cavity contained a quantity of turbid-yellow fluid; there were fibrinous deposits in the cavity of the pelvis, and upon the abdominal viscera the peritoneum itself was vividly injected, opaque and dry. The wounds of both punctures were completely cicatrized, and the corresponding part of the peritoneum was marked by a halo of gray pigment three or four lines in diameter.

In this case, as would appear from the patient's history before admission, with which the result tallied, the disease commenced as chronic peritonitis, which extended along the retroperitoneal areolar tissue, the pancreas, stomach, and lesser omentum, as far as Glisson's capsule in the *fossa hepatis*, and penetrated, with this, deep into the substance of the liver. This peritonitis accounted for the numerous adhesions of the organ, as well as for its lobulated character. The first effusion into the peritoneal cavity, which partially disappeared after the occurrence of the uterine haemorrhage, also dated from this peritonitis.

The case illustrates the statements made above as to the consequences of peri-hepatitis.

RICHARD BRIGHT

Richard Bright (1789-1858) was born in Bristol. From 1820 to 1843 he was the particular ornament of Guy's Hospital, London. He worked six hours a day in the wards and in the dead house, and lectured on clinical medicine and therapeutics. He was the leading consultant of London. The original description of the appearance of the kidneys in cases now known as Bright's Disease, appeared in *Reports of Medical Cases*, in 1827.

REFERENCE

OSMAN, A. A. Original Papers of Richard Bright on Renal Disease. London, Oxford University Press, 1937.

CASES ILLUSTRATIVE OF SOME OF THE APPEARANCES OBSERVABLE ON
THE EXAMINATION OF DISEASES TERMINATING IN
DROPSICAL EFFUSION

[From *Reports of Medical Cases Selected with a View of Illustrating the Symptoms and Cure of Diseases by a Reference to Morbid Anatomy**]

The morbid appearances which present themselves on the examination of those who have died with dropsical effusion, either into the large cavities of the body or into the cellular membrane, are exceedingly various: and it often becomes a matter of doubt how far these organic changes are to be regarded as originally causing or subsequently aiding the production of the effusion, and how far they are to be considered merely as the consequence either of the effusion or of some more general unhealthy state of the system. If it were possible to arrive at a perfect solution of these questions, we might hope to obtain the highest reward which can repay our labours—an increased knowledge of the nature of the disease, and improvement in the means of its treatment.

One great cause of dropsical effusion appears to be obstructed

* 1827.

circulation; and whatever either generally or locally prevents the return of the blood through the venous system, gives rise to effusions of serum more or less extensive. Thus, diseases of the heart which delay the passage of the blood in the venous system, give rise to general effusion, both into the cavities and into the cellular tissue. Obstructions to the circulation through the liver, by causing a delay in the passage of the blood through the veins connected with the vena portae, give rise to ascites. The pressure of the tumors within the abdomen preventing the free passage of blood through the vena cava, gives rise to dropsical effusion into the cellular tissue of the lower extremities: and not infrequently, the obliteration of particular veins from accidental pressure is the source of most obstinate anasarcous accumulation.

These great and tangible causes of hydropic swellings betray themselves obviously after death, and are often easily detected during life:—yet they include so great a variety of disease, that they still present a wide field for the observation of the Pathologist. The different diseases of the heart and of the lungs on which dropsy depends, and the various changes to which the liver is subject rendering it a cause of impediment to the circulation, are still open to much investigation. In fatal cases of dropsy we likewise find the peritoneum greatly diseased in various ways; frequently covered with an adventitious membrane more or less opaque, and capable of being stripped from the peritoneum, which is then left with its natural shining and glossy appearance. At other times the peritoneum is itself altered in structure, or is affected with tubercular or other diseases presenting an accumulation of morbid growth.

There are other appearances to which I think too little attention has hitherto been paid. They are those evidences of organic change which occasionally present themselves in the structure of the Kidney; and which, whether they are to be considered as the cause of the dropsical effusion or as to the consequence of some other disease, cannot be unimportant. Where those conditions of the Kidney to which I allude have occurred, I have often found the dropsy connected with the secretion of albuminous urine, more or less coagulable on the application of heat. I have in general found that the liver has not in these cases betrayed any considerable marks of disease, either during life or on examination after death, though the occasionally incipient disorganization of a peculiar kind has been traced in that organ. On the other hand, I have found that where the dropsy has depended on organic change in the liver, even in the most aggravated state of such change no diseased structure has generally been discovered in the kidneys, and the urine has not coagulated by heat. I have never yet examined the body of a patient dying with dropsy attended with coagulable urine, in whom some obvious derangement was not discovered in the kidneys.

Where the morbid structure by which my attention was first directed to this subject, is to be considered as having in its incipient state given rise to an alteration in the secreting power, or whether the organic change be the consequence of a long continued morbid action, may admit of doubt: the more probable solution appears to be, that the altered action of the kidney is the result of the various hurtful causes influencing it through the medium of the stomach and the skin, thus deranging the healthy balance of the circulation, or producing a decidedly inflammatory state of the kidney itself:— that when this continues long, the structure of the kidney becomes permanently changed, either in accordance with, and in furtherance of, that morbid action; or by a deposit which is a consequence of the morbid action, but has no share in that arrangement of the vessels on which the morbid action depends.

The observations which I have made respecting the condition of the urine in dropsy, are in a great degree in accordance with what has been laid down by Dr. Blackall in his most valuable treatise.

When anasarca has come on from exposure to cold, or from some accidental excess, I have in general found the urine to be coagulable by heat. The coagulation is in different degrees: it likewise differs somewhat in its character: most commonly when the urine has been exposed to the heat of a candle in a spoon, before it rises quite to the boiling point it becomes clouded sometimes simply opalescent, at other times almost milky, beginning at the edges of the spoon and quickly meeting in the middle. In a short time the coagulating particles break up into a flocculent or a curdled form, and the quantity of this flocculent matter varies from a quantity scarcely perceptible floating in the fluid, to so much as converts the whole into the appearance of curdled milk. Sometimes it rises to the surface in the form of a fine scum, which still remains after the boiled fluid has completely cooled. There is another form of coagulable urine, which in my experience has been much more rare; when the urine on being exposed to heat assumes a gelatinous appearance, as if a certain quantity of isinglass had been dissolved in water. I have indeed met with this in one or two cases only.

During some part of the progress of these cases of anasarca, I have in almost all instances found a great tendency to throw off the red particles of the blood by the kidneys, betrayed by various degrees of haematuria from the simple dingy color of the urine which is easily recognized; or the slight brown deposit; to the completely bloody urine, when the whole appears to be little but blood, and when not infrequently a thick ropy deposit is found at the bottom of the vessel.

Besides these cases of sudden anasarcous swelling being generally accompanied by coagulable urine, I have found another and apparently a very opposite state of the system prone to a secretion of the same character; namely, in persons who have been long the subjects of anasarca recurring again and again, worn out and cachectic in their whole frame and appearance, and usually persons addicted to an irregular life and to the use of spirituous liquors. In these cases the albuminous matter has coagulated, in the more ordinary way in flakes and little curdled clots; but instead of rendering the whole milky, the flocculi often incline to a brown colour, looking like the finest particles of bran more or less thickly disseminated throughout the heated urine. Occasionally in these cases the urine has been much loaded with saline ingredients becoming turbid by standing, but rendered quite clear by the application of a much lower degree of heat, than is necessary to coagulate the albumen.

In all the cases in which I have observed the albuminous urine, it has appeared to me that the kidney has itself acted a more important part, and has been more deranged both functionally and organically than has generally been imagined. In the latter class of cases I have always found the kidney decidedly disorganized. In the former, when very recent I have found the kidney gorged with blood. And in mixed cases, where the attack was recent, although apparently the foundation has been laid for it in a course of intemperance, I have found the kidney likewise disorganized.

It is now nearly twelve years since I observed the altered structure of the kidney in a patient who had died dropsical; and I have still the slight drawing which I then made. It was not however until the last two years that I had an opportunity of connecting these appearances with any particular symptoms, and since that time I have added several observations. I shall now detail a few Cases, beginning with the two first, in which an opportunity of connecting the fact of the coagulation of the urine with the disorganized state of the kidney.

CASE I

John King, aet. 34, was admitted October 12, 1825, into the Clinical Ward of Guy's Hospital, under my care. He had been a sailor till within the last four years, and was accustomed to take considerable quantities of spirits;—but he said that he had since avoided taking them, and had been engaged in turning a cutler's wheel. He was pale, and of an unhealthy appearance.

About three weeks before admission he was seized with pain in his loins, knees and ankles;—his legs soon became much swollen, and his hands and face occasionally oedematous. When admitted, the abdomen was painful on pressure. Pulse 78, rather hard; tongue natural, but pale. Bowels somewhat purged; dejections rather light coloured. Urine scanty, about one pint in twenty-four hours. Appetite good.

> Rx. Hydrarg. Oxydi cinerei gr. j,
> Pilul. Scillae compos. gr. xij,
> Opii purificat. gr. j,
> Contunde et in Pilulas iij divide hora somni quotidie sumendas.

The reports of the five following days represent him as rather improving with regard to the quantity of urine. The oedema little reduced; and he lay easiest when raised in bed to nearly a sitting posture; lying flat however, produced neither cough nor irregularity of pulse. The state of his bowels was improved by an occasional dose of castor oil with tincture of opium.

20th—Attacked with severe febrile and inflammatory symptoms with tenderness of the abdomen, pain in the chest, cough, and difficulty of lying down. Tongue furred. The pulse rose to 112 and even 120, and this accompanied with a red and turgid state of the face as if erysipelas were coming on.

> Mittatur sanguis e brachis ad uncias duodecim.
> Foveatur Abdomen.
> Sumat Mist. effervesc. cum.

Vini Ipecachuanhae mxv; sexta quaque hora, et habeat
Olei Ricini f. 3vf cum
Tinct. Opii mv vespere.

21st—The bleeding gave him relief; the blood, which was taken
in a full stream, was covered with a sizy coat nearly half an inch
thick, but was not the least cupped. In the evening the symptoms
returned with severity.

Repetatur sanguinis detractio ad f. zxij.

24th—The inflammation of the face had put on all the characters
of well marked Herpes labialis of most unusual severity, covering
not only both lips but the alae and the point of the nose. Some
blood had been passed in his motions; his urine had become more
copious, and the sediment which subsided to the bottom had
diminished.

Rx. Pulv. Ipecacuanhae gr. j,
    Hydrarg. cum Creta gr. iij,
Fiat Pulvis ter quotidie sumendus.
Foveatur Abdomen.

25th—Urine much more copious, amounting to three pints; it has
assumed the dingy brown colour which marks an admixture of the
red particles of the blood.

26th—Eruption taking its natural course of scabbing,—has not
extended since the first day of its appearance. He continues improving, but has some pain and weakness of his loins; a little pain occasionally in the shoulders and left side;—he lies down easily;—legs
continue to swell. Pulse 78, soft and of good strength. Tongue moist,
but rather furred. One small dejection with slight trace of blood.
The tenderness of abdomen gone. Urine in good quantity, tolerably
clear, but coagulates by heat.

27th—Gums sore from Mercury.

Continuantur Pulvis Ipecacuanhae et Mistura

28th—Complains of sore throat, but there is scarcely a blush of
redness to be perceived.

29th—Throat still sore.

31st—Is decidedly better—the eruption nearly gone. Legs continue to swell, though less; lies down without inconvenience, and
only complains of weakness and pain of the small of the back.
Bowels confined. Urine pretty copious, slightly turbid. Pulse 86, of
good strength. Tongue moist, very slightly furred.

He was removed on November 2nd to another ward so much improved as to be able to walk about; he was taking a grain of ipecacuanhae three times a day with reference to the disordered secretion of his bowels.

On the evening of the 10th, Mr. Stocker, the skilful and experienced apothecary of the hospital, was called, on account of a sudden attack of dyspnoea with symptoms of inflammation in the chest.

    Mittatur sanguis ad j. 3x. Applicetur Empl.
    Canthardis Sterno.

11th—He had been relieved by the bleeding. Blood covered with sizy buff, slightly coagulated, was quite unable to lie down in bed,— his pulse 120, rather indistinct in the right wrist, but not so in the left. He complains of no particular pain, but the dyspnoea is very urgent and apparently increasing,—the urine scanty.

    Rx. Hydrarg. Oxydi cinerei gr. j,
        Pilul. Scillae comp. gr. xij,
        Opii purificat. gr. j,
    In Pilulas iij divide hora somni sumendas
    Repetatur sanguinis detractio ad eandem
    qua hrei quantitatem et applicatur
    Emplastrum Cantharidis Sterno.

The bleeding gave only temporary relief; the blister was repeated on another part of the chest in two or three days. The oedema increased, and his appearance became more depressed.

    15th—
    Rx. Mistur. Camphorae f. zx.
        Liquor. Ammonae Acet. f. 3iij,
        Spir. Aether. nitr. f. 3fs.;
    Misce, fiat Haustur ter quotidie sumendus
    Repetantur Pililae.

November 18th—The symptoms had suffered little change: he sat erect in bed, leaning a little forward, during the day, and at night always wished to sit by the fire. His countenance pallid, rather shrunken, a little puffy about the eyes, and expressive of great anxiety. Hands and legs oedematous; urine very scanty. Pulse 120, quite regular. Respiration 36, with great effort. From the anxiety of his countenance coupled with the position of his body, I was led to consider the mischief to be in the pericardium.

20th—The symptoms unaltered, but he loses flesh and grows

weaker. Urine very scanty. Pulse 104, quite regular, and of considerable strength.

On the 22nd, a grain of digitalis was added to each dose of his pills.

24th—Still as before, never lying down; he complains of some tenderness in the situation of the liver. Respiration 32, performed with great effort and a slight groan on expiration, which however appears voluntary. Pulse 108, full, strong. On percussion the chest appears quite resonant, except about the region of the heart and pericardium. Dejections reported healthy.

25th—Was lying nearly flat in bed, inclined to the left side. Pulse 104. Respiration 40. Rather more urine.

29th—Lies, slightly raised in bed, rather on his left side. There is considerable oedema of the lower extremities. Respiration 32. Pulse 86, firm, hard, with a bound, perfectly regular. Urine scanty, but clear and of a natural colour. Great tenderness in the upper part of the abdomen, which he says, came on since the morning. And he likewise speaks of a sense of water rolling about in the right side of the chest, as having come on since morning. On percussion the right side of the chest is more sonorous than the left, which is rather dull. By assistance of the stethoscope I thought the sound of the heart's beat was as if performed through fluid. Head perfectly free from anything like delirium or wandering.

He died a few hours after the visit.

SECTIO CADAVERIS—NOVEMBER 30TH

Countenance purplish, bloated; some oedema of legs, which became gelatinous a few minutes after being removed. Both portions of the pericardium had many patches of a villous deposit of fibrin, thrown out recently so as to be easily peeled off in some parts, in others the fibrin was more firmly fixed. This coating of fibrin covered with a thin pellicle some inches of continuous surface on the posterior and lower part of the loose portion of the pericardium; it was also remarkable that it was attached very firmly and thickly on the heart in the course of the coronary vessels; it occurred likewise in patches of half the size of a sixpence on many parts; not forming adhesions, but presenting a rough villous surface. The heart was large and firm; the only valvular disease was in the semilunar valves of the aorta, where, in the angle between two of the valves, a triangular and solid deposit of bone of the size of a pea was found. The left lung adhered very firmly throughout most of its extent, and was in every part converted into a gray hepatized structure, very few portions admitting partially the entrance of air. There was some effusion into such parts of the cavity of the chest on this side as the nature of the adhesions admitted. The right lung was soft, and in structure was unnatural, but oedematous; filled by the effusion of serum, so that the fluid ran out mixed with innumerable fine bubbles of air immediately it was cut into. The whole cavity of the chest on this side was filled with serum, but the lung not compressed by it.

A pint or two of clear and transparent serum was effused into the cavity of the abdomen. The intestines and stomach were greatly distended with flatus, and there was an appearance as if the vessels running along the large curvature of the stomach were distended with air; an oblique hernia was found on the right side; a few of the mesenteric glands were enlarged to the size of horse-beans. The peritoneal coat of the liver covered and rendered somewhat opaque by a very thin coating of fibrin apparently not very recent, and a number of flocculent deposits of the same kind. In the size and substance of this organ no obvious disease; rather pale coloured, of a purplish drab throughout, and not of a firm consistence. The gall bladder full of healthy bile, and larger than natural. The pancreas healthy. The spleen dark coloured, with a slight adventitious covering. The kidneys were completely granulated throughout: externally the surface rough and uneven; internally all traces of the natural organization nearly gone, except in the tubular parts, which were of a lighter and more pink colour than usual.

In this case we have a very well marked example of a granulated condition of the kidneys, connected with the secretion of coagulable urine. If we can form any judgment of the priority of disease from the more advanced state of organic change, we shall be inclined to consider that the disease in the kidney was first established, and had probably laid the foundation for that effusion into the cellular membrane which had taken place previously to his admission.

There was no evidence whatever of organic disease in the liver from the beginning, except the account he gave us of his mode of life. Examination after death afforded no ground for the opinion that either the viscera of the chest or the liver were in the first place materially diseased. On the contrary, the organization of the liver and its functions, as far as any means of judging could be afforded by inspection after death or observation during the progress of the disease, remained unimpaired to the very last; and in the morbid appearances of the heart, though evidently connected with the fatal result of the case, were of a nature to evince recent inflammatory action on the pericardium, and not that state of disease which has commonly been observed in connexion with general dropsical effusion. The diseased state of the left pleura was evidently a matter of longer standing, and the firmness of the adhesion gave ground for supposing that some pleuritic attack must have existed previously to his admission; it is not however at all improbable, that greater part of the mischief done to the substance of the left lung had taken place between the 20th of October, when he suffered the severe inflammatory attack, and the 29th of November when he died. The serous effusion which was found more particularly in the right lung, might have been, and most probably was, one of the last circumstances which took place near the close of life. At the same time this case came under my care, my mind was not made up as to the indications which were to be derived from the albuminous quality of the urine; and therefore, though I noticed the fact, I did not afterwards so regularly mark the progressive changes of this secretion as I have since been in the habit of doing. I have however no reason to suppose that it lost its tendency to coagulate. The dingy colour occasionally communicated to the urine in this case by admixture of blood, serves further to connect it with the other cases of dropsy with diseased kidney which I have seen; and it is worthy of remark, that the patient complained often of pain and weakness in the loins, a symptom which is not infrequently connected with this peculiar disease of the kidneys.

JEAN-MARTIN CHARCOT

Jean-Martin Charcot (1825-1893) was physician to the great Paris hospital of the Salpêtrière, used for patients with nervous disease. His clinics there were world famous. His original descriptions are too numerous to list—the most notable being of the triad in multiple sclerosis, gastric crises in tabes, the atrophic joints in tabes, the signs and psychology of hysteria, and various muscular atrophies.

REFERENCE

BEESON, B. B. Jean Martin Charcot. *Ann. M. Hist.*, Vol. X, No. 2, June, 1928.

ON ARTHROPATHIES OF CEREBRAL OR SPINAL ORIGIN

[From *Lectures on the Diseases of the Nervous System**]

Nutritive disorders consecutive on lesions of the nervous centres not unfrequently take up their seat in the articulations. The varieties presented by these articular affections, according to the nature of the cerebral or spinal lesions from which they arise, have led me to establish two principal categories.

A. The first comprises arthropathies of acute or subacute form, accompanied by tumefaction, redness, and sometimes by pain of a more or less severe character. This form was indicated for the first time, if I mistake not, by an American physician, Professor Mitchel, who observed it in the paraplegia connected with Pott's disease of the vertebrae, in which, however, it is very rare, in my opinion. It happens more frequently as a consequence of a traumatic lesion of the spinal cord, as we find from the sufficient evidence of the cases, above quoted, which have been recorded by MM. Vigues and Joffroy. A case of concussion of the cord, related by Dr. Gull, supplies an analogous demonstration.

Acute or subacute inflammation of the joints of paralysed limbs may supervene also, in spontaneous myelitis; as examples of this class, I may mention a case reported by Dr. Gull, and another case which M. Moynier published in the "Moniteur des Sciences Medcales" for 1859. The second case relates to a young man, aged eighteen, who, after lodging for a long time in a damp place, and undergoing great fatigue, had presented all the symptoms of subacute myelitis. Paralysis of motion began to show itself on the 25th of January; it became complete on the 9th of February. On the 23rd of the same month, the skin of the sacral region presented an erythematous patch which gave place to an eschar, on the 5th of March. On the 6th of this month, there was severe pain in the right knee,

* Translated from the French by George Sigerson, M.D., M.Ch. New Sydenham Society, 1878.

which was swollen, and in which the sensation of fluctuation was perceptible. In addition, there was painful tumefaction of the tibio-tarsal articulation of the same side. On the 9th of March, the knee had decreased in size, and on the same day, eschars made their appearance on the heels. The autopsy revealed a focus of ramollissement situated not quite two inches above the cauda equina.

Finally, in a case of central myelitis in a child, having its origin in the neighborhood of a solitary tubercle situated in the cervical region of the cord, Dr. Gull records the formation of an intra-articular effusion, occupying one of the knees, at the time when the paralysis began to invade the lower extremities.

It is remarkable to see these arthropathies, consecutive on the different acute and subacute forms of myelitis, frequently forming, when the muscles of the paralysed limbs are beginning to waste away, or again when an eschar is being rapidly developed on the breech.

The arthropathy of paraplegic patients, first described I believe in 1846, by Scott Alison, afterwards by Brown-Séquard, and the anatomical and clinical characters of which I have made known, belongs, if I mistake not, to the same category. In this second variety, as well as in the first, the arthropathies are limited to the paralysed limbs and mostly occupy the upper extremities. They supervene, especially, after circumscribed cerebral ramollissement (en foyer), and, more rarely, as a consequence of intra-encephalic haemorrhage.

They usually form fifteen days or a month after the attack of apoplexy, that is to say, at the moment when the tardy contracture that lays hold on the paralysed members appears, but they may also show themselves at a later epoch. The tumefaction, redness, and pain of the joints are sometimes marked enough to recall the corresponding phenomena of acute articular rheumatism. The tendinous sheaths are, indeed, often affected at the same time as the articulations.

I have shown that we have here a true synovitis with vegetation, multiplication of the nuclear and fibroid elements which form the articular serous membrane, and augmentation in number and volume of the capillary vessels which are there distributed. In intense cases, a sero-fibrinous exudation is produced, with which are mingled, in various proportions, white blood-corpuscles that may become abundant enough to distend the synovial cavity. The diarthrodial cartilages and ligamentous parts have not hitherto appeared to present any concomitant lesion perceptible to the naked eye. On the other hand, the tendinous synovial sheaths, in the neighborhood of the affected joints, take part in the inflammatory process, and appear greatly congested.

It is needless to insist upon the interest which pertains to these arthropathies as regards diagnosis,—articular rheumatism, whether acute or subacute, being an affection often connected with certain forms of cerebral softening, and one which, shows itself also, occasionally, after traumatic causes capable of determining shock in the nervous centres. On the other hand, many affections of the spinal cord are erroneously attributed to a rheumatic diathesis in consequence of the coexistence of these articular symptoms. The clinical characters which render it easy to recognise arthropathies correlated with lesions of the nervous centres, and which allow them to be distinguished from cases of rheumatic arthritis, are chiefly these:

1. Their limitation to the joints of the paralysed members.

2. The generally determinate epoch in which, in cases of sudden hemiplegia, they make their appearance on the morbid scene.

3. The coexistence of other trophic troubles of the same order, such as eschars of rapid formation; and (when the spinal cord is involved) acute muscular atrophy of the paralysed members, cystitis, nephritis, &c.

B. The type of the second group is to be found in progressive locomotor ataxia. Allow me to fix your attention for an instant upon this species of articular affection, in which I take a paternal interest, all the more lively because the signification I attached to it has had to encounter many sceptics. And at first, a word as to the clinical characters of the arthropathy of ataxia patients.

This disorder generally shows itself at a determinate epoch of the ataxia, and its appearance coincides, so to speak, in many cases with the setting in of motor incoordination.

Without any appreciable external cause, we may see, between one day and the next, the development of a general and often enormous tumefaction of the member, most commonly without any pain whatever, or any febrile reaction. At the end of a few days the general tumefaction disappears, but a more or less considerable swelling of the joints remains, owing to the formation of hydarthus; and sometimes to the accumulation of liquid in the periarticular serous bursae also. On puncture being made, a transparent lemon-colored liquid has been frequently drawn from the joint.

One or two weeks after the invasion, sometimes much sooner, the existence of more or less marked cracking sounds may be noted, betraying the alteration of the articular surfaces which, at this period, is already profound. The hydarthus becomes quickly resolved, leaving after it an extreme mobility in the joint. Hence consecutive luxations are frequently found, their production being largely aided by the wearing away of the heads of the bones which has taken place. I have several times observed a rapid wasting of the muscular masses of the members affected by the articular disorder.

Ataxia arthropathy usually occupies the knees, shoulders, and elbows; it may also take up its seat in the hip-joint. The anatomo-pathological information which we possess respecting it, is as yet very imperfect. However, one character is apparently constant, namely, the enoromus wearing down which it exhibited in a very short space of time by the articular extremities. At the end of three months, this head of a humerus which I show you, and which belonged to a female patient in whom we were enabled to study the invasion of the arthropathy, was, as you may remark. . . to a great extent destroyed. I would call your attention to the fact that you do not find on this specimen the bony burr around the worn surface, which would not fail to be present if this were a case of common dry arthritis.

I now place before you in order to establish the contrast, a knee-joint also taken from a woman who presented the symptoms of ataxic arthropathy, but in whom the articular affection was of much older date. Besides the wearing down of the articular surfaces which, as in the preceding case, is carried very far, you notice here the presence of foreign bodies, of bony stalactites, and, in a word, of all the customary accompaniments of arthritis deformans. These latter alterations, I repeat, were absolutely wanting in the first case. On this account, I am led to believe that they are nowise necessary, and that they are produced in an accidental manner, and to all appearance chiefly by the more or less energetic movements to which the patient sometimes continues to subject the affected members.

I wish to confine myself at present to this indication of the most general features of the arthropathies of ataxic patients, for this is a subject which I propose to treat hereafter in more detail. What I have to say will suffice, I hope, to show that the articular affection in question, itself also, the expression of trophic disorders directly dependent on the lesion of the spinal nerve-centre. But here are the principal arguments upon which I base my opinion.

I would point out, in the first place, the absence of all traumatic or diathetic cause of rheumatism or of gout, for instance, which might explain the appearance of the articular disease in the cases which I have studied. Herr R. Volkmann has said that the arthropathy of ataxic patients is simply the result of the distension of the articular ligaments and capsules, in consequence of the awkward manner of walking peculiar to this class of persons. The cases, which are now numerous, in which our arthropathy affected the upper extremities, and occupied either the shoulder or the elbow, are sufficient to prove that the interpretation proposed by Volkmann could have but a very narrow bearing. The influence of a mere mechanical cause cannot be invoked, at least not as a principal agency, even in cases where the arthropathy occupies the lower extremities. I have, in fact, taken care to point out, supporting my words by oft repeated clinical observations, that the articular affection in question is developed at a comparatively early epoch of the sclerosis of the posterior columns, and at a time when motor incoordination is as yet null, or scarcely manifest.

The clinical characters of our arthropathy are, besides, really special. Its sudden invasion, marked by the general tumefaction of the member; the rapid alterations of the articular surfaces; finally, its appearance at, as it were, a determinate epoch of the spinal disease with which it is connected, constitute so many peculiarities which are, if I err not, found together in no other articular affection.

But here is a more direct argument. Holding as we did that the arthropathy in question is a trophic lesion consecutive on the disease of the spinal cord, we yet could not think of connecting it with any of the common alterations of progressive locomotor ataxia—with sclerosis of the posterior columns, posterior spinal meningitis, or atrophy of the posterior roots of the spinal nerves. On the other hand, a minute examination of many cases had taught us that it was impossible to invoke a lesion of the peripheral nerves. It is in the grey matter of the anterior cornua of the cord that the starting point of this curious complication of the ataxia is to be found according to our belief. It is not very rare to find the spinal grey matter affected in locomotor ataxia; but the lesion is then generally found in the posterior cornua. Now, it was quite different in two cases of locomotor ataxia, complicated with arthropathy, in which a careful examination of the cord has been made; the anterior cornua were, in both cases, remarkably wasted and deformed, and a certain number of the great nerve-cells, those of the external group especially, had decreased in size, or even disappeared altogether without leaving any vestiges. The alteration, besides, showed itself exclusively in the anterior cornu corresponding to the side on which the articular lesion was situated. . . . It affected the cervical region, in the first case, where the arthropathy occupied the shoulder; it was observed, a little above the lumbar region, in the second case which presented an example of arthropathy of the knee. Above and below these points, the grey matter of the anterior cornua appeared to be exempt from alteration.

It was asked whether this alteration of one of the anterior cornua of the cord, which microscopical examination reveals, may not be a result of the functional inertia to which the corresponding member has been condemned on account of the articular lesion. This hypothesis must be rejected because, on the one hand, in both of our cases, the members affected by the arthropathies had preserved to a great degree their freedom of motion; and, on the other hand, the lesion of the grey matter differed essentially here from that which is produced after the amputation of a member, or the section of the nerves supplying it.

From what precedes, I hope to have made it appear at least highly probable that the inflammatory process, first developed in the posterior columns, by gradually extending to certain regions of the anterior cornua of the grey matter was able to occasion the development of the articular affection in our two patients. If the results obtained in these two cases are confirmed by new observations, we should be naturally led to admit that arthritic affections connected with myelitis, and those observed to follow on cerebral softening, are likewise due to the invasion of the same regions of grey matter of the spinal cord. In cases of brain softening, the descending sclerosis of one of the lateral columns of the cord might be considered as the starting point of the diffusion of inflammatory work.

MM. Paturban, Remak, and quite recently, Herr Rosenthal have observed in progressive muscular atrophy, arthropathies which by their clinical characters are closely allied with those of ataxic patients. This is nothing surprising, if we remember that a primary or secondary irritative lesion of the nerve-cells of the anterior cornua of the spinal grey matter appears, in the majority of cases, to be the starting-point of the amyotrophy which, in clinical practice, is usually designated by the name of progressive muscular atrophy. For to-day, gentlemen, I shall stop here in this investigation, which I expect to bring to a conclusion at our next conference.

JAMES PARKINSON

James Parkinson (1755-1824), a London practitioner, published his *Essay on the Shaking Palsy* in 1817. He reported the symptoms and the pathologic appearance of the appendix in a case of appendicitis, in 1812, ascribing the symptoms to the true cause. He was also a geologist and a social reformer.

REFERENCE

JAMES PARKINSON. *Medical Classics,* Vol. II, No. 10, June, 1938.

[From *An Essay on the Shaking Palsy**]

The advantages which have been derived from the caution with which hypothetical statements are admitted, are in no instance more obvious than in those sciences which more particularly belong to the healing art. It therefore is necessary, that some conciliatory explanation should be offered for the present publication: in which, it is acknowledged, that mere conjecture takes the place of experiment, and that analogy is the substitute for anatomical examination, the only sure foundation for pathological knowledge.

When, however, the nature of the subject, and the circumstances under which it has been here taken up, are considered, it is hoped that the offering of the following pages to the attention of the medical public, will not be severely censured. The disease, respecting which the present inquiry is made, is of a nature highly afflictive. Notwithstanding which, it has not yet obtained a place in the classification of nosologists; some have regarded its characteristic symptoms as distinct and different diseases, and others have given its name to disease differing essentially from it; whilst the unhappy sufferer

* 1817.

has considered it as an evil, from the domination of which he had no prospect of escape.

The disease is of long duration: to connect, therefore, the symptoms which occur in its later stages with those which mark its commencement, requires a continuance of observation of the same case, or at least a correct history of its symptoms, even for several years. Of both these advantages the writer has had the opportunities of availing himself: and has hence been led particularly to observe several other cases in which the disease existed in different stages of its progress. By these repeated observations, he hoped that he had been led to a probable conjecture as to the nature of the malady, and that analogy had suggested such means as might be productive of relief, and perhaps even of cure, if employed before the disease had been too long established. He therefore considered it to be a duty to submit his opinions to the examination of others, even in their present state of immaturity and imperfection.

To delay their publication did not, indeed, appear to be warrantable. The disease had escaped particular notice; and the task of ascertaining its nature and cause of anatomical investigation, did not seem likely to be taken up by those who, from their abilities and opportunities, were most likely to accomplish it. That these friends to humanity and medical science, who have already unveiled to us many of the morbid processes by which health and life is abridged, might be excited to extend their researches to this malady, was much desired; and it was hoped, that this might be procured by the publication of these remarks.

Should the necessary information be thus obtained, the writer will repine at no censure which the precipitate publication of mere conjectural suggestions may incur; but shall think himself fully rewarded by having excited the attention of those, who may point out the most appropriate means of relieving a tedious and most distressing malady.

DEFINITION-HISTORY-ILLUSTRATIVE CASES SHAKING PALSY
(PARALYSIS AGITANS)

[Chapter I]

Involuntary tremulous motion, with lessened muscular power, in parts not in action and even when supported; with a propensity to bend the trunk forwards, and to pass from a walking to a running pace: the sense and intellects being uninjured. The term Shaking Palsy has been vaguely employed by medical writers in general. By some it has been used to designate ordinary cases of Palsy, in which some slight tremblings have occurred; whilst by others it has been applied to certain anomalous affections, not belonging to Palsy. The shaking of the Limbs belonging to this disease was particularly noticed, as will be seen when treating of the symptoms, by Galen, who marked its peculiar character by an appropriate term. The same symptom, it will also be seen, was accurately treated of by Sylvius de la Boë. Juncker also seems to have referred to this symptom: having divided tremor into active and passive, he says of the latter, "ad affectus semiparalyticos pertinent; de qualibus hic agimus, quique *tremores paralytoidei* vocantur." Tremor has been adopted, as a genus, by almost every nosologist; but always unmarked, in their several definitions, by such characters as would embrace this disease. The celebrated Cullen, with his accustomed accuracy observes, "Tremorem, utpote semper symptomaticum, in numerum generum recipere nollem; species autem a Sauvagesio recensitas, prout mihi vel astheniae vel paralysios, vel convulsionis symptomata esse videntur," his subjungam. Tremor can indeed only be considered as a symptom, although several species of it must be admitted. In the present instance, the agitation produced by the peculiar species of tremor, which here occurs, is chosen to furnish the epithet by which this species of Palsy, may be distinguished.

HISTORY

So slight and nearly imperceptible are the first inroads of this malady, and so extremely slow is its progress, that it rarely happens, that the patient can form any recollection of the precise period of its commencement. The first symptoms perceived are, a slight sense of weakness, with a proneness to trembling in some particular part; sometimes in the head, but most commonly in one of the hands and arms. These symptoms gradually increase in the part first affected; and at an uncertain period, but seldom in less than twelve months or more, the morbid influence is felt in some other part. Thus assuming one of the hands and arms to be first attacked, the other, at this period becomes similarly affected. After a few more months the patient is found to be less strict than usual in preserving an upright posture: this being most observable whilst walking, but sometimes whilst sitting or standing. Sometimes after the appearance of this symptom, and during its slow increase, one of the legs is discovered slightly to tremble, and is also found to suffer fatigue sooner than the leg of the other side; and in a few months this limb becomes agitated by similar tremblings, and suffers similar loss of power.

Hitherto the patient will have experienced but little inconvenience; and befriended by the strong influence of habitual endurance, would perhaps seldom think of his being the subject of disease, except when reminded of it by the unsteadiness of his hand, whilst writing or employing himself in any nice kind of manipulation. But as the disease proceeds, similar employments are accomplished with considerable difficulty, the hand failing to answer with exactness to the dictates of the will. Walking becomes a task which cannot be performed without considerable attention. The legs are not raised to that height, or with that promptitude which the will directs, so that the utmost care is necessary to prevent frequent falls.

At this period the patient experiences much inconvenience, which unhappily is found daily to increase. The submission of the limbs to the directions of the will can hardly ever be obtained in the performance of the most ordinary offices of life. The fingers cannot be disposed of in the proposed directions, and applied with certainty to any proposed point. As time and the disease proceed, difficulties increase: writing can now be hardly at all accomplished with some difficulty. Whilst at meals the fork not being duly directed frequently fails to raise the morsel from the plate: which, when seized, is with much difficulty conveyed to the mouth. At this period the patient seldom experiences a suspension of the agitation of his limbs. Commencing, for instance in one arm, the wearisome agitation is borne until beyond sufferance, when suddenly changing the posture it is for a time stopped in that limb, to commence, generally, in less than a minute in one of the legs, or in the arm of the other side. Harassed by this tormenting round, the patient has recourse to walking, a mode of exercise to which the sufferers from this malady are in general partial; owing to their attention being thereby somewhat diverted from their unpleasant feelings, by the care and exertion required to ensure its safe performance.

But as the malady proceeds, even this temporary mitigation of suffering from the agitation of the limbs is denied. The propensity to lean forward becomes invincible, and the patient is thereby forced to step on the toes and fore part of the feet, whilst the upper part of the body is thrown so far forward as to render it difficult to avoid falling on the face. In some cases, when this state of the malady is attained, the patient can no longer exercise himself by walking in his usual manner, but is thrown on the toes and fore part of the feet; being, at the same time, irresistibly impelled to take much quicker and shorter steps, and thereby to adopt unwillingly a running pace. In some cases it is found necessary entirely to substitute running for walking; since otherwise the patient, on proceeding only a very few paces, would inevitably fall.

In this stage, the sleep becomes much disturbed. The tremulous motion of the limbs occurs during sleep, and augments until they awaken the patient, and frequently with much agitation and alarm. The power of conveying the food to the mouth is at length so much impeded that he is obliged to consent to be fed by others. The bowels, which had been all along torpid, now, in most cases, demand stimulating medicines of very considerable power: the expulsion of the faeces from the rectum sometimes requiring mechanical aid. As the disease proceeds towards its last stage, the trunk is almost permanently bowed, the muscular power is more decidedly diminished, and the tremulous agitation becomes violent. The patient walks now with great difficulty, and unable any longer to support himself with his stick, he dares not venture on this exercise, unless assisted by an attendant, who walking backwards before him, prevents his falling forwards, by the pressure of his hands against the fore part of his shoulders. His words are now scarcely intelligible; and he is not only no longer able to feed himself, but when the food is conveyed to the mouth, so much are the actions of the muscles of the tongue, pharynx, &c., impeded by impaired action and perpetual agitation, that the food is with difficulty retained in the mouth until masticated; and then as difficultly swallowed. Now also, from the same cause, another very unpleasant circumstance occurs: the saliva fails of being directed to the back part of the fauces, and hence is continually draining from the mouth, mixed with the particles of food, which he is no longer able to clear from the inside of the mouth.

As the debility increases and the influence of the will over the muscles fades away, the tremulous agitation becomes more vehement. It now seldom leaves him for a moment; but even when exhausted nature seizes a small portion of sleep, the motion becomes so violent as not only to shake the bed-hangings, but even the floor and sashes of the room. The chin is now almost immovably bent down upon the sternum. The slops with which he is attempted to be fed, with the saliva, are continually trickling from the mouth. The power of articulation is lost. The urine and faeces are passed involuntarily; and at the last, constant sleepiness, with slight delirium, and other marks of extreme exhaustion, announce the wished-for release.

CASE I

Almost every circumstance noted in the preceding description, was observed in a case which occurred several years back, and which, from the particular symptoms which manifested themselves in its progress; from the little knowledge of its nature, acknowledged to be possessed by the physician who attended; and from the mode of its termination; excited an eager wish to acquire some further knowledge of its nature and cause.

The subject of this case was a man rather more than fifty years of age, who had industriously followed the business of a gardener, leading a life of remarkable temperance and sobriety. The commencement of the malady was first manifested by a slight trembling of the left hand and arm, a circumstance which he was disposed to attribute to his having been engaged for several days in a kind of employment requiring considerable exertion of that limb. Although repeatedly questioned, he could recollect no other circumstance which he could consider as having been likely to have occasioned his malady. He had not suffered much from Rheumatism, or been subject to pains of the head, or had ever experienced any sudden seizure which could be referred to apoplexy or Hemiplegia. In this case, every circumstance occurred which has been mentioned in the preceding history.

MORITZ HEINRICH ROMBERG

Moritz Heinrich Romberg (1795-1873) was professor of neurology at Berlin. According to Garrison his *Lehrbuch der Nervenkrankheiten* (1840-46) was the first formal treatise on nervous diseases. It contains the description of Romberg's sign in ataxia.

TABES DORSALIS

[From *A Manual of the Nervous Diseases of Man**]

The spinal cord viewed as a central organ, not only serves as an agent for the mutual transmission of stimuli, but also as a source of nervous power, of the principle of motor and sensory tension, by which the continuance and vigour of motion and sensation is secured, and a general stimulus for the entire organism provided. The disease, which is characterised by a diminution of this power, is termed tabes dorsalis.

The first symptom by which it is manifested is reduction of the motor power in the muscles, first and foremost in the inferior extremities; at the commencement one leg may be affected more than the other, but in the progress of the disease both suffer. The patient complains of weakness and inability to perform any movement or endure any position for a continuance. If he is required to attempt any act demanding a larger consumption of motor power, *e.g.* to bend down or to stand on one foot, his strength at once fails; the practised rider is unable to hold on to his horse as long as usual. Early in the disease we find the sense of touch and the muscular sense diminished, while the sensibility of the skin is unaltered in reference to temperature and painful impressions. The feet feel numbed in standing, walking, or lying down, and the patient has the sensation as if they were covered with a fur; the resistance of the ground is not felt as usual, its cohesion seems diminished, and the patient has a sensation as if the sole of his foot were in contact with wool, soft sand, or a bladder filled with water. The rider no longer feels the resistance of the stirrup, and has the strap put up a hole or two. The gait begins to be insecure, and the patient attempts to

* Lehrbuch der Nervenkrankheiten, 1840-1846. Translated and edited by Edward H. Sieveking, M.D. Sydenham Society, 1853.

improve it by making a greater effort of the will; as he does not feel the tread to be firm, he puts down his heels with greater force. From the commencement of the disease the individual keeps his eyes on his feet to prevent his movements from becoming still more unsteady. If he is ordered to close his eyes while in the erect posture, he at once commences to totter and swing from side to side; the insecurity of his gait also exhibits itself more in the dark. It is now ten years since I pointed out this pathognomonic sign, and it is a symptom which I have not observed in other paralyses, nor in uncomplicated amaurosis; since then I have found it in a considerable number of patients, from far and near, who have applied for my advice; in no case have I found it wanting. Some patients mention the circumstance without being asked about it; one gentleman, a foreigner, whose eyesight was unimpaired, told me that he was at present unable to wash himself in his dark bedroom while standing; and that if he wished to keep his balance he was obliged to have a light while performing his toilet. Another, whose business rendered it necessary for him to go out at six o'clock in the morning, complained that he required some one to support him in the house and out of doors, but that he could dispense with assistance in full daylight.

Independently of this peculiarity, there is also a difference in the movements themselves; the patient experiences a greater difficulty in executing forced and limited movements, than those in which he merely follows the impulse of his inclinations; he finds it much more laborious to walk slowly with a measured step in a given direction, than to let his feet take their own course; rising from the chair, or going up stairs, is more difficult than sitting down or descending; the most difficult matter is to turn round in walking. After prolonged rest, walking and standing are more laborious and insecure than when once begun. The loss of muscular power is also manifested in organs provided with a sphincter, and especially in the bladder. At the commencement of the disease, the desire to micturate occurs more frequently, and cannot be gratified soon enough, for the patient is unable to retain his urine till the utensil is brought to him. Enuresis not infrequently occurs during sleep. The urine is not discharged in an arched jet, as in health, but falls more perpendicularly; nor is the bladder entirely emptied. Costiveness prevails almost universally; the patient feels that he is unable to strain as long and as forcibly as before. Painful sensations of different kinds almost invariably accompany the affection; the most common is a sense of constriction, which proceeds from the dorsal or lumbar vertebrae, encircles the trunk like a hoop, and not unfrequently renders breathing laborious. Several of my patients have described this sensation as particularly troublesome during sleep, causing them suddenly to start up and scream out. Others complain of a heavy weight pressing upon the rectum and the bladder, others again of colic and gastric pains; the majority suffer from pain shooting through the legs, and a sense of pricking, itching, burning, or cold in the skin of the lower as well as of the upper extremities; the face alone is an exception. Formication very rarely occurs in the back. These symptoms may endure for a considerable time, and at first they attract little attention. After an uncertain period the weakness of the legs diminishes visibly. The patient, owing to the threatening loss of balance, is obliged to evert his feet, and walk with his legs apart; he leaves his heels as long as possible in contact with the ground, and keep his knees bent; he is still able to propel himself, (one of my patients stated to me, that he found it necessary to think of every one of his movements), and to totter along the streets, but if arrested in his progress, he is unable to stand still without clinging to some support. The patient's own strength soon fails to support him, and he is obliged to have recourse to assistance. The necessity of employing his eyes becomes more and more urgent; if he closes his eyes, even while sitting, his body begins to sway to and fro; in one case the patient was unable to maintain himself erect in his chair, and slid down to the ground; when in an horizontal posture, the patient is no longer able to recognise the position of his own limbs, and cannot tell whether the right leg is crossed over the left or the reverse. A foreigner, who was a patient of mine, told me that in visiting the diorama he had not the slightest sensation of his progression when led from the light to the dark apartment. The condition of these unfortunate individuals is rendered the more distressing by the circumstances that amblyopia often supervenes; in many cases it is associated with the disease from the commencement. Even when the optic nerve was not implicated, I have repeatedly found a change in the pupils of one or both eyes, consisting in a contraction with loss of motion, which in one case, that of a man aged 45, attained to such a height that the pupils were reduced to the size of a pin's head. In one case, where there was no cerebral affection, a strabismus towards the inner angle took place, the patient at the same time being able to move his eye outwards at will. As the disease progresses, the loss of power also extends to the superior extremities, though they are not affected to the same degree as the inferior. The sphincter of the bladder becomes completely paralysed; erections cease, and the virile power becomes extinct. The intellect of these patients generally remains unimpaired; the majority do not complain much, and they are inclined to represent their condition, especially to the medical man, in a too favorable light; if they are members of the higher classes, they anxiously endeavour to conceal their loss of motor power, in order to avoid the evil reputation of being affected with tabes dorsalis. Nutrition is not impaired in a measure corresponding to the dimunition of motor and sensory power. Such patients may even retain their embonpoint for a considerable time, so that the term tabes does not apply to this feature. At a later period the muscles become flaccid and atrophied, especially about the nates, the legs, and the back. Towards the termination of the disease the patient becomes utterly incapable of holding himself erect or moving; still he continues able to execute movements with his feet at will when the trunk is supported. Diuresis alternates with ischuria; the faeces pass off involuntarily. Gangrene at the sacrum and trochanters, accompanied by febricitations, ushers in death.

Tabes dorsalis is a chronic disease, which may extend over several —as many as ten and fifteen—years. It is only shortened by complication with other more rapidly fatal diseases, especially pulmonary and intestinal phthisis. Intercurrent diseases may also accelerate its progress.

Although the post-mortem records of this disease may present considerable variations, they almost without exception show the existence of partial atrophy of the spinal cord; the lumbar portion and the nerves given off from it are the parts generally affected. The loss of substance, which may amount to one half or two thirds of the healthy spinal cord, either affects the grey and the white substance, or only one of them. It would be well always to have a fresh, healthy specimen at hand for the sake of comparison, and to render the examination more satisfactory. As yet we possess no microscopic investigation of the atrophied portion. The contents of the cords of the cauda equina have often been found to have disappeared to such an extent that nothing but the empty nurilemmatous sheaths seemed to remain. The roots of nerves inserted at a higher segment of the cord also suffer from atrophy; and it is a point of especial interest to observe that the posterior, sensory roots, are occasionally alone affected in conjunction with the posterior columns of the spinal cord, the anterior motor columns and nerves retaining their normal structure. A remarkable instance of this occurred to me in the person of a medical practitioner of a provincial town, of 52 years of age; after violent emotions and severe colds, caught in the prosecution of his profession, he had in his fortieth year been attacked with partial paralysis of the lower extremities and amblyopia, for which, at the suggestion of myself and the late Professor Rust, he went through a course of the Marienbad waters, but to no purpose; the amblyopia passed into complete amaurosis, and tabes dorsalis became fully developed in spite of all the remedies employed. I did not see the patient again; but I ascertained that the insensibility of the skin was maintained to the last and that he correctly appreciated variations of temperature. I was present at the post-mortem examination which was made by Professor Froriep, and the spinal cord, compared with the fresh cord of a man of the same age, only amounted to two thirds of its normal size; I was not a little surprised to find that the atrophy was confined to the lower part of the posterior columns and nerves. The medullary tissue of the former had almost entirely disappeared, so that they were translucent, and of a greyish-yellow colour. The posterior roots of the nerves were deprived of their nerve matter, and presented a watery appearance. From the middle of the dorsal nerves upwards, the atrophy passed into a healthy condition. The anterior columns and roots of the nerves presented no abnormity. Froriep has observed the same in another case of tabes dorsalis. When the disease has been accompanied by amaurosis, we almost invariably find the optic nerve, the chiasma, and the optic tracts atrophied; one or both optic thalami are also either atrophied, or they exhibit changes of texture and colour. The other morbid changes found in tabes dorsalis vary; sometimes the white substance presents a coriaceous condensation, but it is more usual to find softening of the grey matter. In 1832, I examined a man, aged 42, who had been under my care for three years for tabes dorsalis, and found the lumbar, the cervical, and a portion of the dorsal region of the cord of an almost fluid consistency; it was traversed by a number of white longitudinal fibres, as if a delicate cauda equina passed through the cord. The meninges rarely retain their healthy condition; the arachnoid is thickened and beset with cartilaginous and osseous plates, and contains more or less serum. It is exceptional to meet with morbid changes in the osseous envelopes.

Two circumstances that have been shown with certainty to pre-dispose to tabes dorsalis are the male sex and the period between the thirtieth and fiftieth year of life. Scarcely one eighth of the cases are females. The loss of semen has always been looked upon as one of the most fruitful sources of the complaint; but this in itself does not appear to be a matter of much consequence as influencing the dis-ease, as patients who have been labouring under spermatorrhoea for a series of years, are much more liable to hypochondriasis and cere-bral affection than to tabes dorsalis; but when combined with hyper-stimulation of the nerves to which sensual abuses give rise, it not unfrequently favours the origin and encourages the development of the disease after it has commenced. When the strength is much taxed by continued standing in a bent posture, by forced marches, and the catarrhal influences of wet bivouacs, followed by drunkenness and debauchery, as is so often the case in campaigns, the malady is rife; this is the reason why tabes dorsalis was so frequent during the first decennia following the great wars of the present century. Rheu-matism appears to be the morbid process which most frequently gives rise to it. The cases are not rare in which the most careful examination fails to establish an exciting cause.

There is no prospect of recovery for patients of this class; the fatal issue is unavoidable; the only consolation that can be offered to those fond of existence is the long continuance of the disease. If in any case the busy activity of the physician increases the sufferings of the patient, it is in tabes dorsalis. When one of these unfortunate indi-viduals presents himself to us, we generally find his back seamed with cicatrices, he brings us a heap of prescriptions, and gives a long list of the watering-places he has visited in search of health. It is but common humanity to inform him at once that therapeutic interfer-ence can only injure, and that nothing but the regulation of his diet can retard the calamitous issue. Every unnecessary tax made upon the motor powers, as well as sexual excitement, ought to be strictly prohibited. The best remedy for the obstinate costiveness is to be found in cold water enemata; the careful use of cold water in wash-ing the trunk and spinal cord, and in the shape of affusion to the latter, may be recommended. I have employed an ointment contain-ing veratrine with benefit against the painful sensations in the back and extremities. The thing most to be avoided is the frequent application of cupping and issues; nor are long journeys to watering-places advisable, because the driving itself is injurious, and the baths will only afford temporary relief, which will disappear on the return of the patient. Incurable patients should be allowed to spend their lives quietly in their family circle, that their last moments may be soothed by the fond cares of those whom they love.

GUILLAUME-BENJAMIN-AMAND DUCHENNE

Guillaume-Benjamin-Amand Duchenne (1806-1875) was born in Boulogne, France, and practiced there and in Paris. His investigations of nervous disorders were so extensive that as Collins said, he found "neurology a sprawling infant of unknown parentage, which he succored to a lusty youth."

REFERENCE

Selections from the Clinical Works of Dr. Duchenne (De Boulogne). Translated, edited, and condensed by G. V. Poore, M.D. London, The New Sydenham Society, 1883.

THE SITUATION AND PROGRESS OF MUSCULAR ATROPHY

[From *Selections from the Clinical Works of Dr. Duchenne,**
Chapter II]

Commencement in the upper limbs.—As a rule before becoming general, progressive muscular atrophy attacks the upper limb, and destroys its muscles in an irregular fashion. It begins in such cases by attacking one after another the muscles of the thenar eminence, spreading from the superficial to the deep layer. As soon as the abductor pollicis is wasted, its absence is marked by a depression, and by the attitude, during repose, of the first metacarpal bone, which lies too close to the second. When the deep muscles are affected the thenar eminence gets quite flat, and the first metacarpal bone always lies in the same plane as the second. Depressions of the hypothenar eminence and interosseal spaces next announces the atrophy of the muscles of those regions. The loss of the interossei muscles is shown by the claw-like attitude of the fingers during the extension of the hand. The functional troubles caused by these atrophies are considerable, and are described in my work, *Physiologie des Mouvements* (1861).

* L'Electrisation Localisée, 3rd ed., pp. 486-563. Translated by G. V. Poore, 1883.

If the flexors and extensors of the fingers become in their turn atrophied, the hand (which was "clawed" while the atrophy was limited to the interossei) assumes the death-attitude, and the forearm is literally dissected.

Occasionally, but not often, the muscles on the back of the forearm are first attacked, but even in these cases the preceding muscles are soon involved. The atrophy may remain thus localised for many years, as was the case with the patient whose hand is represented in figs. 7 and 8.

I have always been careful to examine the state of the muscles of all parts of the body, and have ascertained that generally the atrophy was limited in the first instance to the hand and forearm. But this limit once passed the muscles of the arm and trunk waste irregularly and partially.

The flexors of the elbow and the deltoid are the first to atrophy, sometimes the one and sometimes the other taking precedence. The triceps extensor cubiti is the last of the muscles of the upper limb to become affected. Once, indeed, I have seen it completely destroyed, while the rest of the muscles of the shoulder-joint were but little affected.

Whenever all the muscles of the arm have been atrophied, I have found a greater or less number of the muscles of the trunk in the same condition. The patients are nearly always in ignorance of the time when these muscles began to waste, because, not being much hindered in their movements, their attention was not forcibly drawn to them, unless the serratus magnus, one of the most necessary muscles for the movement of the upper limb, has been affected early, as was the case in the patient represented in fig. 9.

The following is the order in which the trunk muscles usually waste. First, the lower half of the trapezius disappears, and then the spinal border of the scapula is further removed from the median line than on the sound side. The clavicular portion of this muscle is generally the ultimum moriens of all the muscles of the trunk and neck. Next in succession atrophy attacks the pectorales, the latissimi dorsi, the rhomboidei, the levatores anguli scapulae, the flexors and extensors of the head, the erectores spinae (*les sacro-spinaix*), and the muscles of the abdomen. At this time I have usually seen the muscles of breathing and swallowing become affected. The atrophy equally invades the lower limbs, but only when the muscles of the upper limbs and trunk are in great destroyed. It is most marked in the flexors of the ankle and hip. The other muscles of the lower limbs may become atrophied in the long run.

I have not seen atrophy attack both sides at once, but when one muscle or group of muscles is affected the corresponding muscles are usually attacked at no distant time, and before the disease spreads to other regions. Thus I have seen a certain number of cases in which the same muscles were successively attacked in the two hands. The same has also occurred with the deltoids and the serrata mangi. Such is the usual mode of development of progressive muscular atrophy.

It must be borne in mind, however, that the course of development may be different. Of this the following is an example.

Case No. 4.—In a patient from Barcelona, aged 32, whom I treated in consultation with Professor Trousseau, and in whom the atrophy became general in two years, the muscles were attacked in the following order. The muscles of the right hand were first attacked, then the flexors of the left ankle; next the left hand was attacked, and this was followed by atrophy of the flexors of the right ankle and right hip; then, in different degrees, the atrophy affected the biceps, deltoids, and muscles of the trunk, neck, and face. When I first saw him the diaphragm, tongue, and muscles of swallowing were so seriously affected as to endanger the patient's life by asphyxia or starvation.

ROBERT WILLAN

Skin diseases were the subject of treatises by Daniel Turner (1667-1740), Joseph Pleuck (1732-1807), and Antoine Charles Larry (1726-1783), but the first systematic and scientific classification was that of Robert Willan (1757-1812), of London, whose *On Cutaneous Diseases* appeared in 1808. "His greatest feat," according to Dr. William Allen Pusey in his *History of Dermatology*, "was his grouping of a great number of various forms of dermatitis under the generalization eczema."

CLASSIFICATION OF SKIN DISEASES

[From *On Cutaneous Diseases*,* Vol. 1]

INTRODUCTION

. . . 1. To fix the sense of the terms employed, by proper definitions.

2. To constitute general divisions or orders of the diseases, from leading and peculiar circumstances in their appearance; to arrange

* Philadelphia, Kimber and Conrad, 1809.

them into distinct genera; and to describe at large their specific forms, or varieties.

3. To class and give names to such as have not been hitherto sufficiently distinguished.

4. To specify the mode of treatment for each disease.

To complete adequately a plan so extensive must be considered an undertaking of much difficulty; and perhaps exceeding the powers of any individual. My own observations are principally founded on the Cutaneous Diseases occurring in London, and its vicinity. I intend, however, to compare them with the accounts of similar complaints in ancient and modern writers.

It is proper here to mention that an outline of my plan for the arrangement and description of Cutaneous Diseases, formerly presented to the Medical Society of London, was honoured with the Fothergillian Gold Medal for the year 1790. I beg leave to express my obligation to the Society for this distinguished testimony of their favour; and shall think myself happy, if the more enlarged view which I shall now endeavour to give of the subject should afford satisfaction to them, and to the public.

Consistently with the principles above laid down, I proceed, in the first place, to define the sense of several technical terms employed in the following pages.

DEFINITIONS

I. *Scurf* (Furfura): small exfoliations of the cuticle, which take place after slight inflammation or irritation of the skin, a new cuticle being formed underneath during the exfoliation.

II. *Scale* (Squama): a lamina of morbid cuticle, hard, thickened, whitish and opaque. Scales have at first the figure and extent of the cuticular lozenges, but they afterwards often increase into irregular layers, denominated *Crusts*. Both Scales and Crusts repeatedly fall off, and are reproduced in a short time.

III. *Scab:* a hard substance covering superficial ulcerations, and formed by a concretion of the fluid discharged from them.

IV. *Stigma:* a small bright red speck in the skin, without any elevation of the cuticle. Stigmata are generally distinct or apart from each other. When they coalesce, and assume a dark red, or lived colour, they are termed *Petechiae*.

V. *Papula:* a very small and accuminated elevation of the cuticle, with an inflamed base, not containing a fluid, nor tending to suppuration. The duration of Papulae is uncertain, but they terminate for the most part in Scurf.

VI. *Rash* (Exanthema): consists of red patches on the skin, variously figured, in general confluent, and diffused irregularly over the body, leaving interstices of a natural colour. Portions of the cuticle are often elevated in a rash, so as to give the sensation of an uneven surface. The eruption is usually accompanied with disorder of the constitution, and terminates, in a few days, by cuticular exfoliations.

VII. *Macula:* a permanent discolouration of some portion of the skin, often with a change of its texture, but not connected with any disorder of the constitution.

VIII. *Tubercle:* a small, hard, superficial tumor, circumscribed, and permanent, or proceeding very slowly to suppuration.

IX. *Wheal:* a rounded, or longitudinal elevation of the cuticle, with a white summit, hard, but not permanent, not containing a fluid, nor tending to suppuration.

X. *Vesicle* (Vesicula): a small, orbicular elevation of the cuticle, containing lymph, which is sometimes clear and colourless, but often opaque, and whitish, or pearl-coloured. Vesicles are succeeded either by Scurf, or laminated Scabs.

XI. *Bleb* (Bulla): a large portion of the cuticle detached from the skin by the interposition of a transparent watery fluid. Soon after the water is discharged, the excoriated surface is covered with a flat, yellow, or blackish Scab, which remains till a new cuticle is formed underneath. Both Vesicles and Blebs, when they have a dark red, or livid base, are by medical and chirurgical writers denominated *Phlyctaenae.*

XII. *Pustule:* an elevation of the cuticle, with an inflamed base, containing Pus. Pustules are various in their size, but the diameter of the largest seldom exceeds two lines.

Some forms of Pustules have been distinguished by specific appellations; as

1. *Phlyzacium:* a Pustule, raised on a hard circular base, of a vivid red colour. It is succeeded by a thick, hard, dark coloured Scab.

2. *Psydracium:* a minute Pustule, irregularly circumscribed, producing but a slight elevation of the cuticle, and terminating in a laminated Scab. Many of these pustules usually appear together, and become confluent. After the discharge of Pus, a thin watery humour exudes, which often forms an irregular incrustation.

3. *Achor:* an acuminated Pustule, of intermediate size between

the two foregoing, which contains a straw-coloured matter, having the appearance, and nearly the consistence of strained honey. It appears most frequently about the head, and is succeeded by a thin brown or yellowish scab.

4. *Cerion*, or *Favus*: this Pustule is somewhat larger than the Achor, and contains a more viscid matter; its base is but slightly inflamed, and it is succeeded by a yellow semi-transparent, and sometimes cellular Scab, like a honeycomb.

I propose to arrange Cutaneous Diseases in eight Orders, to be characterised by the different appearances of Papulae, Scales, Rashes, Bullae, Vesicles, Pustules, Tubercles, and Maculae. By comparing together the 2d, 5th, 6th, 7th, 8th, 10th, 11th, and 12th Definitions, the distinguishing characters of each Order may be readily understood. They will also be further illustrated in treating of the Orders respectively.

FERDINAND VON HEBRA

Ferdinand Hebra (1816-1880) founded the clinic in dermatology in the Allgemeines Krankenhaus of Vienna. His most important contribution was in describing the histologic pathology of skin lesions. Hebra, according to the late Dr. William Allen Pusey, in his *History of Dermatology*, "made his first studies and his first reputation in the study of scabies. . . . Imbued with the current notions of humeral pathology [he] first took the view that scabies was essentially a systemic disease, . . . but quickly convinced himself . . . that scabies was a local disease. . . . This led him to experiment with other irritants, particularly croton oil, and he demonstrated that all of the changes of inflammation of the skin might be produced by purely local causes. . . . He demonstrated the growth of ringworm fungus on the epidermis. . . . His exposition of the essential nature of eczema is still the last word on the subject and it would be fortunate if it were oftener read by dermatologists today."

ECZEMA

[From *Diseases of the Skin**]

Definition.—The name Eczema[1] is now applied to a disease of the skin, of usually chronic course, characterised either by the formation of aggregated papules and vesicles, or by more or less deeply red

* Lehrbuch der Hautkrankheiten, 1860. Translated for the New Sydenham Society, 1866.

[1] In German, Ekzem, nässende Flechte, Salzfluss; in French, dartre squameuse humide; in English, moist tetter.

patches covered with thin scales, or in other cases by a moist surface; while in any of these forms there may be developed in addition, partly yellow and gummy, partly green or brown crusts. This affection is constantly accompanied by violent itching, which leads to excoriations, and it is not contagious.

As the reader will see from this definition, I understand the term eczema in a different way from that hitherto accepted by former and by most living dermatologists. I do not consider the formation of vesicles, and subsequently of a moist surface deprived of its epidermis, as sufficient to characterise the disease; but take in as varieties of the same malady all the morbid changes seen in the course of development and retrogression of the ordinary vesicular and moist eczema. The justification of this lies in the following facts:—

A. We are able by the action of irritants upon the skin to produce eczema, and may then observe that vesicles or moist places do not follow in each case, but sometimes only redness and desquamation, sometimes papules not bigger than pins' heads, and occasionally the rapid formation of pustules and crusts.

B. There are very many cases in which we may notice on the same patient, in one place minute scales upon a red surface of skin, in another red miliary papules, in a third elevations of the epidermis filled with watery fluid, and in others again spaces partly deprived of their cuticle, moist, infiltrated, and covered here and there with yellow points of suppuration, or with partly green, partly yellowish-brown crusts.

C. Observation of the course of individual cases of eczema teaches us that many begin with the formation of smaller or larger vesicles, some of which develop into pustules, others burst and form moist surfaces, and others become covered with yellow crusts; while the surrounding surface is occupied by a papular eruption, or is simply red and desquamating. Towards the end of the morbid process all the pustules will have turned to scabs, and these, after completely drying up, fall off and leave the affected parts covered with minute scales, and more or less red and infiltrated.

The above facts will surely suffice, with the help of clinical observation, to lead all experienced physicians to the same conviction to which my own studies have brought me, that eczema must be studied under five different forms. Inasmuch, however, as there are still sceptics as to the correctness of this view, and authorities who treat the attempt to simplify the diagnosis of cutaneous diseases as a crotchet of my own,[1] I may be allowed to set forth more fully the above-mentioned three grounds for my opinion.

(A.) A simple experiment which any one may repeat on his own person, or on patients at his disposal, gives almost conclusive evidence of the identity of my several forms of eczema. Let any application which will produce artificial eczema, e.g., croton oil, be rubbed into the same parts upon different individuals, say into the flexor surface of the forearm, and that to the same extent. Or, if the experimenter has only a single patient, this one will suffice; different parts of the surface must be selected—the flexor and extensor surfaces of the extremities, the palm or sole, the face, breast, back or genitals—and any convenient quantity of croton oil, say five drops, should be carefully and equally applied with a camel's-hair brush to all these places. Within a few hours certain changes will appear on such regions as the face, the bend of the joints, and the genitals, which, however, do not in each case exhibit the same appearance. In some, as the scrotum and penis, a well-marked œdema and redness will be observed, and often innumerable *minute vesicles* in addition; in the face the swelling is usually greater, and the formation of vesicles less; while in the extremities the orifices of the hair follicles will be swollen, forming *red papules*, and here and there vesicles will also appear. If now these parts be left without further applications, whether irritating or medicinal, there will, in the great majority of cases, have ensued such a change within a few days that the swelling, the papules, and the vesicles, will all have disappeared, and only a slight redness and desquamation will remain in proof of a previous inflammation of the skin. But if on the day after the first experiment croton oil be again applied to the same places, we shall observe not only fresh eruptions appear where the first application did not take effect, but the previous ones will assume more striking and extensive forms—so that the papules already produced will develop into vesicles, obviously because the amount of the subepidermic exudation has increased, and thus becomes visible beneath the cuticle. As a rule, the effects of the second application likewise pass away in a few days, redness and desquamation of the skin again marking the last stages. But if the affected parts be a third time, and again and again

[1] Cf. Veiel's Mittheilungen über die Behandlung der chronischen Hautkrankheiten, 1862, p. 104.

treated with croton oil, the morbid process will grow in intensity as well as in extent. We shall see in such a case not only the parts actually brought in contact with the irritant assume a greater number and development of papules or vesicles, but also the surrounding skin which was untouched by the brush will be included in the extending circle, and will exhibit the same primary changes which were observed in the part originally attacked.

An artificial eczema of this intensity will but rarely terminate as above described with desquamation. In most cases the quantity of exudation will not merely allow the formation of vesicles, but will break through the layers of epidermis which form these, and thus oozing out will form moist or weeping patches.

While this form is being developed at some points, the contents of other vesicles which do not burst undergo metamorphosis into pus. In this way the original vesicles become *pustules*, and thus a fresh form of the disease is presented. Moreover, the influence of the pus upon the surrounding parts is not without result in the change of symptoms. The redness and swelling of the skin surrounding each pustule is increased, and the itching which accompanied the formation of vesicles is changed to a sensation of pain. But the difference between the contents of vesicles and pustules shows itself again in their further metamorphosis, for while the former appeared disposed to dry up, but not to form thick yellow or brown crusts, the pus contained in the latter, even without direct contact with the air, dries up within its epidermic chamber, and develops the firm concretions of various form and colour, which are known as scabs and crusts.

After the artificial eczema has thus arrived at its climax, it sooner or later begins to go through retrograde changes. As soon as progress stops, and no fresh eruption appears, the whole of the vesicles and pustules gradually dry up, the scabs thus formed are pressed upon by the encroaching healthy epidermis, and falling off at last, exhibit the skin beneath red, more or less infiltrated, and covered with scales—the dead remains of diseased epidermis. This appearance is in fact what I have already described as occurring in the slighter forms of eczema.

If now we analyze the morbid conditions presented to us in the complete course of eczema thus artificially produced, we shall have no difficulty in reducing these appearances to five primary ones.

First, that observed immediately after the first application of croton oil, and characterised by the eruption of *papules* and of *vesicles*; next that which is produced by the continued action of the same irritant, the formation of *red, weeping patches*; then the further condition resulting from the development of *pustules* and *scabs* from the papules and vesicles; and lastly, the stage of *redness* and *desquamation*.

If we give special names to these forms, we shall be abundantly justified in laying down as a law that eczema appears and runs its course in five different varieties, which, arranged according to their relative intensity, will be—

(1.) *Eczema squamosum* = *Pityriasis rubra*.

(2.) *E. papulosum*, also called *E. lichenoides* or *Lichen eczematodes*.

(3.) *E. vesiculosum* = *E. solare* of Willan.

(4.) *E. rubrum* seu *madidans*.

(5.) *E. impetiginosum*, or *E. crustosum* of some writers.

Which of these varieties will follow the application of an irritant depends upon its quantity and strength; thus, tartar-emetic ointment, turpentine, cantharides, croton oil or mezereum, will produce more intense forms of eruption than preparations of sulphur, salts of copper, iron or zinc, or than potash and soda soaps. The length of application is another cause of difference in the effects. Transitory irritants are more easily borne, and cause slighter injury than those which last longer, and especially those which are uninterrupted. Moreover we must take into account the specific vulnerability of the patient, which may vary greatly. For while in some cases the skin is as sensitive as a daguerreotype plate, and breaks into an eczematous eruption under the slight stimulus of light, in others it will bear severe irritation before showing the least reaction. The state of health of the patient at the time must also be remembered. For persons who have borne a cutaneous irritant with impunity while in good health, may, as soon as illness supervenes, be at once attacked with eczema. Lastly, the several regions of the skin differ in their sensibility to the irritants above mentioned. Thus the integument of the genitals, the face and the flexor surfaces of joints, shows less power of resistance, and is more easily attacked by eczematous eruptions than that of the extensor surfaces of the limbs and of the back; while we find most indisposition to reaction in the skin of the palms and soles, which is destitute of sebaceous follicles.

(B.) When eczema spreads over large surfaces so as to include at the same time the scalp, the face, and the trunk and extremities, either continuously or in different patches, it is rare for all the affected regions to present the same form of the disease; we much more often find its several varieties of forms represented in the different parts. The scalp and face are more frequently the seat of an impetiginous eczema, while the skin of the external ear, the back of the neck, the axilla, and flexures of joints exhibits *E. rubrum*, the extremities *E. papulosum* and *E. vesiculosum*, and the trunk, *E. squamosum*. Now it will, I think, be obvious to every one, that when we find the whole surface of the body attacked at once, it is more natural to assume the disease to be one and the same, than, according to the rules of diagnosis hitherto recognised, to call the affection of the scalp porrigo, *Tinea mucosa,* or *T. granulata,* or achor; that of the face, *Porrigo larvalis* or *Impetigo faciei rubra,* or *Crusta lactea s. serpiginosa,* or *Melitagra flavescens,* while the title of eczema is only allowed to the moist and vesicular patches on the trunk or extremities. Nay, writers have even chosen to call the crusts which appear at the same time from the eczematous exudation drying up, by the special name impetigo, and the red and desquamating parts, *Pityriasis rubra,* thus separating them from the other phenomena of the disease.

Any one who has ordinary opportunities of observing such extensive eruptions of eczema in their rarer acute or more frequent chronic form upon either adults or children, must certainly agree with my view, that all these appearances are symptoms of a single disease, of eczema, and cannot endorse the notion of the public and of many even in the profession who pronounce the patient to be suffering from a scald head, a tetter of the body, and an itch of the hands.

(C.) But observation of the natural course of an attack of eczema furnishes the most unassailable proof of the connection between its various forms. In one case an eruption of vesicles begins the series of symptoms, in another it is preceded by the appearance of red, scaly patches, or groups of papules; or vesicles and papules are developed together, some of the former rapidly changing to pustules, and forming yellow, gum-like crusts by the drying up of their contents. It is

evident from this that even the primary form of eczema is not necessarily vesicular, but that papules and pustules mingled with vesicles,
or red and infiltrated patches, may mark the onset of the disease.
Moreover, as the eruption advances, it frequently undergoes a change
of form. Not only does the red moist surface of *E. rubrum* appear
after the removal of the crusts, but also the parts first affected with
vesicles often undergo repeated relapses, and the superficial layers
of epidermis being thus undermined, display the deeper ones of the
rete Malpighii. These last moist surfaces shortly exhibit the crusts
described above, and so the form of eczema changes with the progress
of the disease, until at length, when the more active exudative
process has ceased, and the period of involution been reached, we
observe a more or less infiltrated, reddened and desquamating surface, which may be called *E. squamosum (Pityriasis rubra)*. Here
again it will, I believe, be admitted as more probable that these
forms are all metamorphoses of eczema, than that, after the doctrine
of former and even certain living authorities, one disease has passed
into another; eczema into impetigo, or porrigo, or tinea, or *Pityriasis
rubra*, or *Melitagra flavescens*.

I have thus demonstrated that certain affections of the skin, hitherto described under different names, must be regarded as only parts
of one and the same disease, namely, eczema. It remains for me to
describe more particularly the forms of eczema which are partly the
result of its greater or less acuteness, partly of the secondary changes
in the skin and subcutaneous tissue caused by the persistence of the
affection, and partly of its localization in certain regions.

Although attacks of eczema usually follow a chronic and even
tedious course, they also assume in many cases an acute type, and
quickly result in recovery; we must, therefore, in the first place,
separate acute from chronic eczema. And if the admission of an acute
eczema appear to contradict the characteristics of the disease in general, and its place seem rather to be among acute inflammations of
the skin, as dermatitis and *Erysipelas vesiculosum*, I must remark
that there appears in practice a far less intimate relation and slighter
likeness between these inflammatory disorders and acute eczema,
than between the latter and its chronic variety. The simple fact that
every chronic attack of eczema exhibits the acute form in the first
days of its course, and only differs from this by repeated relapses each
of which again repeats the acute type, together with the observation

that an originally acute attack of eczema often ends by assuming a chronic character, is sufficient proof that we have to do with one and the same disease in an acute or in a chronic form.

ACUTE ECZEMA

Acute eczema is characterised in general by inflammatory redness and swelling of the skin; yet not to that degree which we are accustomed to in inflammations of the skin κατ'ἐξοχήν, as erysipelas, where the swelling of the integuments causes such an extreme tension as to produce a smooth, glistening surface. On the contrary, we never find the skin in even the most severe cases of eczema, to be stretched and shining, but observe it œdematous and swollen, and covered with a greater or less number of *minute vesicles*, containing either a clear watery serosity, or, sometimes even from the first, a yellowish fluid. The size of these vesicles also varies, the yellow ones being usually much smaller, while those containing a transparent colourless secretion may attain the size of hemp seeds or peas.

The eruption of these vesicles usually takes place suddenly, and is completed in a space of time not exceeding forty-eight hours. In the most favorable cases they dry up after lasting from six to eight days; the layers of epidermis which cover them fall off in the shape of minute whitish or sometimes darker scales, and leave a perfectly normal surface behind. The sensations accompanying this eruption are those of burning and tension, and it is only towards the end of desquamation that the patient complains of a slight itching. The participation of the system generally in the morbid process is shown either by irritation, sleeplessness, and a feeling of diffused chilliness, without increased frequency of the pulse, or by slight febrile symptoms, which accompany the eruption of vesicles and disappear in a few days at most. An attack of eczema following this course has been often and is still occasionally mistaken for erysipelas, but is, as I have shown, essentially distinct from it.

It is more often the case that within the second week of the attack fresh outbreaks of eczema appear either all around the previously affected parts, or only at certain points, or in quite different regions of the body. These follow a precisely similar course, and may themselves be succeeded after an interval by a third or a fourth attack, each running its own acute course; or relapses follow each other so quickly upon the spots first affected, that they produce the characters of chronic eczema. In these last cases a different set of sensations is experienced by the patient; a violent itching leads irresistibly to scratching, and from this fresh morbid appearances result—excoriations in all their varieties, and still deeper changes in the skin which, in their turn, pass through their own series of transformations.

Acute eczema is, moreover, frequently followed by the form of disease produced by the rapid change of the clear watery contents of the primary vesicles to pus, and the desiccation of the latter into yellow gummy crusts. This has been separated from eczema by certain writers, and described by Fuchs as *Impetigo faciei rubra*, when it occurs in the face; by Alibert, as *Melitagra flavescens*, and *M. nigricans*, and by other authors as porrigo, *Crusta lactea*, etc.

As I have shown above, these are but varieties of eczema, and do not only occur in chronic cases, but have often so rapid a course that we must include them under the acute form of the disease. When the serum exuded in an attack of eczema has accumulated under the epidermis, or has broken it through and reached the surface, it dries into the previously described dark and thin or yellow and gummy crusts, under which the epidermis repairs itself, and after the scabs have fallen off appears with only a slightly heightened colour. We may observe eczema running this course, not only on certain parts of the body, but spreading over the entire surface from the crown of the head to the sole of the foot. Not a single region appears to be exempt from an acute attack of this malady; but there are certain parts most frequently affected, and thus we may notice as the more important local varieties of acute eczema the following:—

1. *E. acutum faciei.*
2. *E. acutum genitalium (penis et scroti).*
3. *E. acutum manuum et pedum.*
4. *E. acutum universale.*

XXXVII

INSTRUMENTS AND METHODS OF PRECISION

I. INSTRUMENTS OF PRECISION

NO SCIENCE ATTAINS MATURITY UNTIL IT ACQUIRES METHODS OF MEASUREMENT. Clinical medicine could not advance until it acquired objective methods of determining or measuring the nature and extent of organic disease present in a patient's body.

Numerous sporadic efforts were made from the earliest times to determine the temperature of the body under different circumstances, to establish conclusions from the pulse, to weigh the body (Sanctorius), but the first contribution which placed objective observations on a scientific basis and turned men's minds towards such an ideal was Auenbrugger's invention of percussion (*q.v.*). Auscultation, inaugurated by Laënnec, followed.

Sir John Floyer (1649-1734) published his *Physician's Pulse Watch*, in 1707. He invented a watch with a second hand, and second hands have been on watches ever since. Our modern "pulse lore"—very instructive and valuable in that it is based on physiologic and pathologic knowledge —has been a gradual accumulation from many sources.

Clinical thermometry, projected by George Martine (1702-1741) in his *Essays and Observations*, was firmly established by Carl Reinhold August Wunderlich (1815-1877), professor of medicine at Leipzig, who published *Das Verhalten der Eigenwärme in Krankheiten* in 1868. Before his time fever was regarded as a disease. Ludwig Traube (1818-1876) was the first, we believe, to make a continuous temperature chart.

REFERENCES

CLENDENING, L. History of Certain Medical Instruments. *Ann. Int. Med.*, Vol. 4, No. 2, August, 1930.

MAJOR, R. H. The History of Taking the Blood Pressure. *Ann. M. Hist.*, New Series, Vol. 2, No. 1, 1930.

MITCHELL, S. W. The Early History of Instrumental Precision in Medicine. *Transactions Congress American Physicians & Surgeons*, Vol. II, 1891.

SIR JOHN FLOYER

COUNTING THE PULSE

[From *The Physician's Pulse-Watch: or, an Essay to Explain the Old Art of Feeling the Pulse, and to Improve it by the Help of a Pulse-Watch**]

I Have for many years try'd Pulses by the Minute in Common Watches, and Pendulum Clocks, when I was among my Patients; after some time I met with the common Sea-Minute-Glass, which I used for my Cold Bathing, and by that I made most of my Experiments; but because that was not portable, I caused a Pulse-Watch to be made which run 60 Seconds, and I placed it in a Box to be more easily carried, and by this I now feel Pulses; and since the Watch does run unequally, rather too fast for my Minute-Glass, I thereby regulate it; and add 5 or 6 to the Numbers told by the Watch: I also made a half Minute Glass, whose Case turns like a dark Lanthorn, and that was portable, and useful in feeling of my Patients Pulses, but that differed 4 beats from the Minute Glass, which I always kept at home as my Standard. After I had found this useful Measure for Pulses, I read over all that GALEN had writ about the Pulse, which I epitomis'd, and Corrected many of the old Errors, especially the old Notions about the Causes of the Pulse. After by my Pulse Watch I had found the most healthful Pulses, I easily discern'd what were the exceeding and deficient Pulses. 'Twas easie for me to take Indications from the hot or cold Pulses, and Cacochymias found out by the Pulse Watch, and to cure them by a contrary Regimen; the exceeding Pulses I learnt to reduce to the healthful state by the cold Regimen, and cool Tastes, and the deficient Pulses, by the hot Regimen and hot Tastes of Medicines. After I had reflected on what I had done I found my Notions hit with the CHINESE Practice, about which I consulted many Printed Travels, but could never procure CLEYER'S SPECIMEN MEDICINAE SINICAE, till these Papers were Printing: I found in GALEN all the useful and sensible PHOENOMENA about the Pulses, which I Collected and Explain'd according to the new Anatomy, and our present Philosophy.

Tho' neither the GREEKS nor the CHINESE knew the true Fabrick of the Organs of the Pulse, nor their true action and uses, nor the Cir-

* London, 1707.

culation of Humours, and the causes of it; yet the GREEKS discovered the Pulses of all Diseases and Humours, and Passions: And the CHINESE founded their Art of Physic on the Pulse and its differences; when more quick, great, frequent, was obvious to, the touch; and this produces the hot Diseases, and the contrary Pulses were evident which produced the Cold. The Cacochymias were the causes of all Diseases with the Greeks, but because those cannot explain all Diseases, and they are sometimes very obscure, or much mixed with one another: I shall endeavour to adjust the Cacochymia to several members of the Pulse, by which they may be known, and will prefer the CHINESE Practice to that of the Greeks as most obvious and certain, and short, and after that upon that we may build all the Practice of Physick. I can find by the Index in my Clock that it goes too fast, or too slow, without knowing the Mechanism of it, and I can add to, or take off the Weights, to regulate its Motion when it exceeds or is deficient; so it happens in the practice of Physick, our Life consists in the Circulation of blood, and that running too fast or slow, produces most of our Diseases. The Physician's Business is to regulate the Circulation, and to keep it in a moderate degree, suppose once in three Minutes; if it run oftner or slower, our Mechanism is out of order; but 'tis not necessary for us to understand the Motions of the Particles in the Blood, nor the Texture of the VISCERA and Organs; 'tis enough that I know by a hot Regimen and hot Tastes I can raise deficient Pulses, and by a cold Regimen and Medicines of a cool Taste, I can depress and sink the number of exceeding Pulses. By this Method all fine Hypotheses will be excluded from Practice, and a more certain and sensible Foundation will be laid for it; and we may give liberty to every Physician to talk what Philosophy pleases him best, we can never disagree in this, whether the Pulse exceeds or is deficient, and whether a hot or cool method must be pursu'd; the greatness of the Disease, the strength of the Patient, and the quantity of the Medicines, as well as their qualities, will be known by the Pulse, and all the old Method of Practice and Rules for cure by contraries, will be comprehended under these two general Indications of stopping the Pulse or Circulation when they run too fast, or promoting them when they move too slow.

I have long since imbibed this Notion about Physical Matters, that our Senses can sufficiently inform us of all the most useful

PHOENOMENA whereby we know or cure our Diseases, or prognosticate concerning them.

CARL REINHOLD AUGUST WUNDERLICH

BODY TEMPERATURE IN HEALTH AND DISEASE
[From *Medical Thermometry*, Chapter 1]

1. There are two well-ascertained facts, which not only justify us in endeavoring to determine the temperature of the body in diseases, and render the use of the thermometer both a duty and a valuable aid to diagnosis, but form the basis of all our investigations. The first fact is the *constancy of temperature in healthy persons*, or, in other words, that healthy human beings of every age and condition, in all places and in all circumstances, and exposed to all kinds of influences, provided these do not impair health, have an almost identical temperature.

The second fact is *the variation of temperature in disease*, for in sick persons we are constantly meeting with deviations from the normal temperature of the healthy.

2. The *average normal temperature* of the healthy human body in its interior, or in carefully covered situations on its surface, varies, according to the plan of measurement, from 98.6° to 99.5° Fahr. (37° to 37.5° C.). It is about 98.6° in the well-closed axilla, and a few tenths of a degree higher (.5 to 1½ or 2° Fahr., .7° Fahr. average) in the rectum and vagina.

3. The temperature of healthy persons is almost constantly the same, although not absolutely so. Indeed, there are spontaneous variations in the course of every twenty-four hours, but these seldom exceed half a degree of the Centigrade scale (.9° Fahr.) for each individual. Unusual conditions, and external influences, may indeed cause variations of temperature, but these are never very great, as long as they produce no disturbance of health. Any elevation of the axillary temperature above 99.5° (37.5° C.), or any depression below 97.2° (36.5° C.) is always very suspicious, and whether it appears to be spontaneous or induced by external circumstances, can only be considered normal when all the facts of the case are known, or in very exceptional cases.

The maintenance of a normal temperature under varying conditions, or, in other words, a constant temperature of the body in any individual, is a proof of a sound constitution.

4. A normal temperature does not necessarily indicate health, but all those whose temperature either exceeds or falls short of the normal range, are unhealthy.

5. There are certain limits, which are rarely exceeded, in the range of temperature observed in disease. The highest temperature yet met with in a living man, noted by a trustworthy observer, amounted to 112.55° Fahr. (44.75° C.), whilst the range of lower temperatures is less accurately determined. But if we put aside cases which are quite exceptional, the range of temperature in the most severe diseases is between 95° Fahr. (35° C.) and 108.5° F. (42.5° C.), and it is very seldom that it exceeds 109.4° Fahr. (43° C.), or sinks below 91.4° F. (33° C.).

6. Deviations from the normal course of temperature are certainly to be regarded as significant, and as never occurring without due cause, whether we regard their origin, their amount, the course which they pursue, or their cessation. Many of these deviations may be referred to fixed laws or rules, even now (which I may call pathological thermonomy), but we sometimes fail to discover these, because in disease even much more than in health, animal heat or the temperature of the body is the result of many different, and, in fact, mutually antagonistic, factors. Besides the essential phenomena of disease, many accidental and collateral influences may alter the sick man's temperature.

7. Influences which in no ways disturb the temperature of a healthy man, have often a very remarkable effect in causing variations of temperature in diseased conditions of body, although the diseased condition itself may affect this but slightly. Mobility of temperature as the result of external influences is, therefore, a sign of some diseased condition of body. The discovery of abnormal temperatures in men who have previously exhibited a normal degree of heat, is therefore, a means of discovering or confirming the existence of latent disease.

8. Alterations of temperature may be confined to special regions of the body, which are the seat of diseased actions (local inflammations), whilst the general temperature remains more or less normal. These circumscribed variations, in topical diseases, are of very little practical moment. They consist for the most part of elevations or

depressions of temperature of very moderate extent, seldom exceeding a degree Centigrade (1.8°, or less than 2° Fahrenheit), over a larger or smaller area. These local changes are almost invariably accompanied with other obvious phenomena, which, in a practical point of view, are far more useful for diagnostic purposes than the locally abnormal temperature.

9. The general temperature of the body (blood-heat), registered by the thermometer in interior parts, or in perfectly sheltered spots on the surface, not locally affected, is the expression of the result of a number of processes, which on the one hand tend to the production of heat (chemical processes, so-called tissue changes), and on the other hand promote the giving up of heat (cooling by various means and apparatus, changes of heat into motion). However varied the combinations of these processes, and however their several values may change almost momentarily, so that they appear dependent on almost countless accidental circumstances; yet experience shows, not only that the final result (the animal heat, or specific heat of the body) remains almost always the same, in health; but also that in disease the variations of temperature, if not absolutely trustworthy, are yet the safest standard for estimating the condition of the whole body. Variations of temperature coincide with other functional and structural disturbances of the diseased organism, but none of them can be determined and measured with such accuracy as the temperature. None of them are so independent (comparatively speaking) of trifling and subordinate surrounding influences as the temperature. Very often these variations of temperature are conspicuous long before either functional or structural changes can be recognized.

10. The average temperature or specific heat of the whole body may be normal in disease, or increased or diminished, whilst the distribution of heat is unequal as regards various regions of the body. A normal temperature in sickness is only to be considered as a relative sign, as a symptom which may exclude certain forms of disease, and may justify, but never by itself lead to a positive diagnosis. A fall of temperature below the normal range is persistent in very few diseases only, but occurs as a temporary phenomenon in many favorable and unfavorable circumstances. Precisely parallel is the case of an unequal distribution of animal heat. In a majority of cases, however, this must be considered an unfavorable symptom.

Abnormal elevations of temperature furnish the most important material for purposes of diagnosis and prognosis.

11. Abnormal variations of temperature, except such as are only momentary, are generally associated with certain common typical states of (ill) health (modalities, or typical forms).

A rapid increase of the temperature of the body from a chill, or in the normal warmth of the hands, feet, nose, or forehead, is commonly associated with strong feelings of chilliness and convulsive movements ("cold shivers"; rigors; "fever-frost").

Any considerable diminution of warmth in the extremities or in the face, or in separate exposed parts; with a high or simultaneously falling temperature of the trunk, is generally associated with a small pulse, sunken features, feelings of weakness, and nausea (Unlust), with much sweating, especially local, principally on the cold parts of the skin (Collapse).

12. The amount of temperature changes, the relation of these changes to one another, and their alterations in the course of the disease (quantity, type, and relation), although often modified by accidental influences, are commonly determined by the nature of the disease: and, indeed, the more typical and well-developed the diseased processes are, the more certainly is this the case. Many separate kinds of disease correspond to well-marked types of altered temperature. These answer to well-known varieties of disease.

In opposition to these there are certain atypical or irregular forms of disease, in which the temperature also is irregular. The contrast between typical and atypical forms is, however, not always sharply defined, so that many affections may be considered as standing on a sort of neutral ground, between typical and ill-defined forms.

True typical states of disease, that is, those which almost invariably show more or less clearly a characteristic type, and in which there is seldom if ever a complete deviation from the typical form, are illustrated by enteric fever (abdominal typhus), true exanthematic typhus, and apparently by relapsing fever, smallpox, measles, and scarletina, primary (croupous or lobar) pneumonia, and recent malarious fevers.

The group of approximately typical forms of disease, in which, indeed, characteristic types may be certainly recognized in the abstract, but which, although in certain stages they exhibit great regularity, yet occasionally deviate very widely from the typical, and almost constantly display a great breadth and laxity of behaviour is less easily defined. Yet we may include under it febricular pyaemia, and septicaemia, varicella and rubeola notha, facial erysipelas, acute catarrhal inflammation, tonsillitis (cynanche tonsillaris), acute rheumatism (rheumatic fever), basilar meningitis, and meningitis of the superior convolutions; cerebro-spinal meningitis, parotitis (mumps), pleurisy, acute tuberculosis, fatal neuroses in their last stages, and the trichina disease.

Another group is formed by those diseases which in certain circumstances conform to a regular type, but which generally run their course without fever: when, however, fever supervenes a regular type is generally displayed. To this group cholera, acute phosphorus-poisoning, acute general fatty degeneration, and syphilis especially belong. Even diseases which we are forced to include under the designation of atypical or irregular do occasionally, in exceptional cases, show a close approximation to typical forms in their progress. Of these we may mention diphtheria, dysentery, pericarditis, peritonitis, acute and chronic suppurations (abscesses), and phthisis.

13. The course of the temperature in many special diseases almost invariably follows a single typical form (monotypical or uniform diseases).

Other maladies, according to their intensity, or from other special causes, follow various types of temperature (multiform, or pleotypic diseases). The study of thermometry can define these variations of disease far more accurately than has yet been done, and thus enable us to discover and differentiate varying types of the same disease. Smallpox, enteric fever, scarletina, pneumonia, and malarious fever, are diseases which occasionally assume the multiform type (pleotypism), although as a rule they decidedly follow a single pure type. Those diseases which usually exhibit only an approximately typical course of temperature, show still greater tendencies to assume a multiplicity of ill-defined types.

14. Any disease, however fixed may be its typical form, may exhibit deviations from this in special cases (irregularities). They are determined by more or less lasting individual peculiarities and circumstances (idiosyncrasies), by external conditions, or therapeutical influences, whether favorable or unfavorable, and by the supervention of complications. These irregularities are circumscribed within

certain limits, and their form and extent are more or less determinate. By means of the thermometer it will be possible to learn more of these irregularities than is yet known, to assign them to their proper causes, and give them their due weight in prognosis. And it will help us better to fix the time when a patient's disease, which has appeared to run an irregular course, reassumes a typical form.

15. A single observation of an abnormal temperature, however great, or however small the deviation from the normal may be, is not by itself conclusive as to the kind of disease from which the patient suffers. All we learn from it is—

1. That the patient is really bodily ill.
2. When there is considerable elevation of temperature, we know that there is fever.
3. When there are extremes of temperature, we know that there is great danger.

We may indeed assign the following general significance to single observations of temperature (in a conventional sense).

A. Temperatures much below normal (collapse temperatures), below 96.8° F. (36° C.).
  (a) Deep, fatal algide collapse, below 92.3° F. (33.5° C.).
  (b) Algide collapse, 92.3° F. to 95° F. (33.5° C.–35°), in which it is possible for life to be saved, but which indicates the greatest danger.
  (c) Moderate collapse, 95°–96.8° F. (35°–36° C.), in itself without danger.

B. Normal, or almost normal temperatures.
  (a) Sub-normal temperatures, 96.8° to 97.7° F. (36°–36.5° C.).
  (b) Really normal temperatures = 97.88° to 99.12° F. (36.6°–37.4° C.).
  (c) Sub-febrile temperatures = 99.5°–100.4° F. (37.5°–38° C.).

C. Febrile temperatures.
  (a) Slight febrile action = 100.4° to 101.12° F. (38°–38.4° C.).
  (b) Moderate degree of fever, 101.3° to 102.2° F. (38.5°–

39° C.) in the morning, and rising to 103\1° (39.5° C.) in the evening.

(c) Considerable fever, about 103.1° F. (39.5° C.) in the morning, and about 104° in the evening (40.5° C.).

(d) High fever is indicated by temperatures above 103.1° (39.5° C.) in the morning, and above 104.9° (40.5°) in the evening.

D. Temperature which in every known disease, except relapsing fever, in all probability indicate a fatal termination = 107.6° F. (42° C.) or more. (Hyperpyretic temperatures.)

II. EXAMINATION OF THE BLOOD

The first blood count was made by Vierordt and announced in an article in *Archiv für physiologische Heilkunde*, Stuttgart, 1852, entitled "Zählungen der Blutkorperchen des Menschen." He used a capillary tube to collect the blood and counted the cells in a complete cubic millimeter of his own blood. Hermann Welcher, on April 25, 1853, first counted the white corpuscles, and found 12,133 per cubic millimeter. Potain invented the diluting pipette, in 1867; and Gowers the modern counting chamber, in 1877. Welcher also recorded the first hemoglobin estimations, in 1854, describing two scales—one fluid and one a colored paper series.

REFERENCE

HADEN. *American Journal of Clinical Pathology*, Vol. 5, No. 5, September, 1935.

KARL VIERORDT

[From *Counting Human Blood Corpuscles**]

Before taking up the blood corpuscle counts which I have made hitherto I must present a few comments which, as supplementing the methods described in the previous number, I can properly commend to the attention of all who wish to make similar investigations of blood or other fluids.

First, I recommend that the fine glass capillary tube should not

* Zählungen der Blutkorperchen des Menschen, *Arch. f. physiol. Heilk.*, 2:327, 1852. Translated by Julian F. Smith, Hooker Scientific Library, Central College, Fayette, Mo., 1941.

be longer than about an inch. For convenient manipulation of such capillaries, however, it is necessary to attach them firmly in a wider glass tube (diameter about 2 lines, length 2 or 3 inches) in such a way that the small tube projects about ¾ inch beyond the wider tube. The shortness of the capillary and the attached wider tube make it easier to blow out the blood and in general permit more thorough cleaning of the capillary.

I still use egg white to dilute the blood, but now I thin the egg white with pure water. Three things are accomplished thereby: (1) the mixture of blood with this diluent is not viscous and is very easily drawn out on the glass in the form of a long streak; (2) cracks seldom appear in the solidified mass, or at least do not become numerous; (3) the solidified streak of material forms such a thin film that every corpuscle is visible in perfect clarity without changing the focus. Indeed, counts can be made even at magnifications of 200 and more and the micrometer divisions and corpuscles can still be observed at the same time with sufficient clarity.

My procedure in blowing blood out of the capillary is to place on the glass first a tiny drop (about as big as a medium-sized pinhead) of diluted egg white. The blood is then blown into this droplet and then mixed with the egg white solution by stirring with a very fine needle. This makes it certain that every blood corpuscle is free and separate. Only then is the whole drop drawn out to a long, narrow streak. To prevent gaps in the streak it is necessary that the slide on which the streak is to be made should first have been cleaned with alcohol.

When the count has been made in the blood area covered by the glass micrometer the micrometer must be moved forward. In doing so it is of course strictly necessary to adjust the micrometer accurately so that the counted area and the area still to be counted on the blood chart can be differentiated with certainty. This I did at first by marking the boundary points on the chart each time before advancing the micrometer, a very tedious method, and especially when the corpuscles happen to lie close together there may be a small error in the count. I have now fully overcome this difficulty in the following manner: A few very fine lines are drawn with a diamond across the slide on which the streak is to be made; these lines are equidistant, parallel to the streak, and their spacing is equal to the length of the micrometer divisions. I find 5 such lines necessary because the micrometer which I use for these counts is very long (see the preceding paper). By these lines the boundaries are automatically marked and when a boundary is reached the micrometer can be advanced without further ado. It is also advisable to draw a line from end to end of the slide perpendicular to the boundary markers, i.e. in my case 6 times the length of the micrometer division. The blood streak is drawn along this line; by this expedient the width of the blood streak drawn with the needle is not too strictly limited. By this arrangement the blood area is divided, independently of the micrometer divisions, in 12 parts. It is at once apparent that this method offers highly essential advantages and is thoroughly necessary.

I must not neglect to mention that a movable slide holder (a device which can be dispensed with in most microscopic investigations) was indeed of the utmost convenience for our purpose, i.e. for passing the micrometer rectangle across the field of vision, and especially since the object is then always kept exactly in focus as it passes.

In my first experiments I was never able to obtain thoroughly satisfactory streaks for blood counts from the first filling of the capillary with blood. Then I extracted the capillary with water and on examination under the microscope I felt sufficiently certain that the inner wall of the capillary was not covered with even the thinnest film of water. This, however, is not always the case in such a procedure. Accordingly it is strictly necessary to dry the capillary by warming it gently after every cleaning with water, etc. Unfortunately my first counts had to be rejected because this circumstance was disregarded.

In breaking off the fine capillaries I have hitherto been unable to obtain a smooth fracture. The broken ends are always more or less irregular. Since the blood column rising in the capillary tube extends to the end of the tube at the bottom the shape of the lower blood meniscus might show some deviation, although I have not been able to observe any. Undoubtedly a method will soon be found for breaking or grinding the tube (as is done with larger glass tubes) in such a way that the ends will be cut off clean. I hope later to accomplish this perhaps by embedding one end of the glass capillary in sealing wax and then cutting cleanly through sealing wax and tube. It would then be necessary to remove residual sealing wax with warm alcohol.

The capillaries used for most of my counts were from 0.08295 to 0.08327 mm. in diameter and I would have used still finer ones if they had been obtainable. A quantity of blood even smaller than 0.002 mm. suffices, as I am now thoroughly convinced; but as a rule I have used a little more.

Finally I asked myself what is probably a hypercritical question, namely is it not likely that a certain thickness of the blood film along the wall will be poorer in corpuscles? If that is the case the error would be avoided, though only at the expense of a longer counting time, by making the count in a larger sample. But why trifle with objections which are probably mere subtilities!

In the counts reported here my problem was simply to ascertain in a general way how many corpuscles there are in ordinary blood. I was under the necessity of making some of these counts in light in which it was impossible to distinguish colored cells from colorless cells. I can solve this problem, like many other individual problems, only little by little and I hope, at least in the favorable season when I can work exclusively by daylight, to be able to make at least *one* such count each week.

During the past winter, after many preliminary experiments (which are indispensable for any one who wishes to engage in this type of investigation), I made 9 blood corpuscle counts. The observed blood was my own and I must not neglect to remark that I am pale and thin, with only a moderate appetite, and for many years I have been spared from serious illness of any sort.

Experiment 1. October 6, 1851, 9 A.M.: 0.0020132 mm³ of blood contained 10,086 corpuscles, i.e. 5,010,000 corpuscles per cubic millimeter.

Experiment 2. October 18, 1 A.M.: 0.0028501 mm³ of blood contained 15,342 corpuscles, or 5,357,300 per mm.³

Experiment 3. November 9, 9:30 A.M.: room temperature 14° R. (63.5° F.); 0.0024610 mm³ of blood contained 12,591, or 5,116,200 corpuscles per mm.³

Experiment 4. November 18, 11 A.M.: room temperature 14° R.; 0.0025151 mm³ contained 12,980 corpuscles, or 5,160,800 per mm³ of blood.

Experiment 5. December 29, 3:30 P.M.; temperature 11° R.; 0.0071665 mm³ of blood contained 37,962 corpuscles, or 5,297,700 per mm³ of blood.

Experiment 6. January 6, 1852, 10 A.M.: room temperature 13° R.; 0.0029942 mm³ contained 15,791 corpuscles, or 5,273,800 per mm³ of blood.

Experiment 7. January 18, 12:10 P.M. (just before lunch): room temperature 14° R.; 0.00647951 mm³ of blood contained 31,989 corpuscles, or 4,936,900 per mm³ of blood.

Experiment 8. February 28, 2:30 P.M.: room temperature 13° R.; 0.001600 mm³ of blood contained 9,310, of 5,818,700 corpuscles per mm³ (there may have been an error in measuring the length of the blood column in Experiment 8).

Experiment 9. March 7, 11:30 A.M.: temperature 13° R.; 0.0026-160 mm³ contained 12,028 corpuscles, or 4,597,800 per mm³.

RESULTS

At 13° R. my blood contains, as the average of 9 counts, 5,714,400 corpuscles per cubic millimeter; referred to the Parisian cubic line (11.4789 cubic millimeters) this amounts to 59,396,100 corpuscles.

III. EXAMINATION OF THE URINE

Albumin in the urine was described by several observers before Bright's time: Frederik Dekker, in *Exercitations practicae circa methodum medendi* (1673); Domenico Cotugno, in *De ischiade nervosa commentarius* (1765); William Charles Wells, in *On the presence of red matter and serum of blood in Urine of Dropsy, which has not originated from scarlet fever* (1811); and John Blackall, in *Observations on the Nature and Cure of Dropsies* (1813), all observed albumin in the urine. But of course Bright's account in 1827 (p. 530) was the event that put the procedure of urinalysis into routine clinical practice.

John Charles Weaver Lever (1811-1859), born in Plumstead, entered Guy's Hospital as a student in 1832. After graduation he acquired a large obstetrical practice, and in 1849 was appointed Lecturer in Midwifery at Guy's. His paper on albumin in the urine of eclamptic women made a practical application of the knowledge to obstetrics.

William Charles Wells (1757-1817), one of the most remarkable scientific geniuses of his time, was born in Charleston, South Carolina. He studied at Edinburgh, practiced in London, and was physician to St. Thomas Hospital from 1800 to 1817. His *Essay on Dew*, his *Essay on Vision*, his description of the cardiac complications of rheumatic fever, and his statement of the theory of natural selection (acknowledged by Darwin), show the sweep of his thought.

JOHN CHARLES WEAVER LEVER

Albumin in the Urine of Eclamptic Women

[From *Cases of Puerperal Convulsions with Remarks**]

Condition of the Urine.—In the first four cases here recorded no mention is made of the condition of the urine, for our attention was not at that time directed to the investigation of this secretion. In the fifth case, Mr. Woolnough, my late colleague Mr. Tweedie, Dr. Gull, as well as myself, particularly noticed the great similarity that presented in her appearance and that of patients labouring under anasarca with the Morbus Brightii; and it was with this view that we proceeded to examine the condition of her urine.

At first, I was induced to believe that it was merely a case of pregnancy occurring in a woman affected with granular degeneration of the kidney; but as the traces of albumin became daily more faint, until they entirely disappeared, I was led to suppose that the albuminous condition of the urine depended upon some transient cause probably connected with the state of gestation itself.

To settle this point, I have carefully examined the urine in every case of puerperal convulsions that has since come under my notice, both in the lying-in-Charity of Guy's Hospital and in private practice; and in every case, but one, the urine has been found to be albuminous at the time of the convulsions. In the case (10) in which the albumin was wanting, inflammation of the membranes of the brain, with considerable effusion, was detected after death. I further have investigated the condition of the urine in upwards of fifty women, from whom the secretion has been drawn, during labour, by the catheter; great care being taken that none of the vaginal discharges were mixed with the fluid: and the result has been *that in NO cases have I detected albumin, except in those in which there have been convulsions, or in which symptoms have presented themselves, and which are readily recognized as the precursors of puerperal fits.*

* *Guy's Hosp. Rep.*, London, 1843.

WILLIAM CHARLES WELLS

[From *On the Presence of the Red Matter and Serum of Blood in the Urine of Dropsy, Which Has Not Originated from Scarlet Fever**]

I have examined by means of one, or other, or both, of the tests which have been mentioned, the urine of one hundred and thirty persons, affected with dropsy from other causes than scarlet fever, of whom ninety-five were males, and thirty-five females; and have found serum in that of seventy-eight, sixty of whom were males and eighteen females.

In about a third of the cases in which serum was detected in the urine, its quantity was small, the bulk of the coagulum produced by heat and nitrous acid, after remaining undisturbed twenty-four hours, being only from one-tenth to one-fortieth of that of the urine, which contained it. On the other hand, the urine, after being exposed to the heat of boiling water, in five cases, became firmly solid, and in seven became soft solid, which separated, from the sides of the glass vial in which it had been formed, when the bottom of the vial was placed uppermost. In one of these cases the urine became solid at every trial during six weeks; in the other eleven, it was sometimes rendered only considerably turbid. In the remaining cases with serous urine, amounting to about a half of the whole number, all the distinguishable intermediate quantities of coagulated matter were formed in that fluid by heat and nitrous acid.

Urine in dropsy, when it contains serum, is often more abundant than in health. It is sometimes discharged, though not for any long time, in the quantity of six pints daily; in one person the daily quantity was for a short time ten pints. It must be mentioned, however, that a great part of my information upon the subject has been derived from the reports of the patients themselves, and their nurses in St. Thomas's Hospital.

* *Tr. Soc. Improve. Med. & Chir. Knowledge*, London, 1812.

XXXVIII

LATER PHYSIOLOGY

"PHYSIOLOGY," WRITES DR. JOHN F. FULTON, AT THE BEGINNING OF HIS excellent history of the subject, "arose relatively late in the evolution of knowledge, for, unlike many other sciences, its growth depends ultimately upon analysis and reasoning rather than upon observation and classification." Harvey's experimental proof of the circulation of the blood founded physiology as a science and gave it a method. The doctrine of the circulation is the fundamental conception of physiology. The landmarks of its further elaboration are Malpighi's and Leeuwenhoek's demonstration of the capillaries; Richard Lower's (1631-1691) feat, described in *Tractatus de Corde* (1669), of transfusing dark venous blood into the lungs and proving that it became bright red; Stephen Hale's (1677-1761) work on blood pressure.

The physiology of respiration was developed by the studies of John Mayow (1643-1670), who proved that when blood is aerated, it changes appearance, and that breathing is an exchange of gases; of Joseph Priestley (1733-1804), who isolated oxygen and proved it is necessary for the burning of a flame, for animal life, and that plants give it to the air. But it was Antoine-Laurent Lavoisier (1743-1794), who, according to Sir Michael Foster, "alone discovered oxygen," and proved all its properties especially its physiological properties. We have selected his masterly *Experiences sur la respiration des animaux, et sur les changements qui arrivent a l'air en passant par leur poumon* to represent him in this collection. Lavoisier by quantitative chemical methods overthrew the doctrine of phlogiston and in many other ways initiated the modern note in chemistry. He cleared up Boyle's rather vague concept of elements, and made a table of elements. He devised the first practical method of weighing gases. He knew that organic compounds were made up of carbon, oxygen and hydrogen, with, in the case of animal tissues, sulphur, phosphorus and lime. He was guillotined by the Tribunal of the French Revolution. "It took but a moment to cut off that head," said his friend La Grange, "although it will require another hundred years to produce another like it."

Muscle-nerve physiology may be said to have originated with Aloysio Galvani (1737-1798), who published *De viribus electricitatis in motu musculari*, in 1792. Other landmarks in the history of physiology of the nervous system are:

René Descartes (1596-1650) published *De homine*, in 1662. Among other things this work contains an account of the reciprocal innervation of the eye muscles.

Sir Charles Bell (1774-1842), in 1811, described the functions of the anterior and posterior roots of the spinal cords, thus suggesting localization of nervous function.

François Magendie (1783-1855) described the reflex arc.

Marie-Jean-Pierre Flourens (1794-1867) analyzed the functions of the cerebellum.

Marshall Hall (1790-1857) systematized clinical and experimental observation on the reflexes of the cord.

Johannes Müller (1801-1858), one of the great general physiologists, announced the theory of specific nerve energies.

Ernst Heinrich (1795-1878) and Eduard Friedrich Webber (1806-1871), of Leipzig, proved the inhibitory action of the vagus nerve, and by the study of touch and temperature sense, laid the foundation of the modern mechanistic concept of psychology—since elaborated by Wundt and Fechner.

In the study of the physiology of the digestion Regner de Graaf (1641-1673) successfully obtained pure pancreatic juice, but he made no tests of its properties. René-Antoine-Ferschault de Réaumur (1683-1757) obtained gastric juice by giving his pet kite sponges to swallow, which the bird ejected. He found this fluid caused the liquefaction of meat. Lazaro Spallanzani (1729-1799) studied the action of saliva.

William Beaumont (1785-1853), an American Army surgeon, really aroused interest in the subject of digestion by his dramatic researches on his human subject Alexis St. Martin, who had a gastric fistula. The story is reproduced here from the original work, *Experiments and Observations on the Gastric Juice* (1833). William Osler attributed to Beaumont's experiments the contributions to the knowledge of the digestive processes listed below:

First—The accuracy and completeness of description of the gastric juice itself.

Second—The important acid of the gastric juice was hydrochloric acid.

Third—The establishment by direct observation of the profound influence of mental disturbances on the secretion of the gastric juice and on digestion.

Fourth—A fuller comparative study of the digestion in the stomach with digestion outside the body, confirming the older observations of Spallanzani and Stevens.

Fifth—The refutation of many erroneous opinions relating to gastric digestion and the establishment of a number of minor points such as the rapid disappearance of water from the stomach through the pylorus.

Sixth—The first comprehensive study of the motions of the stomach.

Seventh—A study of the digestibility of different articles of diet in the stomach.

Ivan Petrovich Pavloff (1849-1940) introduced the concept of conditioned reflexes in the secretion of the digestive juices.

Walter Bradford Cannon (1871-   ), of Harvard University, made early researches on the movements of the stomach and intestines, work which was advanced by A. J. Carlson (1875-   ).

Early studies in nutrition were made by Justus von Liebig (1803-1873), Friedrich Wöhler (1800-1882), Felix Hoppe-Seyler (1825-1895), and Emil Fischer (1852-1919), who founded the chemistry of living matter. Liebig studied urea, hippuric acid, fats, blood, bile, and meat extracts. Wöhler synthesized urea, and proved that benzoic acid taken by mouth appears as hippuric acid in the urine. Hoppe-Seyler ascertained the formulae for hemin, hematin and hematoporphoryn, introduced the word proteid, analyzed milk, bile and urine. Fischer synthesized the sugar groups, thus becoming the "pathfinder of carbohydrate metabolism," demonstrated that the proteins were amino-acid radicals, showed that enzymes were specific in action. Max von Pettenkofer (1818-1901) and Carl von Voit (1831-1908) introduced colorimetric methods into metabolic studies.

Claude Bernard (1813-1878), by his genius for experiment which illuminated great reaches of the subject, deserved front rank among physiologists of all time.

REFERENCES

FOSTER, M. Lectures on the History of Physiology. Cambridge Press, 1901.
FULTON, J. F. Selected Readings in the History of Physiology. Springfield, Ill., C. C. Thomas, 1930.
FULTON, J. F. Physiology. Clio Medica series. New York, Paul B. Hoeber, 1931.
MCKIE, D. Antoine Lavoisier. Philadelphia, J. B. Lippincott, 1935.
OLMSTED, J., M.D. Claude Bernard, Physiologist. New York, Harper & Brothers, 1938.
OSLER, W. A Backwood Physiologist. In: An Alabama Student and Other Biographical Essays, 1908.

ANTOINE-LAURENT LAVOISIER

THE RESPIRATION OF ANIMALS

[From *The Respiration of Animals**]

Of all the phenomena of the animal Economy, none is more striking, none more worthy the attention of philosophers and physiologists than those which accompany respiration. Little as our acquaintance is with the object of this singular function, we are satisfied that it is essential to life, and that it cannot be suspended

* Experiences sur la Respiration des Animaux, *Hist. Acad. roy. d. sc.* Translation by Thomas Henry. 1777.

for any time, without exposing the animal to the danger of immediate death. . . .

The experiments of some philosophers, and especially those of Messrs. Hales and Cigna, had begun to afford some light on this important object; and, Dr. Priestley has lately published a treatise, in which he has greatly extended the bounds of our knowledge; and has endeavored to prove, by a number of very ingenious, delicate, and novel experiments, that the respiration of animals has the property of phlogisticating air, in a similar manner to what is effected by the calcination of metals and many other chemical processes; and that the air ceases not to be respirable, till the instant when it becomes surcharged, or at least saturated, with phlogiston.

However probable the theory of this celebrated philosopher may, at first sight, appear; however numerous and well conducted may be the experiments by which he endeavors to support it, I must confess I have found it so contradictory to a great number of phenomena, that I could not but entertain some doubts of it. I have accordingly proceeded on a different plan, and have found myself led irresistibly, by the consequences of my experiments, to very different conclusions.

Now air which has served for the calcination of metals, is, as we have already seen, nothing but the mephitic residuum of atmospheric air, the highly respirable part of which has combined with the mercury, during the calcination: and the air which has served the purposes of respiration, when deprived of the fixed air, is exactly the same; and, in fact, having combined with the latter residuum, about ½ of its bulk of dephlogisticated air, extracted from the calx of mercury, I re-established it in its former state, and rendered it equally fit for respiration, combustion, &c. as common air, by the same method as that I pursued with air vitiated by the calcination of mercury.

The results of these experiments is, that to restore air that has been vitiated by respiration, to the state of common respirable air, two effects must be produced: 1st. to deprive it of the fixed air (carbon dioxide) it contains, by means of quicklime or caustic alkali: 2dly, to restore to it a quantity of highly respirable or dephlogisticated air, equal to that which it has lost. Respiration, therefore, acts inversely to these two effects, and I find myself in this respect led to two consequences equally probable, and between which my present experience does not enable me to pronounce. . . .

The first of these opinions is supported by an experiment which I have already communicated to the academy. For I have shewn in a memoir, read at our public Easter meeting, 1775, that dephlogisticated air (oxygen) may be wholly converted into fixed air by an addition of powdered charcoal; and, in other memoirs, I have proved that this conversion may be effected by several other methods: it is possible, therefore, that respiration may possess the same property, and that dephlogisticated air, when taken into the lungs, is thrown out again as fixed air. . . . Does it not then follow, from all these facts, that this pure species of air has the property of combining with the blood, and that this combination constitutes its red colour? But whichever of these two opinions we embrace, whether that the respirable portion of the air combines with the blood, or that it is changed into fixed air in passing through the lungs; or lastly, as I am inclined to believe, that both these effects take place in the act of respiration, we may, from facts alone, consider as proved:

1st. That respiration acts only on the portion of pure or dephlogisticated air, contained in the atmosphere; that the residuum or mephitic part is a merely passive medium which enters into the lungs, and departs from them neatly in the same state, without change or alteration.

2dly. That the calcination of metals, in a given quantity of atmospheric air, is effected, as I have already often declared, only in proportion as the dephlogisticated air, which it contains, has been drained and combined with the metal.

3dly. That, in like manner, if an animal be confined in a given quantity of air, it will perish as soon as it has absorbed, or converted into fixed air, the major part of the respirable portion of air, and the remainder is reduced to a mephitic state.

4thly. That the species of mephitic air, which remains after the calcination of metals, is in no wise different, according to all the experiments I have made, from that remaining after the respiration of animals; provided always, that the latter residuum has been freed from its fixed air; that these two residuums may be substituted for each other in every experiment, and that they may each be restored to the state of atmospheric air, by a quantity of dephlogisticated air, equal to that of which they had been deprived. A new proof of this last fact is, that if the proportion of this highly respirable air, contained in a given quantity of the atmospheric, be increased or diminished, in such proportion will be the quantity of metal which we shall be capable of calcining in it, and, to a certain point, the time which animals will be capable of living in it.

ALOYSIO GALVANI

CONTRACTION OF MUSCLE FROM NERVE STIMULATION

[From *On the Effect of Electricity on the Motion of the Muscles**]

I dissected and prepared a frog and placed it with everything else at hand on a table on which was an electric machine, but the frog was completely removed from its conductor and by a considerable interval. One of those who were helping me accidentally lightly touched with the point of his scalpel the inner nerve of this frog's leg, and suddenly all of the leg muscles appeared to become so contracted that they seemed to have fallen into fairly violent tonic convulsion. Another who was helping us producing the electricity, seemed to observe that the phenomenon occurred while the spark was obtained from the conductor of the machine. He, marvelling at the novelty of the occurrence, at once drew my attention to this since I was at the time completely occupied with other things, and was absorbed in my own thoughts. I was fired at once with an extraordinary eagerness and desire to perform the same experiment and to bring into the light of day the hidden secret of the phenomenon. Therefore I myself placed the point of the scalpel on one or other of the leg nerves while one of those who were present produced a spark. The whole phenomenon was by this same means repeated, and, at the same time as the sparks were obtained, there were produced wonderfully strong contractions in each muscle of the joints just as if an animal in tetanus had been used.

But fearing that those very movements arose rather from the contact of the point which accidently acted as a stimulus than from the spark, I tested the same nerves with the point of my scalpel by the same method in the other frogs, and indeed with greater pres-

* *De viribus electricitatis in motu musculari*, 1792. Translation by Dr. John F. Fulton: Selected Readings in the History of Physiology. Springfield, Illinois, C. C. Thomas, 1930.

sure, without, however, any spark on that occasion being produced by anyone; but there seemed to be no movements at all. As a result of this I reasoned to myself that perhaps in order for the phenomenon to be established there were needed at the same time both the contact of some body and the application of a spark. Therefore I again placed the edge of the scalpel on the nerves and held it there motionless, both when a spark was produced and when the machine was absolutely still. But the phenomenon appeared only so long as the spark lasted.

We repeated the experiment always using the same scalpel, but to our amazement when the spark was produced sometimes the movements previously described took place, at other times were absent.

Stirred by the novelty of the occurrence, we began to test it by different experimental methods, always, however, using the same scalpel so that we might, if it were possible, discover the reasons for the unexpected difference. This new task was not empty of result, for we discovered that the whole matter was to be attributed to the question of which different part of the scalpel we were holding with our fingers; the scalpel had a bone handle and, if this handle were held in the hand when a spark was produced, no movements resulted, but they did result when the fingers were placed either on the metal blade or on the iron nails securing the blade of the scalpel.

Therefore since the rather dry bone produced an electric force of its own, and the metal blade and iron nails a conducting or "anelectric" force, as they say, we suspected that perhaps it happened that when we held the bone handle with our fingers all approach to the electric current, entering by some means into the frog, was stopped, but that the current was released when we took hold of the blade or the nails setting up the same current.

MARSHALL HALL

REFLEX ACTION

[From *These Motions Independent of Sensation and Volition**]

The animals experimented on were salamanders, frogs and turtles. In the first of these, the tail, entirely separated from the body, moved as in the living animal, on being excited by the point of a needle

* *Proc. Zoological Soc.,* London, Nov. 27, 1832.

passed lightly over its surface. The motion ceased on destroying the spinal marrow within the caudal *vertebrae*. The head of the frog having been removed, and the spine divided between the third and fourth vertebrae, an eye of the separated head was touched: it was retracted and the eye-lid closed,—a similar movement being observed in the other eye. On removing the brain, these phenomena ceased. On pinching the skin, or the toe of one of the anterior extremities, the whole of this portion of the animal moved. On destroying the spinal marrow, this phenomenon also ceased. Precisely similar effects were observed on pinching the skin or toe of one of the posterior extremities; and on removing the last portion of the spinal marrow, this phenomenon ceased. The head of the turtle continues to move long after its separation from the body; on pinching the eye-lid, it is forcibly closed, the mouth is opened, and the membrane expanded under the lower jaw descends as in respiration. On pinching any part of the skin of the body, extremities, or tail, the animal moves. The posterior extremities and tail being separated together, the former were immovable; the latter moved on the application of the flame of a lighted taper to the skin. Those extremities had no connection with the spinal marrow. All movements ceased in the tail also, on withdrawing the spinal marrow from its canal.

Three things are plain from these observations: 1. That the nerves of sensibility are impressible in portions of an animal separated from the rest; in the head, in the upper part of the trunk, in the lower part of the trunk: 2. that motions similar to voluntary motions follow these impressions made upon the sentient nerves: and 3. that the presence of the spinal marrow is essential as the central and cementing link between the sentient and motor nerves.

WILLIAM BEAUMONT

EXPERIMENTS ON DIGESTION

[From *Experiments and Observations on the Gastric Juice*]

Alexis St. Martin, who is the subject of these experiments, was a Canadian, of French descent, at the above mentioned time about eighteen years of age, of good constitution, robust and healthy. He had been engaged in the service of the American Fur Company, as a voyageur, and was accidentally wounded by the discharge of a musket, on the 6th of June, 1822. . . . The whole mass of materials forced from the musket, together with fragments of clothing and pieces of fractured ribs, were driven into the muscles and cavity of the chest. I saw him in twenty-five or thirty minutes after the accident occurred, and, on examination, found a portion of the lung, as large as a Turkey's egg, protruding through the external wound, lacerated and burnt; and immediately below this, another protusion, which, on futher examination, proved to be a portion of the stomach, lacerated through all its coats, and pouring out the food he had taken for his breakfast, through an orifice large enough to admit the fore finger.

Beaumont describes St. Martin's gradual recovery. He was left with a large fistula, which opened under his left breast and to which the internal mucosa and the musculature of the stomach were attached. Beaumont then proceeded to make experiments of which the following descriptions give a general idea.

EXPERIMENT I.

August 1, 1825. At 12 o'clock m., I introduced through the perforation, into the stomach, the following articles of diet, suspended by a silk string, and fastened at proper distances, so as to pass in without pain—viz.: a piece of high seasoned *a la mode beef*; a piece of *raw, salted, fat pork*; a piece of *raw, salted, lean beef*; a piece of *boiled, salted beef*; a piece of *stale bread*; and a bunch of *raw, sliced cabbage*; each piece weighing about about two drachms; the lad continuing his unusual employment about the house.

At 1 o'clock p.m., withdrew and examined them—found the *cabbage* and bread about half digested: the pieces of *meat* unchanged. Returned them into the stomach.

At 2 o'clock, p.m., withdrew them again—found the *cabbage, bread, pork,* and *boiled beef*, all cleanly digested, and gone from the string; the other pieces of meat but very little affected. Returned them into the stomach again.

At 2 o'clock, p.m. [*sic*], examined again—found the *a la mode beef* partly digested: the *raw beef* was slightly macerated on the surface, but its general texture was firm and entire. The smell and taste of the fluids of the stomach were slightly rancid; and the boy complained of some pain and uneasiness at the breast. Returned them again.

The lad complaining of considerable distress and uneasiness at the stomach, general debility and lassitude, with some pain in his head, I withdrew the string, and found the remaining portions of aliment nearly in the same condition as when last examined; the fluid more rancid and sharp. The boy still complaining, I did not return them any more.

August 2. The distress at the stomach and pain in the head continuing, accompanied with costiveness, a depressed pulse, dry skin, coated tongue, and numerous white spots, or pustules, resembling coagulated lymph, spread over the inner surface of the stomach, I thought it advisable to give medicine; and accordingly, dropped into the stomach, through the aperture, half a dozen *calomel pills*, four or five grains each; which, in about three hours, had a thorough cathartic effect, and removed all the foregoing symptoms, and the diseased appearance of the inner coat of the stomach. The effect of the medicine was the same as when administered in the usual way, by the mouth and oesophagus, except the nausea commonly occasioned by swallowing pills.

This experiment cannot be considered a fair test of the powers of the gastric juice. The cabbage, one of the articles which was, in this instance, most speedily dissolved, was cut into small, fibrous pieces, very thin, and necessarily exposed on all its surfaces, to the action of the gastric juice. The stale bread was porous, and, of course, admitted the juice into all its interstices; and probably fell from the string as soon as softened, and before it was completely dissolved. These circumstances will account for the more rapid disappearance of these substances, than of the pieces of meat, which were in entire solid pieces when put in. To account for the disappearance of the fat pork, it is only necessary to remark, that the fat meat is always resolved into oil, by the warmth of the stomach, before it is digested. I have generally observed that when he has fed on fat meat or butter, the whole superior portion of the contents of the stomach, if examined a short time after eating, will be found covered with an oily pellicle. This fact may account for the disappearance of the pork from the string. I think, upon the whole, and subsequent experiments have confirmed the opinion, that fat meats are less easily digested than lean, when both have received the same advantages of comminution. Generally speaking, the looser the texture, and the more tender the fibre, of animal food, the easier it is of digestion. This experiment is important, in a pathological point of view. It confirms the opinion, that undigested portions of food in the stomach produce all the phenomena of fever; and is calculated to warn us of the danger of all excesses, where that organ is concerned. It also admonishes us of the necessity of a perfect comminution of the articles of diet.

CARL VON VOIT

Voit in his necrology of Pettenkofer, writes: "Imagine our sensations as the picture of the remarkable process of the metabolism unrolled before our eyes, and a mass of new facts became known to us! We found that in starvation protein and fat alone were burned, that during work more fat was burned, and that less fat was consumed during rest, especially during sleep; that the carnivorous dog could maintain himself on an exclusive protein diet, and if to such a protein diet fat were added, the fat was almost entirely deposited in the body; that carbohydrates, on the contrary, were burned no matter how much was given, and that they, like the fat of the food, protected the body from fat loss, although more carbohydrates than fat had to be given to effect this purpose; that the metabolism in the body was not proportional to the combustibility of the substances outside the body, but that protein, which burns with difficulty outside, metabolizes with the greatest ease, then carbohydrates, while fat, which readily burns outside, is the most difficultly combustible in the organism."

METABOLISM

[From *The Elements of the Science of Nutrition**]

The unknown causes of metabolism are found in the cells of the organism. The mass of these cells and their power to decompose materials determine the metabolism. It is absolutely proved that protein fed to the cells is the easiest of all the food-stuffs to be destroyed, next carbohydrates, and lastly fat. The metabolism continues in the cells until their power to metabolize is exhausted. All kinds of influences may act upon the cells to modify their ability to

* Handbuch der Physiologie des Allgemeinen Stoffwechsels und der Ernährung. Leipzig, 1881. Translation by G. Lusk, W. B. Saunders Co., Philadelphia, 1928. By permission of W. B. Saunders Co.

metabolize, some increasing it or others decreasing it. To the former category belong muscular work, cold of the environment (in warm-blooded animals), abundant food, and warming the cells. To the latter, cooling the cells, certain poisons, etc.

In speaking of the power of the cells to metabolize, I have not meant thereby, as may be seen from all my writings, that the cells must always use energy in order to metabolize, but rather I have understood thereby the sum of the unknown causes of the metabolic ability of the cells—as one speaks of the fermentative "power" of yeast cells.

The metabolism of the different food-stuffs varies with the quality and quantity of the food. Protein alone may burn, or little protein and much carbohydrate and fat. I have determined the amount of the metabolism of the various food-stuffs under the most varied conditions. All the phases of metabolism originate from processes in the cells. In a given condition of the cells available protein may be used exclusively if enough be furnished them. If the power of the cells to metabolize is not exhausted by the protein furnished, then carbohydrates and fats are destroyed up to the limit of the ability of the cells to do so.

From this use of materials arise physical results, such as work, heat, and electricity, which we can express in heat units. This is the power derived from metabolism.

It is possible to approach the subject in the reverse order, that is, to study the energy production (Kraftwechsel) and to draw conclusions regarding the metabolism (Stoffwechsel). It is perfectly possible to say that the requirement of energy in the body or the production of the heat necessary to cover heat loss, or for energy to do work, are controlling factors of the metabolism; since on cooling the body or on working correspondingly more matter is destroyed. But one must not conclude that the loss of body heat and muscular work are the immediate causes of this increased metabolism. The causes lie in the peculiar conditions of the organism, and muscle work and loss of heat are merely factors acting favorably upon those causes, raising the power of the cells to metabolize. In virtue of this more is destroyed, and secondly the power to work and increased heat production are determined.

The requirement for energy cannot possibly be the cause of metabolism, any more than the requirement for gold will put it into one's pocket. Hence the production of energy has a very definite upper limit, which is afforded by the ability of the cells to metabolize. If the cells will metabolize no more, then further increase of work ceases even in the presence of direct necessity; and this is also the case with the heat production, even though it were very necessary, and we were likely to freeze.

I therefore maintain my "older" point of view, that of pure metabolism, in order to explain the phenomena of nutrition. I am convinced that it is the right way, and that the clearest and most unifying development will be possible as one investigates what substances are destroyed under different circumstances, such as the performance of work, and loss of heat, and how much of the different materials must be fed to maintain the body in condition.

CLAUDE BERNARD

NUTRITION: UTILIZATION OF SUGAR IN THE BODY

[From *The Origin of Sugar in the Animal Body**]

CONCLUSIONS AND REFLECTIONS

The conclusions which seem to me to derive from the facts contained in this memorandum are:

1. That in the physiologic state, there exists constantly and normally the sugar of diabetes in the blood of the heart and in the liver of man and animals.

2. That the formation of this sugar takes place in the liver, and that it is independent of a sweet or starchy alimentation.

3. That this formation of sugar in the liver begins to operate in the animal before birth, and consequently before the direct ingestion of foodstuffs.

4. That this production of sugary material, which would be one of the functions of the liver, appears to be bound to the integrity of the pneumogastric nerves.

It is evident that before these facts, that law, that animals do not create any immediate principle, but only destroy those furnished them by the vegetables, must cease to be true, because animals in

* De L'Origine Du Sucre Dans L'Èconomie Animale. *Arch. Gén. de Méd.*, 4 S., 18: 303-319, 1848. Translation by Dorothy H. Clendening.

the physiologic state are able, like the vegetables, to create and destroy sugar.

Although the animal organism produces sugar without starch, something which the known chemical means do not permit us to do, I would not conclude therefore that the importance of chemical knowledge should be diminished in the study of the phenomena of life. I am, on the contrary, one of those who appreciate the more all the progress that organic chemistry has caused physiology to make. Only, I believe, as I have already had occasion to state, that in order to avoid error and to render all the service of which it is capable, chemistry should never go adventuring alone in the examination of animal functions; I believe that it alone can solve, in many cases, the difficulties which deter physiology, but it can not precede the latter, and I think, finally, that in no case may chemistry believe itself authorized to restrain the resources of nature, which we know not, to the limits of the facts or procedures which constitute our knowledge of the laboratory.

The question of the origin of sugar in the animals, which we have just examined in this work, is still far from being known to us in all its elements. Indeed, if we already possess very positive results, there are, in another direction, facts to elucidate. We should indicate these facts, in order to point them out for study and to show the extent of our subject, which we have only skirted in this first work.

According to what we have said about the existence of sugar in the liver, it should not be believed that in going into an amphitheater and in taking the liver of a cadaver, one would surely find sugar in it. There are, indeed, a large number of diseases in which the sugar disappears and is no longer found in the liver after death. In diabetics, it is known that the sugar disappears from the urine in the last period of the life; it disappears equally from the liver, because the liver of a diabetic which I had occasion to examine in this connection contained no sugar. I have investigated the sugar in the cadavers of 18 subjects who died of different diseases; there were some who showed different proportions of sugar, and there were others who did not contain a trace. My observations on this point are not numerous enough so that I may decide if there are diseases in which the sugar constantly disappears, while it persists in others. In animals weakened by a very long abstinence, which have become sick or have died of disease the sugar often diminishes considerably, and even disappears completely. All the livers of butchered animals, however, should contain much sugar, if they were killed under proper circumstances. The livers obtained at the butcher's have always given me large quantities of sugar. Finally, there is a question which we should examine with care: it is to know whether sugar exists in the same proportion in all classes of animals, taken in conditions as similar as possible. I may already affirm that there seem to be differences in this regard: 1) in birds (chicken, pigeon) the proportion of sugar is quite considerable; 2) in mammals (dog, rabbit, pig, beef, calf, horse) the proportion of sugar is also quite considerable; 3) in reptiles (frogs, lizards) the quantity of sugar found in the liver is very small; 4) in fish, in the ray and the eel, the livers of which I examined in as fresh a state as possible, I did not find the least trace of sugar. Whence comes this disappearance of sugar in certain cold-blooded animals? Is that related to the lesser energy of the respiratory phenomena, which, as we shall see later on, are in very intimate relation with the formation of sugar in the liver?

XXXIX

ASEPSIS

THE PRINCIPLE OF ASEPSIS, THE APPLICATION OF WHICH HAS (WITH ANESTHESIA and hemostasis) created modern surgery, was preached first by the obstetricians. It is, of course, today quite as fundamental a principle of obstetrics as it is of surgery. Two eighteenth century British obstetricians, Gordon, of Aberdeen (1752-1799), and Charles White, of Manchester (1728-1813), wrote of the contagious nature of puerperal fever. White gave instructions for its treatment, and Gordon said very clearly that it was of the nature of erysipelas and was carried from patient to patient by midwives and physicians. Oliver Wendell Holmes (1809-1894), however, first ventured to declare clearly methods of prevention—that obstetricians should not take part in postmortems on puerperal fever cases, etc. Holmes was not a practicing physician and his warnings had little effect, partly for that reason. His essay, published in 1843, was greeted with scorn by the two leading practitioners of obstetrics in America, Hodge and Meigs, of Philadelphia—Hodge with heavy-handed authority and Meigs with what he considered a withering scorn that has made his name ridiculous ever since. In his second essay, 1855, Holmes stated that he understood one "Senderein" had lessened the mortality of puerperal fever by disinfecting his hands with chloride of lime and a nail-brush. This reference was to Ignaz Phillip Semmelweiss (1818-1865), of Vienna, who published his great work, *The Cause, Concept and Prevention of Child-Bed Fever*, in 1861. Semmelweiss was an active practicing obstetrician and for that reason and also because he offered a practical method of prophylactic procedure he deserves more credit than Holmes. As assistant professor in midwifery and attendant in the wards of the Allgemeines Krankenhaus he noticed that puerperal fever was less prevalent in the wards attended by the midwives than in the wards where the students practiced. One day he found a woman in the waiting room crying: on inquiring the reason for the tears, he found that she had been assigned to the students' ward, which she considered certain death. In 1847, his friend Kolletschka, Rokitansky's assistant, died of a post-mortem dissection wound, and at his autopsy Semmelweiss noticed that the changes in Kolletschka's body were the same as those in the bodies of women who died of puerperal sepsis. The connection of the death rate in the student ward became immediately clear: the students attended postmortems, the midwives did not. The students brought the contagion directly from the post-mortem room to the obstetric ward and examined women in labor without even washing their hands. Semmelweiss instituted the practice of washing and scrubbing in chlorine water before attending a patient and the mortality from puerperal sepsis dropped. His idea, however, met with as little acceptance as did that of Holmes. Indeed, so fierce was the opposition to his views from such pillars of orthodox obstetrical doctrine as Scanzoni and his own chief Klein, that he was driven from his hospital position. He went to Budapest and was appointed professor of obstetrics at the university there. But his life was embittered by the attitude of the profession to the doctrine he so passionately espoused, his health and mentality were undermined, and he succumbed to a dissection wound of the right hand, the disease he had himself studied so fully, a true medical martyr.

All this was before the days of bacteriology. When Koch, in 1878, proved the etiology of wound infection and Pasteur (Septicemie puerperale. *Bulletin de l'Académie de Médicine,* 1879) proved that organisms were present in the bodies and blood of women with puerperal sepsis, the way was opened for a truly scientific presentation of the principles of asepsis. This was the contribution of Joseph Lister (1827-1912), born in Upton, Essex, of Quaker parents. His father was the founder of modern microscopy. After graduating in London, Lister went to Edinburgh to study under Syme, the greatest surgeon of his day in Britain, if not in Europe. After six years as Syme's house surgeon Lister went to Glasgow, in 1860, as professor of surgery. His deep concern over septicemia in amputations and compound fractures, conditioned his mind to the importance of Pasteur's doctrines. He heard of the antiseptic powers of carbolic acid, and in August, 1865, employed it with complete success in a compound fracture. Lister's principle was antisepsis, be it noted, not asepsis. He used a carbolic acid spray in the operating room to disinfect the air. But other surgeons and gynecologists—Lawson Tait, Spencer Wells, Trendelenburg, von Bergmann, Billroth, and Halsted, realized that the emphasis should be placed on asepsis and developed the technique more or less as it is followed today.

REFERENCES

GODLEE, J. R. Life of Lister. London, 1917.
HOLMES, O. W. *Medical Classics,* Vol. 1, No. 3, November, 1936.
LISTER. *Medical Classics,* Vol. 2, No. 1, September, 1937.
SEMMELWEISS. *Medical Classics,* Vol. 5, No. 5, January, 1941.
SINCLAIR, *Sir* W. J. Semmelweiss: His Life and His Doctrine. Manchester, The University Press, 1909.

OLIVER WENDELL HOLMES

PUERPERAL FEVER

[From *The Contagiousness of Puerperal Fever**]

The practical point to be illustrated is the following: The disease known as puerperal fever is so far contagious as to be frequently carried from patient to patient by physicians and nurses. Let me begin by throwing out certain incidental questions, which, without being absolutely essential, would render the subject more complicated, and by making such concessions and assumptions as may be fairly supposed to be without the pale of discussion.

1. It is granted that all the forms of what is called puerperal fever may not be, and probably are not, equally contagious or infectious. I do not enter into the distinctions which have been drawn by authors, because the facts do not appear to me sufficient to establish any absolute line of demarcation between such forms as may be propagated by contagion and those which are never so propagated. This general result I shall only support by the authority of Dr. Ramsbotham, who gives, as the result of his experience, that the same symptoms belong to what he calls the infectious and the sporadic forms of the disease, and the opinion of Armstrong in his original essay. If others can show any such distinction, I leave it to them to do it. But there are cases enough that show the prevalence of the disease among the patients of a single practitioner when it was in no degree epidemic in the proper sense of the term. I may refer to those of Mr. Roberton and of Dr. Peirson, hereafter to be cited for examples.

2. I shall not enter into any dispute about the particular mode of infection, whether it be by the atmosphere the physician carries about him into the sick-chamber, or by the direct application of the virus to the absorbing surfaces with which his hand comes in contact. Many facts and opinions are in favor of each of these modes of transmission. But it is obvious that, in the majority of cases, it must be impossible to decide by which of these channels the disease is con-

* Paper read at the Boston Society for Medical Improvement, 1843, and published in the *New England Quarterly Journal for Medicine & Surgery*, which discontinued publication about a year later.

veyed, from the nature of the intercourse between the physician and the patient.

3. It is not pretended that the contagion of puerperal fever must always be followed by the disease. It is true of all contagious diseases that they frequently spare those who appear to be fully submitted to their influence. Even the vaccine virus, fresh from the subject, fails every day to produce its legitimate effect, though every precaution is taken to insure its action. This is still more remarkably the case with scarlet fever and some other diseases.

4. It is granted that the disease may be produced and variously modified by many causes besides contagion, and more especially by epidemic and endemic influences. But this is not peculiar to the disease in question. There is no doubt that smallpox is propagated to a great extent by contagion, yet it goes through the same periods of periodical increase and diminution which have been remarked in puerperal fever. If the question is asked how we are to reconcile the great variations in the mortality of puerperal fever in different seasons and places with the supposition of contagion, I will answer it by another question from Mr. Farr's letter to the Registrar-General. He makes the statement that "five die weekly of smallpox in the metropolis when the disease is not epidemic," and adds, "The problem for solution is, Why do the five deaths become 10, 15, 20, 31, 58, 88, weekly, and then progressively fall through the same measured steps?"

5. I take it for granted that, if it can be shown that great numbers of lives have been and are sacrificed to ignorance or blindness on this point, no other error of which physicians or nurses may be occasionally suspected will be alleged in paliation of this; but that whenever and wherever they can be shown to carry disease and death instead of health and safety, the common instincts of humanity will silence every attempt to explain away their responsibility.

IGNAZ PHILLIP SEMMELWEISS

[From *The Concept of Child-Bed Fever**]

Supported by the experiences which I have collected in the course of fifteen years in three different institutions all of which were

* Die Aetiologie, Der Begriff, Und Die Prophylaxis des Kindbettfiebers, Pest, Wien, u. Leipzig, 1861. Translation from Herbert Thoms' Selected Readings in Obstetrics & Gynaecology. Springfield, Illinois, Charles C. Thomas. By permission of Dr. Thoms and Charles C. Thomas.

visited from time to time by puerperal fever to a serious extent, I maintain that puerperal fever, without the exception of a single case, is a resorption fever produced by the resorption of decomposed animal organic material. The first result of this resorption is a blood-dissolution; and exudations result from the blood-dissolution.

The decomposed animal organic material which produces child-bed fever is, in the overwhelming majority of cases, brought to the individual from without, and that is the infection from without; these are the cases which represent child-bed fever epidemics; these are the cases which can be prevented.

In rare cases the decomposed animal matter which when absorbed causes child-bed fever, is produced within the limits of the affected organism. These are the cases of self-infection, and these cases cannot all be prevented.

The source whence the decomposed animal organic material is derived from without is the cadaver of any age, of either sex, without regard to the antecedent disease, without regard to the fact whether the dead body is that of a puerperal or non-puerperal woman. Only the degree of putrefaction of the cadaver has to be taken into consideration. . . .

At the Obstetric Clinic of the Faculty of Medicine at Pesth, it was physiologic human blood and normal lochia which were the etiological factor of a puerperal fever, inasmuch as they were left for a long time soaking the bed-linen and undergoing decomposition.

The carrier of the decomposed animal organic material is the examining finger, the operating hand, the bed-clothes, the atmospheric air, sponges, the hands of midwives and nurses which come into contact with the excrementa of sick lying-in-women or other patients, and then come again into contact with the genitals of women in labour or just confined; in a word the carrier of the decomposed animal organic material is everything which can be rendered unclean by such material and then come into contact with the genitals of the patient.

The site of infection by the decomposed animal organic material is the internal os uteri and upward from there. The inner surface of the uterus . . . is robbed of its mucosa and presents an area where absorption occurs with extreme readiness. The other parts of the mucosa are well clad with epithelium and do not absorb unless they are wounded. If it is injured any portion of the genitals becomes capable of absorption.

With regard to the time of infection, it seldom occurs during pregnancy because of the inaccessibility of the inner absorbing surface of the uterus by reason of the closure of the os internum. In cases in which the internal os uteri is open during pregnancy infection may occur then, but these cases are rare because there is seldom any need for passing the finger within the cervix uteri.

I neglected to take notes of the cases in which puerperal fever began during pregnancy at the First Obstetric Clinic of Vienna but I believe it to be near the truth if I put down the number of cases as about twenty. By puerperal infection the pregnancy was always interrupted.

The time within which infection most frequently occurs is during the stage of dilatation. This is owing to frequent examinations made with the object of ascertaining the position of the foetus.

A proof of this is that before the introduction of chlorine disinfection nearly all the patients after labour, protracted in the dilatation period, died of puerperal fever.

Infection seldom takes place during the expulsion stage because the surface of the uterus cannot then be reached.

In the third stage, or after-birth period, and during the puerperium, the inner surface of the uterus is accessible, and at this time especially, the atmospheric air loaded with decomposed animal organic materials may gain access to the internal genitals and set up infection. . . .

In the after-birth period and during lying-in, the infection may be produced by the bed-linen coming into contact with the genitals which have been injured in the process of parturition. . . .

Self-infection: The decomposed animal organic material which when absorbed brings on puerperal fever is in rare cases not conveyed to the individual from without but originates within the affected individual owing to the retention of organic material which should have been expelled in child-bed. Before its expulsion decomposition has already begun, and when absorption occurs puerperal fever is produced by Self-infection. These organic materials are the lochia, remnants of decidua, blood coagula which are retained within the cavity of the uterus. Or the decomposed animal organic material is the product of a pathological process, for example, the result of a forcible use of the midwifery forceps causing gangrene of bruised

portions of the genital organs and consequent child-bed fever by
Self-infection.

When we declare that child-bed fever is a resorption fever in
which as the result of absorption a blood-poisoning occurs, and then
exudation follows, we do not imply that puerperal fever is peculiar
to the lying-in woman and restricted in its incidence to lying-in
women. We have met with the disease in pregnant women and in
new-born infants without regard to sex. This is the disease which
was fatal in the case of Kolletschka; and we find it affecting anatomists, surgeons, and patients who have undergone surgical operations.

Puerperal fever is therefore not a species of disease: puerperal
fever is a variety of Pyaemia.

With the expression pyaemia different meanings are bound up: it
is therefore necessary to explain what I mean by pyaemia. I understand by pyaemia a blood-poisoning produced by a decomposed
animal-organic matter.

A variety of pyaemia I call child-bed fever, because special forms
of it occur in the genital sphere of pregnant parturient and puerperal
women. . . .

Puerperal fever is not a contagious disease. By contagious disease
we understand the sort of disease which itself produces the contagion by which it is propagated, and this contagion again produces
in another individual the same disease. Smallpox is a contagious
disease because smallpox produces the contagion by which smallpox
can be reproduced in another individual. Smallpox produces in another individual smallpox and no other disease. . . . For example,
a person suffering from scarlet fever cannot cause smallpox in another individual.

Such is the position with child-bed fever; this disease can be produced in a healthy normal puerperal by a disease which is not
puerperal fever. . . .

Puerperal fever is not conveyed to a healthy puerpera unless a
decomposed animal-organic material is carried to her. For example,
a patient becomes seriously ill with puerperal fever, and when this
puerperal fever runs its course without the production of a decomposed animal-organic matter, which appears externally, then is the
disease not conveyable to a healthy normal puerpera. But when

puerperal fever runs its course in such a way as to produce a decomposed matter appearing externally, then is child-bed fever capable of being conveyed to a normal healthy puerpera. For example, a puerpera is suffering from the malady in the form of septic endometritis . . . from such a patient is puerperal fever capable of being carried.

Puerperal fever is not a contagious disease, but puerperal fever is conveyable from a sick to a sound puerpera by means of a decomposed animal organic materia.

After death the body of every lying-in woman becomes a source of decomposed material which may produce puerperal fever; in the cadaver of the puerpera we consider only the degree of putrefaction. When we have reflected that the overwhelming majority of cases of puerperal fever are produced by infection from outside, and that these cases can be prevented, and that in only a small minority of cases puerperal fever is the result of unavoidable self-infection, the question arises: if all fatal cases, not resulting from puerperal fever, and if all cases of infection from without are prevented by suitable measures, how many lying-in women die as the consequence of self-infection?

It is not possible to answer this question for want of statistics, and we must attain complete control of material and environment so as to banish conveyed infection from our hospitals before we can obtain reliable statistics of self-infection.

JOSEPH LISTER

Observations on the Condition of Suppuration

[From *On a New Method of Treating Compound Fracture**]

The frequency of disastrous consequences in compound fracture, contrasted with the complete immunity from danger to life or limb in simple fracture is one of the most striking as well as melancholy facts in surgical practice.

If we inquire how it is that an external wound communicating with the seat of fracture leads to such grave results, we cannot but conclude that it is by inducing, through access of the atmosphere, decomposition of the blood which is effused in greater or less amount

* *Lancet*, London, 1867, I, II.

around the fragments and among the interstices of the tissues, and losing by putrefaction its natural bland character, and assuming the properties of an acrid irritant, occasions both local and general disturbance.

We know that blood kept exposed to the air at the temperature of the body, in a vessel of glass or other material chemically inert, soon decomposes; and there is no reason to suppose that the living tissues surrounding a mass of extravasated blood could preserve it from being affected in a similar manner by the atmosphere. On the contrary, it may be ascertained as a matter of observation, that, in a compound fracture, twenty-four hours after the accident the coloured serum which oozes from the wound is already distinctly tainted with the odour of decomposition, and during the next two or three days, before suppuration has set in, the smell of the effused fluids becomes more and more offensive.

This view of the cause of the mischief in compound fracture is strikingly corroborated by cases in which the external wound is very small. Here, if the coagulum at the orifice is allowed to dry and form a crust, as was advised by John Hunter, all bad consequences are probably averted, and, the air being excluded, the blood beneath becomes organized and absorbed, exactly as in a simple fracture. But if any accidental circumstance interferes with the satisfactory formation of the scab, the smallness of the wound, instead of being an advantage, is apt to prove injurious, because, while decomposition is permitted, the due escape of foul discharges is prevented. Indeed, so impressed are some surgeons with the evil which may result from this latter cause, that, deviating from the excellent Hunterian practice, they enlarge the orifice with the knife in the first instance and apply fomentations, in order to mitigate the suppuration which they render inevitable.

Turning now to the question how the atmosphere produces decomposition of organic substances, we find that a flood of light has been thrown upon this most important subject by the philosophic researches of M. Pasteur, who has demonstrated by thoroughly convincing evidence that it is not to its oxygen or to any of its gaseous constituents that the air owes this property, but to minute particles suspended in it, which are the germs of various low forms of life, long since revealed by the microscope, and regarded as merely accidental concomitants of putrescence, but now shown by Pasteur to be its essential cause, resolving the complex organic compounds into substances of simpler chemical constitution, just as the yeast plant converts sugar into alcohol and carbonic acid. . . . A beautiful illustration of this doctrine seems to me to be presented in surgery by pneumothorax with emphysema, resulting from puncture of the lung by a fractured rib. Here, though atmospheric air is perpetually introduced into the pleura in great abundance, no inflammatory disturbance supervenes; whereas an external wound penetrating the chest, if it remains open, infallibly causes dangerous suppurative pleurisy.

Applying these principles to the treatment of compound fracture, bearing in mind that it is from the vitality of the atmospheric particles that all the mischief arises, it appears that all that is requisite is to dress the wound with some material capable of killing these septic germs, provided that any substance can be found reliable for this purpose, yet not too potent as a caustic.

In the course of the year 1864 I was much struck with an account of the remarkable effects produced by carbolic acid upon the sewage of the town of Carlisle, the admixture of a very small proportion not only preventing all odour from the lands irrigated with the refuse material, but as it was stated destroying the entozoa which usually infest cattle fed upon such pastures.

My attention having for several years been much directed to the subject of suppuration, more especially in its relation to decomposition, I saw that such a powerful antiseptic was peculiarly adapted for experiments with a view to elucidating that subject, and while I was engaged in the investigation the applicability of carbolic acid for the treatment of compound fracture naturally occurred to me.

My first attempt of this kind was made in the Glasgow Royal Infirmary in March 1865, in a case of compound fracture of the leg. It proved unsuccessful, in consequence, as I now believe, of improper management; but subsequent trials have more than realized my most sanguine anticipations.

Carbolic acid proved in various ways well adapted for the purpose. It exercises a local sedative influence upon the sensory nerves; and hence is not only almost painless in its immediate action on a raw surface, but speedily renders a wound previously painful entirely free from uneasiness. When employed in compound fracture its caustic properties are mitigated so as to be unobjectionable by admixture with the blood, with which it forms a tenacious mass that hardens into a dense crust, which long retains its antiseptic virtue, and has also other advantages, as will appear from the following cases, which I will relate in the order of their occurrence, premising that, as the treatment has been gradually improved, the earlier ones are not to be taken as patterns.

Case 1. James G——, aged 11 years, was admitted into the Glasgow Royal Infirmary, on the 12th of August, 1865, with compound fracture of the left leg, caused by the wheel of an empty cart passing over the limb a little below its middle. The wound, which was about an inch and a half long, and three-quarters of an inch broad, was close to, but not over, the line of fracture of the tibia. A probe, however, could be passed beneath the integument over the seat of fracture and for some inches beyond it. Very little blood had been extravasated into the tissues.

My house-surgeon, Dr. Macfee, acting under my instructions, laid a piece of lint dipped in liquid carbolic acid upon the wound, and applied lateral pasteboard splints padded with cotton wool, the limb resting on its outer side, with the knee bent. It was left undisturbed for four days, when, the boy complaining of some uneasiness, I removed the inner splint and examined the wound. It showed no signs of suppuration, but the skin in its immediate vicinity had a slight blush of redness. I now dressed the sore with lint soaked with water having a small proportion of carbolic acid diffused through it; and this was continued for five days, during which the uneasiness and the redness of the skin disappeared, the sore meanwhile furnishing no pus, although some superficial sloughs caused by the acid were separating. But the epidermis being excoriated by this dressing, I substituted for it a solution of one part carbolic acid in from ten to twenty parts of olive oil, which was used for four days, during which a small amount of imperfect pus was produced from the surface of the sore, but not a drop appeared from beneath the skin. It was now clear that there was no longer any danger of deep-seated suppuration, and simple water dressing was employed. Cicatrization proceeded just as in an ordinary granulating sore. At the expiration of six weeks I examined the condition of the bones, and, finding them firmly united, discarded the splints; and two days later the sore was entirely healed, so that the cure could not be said to have been at all retarded by the circumstance of the fracture being compound.

This, no doubt, was a favourable case, and might have done well under ordinary treatment. But the remarkable retardation of suppuration, and the immediate conversion of the compound fracture into a simple fracture with a superficial sore, were most encouraging facts.

Case 2. Patrick F——, a healthy labourer, aged 32, had his right tibia broken on the afternoon of the 11th of September, 1865, by a horse kicking him with its full force over the anterior edge of the bone about its middle. He was at once taken to the infirmary, where Mr. Miller, the house surgeon in charge, found a wound measuring about an inch by a quarter of an inch, from which blood was welling profusely.

He put up the fracture in pasteboard splints, leaving the wound exposed between their anterior edges, and dressing it with a piece of lint dipped in carbolic acid, large enough to overlap the sound skin about a quarter of an inch in every direction. In the evening he changed the lint for another piece, also dipped in carbolic acid, and covered this with oiled paper. I saw the patient next day, and advised the daily application of a bit of lint soaked in carbolic acid over the oiled paper; and this was done for the next five days. On the second day there was an oozing of red fluid from beneath the dressing, but by the third day this had ceased entirely. On the fourth day, when, under ordinary circumstances, suppuration would have made its appearance, the skin had a nearly natural aspect, and there was no increase of swelling, while the uneasiness he had previously felt was almost entirely absent. His pulse was 64, and his appetite improving. On the seventh day, though his general condition was all that could be wished, he complained again of some uneasiness, and the skin about the still adherent crust of blood, carbolic acid, and lint was found to be vesicated, apparently in consequence of the irritation of the carbolic acid. From the seventh day the crust was left untouched till the eleventh day, when I removed it, disclosing a concave surface destitute of granulation, and free from suppuration. Water dressing was now applied, and by the sixteenth day the entire sore, with the exception of one small spot where the bone was bare, presented a healthy granulating aspect, the formation of pus being limited to the surface of the granulations.

The Antiseptic Principle in Surgery

[From *On the Antiseptic Principle of the Practice of Surgery**]

In the course of an extended investigation into the nature of inflammation, and the healthy and morbid conditions of the blood in relation to it, I arrived several years ago at the conclusion that the essential cause of suppuration in wounds is decomposition, brought about by the influence of the atmosphere upon blood or serum retained within them, and, in the case of contused wounds, upon portions of tissue destroyed by the violence of the injury.

To prevent the occurrence of suppuration with all its attendant risks was an object manifestly desirable, but till lately apparently unattainable, since it seemed hopeless to attempt to exclude the oxygen which was universally regarded as the agent by which putrefaction was effected. But when it had been shown by the researches of Pasteur that the septic properties of the atmosphere depended not on the oxygen, or any gaseous constituent, but on minute organisms suspended in it, which owed their energy to their vitality, it occurred to me that decomposition in the injured part might be avoided without excluding the air, by applying as a dressing some material capable of destroying the life of the floating particles. Upon this principle I have based a practice of which I will now attempt to give a short account.

The material which I have employed is carbolic or phenic acid, a volatile organic compound, which appears to exercise a peculiarly destructive influence upon low forms of life, and hence is the most powerful antiseptic with which we are at present acquainted.

The first class of cases to which I applied it was that of compound fractures, in which the effects of decomposition in the injured part were especially striking and pernicious. The results have been such as to establish conclusively the great principle that all local inflammatory mischief and general febrile disturbances which follow severe injuries are due to the irritating and poisonous influence of decomposing blood or sloughs. For these evils are entirely avoided by the antiseptic treatment, so that limbs which would otherwise be un-

* Read in the Surgical Section before the Annual Meeting of the British Medical Association, Dublin, Aug. 9, 1867. Published in the *Brit. M. J.*, Sept. 21, 1867.

hesitatingly condemned to amputation may be retained, with confidence of the best results.

In conducting the treatment, the first object must be the destruction of any septic germs which may have been introduced into the wounds, either at the moment of the accident or during the time which has since elapsed. This is done by introducing the acid of full strength into all accessible recesses of the wound by means of a piece of rag held in dressing forceps and dipped into the liquid.[1] This I did not venture to do in the earlier cases; but experience has shown that the compound which carbolic acid forms with the blood, and also any portions of tissue killed by its caustic action, including even parts of the bone, are disposed of by absorption and organisation, provided they are afterwards kept from decomposing. We are thus enabled to employ the antiseptic treatment efficiently at a period after the occurrence of the injury at which it would otherwise probably fail. Thus I have now under my care, in Glasgow Infirmary, a boy who was admitted with compound fracture of the leg as late as eight and one-half hours after the accident, in whom, nevertheless, all local and constitutional disturbance was avoided by means of carbolic acid, and the bones were soundly united five weeks after his admission.

The next object to be kept in view is to guard effectually against the spreading of decomposition into the wound along the stream of blood and serum which oozes out during the first few days after the accident, when the acid originally applied has been washed out or dissipated by absorption and evaporation. This part of the treatment has been greatly improved during the past few weeks. The method which I have hitherto published (see *Lancet* for Mar. 16th, 23rd, 30th, and April 27th of the present year) consisted in the application of a piece of lint dipped in the acid, overlapping the sound skin to some extent and covered with a tin cap, which was daily raised in order to touch the surface of the lint with the antiseptic. This method certainly succeeded well with wounds of moderate size; and indeed I may say that in all the many cases of this kind which have been so treated by myself or my house-surgeons, not a single failure has occurred. When, however, the wound is very large, the flow of blood and serum is so profuse, especially during the first twenty-four

---

[1] The addition of a few drops of water to a considerable quantity of the acid, induces it to assume permanently the liquid form.

hours, that the antiseptic application cannot prevent the spread of decomposition into the interior unless it overlaps the sound skin for a very considerable distance, and this was inadmissible by the method described above, on account of the extensive sloughing of the surface of the cutis which it would involve. This difficulty has, however, been overcome by employing a paste composed of common whiting (carbonate of lime), mixed with a solution of one part of carbolic acid in four parts of boiled linseed oil so as to form a firm putty. This application contains the acid in too dilute a form to excoriate the skin, which it may be made to cover to any extent that may be thought desirable, while its substance serves as a reservoir of the antiseptic material. So long as any discharge continues, the paste should be changed daily, and, in order to prevent the chance of mischief occurring during the process, a piece of rag dipped in the solution of carbolic acid in oil is put on next the skin, and maintained there permanently, care being taken to avoid raising it along with the putty. This rag is always kept in an antiseptic condition from contact with the paste above it, and destroys any germs which may fall upon it during the short time that should alone be allowed to pass in the changing of the dressing. The putty should be in a layer about a quarter of an inch thick, and may be advantageously applied rolled out between two pieces of thin calico, which maintain it in the form of a continuous sheet, which may be wrapped in a moment round the whole circumference of a limb if this be thought desirable, while the putty is prevented by the calico from sticking to the rag which is next the skin. When all discharge has ceased, the use of the paste is discontinued, but the original rag is left adhering to the skin till healing by scabbing is supposed to be complete. I have at present in the hospital a man with severe compound fracture of both bones of the left leg, caused by direct violence, who, after the cessation of the sanious discharge under the use of the paste, without a drop of pus appearing, has been treated for the last two weeks exactly as if the fracture was a simple one. During this time the rag, adhering by means of a crust of inspissated blood collected beneath it, has continued perfectly dry, and it will be left untouched till the usual period for removing the splints in a simple fracture, when we may fairly expect to find a sound cicatrix beneath it.

We cannot, however, always calculate on so perfect a result as this. More or less pus may appear after the lapse of the first week, and the larger the wound, the more likely this is to happen. And here I would desire earnestly to enforce the necessity of persevering with the antiseptic application in spite of the appearance of suppuration, so long as other symptoms are favourable. The surgeon is extremely apt to suppose that any suppuration is an indication that the antiseptic treatment has failed, and that poulticing or water dressing should be resorted to. But such a course would in many cases sacrifice a limb or a life. I cannot, however, expect my professional brethren to follow my advice blindly in such a matter, and therefore I feel it necessary to place before them, as shortly as I can, some pathological principles intimately connected, not only with the point we are immediately considering, but with the whole subject of this paper.

If a perfectly healthy granulating sore be well washed and covered with a plate of clean metal, such as block tin, fitting its surface pretty accurately, and overlapping the surrounding skin an inch or so in every direction and retained in position by adhesive plaster and a bandage, it will be found, on removing it after twenty-four or forty-eight hours, that little or nothing that can be called pus is present, merely a little transparent fluid, while at the same time there is an entire absence of the unpleasant odour invariably perceived when water dressing is changed. Here the clean metallic surface presents no recesses like those of porous lint for the septic germs to develop in, the fluid exuding from the surface of the granulations has flowed away undecomposed, and the result is the absence of suppuration. This simple experiment illustrates the important fact that granulations have no inherent tendency to form pus, but do so only when subjected to preternatural stimulus. Further, it shows that the mere contact of a foreign body does not of itself stimulate granulations to suppurate; whereas the presence of decomposing organic matter does. These truths are even more strikingly exemplified by the fact that I have elsewhere recorded (*Lancet*, March 23rd, 1867), that a piece of dead bone free from decomposition may not only fail to induce the granulations around it to suppurate, but may actually be absorbed by them; whereas a bit of dead bone soaked with putrid pus infallibly induces suppuration in its vicinity. . . .

I left behind me in Glasgow a boy, thirteen years of age, who, between three and four weeks previously, met with a most severe injury to the left arm, which he got entangled in a machine at a fair. There was a wound six inches long and three inches broad, and the skin was very extensively undermined beyond its limits, while the soft parts were generally so much lacerated that a pair of dressing forceps introduced at the wound and pushed directly inwards appeared beneath the skin at the opposite aspect of the limb. From this wound several tags of muscle were hanging, and among them was one consisting of about three inches of the triceps in almost its entire thickness; while the lower fragment of the bone, which was broken high up, was protruding four inches and a half, stripped of muscle, the skin being tucked in under it. Without the assistance of the antiseptic treatment, I should certainly have thought of nothing else but amputation at the shoulder-joint; but, as the radial pulse could be felt and the fingers had sensation, I did not hesitate to try to save the limb and adopted the plan of treatment above described, wrapping the arm from the shoulder to below the elbow in the antiseptic application, the whole interior of the wound, together with the protruding bone, having previously been freely treated with strong carbolic acid. About the tenth day, the discharge, which up to that time had been only sanious and serous, showed a slight admixture of slimy pus; and this increased till (a few days before I left) it amounted to about three drachms in twenty-four hours. But the boy continued as he had been after the second day, free from unfavorable symptoms, with pulse, tongue, appetite and sleep natural and strength increasing, while the limb remained as it had been from the first, free from swelling, redness, or pain. I, therefore, persevered with the antiseptic dressing; and, before I left, the discharge was already somewhat less, while the bone was becoming firm. I think it likely that, in that boy's case, I should have found merely a superficial sore had I taken off all the dressings at the end of the three weeks; though, considering the extent of the injury, I thought it prudent to let the month expire before disturbing the rag next the skin. But I feel sure that, if I had resorted to ordinary dressing when the pus first appeared, the progress of the case would have been exceedingly different.

The next class of cases to which I have applied the antiseptic treatment is that of abscesses. Here also the results have been extremely satisfactory, and in beautiful harmony with the pathological principles indicated above. The pyogenic membrane, like the granulations of a sore, which it resembles in nature, forms pus, not from any inherent disposition to do so, but only because it is subjected to some preternatural stimulation. In an ordinary abscess, whether acute or chronic, before it is opened the stimulus which maintains the suppuration is derived from the presence of pus pent up within the cavity. When a free opening is made in the ordinary way, this stimulus is got rid of, but the atmosphere gaining access to the contents, the potent stimulus of decomposition comes into operation, and pus is generated in greater abundance than before. But when the evacuation is effected on the antiseptic principle, the pyogenic membrane, freed from the influence of the former stimulus without the substitution of a new one, ceases to suppurate (like the granulations of a sore under metallic dressing), furnishing merely a trifling amount of clear serum, and, whether the opening be dependent or not, rapidly contracts and coalesces. At the same time any constitutional symptoms previously occasioned by the accumulation of the matter are got rid of without the slightest risk of the irritative or hectic fever hitherto so justly dreaded in dealing with large abscesses.

In order that the treatment may be satisfactory, the abscess must be seen before it is opened. Then, except in very rare and peculiar cases,[1] there are no septic organisms in the contents, so that it is needless to introduce carbolic acid into the interior. Indeed, such a procedure would be objectionable, as it would stimulate the pyogenic membrane to unnecessary suppuration. All that is requisite is to guard against the introduction of living atmospheric germs from without, at the same time that free opportunity is afforded for the escape of the discharge from within. . . .

Ordinary contused wounds are, of course, amenable to the same treatment as compound fractures, which are a complicated variety of them. I will content myself with mentioning a single instance of this class of cases. In April last, a volunteer was discharging a rifle when it burst, and blew back the thumb with its metacarpal bone, so that it could be bent back as on a hinge at the trapezial joint, which had evidently been opened, while all the soft parts between the metacarpal bones of the thumb and forefinger were torn through. I need not insist before my present audience on the ugly character of such an injury. My house-surgeon, Mr. Hector Cameron, applied carbolic acid to the whole raw surface, and completed the dressing as if for

---

[1] As an instance of one of these exceptional cases, I may mention that of an abscess in the vicinity of the colon, and afterwards proved by postmortem examination to have once communicated with it. Here the pus was extremely offensive when evacuated, and exhibited vibrios under the microscope.

compound fracture. The hand remained free from pain, redness or swelling, and with the exception of a shallow groove, all the wound consolidated without a drop of matter, so that if it had been a clean cut, it would have been regarded as a good example of primary union. The small granulating surface soon healed, and at present a linear cicatrix alone tells of the injury he has sustained, while his thumb has all its movements and his hand a fine grasp.

If the severest forms of contused and lacerated wounds heal thus kindly under the antiseptic treatment, it is obvious that its application to simple incised wounds must be merely a matter of detail. . . .

It would carry me far beyond the limited time which, by the rules of the Association, is alone at my disposal, were I to enter into the various applications of the antiseptic principle in the several special departments of surgery.

There is, however, one point more that I cannot but advert to, viz., the influence of this mode of treatment upon the general healthiness of a hospital. Previously to its introduction the two large wards in which most of my cases of accident and of operation are treated were among the unhealthiest in the whole surgical division of the Glasgow Royal Infirmary, in consequence apparently of those wards being unfavorably placed with reference to the supply of fresh air; and I have felt ashamed when recording the results of my practice, to have so often to allude to hospital gangrene or pyaemia. It was interesting, though melancholy, to observe that whenever all or nearly all the beds contained cases with open sores, these grievous complications were pretty sure to show themselves; so that I came to welcome simple fractures, though in themselves of little interest either for myself or the students, because their presence diminished the proportion of open sores among the patients. But since the antiseptic treatment has been brought into full operation, and wounds and abscesses no longer poison the atmosphere with putrid exhalations, my wards, though in other respects under precisely the same circumstances as before, have completely changed their character; so that during the last nine months not a single instance of pyaemia, hospital gangrene, or erysipelas has occurred in them.

As there appears to be no doubt regarding the cause of this change, the importance of the fact can hardly be exaggerated.

XL

CELLULAR PATHOLOGY

IN SECTION XXII WILL BE FOUND SELECTIONS FROM THE EARLIER PATHOLOGISTS who, by means of the postmortem examination, built up the knowledge of gross pathology. Not, however, until the conception of the cell as the unit and basis of all structural changes was established could pathology reach the dignity of a science. This was the unique work of Rudolph Virchow (1821-1902), whose *Cellular Pathology* was published in 1858. The distinguishing feature of this work was that it stated the conception of cellular pathology clearly and completely and for the first time. Virchow's opportunity, of which he took advantage, was that with the cell doctrine as a starting point he could review every department of pathology and put it on a clearly understandable basis. All the apparently confused phenomena of disease fell into place, and such things as the crases and dyscrases of the Viennese pathologist Rokitansky seemed immediately ridiculous. Virchow's doctrine was *"omnis cellula e cellula"* (where a cell arises, there a cell must have previously existed).

There was little to add to his work except details. In 1866, His invented the microtome. In 1871 and 1875, Weigert invented staining methods for sections. Virchow made some mistakes: all his life he consistently refused to acknowledge the unity of different manifestations of tuberculosis.

Virchow's greatest successor and pupil was Julius Cohnheim (1839-1884). His monograph on inflammation, *Neue untersuchungen über die Entzüngdung* (1873), embodies the modern conception, showing in opposition to Virchow's teaching that the essential feature of the inflammatory process is the passage of white blood cells through the walls of the capillaries. This was dynamic rather than static pathology because the observations were made on the cornea of the eye of the living rabbit and the mesentery of the living frog.

REFERENCES

COHNHEIM, J. Lectures on General Pathology. Memoir of Cohnheim. London, New Sydenham Society, 1889.

KRUMBHAAR, E. B. Pathology. Clio Medica series. New York, Paul B. Hoeber, 1937.

LONG, E. R. A History of Pathology. Baltimore, Williams and Wilkins, 1928.

SIGERIST, H. E. The Great Doctors—Virchow. New York, W. W. Norton, 1933.
WELCH, W. H. Virchow. *Philadelphia M. J.*, 1902.

RUDOLF VIRCHOW

CELLULAR PATHOLOGY

[From *Cellular Pathology*,* Chapter 1]

THE CELL THEORY

The present reform in medicine, of which you have all been witnesses, essentially had its rise in new anatomical observations, and the exposition also, which I have to make to you, will therefore principally be based upon anatomical demonstrations. But for me it would not be sufficient to take, as has been the custom during the last ten years, pathological anatomy alone as the groundwork of my views; we must add thereto those facts of general anatomy also, to which the actual state of medical science is due. The history of medicine teaches us, if we will only take a somewhat comprehensive survey of it, that at all times permanent advances have been marked by anatomical innovations, and that every more important epoch has been directly ushered in by a series of important discoveries concerning the structure of the body. So it was in those old times, when the observations of the Alexandrian school, based for the first time upon the anatomy of man, prepared the way for the system of Galen; so it was, too, in the Middle Ages, when Vesalius laid the foundations of anatomy, and therewith began the real reformation of medicine; so, lastly, was it at the commencement of this century, when Bichat developed the principles of general anatomy. What Schwann, however, has done for histology, has as yet been but in a very slight degree built up and developed for pathology, and it may be said that nothing has penetrated less deeply into the minds of all than the cell-theory in its intimate connection with pathology.

If we consider the extraordinary influence which Bichat in his time exercised upon the state of medical opinion, it is indeed astonishing that such a relatively long period should have elapsed since Schwann made his great discoveries, without the real importance of

* Cellular-Pathologie, 1858. Translated from the second edition of the original by Frank Chance, 1860.

the new facts having been duly appreciated. This has certainly been essentially due to the great incompleteness of our knowledge with regard to the intimate structure of our tissues which has continued to exist until quite recently, and, as we are sorry to be obliged to confess, still even now prevails with regard to many points of histology to such a degree, that we scarcely know in favour of what view to decide.

Especial difficulty has been found in answering the question, from what parts of the body action really proceeds—what parts are active, what passive; and yet it is already quite possible to come to a definitive conclusion upon this point, even in the case of parts the structure of which is still disputed. The chief point in this application of histology to pathology is to obtain a recognition of the fact, that the cell is really the ultimate morphological element in which there is any manifestation of life, and that we must not transfer the seat of real action to any point beyond the cell. Before you, I have no particular reason to justify myself, if in this respect I make quite a special reservation in favour of life. In the course of these lectures you will be able to convince yourselves that it is almost impossible for any one to entertain more mechanical ideas in particular instances than I am wont to do, when called upon to interpret the individual processes of life. But I think that we must look upon this as certain, that, however much of the more delicate interchange of matter, which takes place within a cell, may not concern the material structure as a whole, yet the real action does proceed from the structure as such, and that the living element only maintains its activity as long as it really presents itself to us as an independent whole.

In this question it is of primary importance (and you will excuse my dwelling a little upon this point, as it is one which is still a matter of dispute) that we should determine what is really to be understood by the term cell. Quite at the beginning of the latest phase of histological development, great difficulties sprang up in crowds with regard to this matter. Schwann, as you no doubt recollect, following immediately in the footsteps of Schleiden, interpreted his observations according to botanical standards, so that all the doctrines of vegetable physiology were invoked, in a greater or less degree, to decide questions relating to the physiology of animal bodies. Vegetable cells, however, in the light in which they were at that time universally, and as they are even now also frequently regarded, are structures, whose identity with what we call animal cells cannot be admitted without reserve.

It is only when we adhere to this view of the matter, when we separate from the cell all that has been added to it by an after-development, that we obtain a simple homogeneous, extremely monotonous structure, recurring with extraordinary constancy in living organisms. But just this very constancy forms the best criterion of our having before us in this structure one of those really elementary bodies, to be built up of which is eminently characteristic of every living thing—without the pre-existence of which no living forms arise, and to which continuance and the maintenance of life is intimately attached. Only since our idea of a cell has assumed this severe form—and I am somewhat proud of having always, in spite of the reproach of pedantry, firmly adhered to it—only since that time can it be said that a simple form has been obtained which we can everywhere again expect to find, and which, though different in size and external shape, is yet always identical in its essential constituents.

In such a simple cell we can distinguish dissimilar constituents, and it is important that we should accurately define their nature also.

In the first place, we expect to find a nucleus within the cell; and with regard to this nucleus, which has usually a round or oval form, we know that, particularly in the case of young cells, it offers greater resistance to the action of chemical agents than do the external parts of the cell, and that, in spite of the greatest variations in the external form of the cell, it generally maintains its form. The nucleus is accordingly, in cells of all shapes, that part which is the most constantly found unchanged. There are indeed isolated cases, which lie scattered throughout the whole series of facts in comparative anatomy and pathology, in which the nucleus also has a stellate or angular appearance; but these are extremely rare exceptions, and dependent upon peculiar changes which the element has undergone. Generally, it may be said that, as long as the life of the cell has not been brought to a close, as long as the cells behave as elements still endowed with vital power, the nucleus maintains a very nearly constant form.

The nucleus, in its turn, in completely developed cells, very constantly encloses another structure within itself—the so-called nucleolus. With regard to the question of vital form, it cannot be said of the nucleolus that it appears to be an absolute requisite; and, in a considerable number of young cells, it has as yet escaped detection. On the other hand, we regularly meet with it in fully developed, older forms; and it, therefore, seems to mark a higher degree of development in the cell. According to the view which was put forward in the first instance by Schleiden, and accepted by Schwann, the connection between the three coexistent cell-constituents was long thought to be on this wise: that the nucleolus was the first to shew itself in the development of tissues, by separating out of a formative fluid (blastema, cyto-blastema), that it quickly attained a certain size, that then fine granules were precipitated out of the blastema and settled around it, and that about these there condensed a membrane. That in this way a nucleus was completed, about which new matter gradually gathered, and in due time produced a little membrane (the celebrated watch-glass form). This description of the first development of cells out of free blastema, according to which the nucleus was regarded as preceding the formation of the cell, and playing the part of a real cell-former (cytoblast), is the one which is usually concisely designated by the name of the cell-theory (more accurately, theory of free cell-formation),—a theory of development which has now been almost entirely abandoned, and in support of the correctness of which not one single fact can with certainty be adduced. With respect to the nucleolus, all that we can for the present regard as certain, is, that where we have to deal with large and fully developed cells, we almost constantly see a nucleolus in them; but that, on the contrary, in the case of many young cells it is wanting.

In my first lecture, gentlemen, I laid before you the general points to be noted with regard to the nature and origin of cells and their constituents. Allow me now to preface our further considerations with a review of the animal tissues in general, and this both in their physiological and pathological relations.

The most important obstacles which, until quite recently, existed in this quarter, were by no means chiefly of a pathological nature. I am convinced that pathological conditions would have been mastered with far less difficulty if it had not, until quite lately, been utterly impossible to give a simple and comprehensive sketch of the physiological tissues. The old views, which have in part come down to us from the last century, have exercised such a preponderating influence upon that part of histology which is, in a pathological point of view, the most important, that not even yet has unanimity been arrived at, and you will therefore be constrained, after you have inspected the preparations I shall lay before you, to come to your own conclusions as to how far that which I have to communicate to you is founded upon real observation.

If you read the "Elementa Physiologiae" of Haller, you will find, where the elements of the body are treated of, the most prominent position in the whole work assigned to fibres, the very characteristic expression being there made use of, that the fibre (fibra) is to the physiologist what the line is to the geometrician.

This conception was soon still further expanded, and the doctrine that fibres serve as the groundwork of nearly all parts of the body, and that the most various tissues are reducible to fibres as their ultimate constituents, was longest maintained in the case of the very tissue in which, as it has turned out, the pathological difficulties were the greatest—in the so-called cellular tissue.

In the course of the last ten years of the last century there arose, however, a certain degree of reaction against this fibre-theory, and in the school of natural philosophers another element soon attained to honour though it had its origin in for more speculative views than the former, namely, the globule. Whilst some still clung to their fibres, others, as in more recent times Milne Edwards, thought fit to go so far as to suppose the fibres, in their turn, to be made up of globules ranged in lines. This view was in part attributable to optical illusions in microscopical observation. The objectionable method which prevailed during the whole of the last and a part of the present century—of making observations (with but indifferent instruments) in the full glare of the sun—caused a certain amount of dispersion of light in nearly all microscopical objects, and the impression communicated to the observer was, that he saw nothing else than globules. On the other hand, however, this view corresponded with the ideas common amongst natural philosophers as to the primary origin of everything endowed with form.

These globules (granules, molecules) have, curiously enough, maintained their ground, even in modern histology, and there are but few histological works which do not begin with the consideration of elementary granules. In a few instances, these views as to the globular nature of elementary parts have, even not very long ago, acquired such ascendancy, that the composition, both of the primary tissues in the embryo and also of the later ones, was based upon them. A cell was considered to be produced by the globules arranging themselves in a spherical form, so as to constitute a membrane, within which other globules remained, and formed the contents. In this way did even Baumgartner and Arnold contend against the cell theory.

This view has, in a certain manner, found support even in the history of development—in the so-called investment-theory (*Umhullungstheorie*)—a doctrine which for a time occupied a very prominent position. The upholders of this theory imagined, that originally a number of elementary globules existed scattered through a fluid, but that, under certain circumstances, they gathered together, not in the form of vesicular membranes, but so as to constitute a compact heap, a globe (mass, cluster—*Klumpchen*) and that this globe was the starting point of all further development, a membrane being formed outside and a nucleus inside, by the differentiation of the mass, by apposition, or intussusception.

At the present time, neither fibres, nor globules, nor elementary granules, can be looked upon as histological starting-points. As long as living elements were conceived to be produced out of parts previously destitute of shape, such as formative fluids, or matters (plastic, matter, blastema, cytoblastema), any one of the above views could of course be entertained, but it is in this very particular that the revolution which the last few years have brought with them has been the most marked. Even in pathology we can now go so far as to establish as a general principle, that no development of any kind begins de novo, and consequently as to reject the theory of equivocal (spontaneous) generation just as much in the history of the development of individual parts as we do in that of entire organisms. Just as little as we can now admit that a taenia can arise out of saburral mucus, or that out of the residue of the decomposition of animal or vegetable matter an infusorial animalcule, a fungus, or an alga, can be formed, equally little are we disposed to concede either in physiological or pathological histology, that a new cell can build itself up out of any non-cellular substance. Where a cell arises, there a cell must have previously existed (omnis cellula e cellula), just as an animal can spring only from an animal, a plant only from a plant. In this manner, although there are still a few spots in the body where absolute demonstration has not yet been afforded, the principle is nevertheless established, that in the whole series of living things, whether they be entire plants or animal organisms, or essential constituents of the same, an eternal law of continuous development prevails. There is no discontinuity of development of such a kind that a new generation can of itself give rise to a new series of developmental forms. No developed tissues can be traced back either to any large or small simple element, unless it be unto a cell.

JULIUS COHNHEIM

THE PHENOMENA OF INFLAMMATION

[From *Lectures on General Pathology*,* Vol. I, Chapter v]

You need only expose the vessels of a part to the air by removing its protective coverings; when if you have selected a transparent tissue, there is nothing to hinder microscopic observation. The simplest method is to draw out the intestine of a curarized frog through a laterally placed opening in the abdominal wall, and to bring the mesentery under the microscope, after having carefully spread it out on a slide adapted to the purpose. Or you may wound the papillary surface of the frog's tongue by removing the papillae with a cut of the scissors, carried parallel to the surface; a number of larger and smaller vessels will thus be exposed in the base of the wound. No further violence should be used after this; on the contrary, the more carefully you protect the preparation from all disturbing accidents, as contamination by blood, stretching or loss of moisture, the more regularly will a succession of appearances be developed, which are well calculated to fully engross your attention.

The first thing you notice in the exposed vessels is a dilatation which occurs chiefly in the arteries, then in the veins, and least of all in the capillaries. With the dilatation which is gradually developed, but which during the space of fifteen to twenty minutes has usually attained considerable proportions (often exceeding twice the original diameter) there immediately sets in in the mesentery an acceleration of the blood-stream, most striking again in the arteries, but very apparent in the veins and capillaries also. Yet this acceleration never lasts long; after half an hour or an hour, or sometimes

* Neue Untersuchungen über die Entzündung, Berlin, 1873. Translated from the second German edition by Alexander B. McKee, 1889.

after a shorter or longer interval, it invariably gives place to a decided retardation, the velocity of the stream falling more or less
below the normal standard, and so continuing as long as the vessels
occupy their exposed situation. Such is the course of events in the
mesentery experiment, in which not only the vessels of the mesentery
but their terminal ramifications in the intestine are laid bare. In the
wound of the tongue, on the other hand, the acceleration is often
altogether absent; and from the first there is associated with the
dilatation a retardation of the stream, which increases as the dilation increases. This is the case at least when a number of larger
branches are exposed in the wound, but not their finer ramifications.
Should the latter also be laid bare, a temporary acceleration precedes the slowing of the blood-stream, which never fails finally to
set in in the exposed vessels.

This stage having been reached, the vessels are seen to be all of
them very wide; a multitude of capillaries which were formerly
hardly perceptible can now be clearly distinguished; pulsation is unusually conspicuous on into the finest ramifications of the arteries;
while the flow is everywhere slower than normal, so that the individual corpuscles may easily be recognised not only in the capillaries
but also in the veins, and during diastole even in the arteries. In
consequence of the tardy forward movement the corpuscles accumulate in large numbers in the capillaries, so that the latter appear
redder than usual, and therefore fuller, more voluminous; yet their
cross-section, as just stated, is only very inconsiderably enlarged. But
it is the veins rather than the capillaries that attract the notice of the
observer; for slowly and gradually there is developed in them an
extremely characteristic condition; the originally plasmatic zone becomes filled with innumerable colourless corpuscles. The plasmatic
zone of the veins, you will remember, is always occupied by scattered colourless blood-corpuscles, which, owing to their globular
form and low specific gravity, are driven into the periphery of the
stream, and whose adhesiveness makes it difficult for them to escape
from the wall once they have come into contact with it. It is obvious
that this difficulty will be enhanced in proportion to the slowness of
the blood-stream; and thus it is not surprising that a gradual accumulation of large numbers of colourless corpuscles should take place in
the peripheral zone, and here come to be comparatively motionless.
For that a state of absolute rest, an actual standstill is out of the

question, I need hardly mention expressly; the colourless cells of the plasmatic layer remain stationary at most for a time, they then advance a little and perhaps make another short halt, and so on. Yet this does not lessen the striking contrast presented by the central column of red blood-corpuscles, flowing on in an uninterrupted stream of uniform velocity, and the peripheral layer of resting colourless cells; the internal surface of the vein appears paved with a single but unbroken layer of colourless corpuscles without the interposition at any time of a single red one. It is the separation of the white from the red corpuscles that gives the venous stream in these cases that characteristic appearance, that anything analogous to which you will look for in vain in the other vessels. For in the capillaries, although large numbers of colourless blood-corpuscles adhere to the walls, there is always an admixture of red cells, or rather these are very decidedly in the majority. Lastly, in the arteries there is seen during diastole, almost at the moment of exit of the wave, a number of colourless blood-corpuscles rolling straight towards the periphery; yet these are always swept into the stream at the next systole, so that the development of a resting peripheral layer is here altogether out of the question.

But the eye of the observer hardly has time to catch all the details of the picture before it is fettered by a very unexpected occurrence. Usually it is a vein with the typical peripheral arrangement of the white corpuscles, but sometimes a capillary, that first displays the phenomenon. A pointed projection is seen on the external contour of the vessel wall; it pushes itself further outwards, increases in thickness, and the pointed projection is transformed into a colourless rounded hump; this grows longer and thicker, throws out fresh points, and gradually withdraws itself from the vessel wall, with which at last it is connected only by a long thin pedicle. Finally this also detaches itself, and now there lies outside the vessel a colourless, faintly glittering, contractile corpuscle with a few short processes and one long one, of the size of a white blood-cell, and having one or more nuclei, in a word, a colourless blood-corpuscle. While this is taking place at one spot, the same process has been carried on in other portions of the veins and capillaries. Quite a large number of white blood-cells have betaken themselves to the exterior of the vessels, and these are constantly followed by fresh ones, whose place in the peripheral layer is immediately occupied by others. Like every stage of the entire process on from the moment of exposure, these phenomena may develop either rapidly or slowly; at one time the earliest emigration very quickly succeeds the pavementing; at another an hour or more may pass without anything happening to draw attention to the contour of a single vein or capillary. In any case the final result, after six or eight or more hours have elapsed, will be the enclosure of all the veins, small and large of the mesentery or wound of the tongue with several layers of colourless blood-corpuscles. These fence in the veins, in the interior of which the previously described conditions continue, namely, the peripheral arrangement of the colourless cells and the central unbroken flow of red blood-corpuscles. Nothing analogous has occurred in connection with the arteries, their contour has remained smooth as before, nor can a solitary corpuscle, red or white, be discovered on their outer surfaces, except of course such as may have reached them from the neighbouring veins. On the other hand, the capillaries take, as already mentioned, a very active part in the process, yet these and the capillary veins differ remarkably from the veins proper in that not merely colourless but red corpuscles emigrate from them. This result is completely in harmony with the condition of the stream in these vessels, for I have already called your attention to the fact that in the veins only white corpuscles, in the capillaries both varieties, are in contact with the vessel wall, so that whether a preponderance of white or of red corpuscles passes out of a given capillary depends solely on the numerical relations of the cells accumulated in its interior.

Keeping pace with this exodus, emigration, or, as it is also called, extravasation of corpuscular elements there occurs an increased transudation of fluid, in consequence of which the meshes of the mesentery, or the tissues of the tongue, are infiltrated and swell. But this is not all. The extravasated colourless corpuscles distribute themselves, in proportion as their numbers increase, over a larger area, forsaking the neighbourhood of the vessels from which they were derived. The tissues become more and more densely packed with them, while the red cells, which have not the power of independent locomotion, remain seated in the vicinity of their capillaries, yet these may also be carried off by the stream of transudation. Soon a moment must arrive when the products of exudation and transudation can no longer be accommodated in the tissues. They now gain the free surface of the mesentery, and should the transuded fluid coagulate, as is the rule here, the final result of the processes just described will be the deposition on the mesentery, as well as on the intestine, of a fibrinous pseudo-membrane, densely packed with colourless blood-corpuscles, and interspersed with isolated red cells. The appearances are essentially the same after painting the smooth surface of the frog's tongue with croton-oil. Of course it is absolutely necessary to employ the croton-oil in extreme dilution—about one part to forty or fifty of olive oil—and even then to allow the mixture to act only a very short time. For if you do not soon wipe off the oil, or still more if you make use of a concentrated solution, you at once get an intense corrosive action, as evidenced by the formation of thrombi in the larger vessels, and the occurrence of complete stasis in some of the capillaries, more especially in the superficial ones. The weak solution, on the other hand, provokes an enormous dilatation of all the vessels, which at first is accompanied by a very great acceleration of the blood-stream. After a time, however, the velocity commences to diminish in the dilated vessels, and is converted into a pronounced retardation of the entire circulation through the tongue. With the retardation there is simultaneously developed the peripheral arrangement of the colourless blood-corpuscles in the veins, and the accumulation of blood-corpuscles in the capillaries, which is so extreme as to result in the actual stagnation of the red cells in such of the latter as are superficially situated. And now it will not be long before extravasation from some of the capillaries and veins begins. As might be expected, the veins supply only colourless corpuscles; the capillaries whose blood is becoming stagnant almost exclusively red, while from those capillaries in which the flow, though retarded, is still sustained, coloured and un-coloured cells pass out together, at one point more colourless, at another more red, and these may even collect into small clumps outside the vessels. At the same time the swelling of the tongue gradually increases, it becomes intensely reddened; a multitude of small punctiform haemorrhages are already apparent even to the naked eye, while microscopic examination reveals a no less dense accumulation of colourless corpuscles throughout its tissues.

XLI

RISE OF THE SPECIALTIES

IT IS A CURIOUS HISTORICAL FACT THAT THE SUBJECTS THAT WE NOW CALL the specialties—diseases of the eye, the ear, the nose and throat—were very late in developing, and until 1800, were almost literally relegated by the medical profession to the quacks. The plight of a patient with an eye or ear disease up to the end of the eighteenth century, was pitiable in the extreme. A few surgeons, such as Scarpa and Sir Astley Paston Cooper, performed some eye operations, but when John Cunningham Sanders* (1773-1810) founded Moorfield's Hospital in London (1805) for the exclusive care of eye patients, he apologized to his regular colleagues for his invasion of the field of the quack.

Grassi of Salerno, also called Benvenuto, wrote a work *De Oculis, eorumque egritudinibus et curis,* which was printed in 1475. It was widely used during the Middle Ages as a text on eye diseases. (It has been translated by Dr. Casey Wood, Stanford University Press, 1929.) Georg Bartisch (1536-1606) published, in 1583, his famous *Ophthalmodouleia* or *Augendiènst,* which contains illustrations of some eye operations as well as of the crude anatomy of the eye. It was probably used as a manual of instruction for traveling quacks. Spectacles or reading glasses of some kind were used from about 1270: the credit for their invention is in dispute (see Garrison's *History of Medicine,* pp. 184-185).

More scientific and dignified pioneers in the study of ophthalmology include Jacques Daviel (1696-1762), who described a method of extraction of cataract, which for all practical purposes is the procedure now used; and Thomas Young (1773-1829) who described the mechanism of the eye (1801) and explained visual accommodation and astigmatism. It was not, however, until the fortuitous meeting of Albrecht von Graefe (1828-1870) and Frans Cornelius Donders (1818-1889) in London, in 1851, that modern ophthalmology may be said to have had a beginning. Both men had studied in such eye clinics as existed on the Continent. When they met in London they immediately became friends, immersed in a common enthusiasm. To their studies Donders brought his knowledge of physiology and optics, Graefe, clinical genius. Graefe returned to Berlin in 1857, to become professor of ophthalmology and Donders to Utrecht in

* Sanders was associated in his pioneer work with John Richard Frere and Richard Battley. In Collins' "History of Moorfield's Hospital" he relates that the return of the English sailors and soldiers from Egypt after the campaign against Napoleon with many instances of trachoma, stimulated the three young men to their special study.

a similar chair. In 1851 Helmholtz had invented the ophthalmoscope. Graefe's work included a description of iridectomy for iritis, improvement of the cataract operation, the application of the ophthalmoscope to all eye examinations, descriptions of sympathetic ophthalmia, and ocular paralyses. Donders' greatest work was *The Anomalies of Refraction and Accommodation* (1864), first published in English by the New Sydenham Society. This work is the basis of all scientific fitting of glasses for myopia, strabismus, hypermetropia, and so forth.

Otology did not have quite such a definite point of origin. Jean-Marc-Gaspard Itard (1775-1838) of Provence, may possibly be said to be the pioneer, not only in the physiology of the ear, but in clinical practice. He wrote the first formal treatise on diseases of the ear, in 1821. Previously Guyot and Cleland had catheterized the eustachian tube, and Petit had performed mastoid operations. Adam Politzer (1835-1920) popularized the use of the otoscope and published an atlas showing the appearance of the membrana tympani by illumination.

Laryngology and rhinology developed coincidentally with surgery and clinical medicine. That is to say, knowledge of tonsillitis and diphtheria, nasal polyps, and tuberculosis and carcinoma of the larynx existed in very early times but it was not incorporated into a specialty until the nineteenth century. Several men might be taken to represent its crystallization—Johann Czermak (1815-1851), Horace Green (1802-1866), and Morell Mackenzie (1837-1892). We reproduce here Mackenzie's history of laryngoscopy, because it was with the introduction of the laryngoscope that the specialty really found a point of focus, and the account is a fine bit of history from one of the pioneers. Mackenzie wrote one of the first systematic treatises on diseases of the nose and throat (1880-1881), and in establishing a special hospital for such cases and insisting on the legitimate claims of the specialty, he gave it dignity. A highly dramatic and historic incident of his career was his treatment of the Emperor Frederick of Germany. Mackenzie removed a growth from the larynx, which Virchow pronounced benign. The German consultants—von Bergmann, in particular—insisted that it was malignant, and so it proved, as it caused the death of the Emperor within a year. The German medical profession accused Mackenzie of deliberate delay in the conduct of the case, and the entire incident caused most strained relations between the governments of England and Germany.

REFERENCES

CHANCE, B. Ophthalmology. Clio Medica series. New York, Paul B. Hoeber, 1939.

HIRSCHBERG, J. Geschichte der Augenheilkunde. In: Graefe-Salmisch's Handbuch. Leipzig-Berlin, 1899. 12 vols. Complete bibliographies.

LEBENSOHN, J. E. History of Spectacles. *Bull. Soc. Med. Hist., Chicago,* Vol. IV, No. 2, July, 1930.

PERERA. Albrecht von Graefe, Founder of Modern Ophthalmology. *Arch. Ophth.*, Vol. 14, No. 5, Nov., 1935.

POLITZER, A. Geschichte der Ohrenheilkunde. Stuttgart, 1907-1913.

WRIGHT, J. A History of Laryngology and Rhinology. Philadelphia, Lea & Febiger, 1914.

ALBRECHT VON GRAEFE

IRIDECTOMY IN GLAUCOMA

[From *On Iridectomy in Glaucoma**]

To the report on the curative effects of iridectomy, already published in the "Archiv," I am about to add another of a highly gratifying character, for it refers to a comprehensive category of diseases hitherto incurable. The present communication would much sooner have been issued, had not the insidious nature of the affection demanded extreme care in judging of the results, and long-continued observation. Hence it is perfectly possible that iridectomy, as a remedy in the glaucomatous process, is already well known to most readers of the "Archiv," the subject having been discussed in my clinic through two sessions, communicated to many of my colleagues both orally and by letter, and very widely imitated. The best amends for my delay will be to specify the indications and prognosis as accurately as possible; otherwise, indeed, I would not give publicity to these observations, lest those who may adopt my views should not prove equally successful, and I should bring into discredit a method which is happily becoming naturalised in practice within its proper limitations.

When we recommend medicines or modes of operation, it is especially necessary to define the disease in reference to which they are recommended. The absence of general agreement constitutes an hereditary evil of therapeutical science, only to be cured by slow degrees, just as a sanguine predilection for medicines gradually yields to an intelligent analysis of the indications of treatment. I feel the necessity of agreement the more acutely, because the affection in question has ever seemed one attended by confusion and misunderstanding.

The name glaucoma formerly indicated a vague, expressionless

* Uber die Wirkung der Iridektomie bei Glaukom, *Arch. Ophthal.*, Berlin, 1857. Translation by Thomas Windsor. New Sydenham Society, 1859.

symptom-a sea-green, bottle-green, or dirty-green background of the eye, seen through a fixed, dilated pupil. When greater exactness was required, efforts were made to discover definite, material changes, sufficient to account for this symptom, but they terminated most variously and contradictorily. Whilst some imagined in glaucoma a peculiar degeneration of the refractive media, and especially of the vitreous body, others referred the origin of the disease to the choroid, others again to the retina; and as each of these views was contradicted, some gave up the seemingly 'futile attempt to localize, and considered glaucoma a disease of the whole globe. The latter hypothesis obviously attests only the incompleteness of our knowledge, for, owing to the great variety of the tissues of the eye, and exact pathology requires as accurate localization as in diseases of the abdomen or thorax. No one can doubt that in the course of the glaucomatous process most tissues of the eye become diseased, but it is equally certain that they are attacked at different periods, and that in consequence we have to distinguish the primary from the secondary changes. Of all the opinions brought forward certainly that one had the most numerous and powerful followers, which explained glaucoma as an inflammation of the choroid, with effusion between it and the retina. This view seemed to be favoured by pathologico-anatomical facts first collected by Schröder van der Kolk, and afterwards especially by Arlt. Notwithstanding, it was still open to controversy which of the changes were primary and which secondary or quite accessory. Most of the preparations were taken from far-advanced cases, and hence, at the time of dissection, did not present the typical appearances of glaucoma; and with the exception of a few instances brought forward by Arlt, there had been no examination during life, a point which is almost indispensable for fixing the glaucomatous origin.

Finally, more than five years ago, there appeared Helmholtz's immortal discovery, destined to throw so much light upon many obscure cases, especially upon amaurotic affections. This naturally excited the hope that the question of glaucoma would also be decided. But, unfortunately, this expectation was not so speedily to be gratified; it rather appeared as if glaucoma would remain an insoluble mystery, even when examined by the new instrument. The immediate results were of a purely negative nature, proving the nonexistence of those effusions suspected to lie between the retina and

the choroid. The diagnosis of such effusions, formerly possible only
when they were very extensive, so as to produce the so-called hydrops
subretinalis, had been so much aided by the application of the
ophthalmoscope, that they could not possibly be overlooked; yet
they were never seen, provided typically pure cases of glaucoma—
and such, of course, were necessary—were employed in the investigation. Such cases were, in general, somewhat advanced, for in the
specially acute period of the process, it is seldom possible, owing to
the diffuse opacity of the refractive media, to determine with certainty the details of the back of the eye. Hence it could no longer be
supposed that subretinal effusions caused the glaucomatous blindness; either they had no connexion whatever with the disease, or
were developed at an advanced period, as a secondary affection.
Neither could the diffuse cloudiness of the aqueous and vitreous
humours sufficiently account for the blindness caused by glaucoma;
for the opacity is never so great as to explain the entire loss of perception; besides, at times, if not prevented by the formation of
cataract, it may be seen spontaneously to disappear, without any
corresponding power of vision being restored. The changes in the
internal membranes, apoplexies, and, at a later period, partial atrophies of the choroid and retina, did not by any means constantly
occur, and were also developed to a far less extent than in chronic
retinitis and choroiditis. Hence it must follow that these structural
changes did not directly cause the occurrence of the blindness. Since,
however, a peculiar alteration in the entrance of the optic nerve was
always apparent in well-marked glaucoma, the attention of all investigators was directed to this point, as the probable source of the
disease. . . .

When the changes in the optic nerve were perceived in glaucoma,
there naturally ensued the task of connecting them with the symptoms previously known of circulatory and trophic disturbances. The
solution of this problem presented the greatest difficulties. Even if
the affection of the optic nerve were really the first cause of blindness, still the remaining complex group of symptoms could not result
from it, as a secondary affection. We see glaucoma, in its most typical
variety, sometimes occurring in previously healthy eyes in the form
of acute inflammatory attacks. Besides, a causal relation of the lesion
of the optic nerve with the other alterations, of the refractive media,
&c., could scarcely be imagined. In amaurotic cases we observe the

most advanced metamorphoses of the papilla, extending even to perfect atrophy of the optic nerve and retina, without the other parts of the eye being affected; even the most exquisite granular exudation processes in the papilla usually only lead to changes of the retina. This may be readily explained by the nutrition of the retina, which is independent of the other membranes of the eye. Indeed, this opinion is almost disproved by the fact, that the papilla is not convex, but concave.

There were only two ways of discovering the connexion of the lesion of the optic-nerve with the other glaucomatous symptoms; first by pathological anatomy, and secondly by the most exact clinical observation. Since my opportunities were very insufficient for the anatomical examination of fit cases, I was restricted to the latter method of study. I especially watched those cases in which the appearance of glaucoma was developed after repeated acute internal inflammations (ophthalmia arthritica). For the time I neglected the question as to what internal membrane the origin of inflammation should be referred, and adhered to the supposition of a choroiditis, which was especially favored by Arlt's dissections, the whole appearance of the malady, the sympathy of the iris, and the cloudiness of the vitreous body. The ophthalmoscope had only refuted the occurrence of subretinal effusions, but not of choroiditis, a point to which I shall again refer, especially as in my original note on glaucoma, fettered by the degeneration of the optic nerve, I passed too rapidly over the account of the internal membranes. Now when I compared the general appearance of this glaucomatous inflammation with that of the internal inflammations, for example, of the common iridochoroiditis, it seemed to me that all the characteristic symptoms tended to one point—*increase of the intra-ocular pressure.*

The hardness of the glaucomatous globe has been remarked from the earliest periods of ophthalmology. Since no change in the sclerotic, capable of explaining the altered resistance of the globe, can be justly admitted, it must be founded on the more complete filling of the globe with fluid. The dilatation and immobility of the pupil, are, it is known, not caused by the blindness: were this the case, the pupil must contract on the passage of light into the other and healthy eye as in unilateral anaesthesia of the retina; the diameter of the pupil must also change in rotations of the globe, in alterations of accommodation, and in closure of the lids. Besides, not unfrequently, after the first attack of glaucomatous inflammation, more or less power of vision returns, and yet the pupil preserves its abnormal properties. Evidently we must refer the pupillary affection directly to iridoplegia-paralysis of the nerves passing to the iris. The degree of mydriasis in glaucoma compared with that in paralysis of the oculomotorius, might indeed seem too great for the admission of iridoplegia; but as I have already elsewhere stated, the maximum dilatation, in which sometimes the iris almost peripherically vanishes, does not exist from the commencement, but is developed with the progressive atrophy of tissue, and it sometimes also proceeds from other mechanical causes (see the account of dissections in the previous treatise on Sympathetic Amaurosis). The increase of the intraocular pressure would furnish a further reason for iridoplegia, the power of conduction in the ciliary nerves being thereby annulled. As a phenomenon analogous to iridoplegia, I found *corneal anaesthesia*; this is also explained by compression of the nerves passing to it; in a second note on glaucoma (*vide* A. F. O., Bd. i, Abth. 2, S. 305) I have already given a direct proof of this, by showing that after the aqueous humour has escaped, the sensibility of the cornea is again restored, provided the operation is performed at a sufficiently early period of the disease. With these symptoms may be classed the *flattening of the anterior chamber*; which, in my opinion, depends upon two circumstances. First, the convexity of the cornea is diminished; secondly, the iris is really more arched forwards. The flattening of the cornea, which may be proved by examinations of the reflection from it, compared with that of another and healthy eye, is in itself an invaluable argument for the increase of intraocular pressure. In an inflammatory affection we could scarcely explain this symptom in any other way; perhaps, also, we may be able to found on it an accurate method of measuring the amount of increased pressure, for it is subject to mathematical estimate by means of Helmholtz's ophthalmometer. If the radius of corneal curvature approach that of the sclerotic, the receding angle of the cornea will be pushed outwards, and occasion an alteration in the form of the anterior chamber; at the same time, the iris will be pressed backwards. Since, on the contrary, the iris in glaucoma appears more convex anteriorly, there must, with the demonstrable flattening of the cornea, be a compensating, or more correctly an over-compensating momentum, a far greater increase of pressure acting in the space occupied by the vitreous body than in the anterior chamber, so that the iris is actually pressed forward.

When I published my second notice upon glaucoma in the first volume of this journal (Abth. 2), I expressed the opinion that glaucomatous inflammations essentially depend on increased intraocular pressure; this, indeed, induced me to make the trials of paracentesis, there related. *The form of the papilla,* however, prevented me from coming to a general decision. The supposed convexity continually turned my attention back to a substantial lesion of the optic nerve. When I became convinced of the concavity of the papilla, an explanation was immediately sought in another direction; but it required longer observation to become fully satisfied, the cases in which the lesion of the optic nerve apparently pre-existed (see hereafter) still remaining perfectly inexplicable. A case treated by paracentesis, first showed me that the excavation of the optic nerve only becomes developed secondarily in the so-called acute glaucoma. The patient had come to me during a violent attack of glaucomatous choroiditis. I had performed paracentesis three times in a space of eight days. Clearness of the refractive media was as perfectly obtained as could be wished, and continued also during the following weeks; at the same time there were all the signs of a glaucomatous affection, visible in the iris and pupil. The ophthalmoscope showed the papilla to be perfectly normal, and only after many months, when fresh obscurations occurred, did the optic nerve progressively degenerate, advancing with the other symptoms of increased pressure. I only attained a more general conviction as to the point in question, after the employment of iridectomy had enabled me to decide more accurately upon the condition in the earliest periods of disease. It was constantly found that the lesion of the optic nerve did not yet exist after the first inflammatory attacks, but was gradually developed at the same time as the other symptoms of increased pressure. Hence, I felt obliged to conclude that the excavation of the papilla arises in the same way as many of the sclerotic ectasiae which appear in the later stages of the glaucomatous process. The entrance of the optic nerve is, in respect to resistance, the weakest part of the envelopes of the bulb, and it may easily be imagined that this part would also be the first to be pressed outwards when the intra-ocular pressure is increased. . . .

The question now occurred, whether iridectomy also causes a

diminution of intra-ocular pressure in the healthy eye, or whether this only takes place under certain conditions of disease. I think that I may answer this question affirmatively, although I cannot at present adduce any more exact investigations. The eyes of animals, from which I had excised large pieces of iris, generally appeared to me a little softer to the touch, and when I carefully introduced the cannula of Anel's syringe into the anterior chamber of such eyes, the whole of the aqueous humour did not ascend through the action of the intra-ocular pressure, as is usually the case, but only a part. I think I have also perceived a slight permanent diminution of tension in patients, where an artificial pupil has been made for leucoma adhaerens.

Supported by these facts and considerations, I considered myself perfectly justified in performing iridectomy in glaucoma; for I knew the favorable action of the operation on the condition of the choroid in regard to its circulation; and everything seemed to favour the opinion that the operation probably possessed a physiological, and certainly, in many cases, a therapeutical pressure-diminishing action. The first trials were extremely uncertain, for I had no fixed principles, either in regard to the choice of cases or the manner of making the trial. I first employed this method in June, 1856, and from that time have continued it, especially in the cases which I have already described as acute glaucoma. The immediate effects appeared from the first very favorable; but remembering how my hopes had been frustrated in paracentesis, I was extremely mistrustful, and remained so until, in time, distinct differences appeared between the present results and those formerly obtained. A continued improvement was generally apparent exactly in proportion as the observation was prolonged; the signs of glaucoma retrograded in the manner hereafter to be described, and now, having followed some cases more than a year, and a considerable number more than nine months, I think I cannot be mistaken in regarding iridectomy as a true curative treatment of the glaucomatous process. It has its natural limits, like every therapeutical method; and in some degree to define these is the object of the following communication.

Were I to discuss the application of iridectomy to the various stages and groups of glaucoma in the same order in which they were successively presented for trial, I should have to begin with the old, and in part with cases that had completely run their course; for it is obvious that, owing to the uncertainty of success, a new method will be first tried in cases where as little as possible can be lost. I think, however, that by such an exposition I should lose sight of the indications on which I especially depend; and therefore prefer to leave entirely out of question the historical development, and to discuss the relative results in accordance with the previous nosological divisions. Accordingly I shall communicate my results in regard to iridectomy—

1. In the premonitory stage of glaucoma.
2. In the acute period of inflammatory glaucoma.
3. In the later period of inflammatory glaucoma.
4. In chronic glaucoma.
5. In amaurosis with excavation of the optic nerve.

Vision was perfectly restored in all cases in which the operation was performed before the termination of two weeks from the occurrence of inflammation. Some of these cases seemed perfectly desperate; for every trace of the qualitative perception of light had been already extinguished. I need scarcely mention that at first I promised but little to these patients. I frequently undertook the operation . . . altogether on account of the violent ciliary neuroses, and the effect was, in both respects, very surprising. It is only within the last half-year that I have ventured to predict complete restoration, even where the power of distinguishing has been perfectly lost, it being presupposed that less than two weeks have passed since the commencement of the inflammation, and that a moderate quantitative perception of light exists.

FRANS CORNELIUS DONDERS

Accommodation in the Eye

[From *The Anomalies of Accommodation and Refraction of the Eye*,* Chapter 1]

PROOFS OF THE EXISTENCE OF ACCOMMODATION IN THE EYE

The media of the eye form a compound dioptric system, wherein we can accurately and easily trace the course of the rays of light only

* Translated from the author's manuscript by William Daniel Moore. New Sydenham Society, 1864. First published in book form in English with the above title.

by being acquainted with its cardinal points. But, to clear up a number of questions, it is satisfactory to consider the whole system as a single lens, with a definite focus, and the action of such a lens is then sufficient to give an idea of the accommodation.

It is well known that when *parallel* rays of light fall upon a convex lens, . . . these, at a certain distance behind the lens, unite nearly into a point, called the *principal focus*. The distance between a particular point h'' of the lens and the focus I'' is termed the *focal distance F*. Parallel rays of light proceed from infinitely distant objects. From each point of an object, placed at a finite distance, proceed rays, which have a *diverging* direction. When such rays fall on the lens, they unite likewise almost into a point j, but this point lies further behind the lens than the principal focus. Such a point is in general called a focus. The principal focus is the focus for parallel rays. It is evident that the rays may fall on the lens in such a diverging direction as to maintain a degree of divergence behind the lens. This is the case when the point (Fig. 4i), whence the rays $i\ a$ and $i\ a'$ proceed, is at a less distance from the lens than amounts to the focal distance F of the lens. The rays then acquire after refraction through the lens, as $b\ c$ and $b'\ c'$, a direction as if they had come from a point more distant from the lens. In the explanation of ordinary vision this last is, however, of no importance, inasmuch as objects are always held far enough from the eye to allow of the rays proceeding from them being brought, if not into union, at least into a converging direction.

In the normal eye, the retina is placed precisely at the focal distance of the dioptric system. Parallel rays, derived from infinitely distant objects, are therefore brought into union exactly in the retina. The objects are accurately perceived. From near objects, as we have observed, the rays proceed in a diverging direction, and their point of union in the normal eye, consequently, lies behind the retina, and yet the organ is capable of perceiving near objects also accurately. It therefore has the further power of bringing divergent rays into union on the retina. Now, this power of bringing at will rays of different directions into union on the retina is the *power of accommodation of the eye*.

We can easily convince ourselves that the normal eye possesses such a power. That we are able clearly and accurately to distinguish objects at different distances, everyone knows by experience. We need, therefore, only assure ourselves that we *cannot at the same time* plainly distinguish remote and proximate objects, to obtain a proof that an accommodating power exists; in other words, that in the eye a change is produced in connexion with the distance at which we can see accurately. It is almost superfluous to adduce a direct proof in support of this statement. Ordinary observation will abundantly demonstrate it. It is well illustrated by holding a veil at some inches from the eye, and a book at a greater distance; we can then at will see accurately either the texture of the veil, or the letters of the book, but never both together. If we see the texture of the veil, we cannot distinguish the letters of the book; if we read, the veil produces only a feeble, almost uniform obscuration of the field of vision; of the separate threads we see scarcely anything; the circle of diffusion in imperfect accommodation can be most distinctly seen at an illuminated point, or at a darker spot on a piece of ordinary window-glass. The latter is held close to one eye (while the other eye is shut), but so that the point can still be accurately perceived— the objects situated at a certain distance on the other side of the glass are then observed without defined contours. We can now, however, at will, immediately see, in the direction of the point, the objects at the remote side of the glass distinctly, whereupon the point appears as a larger, diffused spot. A change has consequently taken place in the eye. Of this we are ourselves distinctly conscious. When we looked at distant objects through the glass, the eye was adjusted for almost parallel rays; the diverging rays proceeding from the point had therefore their point of union behind the retina. When the point was accurately seen, the eye was accommodated to the diverging rays proceeding from it, and the almost parallel rays derived from the distant objects had already united in front of the retina, and had decussated in a focus. In uniting, whether before or behind the retina, the rays proceeding from each separate point formed a *round* spot on the retina, instead of a point. The section of these rays has, in fact, nearly the *form of the pupil*, and, if the rays of the cone have not yet been brought into union, or if they have already decussated, they form on the retina a little spot of the form of the pupil. All the little spots, which represent the several points of the object in the retinal picture, are now like so many blotted points of an accurate image covering one another, and it is evident, that the former must, therefore, lose its sharp contour and be diffused on the surface. But as the retinal, so is the projected picture, and we therefore say, that we see the object diffused. In such a state do all objects appear, for which the eye is not accommodated.

CHANGE OF THE DIOPTRIC SYSTEM OF THE EYE IN ACCOMMODATION

That in the eye, in accommodation, a change is produced, has in the preceding section been placed beyond a doubt. The question now is, in what that change consists? Since Kepler first attempted to answer it, the inquiry has been the constant source of much difference of opinion among natural philosophers and physiologists. All imaginable hypotheses have been advanced. Alteration of situation of the lens, elongation of the axis of vision, contraction of the pupil, change of form of the lens, have all in turns been made use of in the explanation, and those who were satisfied with none of these theories, were sometimes bold enough altogether to deny the existence of an accommodating power. The ophthalmoscope, which enables us to see, in the fundus of the eye, the diffused images of objects for which the eye is not accommodated, effectually silences these last.

It is not my intention to subject anew to criticism the long series of incorrect views upon the subject. I am not writing a history of errors. We now know what change the dioptric system undergoes in accommodation, and the source of this knowledge alone can here be sketched in its leading features. *The change consists in an alteration of form of the lens: above all, its anterior surface becomes more convex and approaches to the cornea.*

It is now nearly sixty years since Thomas Young[1] had satisfied himself that the power of accommodation depends upon a change of form in the lens. Nor was he led to this conviction merely by the exclusion of other hypotheses; he adduced reasons which, properly understood, should be taken as positive proofs. As an hypothesis the idea had already existed; but previously to the time of Young it could be considered as little more than a loose assertion, to which no value was to be attached. The force of Young's experiments was, however, not understood, and his doctrine scarcely found a place in the long list of incorrect opinions and hazardous suggestions, which were constantly anew brought forward. Perhaps the necessary atten-

[1] Philosophical Transactions, 1801, vol. xcii, p. 23. Conf. Miscellaneous Works of the late Th. Young, edited by George Peacock, Vol. I, p. 12. London, 1855.

tion was not paid to Young's demonstration, because physiologists, not being acquainted with any muscular elements in the eye, could scarcely imagine by what mechanism the crystalline lens should change its form, and they were little inclined to believe with Young[1] in the contractility of the fibres of the lens. It was not until after direct proofs (within the reach of every one's observation and comprehension) of the change of form of the lens had been brought forward by others, that Helmholtz[2] placed the able investigation of Thomas Young in its proper light. The direct proofs were given a few years ago, and to our fellow-countryman Cramer,[3] too early snatched from science, belongs the highest honour in the matter.

For many years the reflected images of the anterior and posterior surfaces of the lens were generally known. Purkinje had discovered them in 1823, and Sanson had made them available in the diagnosis of cataract (1837). If some doubt still remained respecting the origin of the two reflected images, observed in the eye behind that of the cornea, the doubt was removed by the experiments of Meyer.[4] For the recognition of cataract they lost their value, when more decisive means of attaining it were discovered. But it was they which could give an infallible answer to the question, whether the lens in the accommodation of the eye undergoes a change, either in form or in situation.

Maximilian Langenbeck[5] was the first to whom it occurred to investigate the reflected images of the lens with reference to this important question. He examined them, however, only with the naked eye, moreover at a very unfavourable angle, almost solely with respect to the depth of their situation in the eye, and we can, therefore, scarcely assume that this investigation was sufficiently decisive to produce conviction. Nevertheless, he announced the most important fact: namely, that *in accommodation for near objects the anterior surface of the lens becomes more convex.* This statement lay hidden in a work, whose title was little adapted to attract the attention of physiologists. Accidentally the book fell into my hands. Struck with Langenbeck's fortunate idea, I immediately endeavoured to satisfy

[1] Miscellaneous Works, Vol. I, pp. 1 *et seq.*
[2] Allgemeine Encyclopaedie der Physik, herausgegeben von G. Karsten. Erste Lieferung, B. I., pp. 112 *et seq.*
[3] Het accommodatie-vermogen, physiologisch toegelicht. Haarlem, 1853.
[4] *Zeitschrift f. ration. Medizin*, Band V, p. 262. 1846.
[5] Klinische Beiträge aus dem Gebiete der Chirurgie u. Ophthalmologie. Göttingen, 1849.

myself of the correctness of his assertion; but owing to defects in the means I employed, no satisfactory result was obtained. That on examination with a magnifier the reflected images should show with certainty, whether in accommodation a change of the crystalline lens arises, I did not hesitate to predict.[1] I soon heard that Cramer, led by this prediction, had taken up the question.[2] He comprehended its full importance, solved it in the manner pointed out by me, and so put forward his result, that its correctness was in a very short time universally admitted.

I have above observed, that from the reflected images of the lens we may learn both the *curvature* and the *situation* of its surfaces. Cramer had already deduced both from his investigations.

In the first place, as relates to the curvature, we know that convex mirrors produce a diminished image behind, concave mirrors before the reflecting surface, and that the images are smaller in proportion as the radius of curvature is less. This is easily seen by comparing the reflected image of a flame formed by biconvex spectacles ground with different radii. We see an erect reflected image behind the anterior surface of the glass, and an inverted image before the glass, and both are smaller in proportion to the convexity of the surfaces of the glass employed. The posterior erect image is formed by reflexion on the anterior surface of the glass; the anterior inverted image is formed by reflexion on the posterior surface, or, to speak more correctly, on the concave surface of the air contiguous with the posterior convex surface. Now, the anterior surface of the crystalline lens is a convex mirror; the posterior surface, or rather the anterior surface of the vitreous humour corresponding thereto, represents a concave mirror. The reflected images are feebly illuminated, because the difference in refraction between the fluids of the eye and the lens being small, the reflexion is not considerable. They are, however, clearly discernible, when we hold a bright flame at one side of the eye, and look into the organ at the other side. If a line, drawn from the flame to the eye, forms an angle of about 30° with the axis of vision, and if we look at the other side, likewise at an angle of about 30° with the axis of vision, into the eye, the three little images appear flat, close to one another, in the pupil. (Fig. 5.) A represents their situation in the eye accommodated for distance;

[1] Nederl, Lancet, 2 Sér., D. V., pp. 135 and 147.
[2] Loc. cit.

B in the eye accommodated for near objects. In both a is the reflected image of the cornea; b that of the anterior surface, and c that of the posterior surface of the lens. Cramer viewed them magnified 10 or 20 times. He thus convinced himself that the image b reflected by the anterior surface of the lens is, in accommodation for near objects, considerably smaller, and he thence correctly inferred that the anterior surface of the lens increases in convexity, that the radius of curvature diminishes. Subsequently Helmholtz,[1] who, independently of Cramer, had discovered the true principle of accommodation,[2] has stated that also of the little inverted image c formed by reflexion on the surface of the vitreous humour, not only the apparent, but the actual size diminishes a little in accommodation for near objects, and that, consequently, the posterior surface of the lens, too, increases in convexity, although this increase is very slight.

ON THE MECHANISM OF ACCOMMODATION

So soon as the changes which the dioptric system undergoes in accommodation had become known, physiologists were in a position to investigate, with some hope of success, the mechanism whereby those changes are produced. Many modes of solving this question have been tried. By some, experiment has been resorted to; others have instituted an accurate examination of the anatomy of the parts which appear to be concerned in the mechanism referred to; pathology, finally, has been made use of in prosecuting the inquiry. But, notwithstanding all these efforts, it cannot be said that any theory brought forward has as yet been fully proved: the utmost we have attained to is, that by exclusion, the limits wherein our views may range have been much restricted.

It has been in general tacitly assumed that the accommodation for distance, and even for the farthest point of distinct vision, is purely passive,—that in it only relaxation of the parts which actively produce accommodation for near objects takes place. I believe that this idea was in all respects fully justified. But, if we endeavour to explain the mechanism of accommodation, it is, as a preliminary question, so important, that it may well be specially treated of, the more so, because some advocate an active accommodation also for distant objects. The grounds on which it may, in my opinion, be maintained

[1] Arch. f. Ophthalmologie, herausgegeben von Artl, Donders und von Graefe, Bd. I, Abth. 2, p. 1.
[2] Monatsberichte der Akademie zu Berlin, Febr. 1853, p. 137.

that accommodation for near objects only is active, while that for distant objects is passive, are the following:—

1. The subjective sensation;—for myself this is conclusive.

2. The phenomena produced by mydriatics. If we drop into the eye, a solution of one part of sulphate of atropia in 120 of water, the pupil, after ten or fifteen minutes, begins to dilate, and soon afterwards the nearest point of distinct vision removes farther and farther from the eye. At the end of forty minutes all action is destroyed, and the eye remains accommodated to its farthest point. The muscular system for accommodation is now paralysed, and paralysis, that is, the highest degree of relaxation, is thus proved to be equivalent to accommodation for the farthest point. Now, did we assume the existence of a distinct system, working actively in accommodation to the farthest point, we ought to maintain—1st, That this system is not paralysed by atropia; 2nd, That it is by this agent brought into a condition in which it is incapable of relaxation. This supposition would not be quite absurd. Something of the kind is said to occur in the action of atropia upon the iris: the circular fibres of the latter are thereby paralysed, but at the same time its radiating (?) fibres are said to be brought into the condition of spasm, so that the pupil becomes much wider than in cases of paralysis of the sphincter, and is also not at all or scarcely capable of further dilatation by irritation of the sympathetic nerve in the neck.[1] But though the supposition is not absurd, it is nevertheless far-fetched and little admissible. That it is incorrect appears further:—

3. From the phenomena attending paralysis of the oculo-motor nerve. In this affection the power of accommodation is not unfrequently wholly lost. This condition may occur with paralysis of some or of all the muscles governed by the oculo-motor nerve; but it may also exist quite independently. In it the refraction corresponds to the original *farthest point*, as cases of recovery have satisfactorily proved to me. The pupil is immovable and dilated, although not highly so. On instillation of atropia, the diameter becomes much greater, but the refraction of the eye remains unaltered. Accommodation for the farthest point corresponds, therefore, to total paralysis. In imperfect paralysis (paresis of accommodation) the nearest point is always removed further from the eye, the farthest remaining un-

[1] See de Ruyter, De actione Atropae Belladonnae in iridem, Trajeni ad Rhenum, 1856. Kuyper, Onderzoekingen betrekkelijk de kunstmatige verwijding van den oogappel. Utrecht, 1860.

altered. Cases of paralysis, where the farthest point should be approximated to the eye, do not occur: they should necessarily occur, did a muscular system exist, actively producing accommodation for remote objects.

4. The lens, enclosed in its capsule, has an important property, which must here be expressly pointed out. It possesses a high degree of elasticity. On gentle pressure its form is easily altered, but it immediately regains its original form when the pressure ceases.

Hence, too, it appears, that only the mechanism of accommodation for near objects is explicable by muscular action, and that the return to accommodation for distant objects occurs spontaneously (with the co-operation of elastic parts) when the active muscular operation ceases. The efforts of myopic individuals to see distinctly at a greater distance, are confined, as we shall subsequently observe, to diminishing the circles of diffusion, by excluding a part of the pupil: they produce no true accommodation—no change of the dioptric system.

Now the accommodation for near objects must take place through the intervention of muscular action. The accommodation is produced voluntarily, and we know no voluntary movement without the intervention of contractile—of muscular elements.

JEAN-MARC-GASPARD ITARD

[From *Treatise on Diseases of the Ear and Hearing**]

DISEASES OF THE EAR

ON RUPTURE OF THE MEMBRANE OF THE TYMPANA

[Chapter XI]

Rupture of the membrane of the tympanum is an inevitable result of internal otitis whenever the products of this inflammation exude through the auditory passage. Thus all the observations which I have given above on this inflammation of the eardrum, in so far as it affects this result, can be regarded as so many histories of rupture of the membrane of the tympanum. It is worth noting that it is almost always at the edges, and rarely at the center, of this membrane that

* Traité des Maladies de L'Oreille et de L'Audition, 1821. Translation by Dorothy H. Clendening.

the rupture occurs; sometimes to the point that it appears to have but one simple discharge from one of the edges. From this results a slight derangement of the functions of the membrane, except in the event this discharge continues to work near the center, near the point where the handle of the hammer is inserted. It is also remarkable with what ease Nature closes this opening in a few days.

Nothing is more commonly seen in cases of severe internal otitis than a large quantity of purulent matter discharging from the eardrum across the tympanic membrane, the discharge drying up and the opening which has given it passage closing in a few days, to the extent that it leaves no trace visible to the eye. This is not so if through the duration, or perhaps by the nature of the drainage, the suppuration attacks the membrane, corrodes it, and destroys a large part of it, and causes destruction of the ossicles. Then the opening, or rather the destruction of the membrane, makes absolutely no attempt to close. But here, as in other experiences in medical practice, one must admit of exceptions. One would think that this membrane could never heal after being totally destroyed. Nevertheless Valsalva assures us that he saw this in the ear of a woman who had become deaf following a prolonged drainage. Behind the tympanic membrane, which had been corroded by the suppuration, another membrane was obliquely raised which left the hammer and anvil outside and covered the base of the stirrup.

In the case, however, of accidental tears in the membrane caused by the action of foreign bodies carelessly invested in the auditory passage or of fractures to the skull caused by a fall or a blow on the head, provided the ossicles have not been injured or destroyed, the wound heals with an ease and speed which one would hardly dare hope for in a part so tenuous and lacking in cellular tissue.

Valsalva proved by experiments this healing power of the membrane. He opened those of several dogs: he opened it by sinking a large probe into the passage; he heard the crepitation of the probe tearing the film; he once dilated the rupture as much as possible, moving the end of the probe about extensively; and when he finally killed the animals he had their ears examined and found the damage caused the membrane by the probe well healed, no trace of it left, not even a scar being perceptible. I repeated the same experiments on the same kind of animals without always getting the same result. When I simply perforated the membrane, either with a blunt probe, or with a sharp one, the opening closed completely at the end of 15 to 20 days. But if, not content with opening the membrane, I tore about with the end of a blunt probe, I thereby started a slight discharge, and the membrane did not heal.

I observed the same thing in people on whom I performed this operation for deafness. At first I used a small sharp, pointed trocar; the resulting opening healed in a few days. When later I used a dull probe at the blunt end it never closed, and air continued to pass through the auditory passage as freely as after the operation.

ON OCCLUSION OF THE EUSTACHIAN TUBE
[Chapter XVI]

Occlusion of the Eustachian tube has several causes. The most common are congestion of the tonsils; the development of polypus tumors near the orifice of this canal; congestion or chronic tumefaction of its membrane; or, finally adhesions of its walls. The latter mode of occlusion is commonly caused by syphilitic ulcers of the soft palate, or gangrenous sore throats which have affected these same parts.

The secretions which obstruct the tube often have a remarkable consistency and odor. They resemble soap, or, rather, soft cheese, and are extremely fetid. Frequently efforts of excretion or sneezing detach some of these concretions of the guttural orifice of this passage, and one immediately notices a most disagreeable odor and taste in the mouth. But the most common occlusion of the tube starts with partial deafness. In treating lesions of hearing from this cause I will show the signs by which you can recognize the occlusion or obstruction of the tube and the remedies to use.

ON DEAFNESS DUE TO ENGORGEMENT OF THE INTERNAL EAR
[Part II, Chapter XIII]

This form of chronic deafness is the one which I have most frequently encountered in my practice and on which I can report a large number of observations. I have little doubt that in this kind of deafness, which I voluntarily designate under the name of catarrhal deafness, the different humors which bathe the inner ear become infectiously increased. It is found most frequently in young people, especially those with a lymphatic tendency, a weak constitution, pale, or with little color, and troubled, according to their story, by *glaires* (white of egg) in the stomach, and fonts of water falling from the skull. When, without draining, without pain, without concretions in the external ear these people become deaf, one can believe they are of the type I have described above. I consider the diagnosis less doubtful if the posterior fauces is filled with a great deal of mucous, if the voice is not clear and is obstructed as it is during a mercurial salivation or in catarrhal quinsy. It is not unusual for these people to talk through the nose, although their nasal cavities are drier than usual, even to the point where you seldom see them blow their noses. Two or three times I have noticed that the root of the nose was much larger than usual.

One common symptom of this type of deafness is its wide range in intensity, and it rarely seems to depend on the climate. Changes for better or worse often occur suddenly, sometimes being caused by efforts at excretion; by sneezing or blowing one's nose, or sniffling; sometimes without any movement of the head, and without known cause.

These sudden changes always become less frequent and even cease to be noticed when the deafness dates back several years. You also notice that people afflicted with this infirmity are much deafer on awakening in the morning, also during the day when the stomach is empty, and always when exposed to cold or to wet feet; that, on the contrary, they hear noticeably better during the heat of summer, or when they use hot inhalants or hot medications; and especially following spontaneous or artificially induced vomiting. The auditory passage instead of being dry and mealy, as in some of the other forms of deafness, is heavily coated with cerumen which is always more liquid; and it is not uncommon to find the tympanic membrane lacking in its usual transparency.

Emetics are the basis of the treatment. They must be repeated frequently and administered mildly, rather to produce nausea than vomiting, and to prolong for the longest possible time the preliminary stimulus of the emetic on the salivary glands and the mucous membrane of the throat.

I prescribe at the same time, in the guise of tobacco, the use of thrush powder, which powerfully provokes sneezing. I have the head shaved and have it rubbed every day with flannel saturated in balsamic vapors. Finally, I sometimes place a seton at the nape of the neck. which I change after it has produced a good hold, by two cauteries inserted one inch into each opening. I finally resort to direct medications to the inner ear after all other methods have failed. You can do this by three different methods—1 through the mastoid bone; 2 through the tympanic membrane; 3 through the Eustachian tube.

SIR MORELL MACKENZIE

HISTORY OF THE LARYNGOSCOPE

[From *The Laryngoscope*,* Chapter 1]

"Honour belongs to the first suggestion of a discovery, if that suggestion was the means of setting some one to work to verify it; but the world must ever look upon this last operation as the crowning exploit."—BAIN.

It may seem strange to some that it was not till the middle of the last century that an instrument was invented for examining the lower part of the throat during life, nor till more than a hundred years later that that instrument was sufficiently improved and simplified to be capable of general application. The dentist's mirror seems to have been used from time immemorial,[1] and polished tubes for passing into the external canals of the body, and thus obtaining an inspection, are of very ancient origin.[2]

A mere transfer of the dentist's mirror from the mouth to the back of the throat was not however sufficient to give birth to the laryngoscope; and the speculum (which is simply a rigid tube meant to press back the flaccid walls of a straight canal, and thus allow luminous rays to pass through it) was not applicable to the examination of a part situated at an angle to the line of vision. It was only by a combination of these two elements (reflexion and illumination) that the interior of the larynx could be seen in the living subject. This fact, together with the circumstance that it was not till comparatively recently, that physicians attempted to discriminate between diseases of the fauces and those of the windpipe, will perhaps ac-

* London, 1865.
[1] In the Augustan age, dental surgery had attained a degree of perfection which implies the employment of mirrors for examining the inner surface of the teeth.—Celsus, lib. vii, cap. xii.
[2] Some of my readers who have been in Italy may have seen the speculum found in excavating the buried city of Pompeii.

656 SOURCE BOOK OF MEDICAL HISTORY

count for the noninvention of the laryngoscope at an earlier date. Whatever the cause may be, however, there is no trace of a laryngoscope before the middle of the eighteenth century.

In the year 1743, and probably some years previously, M. Levret, a distinguished French accoucheur, whose highly inventive genius had led him to contrive surgical instruments of almost every description, occupied himself in discovering means, whereby polypoid growths in the nostrils, throat, ears, and other parts, could be tied by ligatures. It is unnecessary to describe here, the various ingenious instruments which he invented for the purpose, and it is only requisite to observe that in using them he employed a speculum which differed from the various *specula oris* then in use. It consisted mainly of a plate of polished metal (*plaque polie*) which "reflected the luminous rays in the direction of the tumour," and at the same time received the image of the tumour on its reflecting surface. It is evident that this little mirror was regarded as a mere appendix of that which Levret considered much more important—viz., his method of applying ligatures, and that he did not recognize its value as a means of diagnosis in diseases of the larynx. The whole subject was soon lost sight of, and it was not till more than fifty years later that it again excited attention.

Then it was that a certain Dr. Bozzini, of Frankfort-on-the-Maine, made a great sensation throughout Germany, with his invention for illuminating the various canals of the body. About the year 1804, he first made known his ideas, which in the beginning were treated with derision. Gradually, however, the fame of the physician spread, the value of his invention was enormously exaggerated, and not only the professional press, but even political and literary journals, teemed with accounts of the wondrous apparatus. In the year 1807 Dr. Bozzini published a work on the subject of his invention, entitled "The Light-Conductor, or Description of a Simple Apparatus for the Illumination of the Internal Cavities and Spaces in the Living Animal Body. About this time, the public seem to have become still more impressed with the value of Dr. Bozzini's invention, and an absurd idea appears to have got abroad that the apparatus would enable practitioners to inspect, not merely the outlets of the body, but even the internal viscera. There was nothing in the work except perhaps its rather ambitious title to encourage this idea, but this did not save it from incurring the wrath of the profession. It is curious, that the feeling against the invention should have been strongest in the very city, from which so many of the earliest and most valuable laryngoscopic observations afterwards issued. The Faculty of Physicians of Vienna, in concert with the members of the Joseph's Academy, passed a very damaging opinión on Dr. Bozzini's invention. They prefaced their admonition by remarking that "premature conclusions were likely to be arrived at concerning the instrument," and "that perhaps even there might be an outlay of money (!!), which might afterwards be regretted." They then went on to say that "only very small and unimportant parts of the body could be examined," that "the illuminated spot was so small—its diameter being never more than an inch—that if a person did not know beforehand exactly what he was to look at, he would not generally be able to tell what part of the body was presented to view."

This was the spirit in which Bozzini's invention was received, a description of it will show that it deserved a better fate. It consisted of two essential parts: 1st, a kind of lantern, and 2ndly, a number of hollow metal tubes (*specula*) for introducing into the various canals of the body. The lantern was a vase-shaped apparatus made of tin, in the centre of which there was a small wax candle. The top of the apparatus was covered, but a large aperture at the upper part, and some holes in its base, allowed sufficient supply of air for the candle, the latter was fixed in a metal tube and forced upwards by a spring, after the manner of a Palmer's lamp. In the side of the apparatus there were two round holes, a larger and smaller one, opposite each other. To the smaller one an eye-piece was fixed, to the larger the speculum was fitted. The flame of the candle came just below the level of these two apertures. The mouth of the speculum—a tube of polished tin or silver—was always the same size, but the diameter of the tube beyond varied according to the canal in which it had to be introduced. The apparatus was about thirteen inches high, two inches from before backwards, and rather more than three from side to side. These measurements were considered necessary, in order that there should be sufficient space for the candle to burn steadily, and that the lantern should not become too hot. The eye-piece was arranged to fit the eye, so that everything was hidden from view, except the spot seen through the speculum. It may be remarked, that the vase-shaped chamber lined with tin constituted, in fact, two concave mirrors, one behind and the other in front of the reflector,

the posterior reflector (if the expression may be used) being perforated by a hole for the eye-piece, and the anterior by another for
the speculum.

It is not necessary to enter into details concerning the different
canals for which this "simple apparatus" was recommended, but the
following quotation shows that the requisite for making a laryngoscopic examination were fully appreciated by Dr. Bozzini: "If *a person wishes to see round a corner into a part of the throat*,[1] or behind
the plate into the posterior nares, the rays must be broken, and *a
mirror is required for illumination and reflexion*." In employing reflected light, Bozzini had the speculum divided by a vertical partition, so that there were, in fact, two canals and two mirrors. One of
these mirrors was intended to convey the light, the other to receive
the image. We now know that this arrangement is quite unnecessary,
as one mirror is able to serve both purposes.

In the year 1827, twenty years after the publication of Bozzini's
pamphlet, Dr. Senn, of Geneva, tried to examine the larynx of a little girl, suffering from difficulty of breathing, and extreme dys-
phagia. The case was not a favourable one for the trial, and the
attempt failed; but as Dr. Senn did not employ any means for throwing a light into the larynx, it is not likely that his efforts would,
under any circumstances, have proved more successful. He did not,
like Bozzini before and Babington after him, perceive that in laryngoscopy, two factors (illumination and reflexion) must always be
employed. The following are Dr. Senn's remarks upon the subject:—
"I had a little mirror constructed for introduction to the back of
the pharynx; with it I tried to see the upper part of the larynx,—
the glottis; but I gave up its use on account of the small size of the
intrument. However, I believe that this method could be employed
with advantage in the case of adults, and that in certain cases of
laryngeal phthisis, it might assist in diagnosis."[2] Though this attempt
was made in the year 1827, it was not recorded before the end of
the year 1829; even then the account of the employment of the
mirror was not embodied in the text of the report, but was merely
appended as a note to the communication. The case was one of considerable interest, both on account of its general features, and espe-

---

[1] The word used is "Schlund." This term is now employed, anatomically speaking,
for the pharynx, but it is often used to express the throat generally, and by Hilpert
is considered synonymous with "Kehle," the larynx.

[2] Journal des Progrès," 1829, p. 231, note.

cially from its having been one of the first in which a canula had been worn in the trachea for any length of time. It was particularly with reference to this circumstance, that the case had been brought before the Académie des Sciences on the 10th December, 1827.[1] In the published account of the séance there is no mention of any attempt at laryngoscopy.

In the year 1829[2] Dr. Benjamin Guy Babington exhibited at the Hunterian Society of London, an instrument closely resembling the Laryngoscope now in use. Two mirrors were employed by this physician, one the smaller, for receiving the laryngeal image, the other, larger one, for concentrating the solar rays on the first. The patient sat with his back to the sun, and whilst the illuminating mirror (a common hand looking-glass) was held with the left hand, the laryngeal mirror—a glass one coated with quicksilver—was introduced with the right. By a very simple mechanism, a tongue-depressor was united with the laryngeal mirror, and thereby one of the most serious obstacles to laryngoscopy was attempted to be overcome. A spring was fixed between the shanks of the laryngeal mirror and spatula, in such a way, that, by pressing the two handles together, the tongue was depressed. At a later period (between the years of 1829 and 1835) Dr. Babington abandoned the attractive combination of mirror and spatula, and had mirrors made, which closely resemble those now in general use. The mirrors were made of polished steel, and were, like those now in use, inclined to the shanks at an angle of about 120. Though Dr. Babington used his laryngoscope on many patients, there are no cases recorded in which the instrument was employed.

Priority of publication had long been the established touchstone, by which the disputed claims of inventors have been tested. Tried by this criterion, Babington must be regarded as the inventor of the Laryngoscope; for whilst an account of his invention was published in London, in March, 1829, Senn's attempt to examine the larynx was not recorded in Paris till after August in the same year. The claims of Babington, however, rest on something better than a technical basis; for whilst Senn merely attempted to employ a *laryngeal mirror*, Babington invented a *laryngoscope*. With the mirror alone it was impossible to see the interior of the larynx; but when a

[1] "Journal Général de Médecine," tom. CII. January, 1828.
[2] "Lond. Med. Gaz.," Vol. 111, p. 555. London, 1829.

method of illumination was at the same time employed, the inspection became, if not easy, at any rate practicable. The only difference between Dr. Babington's laryngoscope and the one now in general use is, that whilst in the latter the light is thrown into the larynx (or rather on to the laryngeal mirror) by a circular mirror attached to the head of the operator, in the former the illumination was effected by a mirror held in the operator's hand. Dr. Babington, moreover, does not appear to have employed artificial light; and his mirrors were of more clumsy construction than those now used. Those who have learnt to use the laryngoscope, will readily appreciate the difficulties of illuminating the larynx with a hand mirror; whilst in this country where the sun very often does not shine brightly for weeks together, the art of laryngoscopy could never have flourished till artificial light had been substituted for the uncertain solar rays.

In the year 1832, whilst Babington was still working with his "glottiscope," to use the term employed by him at the time, Dr. Bennati, of Paris, asserted his ability to see the vocal cords. A mechanic named Selligue, who was suffering from laryngeal phthisis, had invented "a doubletubed speculum, of which one tube served to carry the light to the glottis, and the other to bring back to the eye the image of the glottis reflected in the mirror, placed at the guttural extremity of the instrument." A complete recovery rewarded the ingenious patient for his clever invention, and with this instrument Bennati professed to be able to see the glottis. Trousseau, however, disbelieved his statements, and devoted several pages of his well-known work[1] to prove, that the epiglottis formed an insuperable impediment to a view of the interior of the larynx. This renowed physician had an instrument constructed after the model of Selligue's, but he does not appear to have attempted its use. It is worthy of note, that Sellique's laryngeal speculum closely resembled that of Bozzini, for whilst the latter was made in one tube divided by a vertical partition, the former consisted of two tubes.

In the year 1838[2] M. Baumês exhibited at the Medical Society of Lyons a mirror about the size of a two-franc piece, which he described as being very useful for examining the posterior nares and larynx.

[1] "Mémoire sur la Phthisie Laryngée." Par MM. Trousseau et Belloc. "Mémoire de L'Académie de Médecine," tome VI, 1837.
[2] "Compte Rendu des Travaux de la Société de Médecine de Lyons, 1836-38," p. 62.

In the year 1840,[1] Liston, in treating of oedematous tumours which obstruct the larynx, observed as follows: "The existence of this swelling may often be ascertained by a careful examination with the fingers, and a view of the parts may sometimes be obtained by means of a speculum,—such a glass as is used by dentists on a long stalk previously dipped in hot water, introduced with its reflecting surface downwards and carried well into the fauces." When the real art of laryngoscopy was founded almost twenty years later, the name of our talented countryman was prominently associated with its invention. But it is obvious from the above passage, that Liston never contemplated an inspection of the vocal cords. It is plain that in his estimation the sense of touch was more to be relied on than that of sight; and the fact that the fingers were to be used, indicates pretty clearly that Liston was referring rather to the epiglottis than the parts below.

In the year 1844,[2] Dr. Warden, of Edinburgh, conceived the idea of employing a prism of flint glass for obtaining a view of the larynx. The success which had attended his efforts to inspect the membrane of the tympanum, induced him to apply the principle of the prism to other canals. He reported two cases[3] in which he considered that he had had "satisfactory ocular inspection of diseases affecting the glottis." The possibility of inspecting the larynx in this way admits of no doubt, but Dr. Warden's method of employing the prism was not calculated to bring about very favourable results. The particulars of one of the cases referred to are given, but in the other "the appearances were so far similar as to render their detail unimportant." The patient whose case is narrated was a lady, "who had been the subject of medical treatment for chronic inflammation of the pharynx of nearly a year's duration"; the inflammation had latterly spread in the direction of the glottis, and painful deglutition and paroxysms of suffocation now supervened. "After the preliminary examination and *quietening the irritability of the parts by touch with the finger*, there was no longer any impediment or inconvenience experienced from the tendency to retching. . . . *The dilator faucium was employed to depress the tongue and expand the isth-*

[1] "Practical Surgery," third edition, p. 417. 1840.
[2] Royal Scottish Society of Arts. Description, with illustrations, of a Totally Reflecting Prism for Illuminating the open cavities of the body, &c., &c., May, 1844. See also "Lond. Med. Gaz.," Vol. XXXIV, p. 256.
[3] "Month. Journ. Med. Science," July, 1845, p. 552.

mus of the fauces." The result of the examination was that the epiglottis was seen to be very much thickened and inflamed, "but it was only when efforts to swallow were made or repeated that the arytaenoid cartilages, in a similar condition of thickening, were raised out of concealment, and brought brilliantly to show their picture in the reflecting face of the mirror." For the purpose of illumination Dr. Warden employed "a powerful argand-lamp, with a large prism attached, so as to throw the full light of the lamp into the fauces and pharynx." That is to say, instead of the two plane mirrors we use (one for illumination, and the other for reflexion), he employed two prisms.

In concluding the report of these cases Dr. Warden remarks that "the experience afforded by both gives ground for the same conclusion, that the instrument made use of can have no farther range than the bottom of the pharynx and mouth of the glottis, and of the latter only so often as it is raised from its natural depth, by the contraction of the muscles employed in the act of deglutition. By this means, therefore, we can obtain no assistance in the investigation or treatment of disease below the pharynx." It is not surprising that Dr. Warden should have expressed himself thus unfavourably concerning his attempts to examine the larynx. What with "quietening the irritability of the throat by touch with the finger, depressing the tongue, dilating the fauces, and encouraging the patient to swallow," it was utterly impossible for him to have succeeded. No disciple of Czermak could hope to see the vocal cords were he to prepare his patient in the way described by Warden; and when we remember how limited was his experience, and how imperfect his instruments, the appearances described by him can scarcely be regarded otherwise than the baseless fabric of a very imperfect vision.

In the year 1854,[1] "the idea of employing mirrors for studying the interior of the larynx during singing" occurred to M. Manuel Garcia. He had often thought of it before, but believing it impracticable, had never attempted to realise the idea. M. Garcia, though long a distinguished singing-master in London, was a Frenchman by birth, and a Spaniard by descent; and though his observations with the laryngoscope were first published in England, they were first made in France.

---

[1] "Notice sur l'Invention du Laryngoscope." Par Paulin Richard. Paris, 1861.—See M. Garcia's letter to Dr. Larrey, dated May 4, 1860. (Page 12 in Richard's Pamphlet.)

In the month of September, 1854, whilst Garcia was spending his holidays in Paris, he determined to clear up his doubts concerning the possibility of inspecting the larynx. His efforts were crowned with success, and the following year he presented a paper to the Royal Society of London, entitled "Physiological Observations on the Human Voice."[1] This paper contained an admirable account of the action of the vocal cords during inspiration and vocalisation; some very important remarks on the production of sound in the larynx, and some valuable reflections on the formation of the chest and falsetto notes. M. Garcia's laryngoscopic investigations were all made on himself; indeed, he was the first person who conceived the idea of an autoscopic examination.

His method, which he believed had never been employed by any one before, consisted in introducing a little mirror, fixed to a long stem, suitably bent, to the top of the pharynx. He directed that the person experimented upon should turn towards the sun, so that the luminous rays falling on the little mirror should be reflected into the larynx;[2] but he added in a foot note, that "if the observer experiments on himself, he ought, by means of a second mirror, to receive the rays of the sun, and direct them on the mirror which is placed against the uvula." In practising auto-laryngoscopy after the manner of Czermak, three mirrors are employed; one for illumination, another for introducing to the fauces, and a third to enable the observer to see the image in the mirror held in his own throat. Garcia employed only two mirrors: a small one at the end of a long stem for introducing to the pharynx, and a large one which served the double purpose of illuminating the little mirror, and enabling the operator to see the image formed on it. It will be seen that Garcia's method was precisely similar to that employed by Babington; the one, however, limited his observations to his own larynx, the other never made an attempt at auto-laryngoscopy. Garcia's communication to the Royal Society, though causing little stir at the time, was destined to experience a fate in many respects similar to that which befell the paper of our countryman, Mr. Cumming.[3] Treated with

---

[1] "Proc. Royal Soc. London," Vol. VII, No. 13, 1855. "Philosoph. Magazine and Journal of Science," Vol. X, p. 218, and "Gaz. Hebdom. de Méd. et Chir.," Nov. 16, 1855, No. 46.

[2] It is worthy of note, that Garcia really never followed this plan, but, in point of fact, always used a second mirror for throwing the solar rays on the laryngeal mirror. In the mirror which he used as a reflector. He also saw the autoscopic image.

[3] "Transactions of Med. Chir. Soc.," 1846.

apathy, if not with incredulity in England, both papers passed into the hands of foreign professors, and whilst Helmholtz matured the ophthalmoscope, Czermak developed the laryngoscope. In the year 1857, during the summer months, Dr. Türck, of Vienna (who had read Garcia's paper), endeavoured to employ the laryngeal mirror in the wards of the General Hospital. He was not successful, however, at first, and at the end of the autumn he seems to have abandoned his fruitless attempts. Trusting entirely to the solar rays, having no apparatus (no second mirror) for concentrating the light on the laryngeal mirror, and the latter being a clumsy hinged instrument, it was scarcely possible for him to succeed. When at a later period, however, Czermak proved the practical value of the laryngoscope, Türck put forth his claims to priority. Nevertheless, in the very communication[1] in which he asserted his pretensions, he observed that "he was very far from having any exaggerated hopes about the employment of the laryngeal mirror in practical medicine." This unfortunate remark shows that he did not even then recognize the value of the laryngoscope.

In the year 1857, in the month of November, Professor Czermak, of Pesth, borrowed from Dr. Türck the little mirrors which that gentleman, in spite of the exhortation of his friends, had thrown aside as useless.[2] In a short time his superior genius, untiring perseverance, and natural dexterity enabled him to overcome all difficulties. When the dentist's mirror passed into the hands of Dr. Czermak, the examination of the larynx was dependent—so to speak —on the clock and the barometer, but he soon relieved it from both these troublesome monitors. Artificial light was substituted for the uncertain rays of the sun; the large ophthalmoscopic mirror of Ruete was used for concentrating the luminous rays; the awkward hinge which united the laryngeal mirrors were made of different sizes. Thus it was that Czermak created the art of laryngoscopy. Others before him had contrived instruments, with which they had sometimes succeeded in inspecting the interior of the larynx, but "the tools fitted for the art" of laryngoscopy were not constructed before his time. His first publication appeared in March, 1858,[3] and a month

---

[1] "Zeitschrift der Ges. der Aerzte zu Wien.," April 26, 1858.
[2] Professor Brücke's Letter to Czermak. "Selected Monographs: New Syd. Soc.," Vol. XI.
[3] "Wien. Medizin. Wochenschrift."

later a very important paper of his was brought before the Academy of Sciences of Vienna.[1] In claiming for Czermak the honour of having so modified the laryngoscope, that its application became comparatively easy, it would not be right to withhold from Dr. Türck the merit of having patiently and productively worked at the subject at a later period. A careful investigation of facts and dates, however, must convince every disinterested person, that Türck's subsequent successful labours were prompted by the proofs which Czermak had given of the value of the laryngoscope.

Czermak's investigations were at first confined to his own larynx, and his success must in part be attributed to his great physical advantages. Possessed of a most capacious pharynx, small tonsils and uvula, and large laryngeal aperture, it would be difficult to find a subject better suited for laryngoscopy. Notwithstanding the beautiful simplicity effected by Czermak in the details of the laryngoscope, the profession might not have become impressed with the value of the instrument, had not his brilliant demonstrations delighted and astonished the medical public throughout Europe. The general employment of the laryngoscope in practical medicine must be attributed not less to his enthusiastic and universal teaching—to his brilliant demonstrations and personal influence, than to his entire remodeling of the instrument itself. The fact that no improvement has been made in the mechanism of the laryngoscope for the last five years, though a great number of practical men in all parts of the world have been constantly working at the subject, is the strongest testimony to the value of Czermak's labours.

[1] "Physiolog. Unters mit Garcia's Kehlkopfspiegel," mit 111. Tafeln. Sitzber d. kk. Akademie d. Wiss. in Wien. vom April, Bd. XXIX, p. 557. (Afterwards reprinted in a separate form.)

XLII

WILHELM CONRAD ROENTGEN

WILHELM CONRAD ROENTGEN (1845-1922) WAS PROFESSOR OF PHYSICS successively at the Universities of Strassburg, Giessen, Würzburg and Munich. He was interested, as most physicists of the time were, in the peculiar phenomena resulting from the discharge of electricity through a vacuum. In 1895, at Würzburg, while working with a Crookes' tube, he accidentally found that shadows were formed on a photographic plate. He found by careful and well planned experiments (it was afterwards irresponsibly stated, to his great indignation, that all his results were accidental) that by making his tube light-proof, a greenish, fluorescent light could be thrown upon a platino-barium screen nine feet away. These rays passed through substances ordinarily opaque, such as the soft parts of the body, revealing the bones. December 28, 1895, before the Würzburg Society he read the paper reprinted in the following pages. Professor Kölliker, who presided, submitted to having his hand photographed with the new ray: the bones were plainly visible (many copies of this original plate are in existence) and all doubts about the authenticity of Roentgen's claims were set at rest. Roentgen suggested that the rays be called X rays but the society, on Kölliker's motion, voted that the name "Roentgen ray" be used. Roentgen published a few more papers on the subject, but his total contribution is relatively small.

REFERENCE

GLASSER, O. Wilhelm Conrad Röntgen and the Early History of the Röntgen Rays. London, John Bale Sons and Danielsson, 1933.

PRELIMINARY COMMUNICATION

[From On a New Kind of Rays*]

(1) If the discharge of a fairly large induction-coil be made to pass through a Hittorf vacuum-tube, or through a Lenard tube, a Crookes' tube, or other similar apparatus, which has been sufficiently exhausted, the tube being covered with thin, black cardboard which fits it with tolerable closeness, and if the whole apparatus be placed in a completely darkened room, there is observed at each discharge

* Eine Neue Art von Strahlen, Würzburg, 1895. Translation by Dr. Otto Glaser, of Cleveland.

a bright illumination of a paper screen covered with barium platino-cyanide, placed in the vicinity of the induction-coil, the fluorescence thus produced being entirely independent of the fact whether the coated or the plain surface is turned towards the discharge-tube. This fluorescence is visible even when the paper screen is at a distance of two metres from the apparatus.

It is easy to prove that the cause of the fluorescence proceeds from the discharge-apparatus, and not from any other point in the conducting circuit.

(2) The most striking feature of this phenomenon is the fact that an active agent here passes through a black cardboard envelope, which is opaque to the visible and the ultraviolet rays of the sun or of the electric arc; an agent, too, which has the power of producing active fluorescence. Hence we may first investigate the question whether other bodies also possess this property.

We soon discover that all bodies are transparent to this agent, though in very different degrees. I proceed to give a few examples: Paper is very transparent;[1] behind a bound book of about one thousand pages I saw the fluorescent screen light up brightly, the printer's ink offering scarcely a noticeable hindrance. In the same way the fluorescence appeared behind a double pack of cards; a single card held between the apparatus and the screen being almost unnoticeable to the eye. A single sheet of tinfoil is also scarcely perceptible, it is only after several layers have been placed over one another that their shadow is distinctly seen on the screen. Thick blocks of wood are also transparent, pine boards 2 or 3 cm. thick absorbing only slightly. A plate of aluminium about 15 mm. thick, though it enfeebled the action seriously did not cause the fluorescence to disappear entirely. Sheets of hard rubber several centimetres thick still permit the rays to pass through them.[2] Glass plates of equal thickness behave quite differently, according as they contain lead (flint-glass) or not; the former are much less transparent than the latter. If the hand be held between the discharge-tube and the screen, the darker shadow of the bones is seen within the slightly dark shadow-image of the hand itself. Water, carbon disulphide, and

[1] By "transparency" of a body I denote the relative brightness of a fluorescent screen placed close behind the body, referred to the brightness which the screen shows under the same circumstances though without the interposition of the body.

[2] For brevity's sake I shall use the expression "rays" and to distinguish them from others of this name, I shall call them "X-rays." (See paragraph No. 14.)

various other liquids, when they are examined in mica vessels, seem also to be transparent. That hydrogen is to any considerable degree more transparent than air I have not been able to discover. Behind plates of copper, silver, lead, gold, and platinum, the fluorescence may still be recognized, though only if the thickness of the plates is not too great. Platinum of a thickness of 0.2 mm. is still transparent, the silver and copper plates may even be thicker. Lead of a thickness of 1.5 mm. is practically opaque, and on account of this property this metal is frequently most useful. A rod of wood with a square cross-section (20 by 20 mm.), one of whose sides is painted white with lead paint, behaves differently according as to how it is held between the apparatus and the screen. It is almost entirely without action when the X-rays pass through it parallel to the painted side; whereas the stick throws a dark shadow when the rays are made to traverse it perpendicular to the painted side. In a series similar to that of the metals themselves their salts can be arranged with reference to their transparency, either in the solid form or in solution.

(3) The experimental results which now have been given, as well as others, lead to the conclusion that the transparency of different substances, assumed to be of equal thickness, is essentially conditioned upon their density; no other property makes itself felt like this, certainly to so high a degree.

The following experiments show, however, that the density is not the only cause acting. I have examined, with reference to their transparency, plates of glass, aluminium, calcite, and quartz, of nearly the same thickness, and while these substances are almost equal in density, yet it was quite evident that the calcite was sensibly less transparent than the other substances which appeared almost exactly alike. No particularly strong fluorescence (see No. 6 below) of calcite, especially by comparison with glass, has been noticed.

(4) All substances with increase in thickness become less transparent. In order to find a possible relation between transparency and thickness, I have made photographs (see No. 6 below) in which portions of the photographic plate were covered with layers of tinfoil, varying in the number of sheets superimposed. Photometric measurements of these will be made when I am in possession of a suitable photometer.

(5) Sheets of platinum, lead, zinc, and aluminium were rolled of such thickness that all appeared nearly equally transparent. The fol-

lowing table contains the absolute thickness of these sheets measured in millimetres, the relative thickness referred to that of the platinum sheet and their densities.

	Thickness	Relative Thickness	Density
Pt 0.018 mm...............		1	21.5
Pb 0.05 "		3	11.3
Zn 0.10 "		6	7.1
Al 3.5 "		200	2.6

We may conclude from these values that different metals possess transparencies which are by no means equal, even when the product of thickness and density are the same. The transparency increases much more rapidly than this product decreases.

(6) The fluorescence of barium platinocyanide is not the only recognizable effect of the X-rays. It should be mentioned that other bodies also fluoresce; such, for instance, as the phosphorescent calcium compounds, then uranium glass, ordinary glass, calcite, rocksalt, and so on.

Of special significance in many respects is the fact that photographic dry plates are sensitive to the X-rays. We are, therefore, in a condition to determine more definitely many phenomena, and so the more easily to avoid deception; wherever it has been possible, therefore, I have controlled, by means of photography, every important observation which I have made with the eye by means of the fluorescent screen.

In these experiments the property of the rays to pass almost unhindered through thin sheets of wood, paper and tinfoil is most important. The photographic impressions can be obtained in a non-darkened room with the photographic plates either in the holders or wrapped up in paper. On the other hand, from this property it results as a consequence that undeveloped plates cannot be left for a long time in the neighbourhood of the discharge-tube, if they are protected merely by the usual covering of pasteboard and paper.

It appears questionable, however, whether the chemical action on the silver salts of the photographic plates is directly caused by the X-rays. It is possible that this action proceeds from the fluorescent light which, as noted above, is produced in the glass plate itself or perhaps in the layer of gelatin. "Films" can be used just as well as glass plates.

I have not yet been able to prove experimentally that the X-rays are able also to produce a heating action; yet we may well assume that this effect is present, since the capability of the X-rays to be transformed is proved by means of the observed fluorescence phenomena. It is certain, therefore, that all the X-rays which fall upon a substance do not leave it again as such.

The retina of the eye is not sensitive to these rays. Even if the eye is brought close to the discharge-tube, it observes nothing, although, as experiment has proved, the media contained in the eye must be sufficiently transparent to transmit the rays.

(7) After I had recognized the transparency of various substances of relatively considerable thickness, I hastened to see how the X-rays behaved on passing through a prism, and to find whether they were thereby deviated or not.

Experiments with water and with carbon disulphide enclosed in mica prisms of about 30° refracting angle showed no deviation, either with the fluorescent screen or on the photographic plate. For purposes of comparison the deviation of rays of ordinary light under the same conditions was observed; and it was noted that in this case the deviated images fell on the plate about 10 or 20 mm. distant from the direct image. By means of prisms made of hard rubber and of aluminium, also of about 30° refracting angle, I have obtained images on the photographic plate in which some small deviation may perhaps be recognized. However, the fact is quite uncertain; the deviation, if it does exist, being so small that in any case the refractive index of the X-rays in the substances named cannot be more than 1.05 at the most. With a fluorescent screen I was also unable to observe any deviation.

Up to the present time experiments with prisms of denser metals have given no definite results, owing to their feeble transparency and the consequently diminished intensity of the transmitted rays.

With reference to the general conditions here involved on the one hand, and on the other to the importance of the question whether the X-rays can be refracted or not on passing from one medium into another, it is most fortunate that this subject may be investigated in still another way than with the aid of prisms. Finely divided bodies in sufficiently thick layers scatter the incident light and allow only a little of it to pass, owing to reflection and refraction, so that if powders are as transparent to X-rays as the same substances are in mass—equal amounts of material being presupposed—

it follows at once that neither refraction nor regular reflection takes place to any sensible degree. Experiments were tried with finely powdered rock-salt, with fine electrolytic silver-powder, and with zinc-dust, such as is used in chemical investigations. In all these cases no difference was detected between the transparency of the powder and that of the substance in mass, either by observation with the fluorescent screen or with the photographic plate.

From what has now been said it is obvious that the X-rays cannot be concentrated by lenses; neither a large lens of hard rubber nor a glass lens having any influence upon them. The shadow-picture of a round rod is darker in the middle than at the edge; while the image of a tube which is filled with a substance more transparent than its own material is lighter at the middle than at the edge.

(8) The question as to the reflection of the X-rays may be regarded as settled, by the experiments mentioned in the preceding paragraph, in favour of the view that no noticeable regular reflection of the rays takes place from any of the substances examined. Other experiments, which I here omit, lead to the same conclusion.

One observation in this connection should, however, be mentioned, as at first sight it seems to prove the opposite. I exposed to the X-rays a photographic plate which was protected from the light by black paper, and the glass side of which was turned towards the discharge-tube giving the X-rays. The sensitive film was covered, for the most part, with polished plates of platinum, lead, zinc, and aluminium, arranged in the form of a star. On the developed negative it was seen plainly that the darkening under the platinum, the lead, and particularly the zinc, was stronger than under the other plates, the aluminium having exerted no action at all. It appears, therefore, that these three metals reflect the rays. Since, however, other explanations of the stronger darkening are conceivable, in a second experiment, in order to be sure, I placed between the sensitive film and the metal plates a piece of thin aluminium-foil, which is opaque to ultra-violet rays, but is very transparent to the X-rays. Since the same result substantially was again obtained, the reflection of X-rays from the metals above named is proved.

If we compare this fact with the observation already mentioned that powders are as transparent as coherent masses, and with the further fact that bodies with rough surfaces behave like polished bodies with reference to the passage of the X-rays, as shown also in the last experiment, we are led to the conclusion already stated that regular reflection does not take place, but that bodies behave towards the X-rays as turbid media do towards light.

Since, moreover, I could detect no evidence of refraction of these rays in passing from one medium into another, it would seem that X-rays move with the same velocity in all substances: and, further, that this speed is the same in the medium which is present everywhere in space and in which the particles of matter are embedded. These particles hinder the propagation of the X-rays, the effect being greater, in general, the more dense the substance concerned.

(9) Accordingly it might be possible that the arrangement of particles in the substance exercised an influence on its transparency; that, for instance, a piece of calcite might be transparent in different degrees for the same thickness, according as it is traversed in the direction of the axis, or at right angles to it. Experiments, however, on calcite and quartz gave a negative result.

(10) It is well known that Lenard came to the conclusion, from the results of his beautiful experiments on the transmission of the cathode rays of Hittorf through a thin sheet of aluminium, that these rays are phenomena of the ether, and that they diffuse themselves through all bodies. We can say the same of our rays.

In his most recent research, Lenard has determined the absorptive power of different substances for the cathode rays, and, among others, has measured it for air from atmospheric pressure to 4.10, 3.40, 3.10, referred to 1 centimetre, according to the rarefaction of the gas contained in the discharge-apparatus. Judging from the discharge-pressure as estimated from the sparking distance, I have had to do in my experiments for the most part with rarefactions of the same order of magnitude, and only rarely with less or greater ones. I have succeeded in comparing by means of the L. Weber photometer—I do not possess a better one—the intensities, taken in atmospheric air, of the fluorescence of my screen at two distances from the discharge apparatus—about 100 and 200 mm., and I have found from three experiments, which agree very well with each other, that the intensities vary inversely as the squares of the distances of the screen from the discharge-apparatus. Accordingly, air absorbs a far smaller fraction of the X-rays than of the cathode rays. This result is in entire agreement with the observation mentioned above, that it is still possible to detect the fluorescent light at a distance of 2 metres from the discharge apparatus.

Other substances behave in general like air; they are more transparent to X-rays than to cathode rays.

(11) A further difference, and a most important one, between the behavior of cathode rays and of X-rays lies in the fact that I have not succeeded, in spite of many attempts, in obtaining a deflection of the X-rays by a magnet, even in very intense fields.

The possibility of deflection by a magnet has, up to the present time, served as a characteristic property of the cathode rays, although it was observed by Hertz and Lenard that there are different sorts of cathode rays, "which are distinguished from each other by their production of phosphorescence, by the amount of their absorption, and by the extent of their deflection by a magnet." A considerable deflection, however, was noted in all of the cases investigated by them, so that I do not think that this characteristic will be given up except for stringent reasons.

(12) According to experiments especially designed to test the question, it is certain that the spot on the wall of the discharge-tube which fluoresces the strongest is to be considered as the main centre from which the X-rays radiate in all directions. The X-rays proceed from that spot where, according to the data obtained by different investigators, the cathode rays strike the glass wall. If the cathode rays within the discharge-apparatus are deflected by means of a magnet, it is observed that the X-rays proceed from another spot— namely, from that which is the new terminus of the cathode rays.

For this reason, therefore, the X-rays, which it is impossible to deflect, cannot be cathode rays simply transmitted or reflected without change by the glass wall. The greater density of the gas outside of the discharge-tube certainly cannot account for the great difference in the deflection, according to Lenard.

I therefore reach the conclusion that the X-rays are not identical with the cathode rays, but that they are produced by the cathode rays at the glass wall of the discharge-apparatus.

(13) This production does not take place in glass alone, but, as I have been able to observe in an apparatus closed by a plate of aluminium 2 mm. thick, in this metal also. Other substances are to be examined later.

(14) The justification for calling by the name "rays" the agent

which proceeds from the wall of the discharge apparatus, I derive in part from the entirely regular formation of shadows, which are seen when more or less transparent bodies are brought between the apparatus and the fluorescent screen (or the photographic plate).

I have observed, and in part photographed, many shadow-pictures of this kind, the production of which has a particular charm. I possess, for instance, photographs of the shadow of the profile of a door which separates the rooms in which, on one side, the discharge-apparatus was placed, on the other the photographic plate; the shadow of the bones of the hand; the shadow of a covered wire wrapped on a wooden spool; of a set of weights enclosed in a box; of a compass in which the magnetic needle is entirely enclosed by metal; of a piece of metal whose lack of homogeneity becomes noticeable by means of the X-rays, etc.

Another conclusive proof of the rectilinear propagation of the X-rays is a pin-hole photograph which I was able to make of the discharge-apparatus while it was enveloped in black paper; the picture is weak but unmistakably correct.

(15) I have tried in many ways to detect interference phenomena of the X-rays; but, unfortunately, without success, perhaps only because of their feeble intensity.

(16) Experiments have been begun, but are not yet finished, to ascertain whether electrostatic forces affect the X-rays in any way.

(17) In considering the question, what are the X-rays—which, as we have seen, cannot be cathode rays—we may perhaps at first be led to think of them as ultra-violet light, owing to their active fluorescence and their chemical actions. But in so doing we find ourselves opposed by the most weighty considerations. If the X-rays are ultra-violet light, this light must have the following properties:—

(a) On passing from air into water, carbon disulphide, aluminium, rock-salt, glass, zinc, etc., it suffers no noticeable refraction.

(b) By none of the bodies named can it be regularly reflected to any appreciable extent.

(c) It cannot be polarized by any of the ordinary methods.

(d) Its absorption is influenced by no other property of substances so much as by their density.

That is to say, we must assume that these ultra-violet rays behave entirely differently from the ultra-red, visible, and ultra-violet rays which have been known up to this time.

I have been unable to come to this conclusion, and so have sought for another explanation.

There seems to exist some kind of relationship between the new rays and light rays; at least this is indicated by the formation of shadows, the fluorescence and the chemical action produced by them both. Now, we have known for a long time that there can be in the ether longitudinal vibrations besides the transverse light vibrations, and, according to the views of different physicists, these vibrations must exist. Their existence, it is true, has not been proved up to the present, and consequently their properties have not been investigated by experiment.

Ought not, therefore, the new rays to be ascribed to longitudinal vibrations in the ether?

I must confess that in the course of the investigation I have become more and more confident of the correctness of this idea, and so, therefore, permit myself to announce this conjecture, although I am perfectly aware that the explanation given still needs further confirmation.

Würzburg, Physikalisches Institut der Universität, December, 1895.

INDEX

Vinci, Leonardo da, 123
  notes of, 124
  on heart and vessels, 124
  on urine, 125
Virchow, Rudolf, 622, 623
Vision, Essay on, 585
Vital statistics, 464
Vitamin C, 464
Voit, Carl von, 590

Warren, John C., 355
Watch, pulse, physician's, 572
Water, in treatment of fever, 428
Water-borne diseases, 464
Waterhouse, Benjamin, 291
  on the history of cowpox, 301
Webber, E. H., and E. F., 589
Weller, Sam, 339
Wells, Horace, 355
Wells, Spencer, 604
Wells, William Charles, 585
  on dropsy, 587
  on scarlet fever, 587
Wepfer, 244

Wet nurse, 178
Wheal, 562
White, Charles, 603
Willan, Robert, 62, 560
  classification of skin diseases, 560
Wine, studies on, of Pasteur, 378
Withering, William, 419
  on the foxglove, 421
Wöhler, Friedrich, 590
Woodville, 304
Worms, in children, 266
Wounds, gunshot, 192
  Paré on, 189
  treatment of Chauliac on, 89
Wright, Sir Almroth E., 418
Wunderlich, Carl R. A., 572, 575
Wurtz, 261

X-ray, 666

Yellow fever, 465, 479
York Retreat, 441
Young, Thomas, 634